Mapping Tree Species Diversity

Mapping Tree Species Diversity

Editors

Markus Immitzer
Onisimo Mutanga
Clement Atzberger

MDPI

Basel • Beijing • Wuhan • Barcelona • Belgrade • Novi Sad • Cluj • Manchester

Editors

Markus Immitzer
University of Natural
Resources and Life Sciences
Vienna, Austria

Onisimo Mutanga
University of Kwazulu-Natal
Durban, South Africa

Clement Atzberger
University of Natural
Resources and Life Sciences
Vienna, Austria

Editorial Office
MDPI
St. Alban-Anlage 66
4052 Basel, Switzerland

This is a reprint of articles from the Special Issue published online in the open access journal *Remote Sensing* (ISSN 2072-4292) (available at: https://www.mdpi.com/journal/remotesensing/special_issues/Mapping_Tree_Species).

For citation purposes, cite each article independently as indicated on the article page online and as indicated below:

Lastname, A.A.; Lastname, B.B. Article Title. *Journal Name* **Year**, *Volume Number*, Page Range.

ISBN 978-3-0365-8526-0 (Hbk)
ISBN 978-3-0365-8527-7 (PDF)
doi.org/10.3390/books978-3-0365-8527-7

Cover image courtesy of Markus Immitzer

Contents

About the Editors

Markus Immitzer

Markus Immitzer works as a Senior Scientist at the Institute of Geomatics, University of Natural Resources and Life Sciences, Vienna (BOKU), Austria. His main fields of interest are remote sensing based on optical data with very high spatial and spectral resolutions for forest mapping, stress monitoring, the modeling of biophysical parameters, and land cover classification and monitoring. His main expertise are in the following fields: classification of EO data from different platforms (drone to satellite), inclusive data fusion, the modeling of forest parameters, and habitat and distribution modeling.

Onisimo Mutanga

Onisimo Mutanga is a Professor of Remote Sensing and South African Research Chair on Land Use Planning and Management. His expertise lie in resource modelling, pattern and condition analysis in the face of global and land use change. This includes the development of innovative remote sensing approaches for the improved monitoring and management of various ecosystem traits. Focus areas include vegetation mapping, ecosystem services quantification, forest fragmentation, disaster risk assessment, rangeland quality and quantity monitoring, as well as invasive species modelling, among others. He integrates ecology, biodiversity conservation and remote sensing in order to model environmental characteristics as an integral component of land use planning and management.

Clement Atzberger

Clement Atzberger currently serves as a Full Professor and is the head of the Institute of Geomatics at the University of Natural Resources and Life Sciences, Vienna (BOKU), Austria. Besides his excellent background in remote sensing applications, his primary areas of expertise include satellite-based natural resource management, time series analysis, radiative transfer modeling, crop growth modeling, data assimilation, and imaging spectroscopy. He has extensive experience in reviewing scientific literature and conducting assessments of new methodologies. Amongst his other roles as a reviewer for scientific conferences and journals, he has also served for more than 5 years as Editor-in-Chief for the section "Agriculture and Vegetation" in the SCI journal *Remote Sensing*, giving him a broad and unbiased overview of new developments in the scientific fields of Earth Observation (EO) and machine learning/AI.

remote sensing

MDPI

Communication

Tree Species Diversity Mapping—Success Stories and Possible Ways Forward

Markus Immitzer * and Clement Atzberger

Institute of Geomatics, Department of Landscape, Spatial and Infrastructure Sciences, University of Natural Resources and Life Sciences Vienna (BOKU), Peter-Jordan-Straße 82, 1190 Vienna, Austria; clement.atzberger@boku.ac.at

* Correspondence: markus.immitzer@boku.ac.at

Abstract: The special issue "Tree species diversity mapping" presents research focused on the remote assessment of tree species diversity, using different sensor modalities and platforms. The special issue thereby recognizes that the continued loss of biodiversity poses a great challenge to humanity. Precise and regularly updated baseline information is urgently needed, which is difficult, using field inventories, especially on a large scale. On such scales, remote sensing methods excel. The work presented in the special issue demonstrates the great potential of Earth Observation (EO) for addressing knowledge gaps, as EO provides rich (spectral) information at high revisit frequencies and spatial resolutions. Many tree species can be distinguished well using optical data, in particular, when simultaneously leveraging both the spectral and temporal dimensions. A combination with other sensor modalities can further improve performance. EO approaches are, however, limited by the availability of high-quality reference information. This complicates the task as the collection of field data is labor and time-consuming. To mitigate this limiting factor, resources should be better shared amongst the community. The reliance on in situ data also highlights the need to focus research on the extraction of more permanent (i.e., species-inherent) properties. In this respect, we identify and discuss some inherent limitations of current approaches regarding tree species discrimination. To this end, we offer a more fundamental view on tree species classification based on physical principles. To provide both a summary of the special issue and some stimulating thoughts about possible future research directions, we structured the present communication into four parts. We first introduce the need for biodiversity information, followed by a summary of all 19 articles published within the special issue. The articles are ordered by the number of species investigated. Next, we provide a short summary of the main outputs. To stimulate further research and discussion within the scientific community, we conclude this communication by offering a more fundamental view on tree species classification based on EO data and its biophysical foundations. In particular, we purport that species can possibly be more robustly identified if we classify/identify them in the biophysical feature space and not in the spectral-temporal feature space. This involves the creation and inversion of so-called physically-based radiative transfer models (RTM), which take hyper/multispectral observations together with their observation geometry (as well as other priors), and project these into biophysical variables such as chlorophyll content and LAI etc. The perceived advantage of such an approach is that the generalizability (and scalability) of EO based classifications will increase, as the temporal trajectory of species in the biophysical parameter space is probably more robust compared to the sole analysis of spectral data, which—amongst other perturbing factors—also depend on site/time specific illumination geometry.

Keywords: tree species; classification; biodiversity; machine learning; feature extraction; optical data; SAR; LiDAR; time series; radiative transfer model

Citation: Immitzer, M.; Atzberger, C. Tree Species Diversity Mapping—Success Stories and Possible Ways Forward. *Remote Sens.* **2023**, *15*, 3074. https://doi.org/10.3390/rs15123074

Academic Editor: Jan Altman

Received: 6 May 2023
Revised: 26 May 2023
Accepted: 6 June 2023
Published: 12 June 2023

1. Introduction

The IPBES report on Biodiversity and Ecosystem Services depicts an alarming and shocking picture of the Earth [1]. With accelerating rates of species extinction, our environment is declining globally at unprecedented rates. Transformative economic and societal change is necessary, and will involve far-reaching alterations and actions at both local and global levels. This issue was recently discussed at UN Biodiversity Conference COP15 in Montréal [2]. To cope with the pace of global change, amongst many other activities, a rapid increase in knowledge about species numbers, compositions, and conditions is required, as well as species interactions and environments [1]. This information should possibly be condensed into quantitative measures of biodiversity [3].

Remote sensing provides the only feasible way to cost-effectively and repeatedly measure and monitor land cover changes on large scale—while at the same time, allowing a user to look as far back as the start of the space age, in the 1970s [4,5]. Today's sensors on platforms such as satellites, aircraft, and UAV provide a wide range of observational capability in terms of spatial, temporal, and spectral resolutions. Machine learning approaches and computational capacity are improving quickly, offering huge potential for improved data analysis, including "big data" and the development of powerful monitoring systems. First efforts were undertaken to bridge gaps between Earth observation (EO) scientists on one hand, and ecologists etc. on the other hand [3]. A first set of comprehensive essential biodiversity variables (EBVs) and how they can be monitored by EO technique were defined [4,6,7].

2. Overview of Contributions

The special issue "Mapping Tree Species Diversity" aimed to collect scientific research covering the following themes:

- biodiversity of forests, with respect to classical species diversity;
- the mapping of changes in diversity;
- the floristic composition of forests;
- invasive species;
- the functional diversity of forests.

In total, the special issue includes one review and 17 research articles. Despite the limited number of studies, a wide range of different ecoregions could be covered. The studies were conducted on all continents except Antarctica. With six studies each, there is a focus on Europe and Asia, followed by Africa with three studies. One study each was performed in North America, South America, and Oceania. As a result of the broad geographical coverage, a relatively wide range of plant species and climatic zones were examined, from boreal and temperate forests to tropical forests. It is also exciting that numerous studies were carried out in forests that are subject to a certain protection category, from biosphere parks to nature reserves and national parks, which shows the increasing interest in monitoring such areas with EO data.

Most studies focused on the separation and classification of tree species or tree species groups, using optical data from different platforms and sensors (Table 1). The review paper summarized the current knowledge of LiDAR for tree species classification. Two studies investigated the advantages of combining optical data with LiDAR and one with SAR data. Most studies used spectral signatures or resultant spectral indices as discriminative features, either derived from single scenes or multiple images. In addition to the well-established Landsat and Sentinel-2 (S2) satellites, which provide high quality data free of charge, commercial satellites with higher spatial resolution such as Worldview (WV)-2/3 or Gaofen-1/2, were also used. Textural measures, and approaches borrowed from computer vision, were used in the case of very high-resolution (sub-meter) data recorded by aircraft and unmanned aerial vehicles (UAV). To separate and classify tree species, Random Forest (RF) and Support Vector Machine (SVM) algorithms were most often used, in particular when using (orbital) remote sensing data with metric to hecto-metric ground sampling distance (GSD). A few neural network approaches were used when UAV data was analyzed.

Table 1. Summary of the contributions to the special issue about tree species diversity mapping describing the used data, location of the study, number of species considered (used for the order of the table), used classifier, and achieved accuracy (GSD: ground sampling distance, MS: multispectral, HS: hyperspectral, LS: Landsat, S1: Sentinel-1, S2: Sentinel-2, WV: WorldView, UAV: unmanned aerial vehicle, ANN: artificial neural network, CNN convolutional neural network, RF: random forest, SVM: support vector machine; OA: overall accuracy).

Data Type	Platform	Sensor	Spectral Bands	GSD	Temporal Coverage	Location	Number of Species *	Classifier	OA (%)	Ref.
LiDAR				Review study					60–80	[8]
MS	Satellite	S2	10	10 m	Multi-temporal	Senegal	1	ANN	91	[9]
RGB	Airplane	Different aerial cameras	3	0.1 cm	Bi-temporal	New Zealand	1	CNN & XGBoost	83	[10]
MS	Satellite	LS	6	30 m	Multi-temporal	Russia	2	RF	90	[11]
MS + LiDAR	Airplane + Helicopter	Xp-w/a I DMCIII; RIEGL LMS Q 680i	4	55 pts/m^2, 0.17/0.095 m	Mono-temporal	Germany	2–4	RF	87–98	[12]
MS	Satellite	LS8	6	30 m	Multi-temporal	China	4	RF	79	[13]
RGB	UAV	RGB camera	3	0.003–0.5 m	Serval flights	Germany	4	CNN	92	[14]
MS	Satellite	Gaofen-1&2, S2, LS8	2 (NDVI)	4, 10, 16, 30 m	Multi-temporal	China	4	RF	85	[15]
MS	Satellite	WV-3	4	0.4 m	Mono-temporal	China	6	CNN, SVM, RF	83	[16]
MS	Satellite	S2	10	10 m	Multi-temporal	Serbia	8	RF	83	[17]
HS	UAV		25	0.1 m	3 flights in 3 years	Brazil	8	RF	50	[18]
MS	Satellite	WV-2	8	0.5 m	Mono-temporal	Kenya	8	RF, SVM	70–73	[19]
MS	Satellite	WV-3	16	0.3 cm	Mono-temporal	Canada	11	SVM, RF	75	[20]
MS + LiDAR	Satellite + Airplane	S2, RGB, RIEGL LMS-Q680	3 + 10	3.6 pts/m^2, 0.12 m, 10 m	Mono-/multi-temporal	China	11	SVM, RF	90–94	[21]
MS + SAR	Satellite	S2, S1	10	10 m	Multi-temporal	Austria	12	RF	84	[22]
MS	Satellite	S2	10	10 m	Multi-temporal	Austria	12	RF	89	[23]
MS	Satellite	Formosat-2	4	8 m	Multi-temporal	France	13	SVM	48–60	[24]
HS	Airplane	AVIRIS	366	4 m	Mono-temporal	India	20	SVM	86	[25]

* Sometimes, species were grouped into broader "classes".

The review paper on LiDAR [8] pointed out that LiDAR data are currently often used for the separation of only a few (2–4) tree species. In addition to geometry, the use of the full-waveform and intensity features improves the classification performances. The authors concluded that combinations with other data sources from the spectral domain are promising, but also challenging.

Lu et al. [9] mapped *Faidherbia albida*, often used in Senegal's agroforestry systems, using multi-spectral S2 time series. The "reversed" phenology of this species, compared to many other species in the region, did not show the expected clear signal. This was explained by the spectral contribution of the herbaceous layer. Nevertheless, especially with data from the senescence phase, the species could be detected very well, and thus, a good data basis for management activities could be established.

Pearse et al. [10] highlighted the potential of deep learning approaches for the mapping of one specific species (*Metrosideros excelsa*) based on RGB data from drones. Distinctive phenological features like red flowers facilitate the recognition of the species but are not mandatory. Higher spectral resolution and image quality would be helpful to distinguish more species.

Tian and Fu [11] showed the possibility of EO data, in particular Landsat scenes, for largescale mapping of the boreal forest cover in Siberia. They analyzed changes in the cover of broadleaved and coniferous trees between 1985 and 2015. The quality of the 2015 map was validated by using very high resolution (VHR) data (Gaofan-2).

Next to the separation of coniferous and broadleaved trees; the main aim in the study of Krzystek et al. [12] was the mapping of standing dead trees which were separated in snags and among trees with crowns. Dead trees are often important habitats for fauna and flora, and therefore, often a driver for biodiversity in forests. The work highlighted the potential of ALS data for individual tree crown (ITC) approaches and the detection of standing deadwood in a nature protected area (national parks) on the border Germany and Czech Republic.

Well-executed topographic corrections are a prerequisite for the optimum analysis of high-quality EO data. Dong et al. [13] found that the overall accuracy of classifications models for four tree species in a mountainous area in the Shandong Province, China can be improved by 4–13 percentage points by using better-suited topographic correction methods. The use of vegetation indices (VI) can further reduce the influence of the topography and can therefore improve the classification.

Egli and Höpke [14] analyzed UAV data acquired at different daytimes for tree species classification with a convolutional neural network (CNN) and found that the models are independent from illumination conditions. In contrast, the spatial resolution of the input data has a strong impact on the classification accuracy. Decreasing the GSD/resolution through resampling degraded the model results, however, the classification accuracy of a human interpreter dropped much faster and earlier than the CNN models.

Xu et al. [15] analyzed the influence of the spatial resolution on tree species classification in northeast China. They analyzed NDVI time series obtained by monthly data from four different sensors with spatial resolution between 4 to 30 m for the separation of four species. The 10 m data sets outperformed the other data sets. The use of additional metrics describing the land surface phenology (LSP) improved the models. The study showed that phenological parameters were not independent from spatial resolution.

Yan et al. [16] used WV-3 data with the four standard bands to distinguish tree species in a park landscape in Beijing, China by using an ITC approach. They pointed out the importance of high quality crown delineation for achieving good accuracies which is more complicated in natural forests. The CNN outperformed traditional classifiers such as RF and SVM while not needing a separate feature extraction step.

Kovačević et al. [17] used multi-temporal S2 data and analyzed different vegetation types for the entire geography of Serbia. Their approach is based on a probability classification model that considers the temporal information of the data sets and a spatial-temporal aggregation. The authors also highlighted the challenges regarding reference data for the classes of interest in terms of occurrence, recognizability, and demarcation.

Takahashi Miyoshi et al. [18] used UAV-borne hyperspectral data for their ITC approach in a study site in the Brazilian Atlantic forest. They found some differences in the interpretability of the three flights, which were done in June/July of three consecutive years. Different phenological stages and illumination conditions have an influence on the models. They found that the added value of multi-temporal data has to be investigated in detail with a more extensive reference data set.

Jackson and Adam [15] analyzed a WV-2 scene in a study area on Mount Kenia. They highlighted the importance of the Red, RedEdge, and NIR bands for the separation of the tree species. They used under and oversampling to generate a more balanced data set which improved the classification accuracy slightly.

Varin et al. [20] combined WV-3 and LiDAR data to analyze ITC. Transferring the canopy delineation done on the LiDAR canopy height model (CHM) to the WV-3 scene was challenging. The different spatial resolutions of the bands, especially the SWIR bands, also complicate the analyses. Therefore, they concluded that the 8-band data is preferable to the 16-band dataset. They also mentioned the problem of trees which were either not seen (too small crown and/or located in the understory) or species which were not considered in the reference data set.

Wan et al. [21] combined LiDAR data, aerial images, and S2 time series data in an hilly forest in China. They found the data with higher spatial resolution were more important for the species classification. However, the highest accuracies were achieved with the fusion approach using all data sets.

Lechner et al. [22] tested the combination of S2 and S1 data for the classification of 12 tree species in a biosphere reserve in Austria. A dense S1 time series (SAR) was able to separate coniferous species quite well and also improved mono-temporal S2 models. However, considerably higher accuracies were achieved by a multi-temporal S2 data set. In this case, additional SAR data could not further increase the accuracy.

Immitzer et al. [23] analyzed, for the same study site, an extensive multi-temporal S2 data set covering several years to identify the most important dates and bands to distinguish 12 tree species. They showed the high importance of the Red band for the separation of species in the coniferous group, the SWIR band for the broadleaved trees, and the NIR for the separation between these two tree species groups. VIs further improved the model performance when added to the spectral signatures. Spectral data sets with at least 5–7 scenes, covering the main phenological stages such as spring, summer, and autumn, led to very good results.

Karasiak et al. [24] used multi-temporal, 4-band Formosat-2 data from several years for tree species classification. The results showed that the usage of multi-temporal data improves the model performance; however, the best dates for optical data acquisition varied between years. They also tested different validation approaches and emphasized the importance of the number and quality of reference samples.

Chaurasia et al. [25] used hyperspectral data to produce species abundance maps and analyzed the intra- and inter-species spectral diversity for three test sites in India. They found higher spectral diversity within abundant species and attributed this to improved adaptation of these species to local conditions. Changes in species diversity over climatic gradients were also observed, which can be important with respect to climate change.

3. Summary of Main Outputs and Findings

The work presented in this special issue re-confirmed a number of findings. For example, no noticeable issues were reported in separating coniferous and broadleaved trees, thus confirming many previous studies [26,27]. The main features permitting this separation were located in the NIR, as in this spectral region the spectral differences are largest (Figure 1).

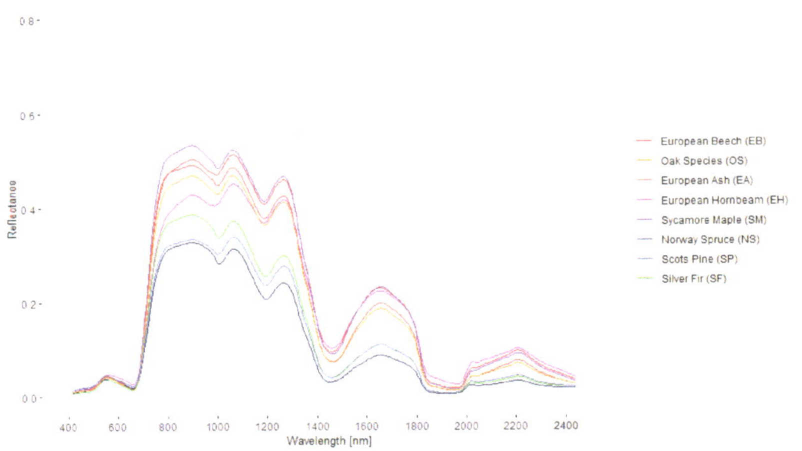

Figure 1. Mean spectral signatures of tree crowns from seven Central European tree species using airborne HySpex imaging spectrometer data. Species with reddish lines are broadleaved species; green and blue lines indicate coniferous species. The data used for this illustration was also used in the work of Maschler et al. [28].

This points to an overwhelming impact of leaf optical properties, and in particular, leaf anatomy (Figure 2a) [29,30]. Indeed, both empirical and physically-based leaf optical properties models have demonstrated that the difference in leaf anatomy is the main trait that makes needle leaves much less reflective in the NIR compared to leaves from broadleaved trees [31]. All other conditions being identical, simulations using the forest radiative transfer model INFORM [32,33] confirm this (Figure 2b).

Within the two broad tree species groups (broadleaved/coniferous trees), some species can be distinguished relatively well, while others are more difficult to separate [26,34,35]. For example, some species are usually sufficiently separable in the VIS-NIR-SWIR optical domain, while others are spectrally too similar to be distinguished. This overlap in spectral signatures persists throughout the growing season and points to three main effects:

- Between several tree species, often only very small inter-class differences in leaf biochemical and structural properties exist, leading to very similar leaf optical properties (Figure 2a).
- Most species exhibit a relatively large intra-class variability of canopy properties [34,36]. Intra-class variability exist, for example in terms of differences in tree age, stem density, growth form, and crown closure [37–39]. Additional intra-class variability results from changes in weather and growth conditions.
- Even for relatively closed forests, the canopy reflectance is heavily impacted by the optical properties of the forest floor, specifically, the type and amount of understory vegetation [38,40,41]. This induces a large range of "background noise", thus "widening" the spectral signature recorded by the sensor [42–44].

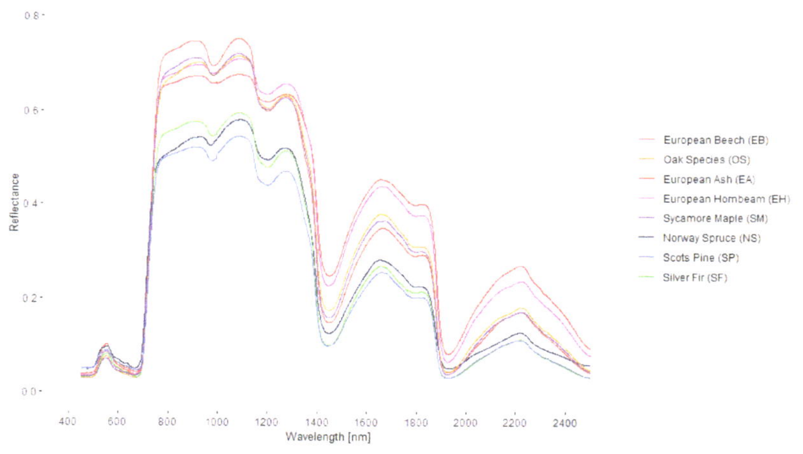

(a)

Figure 2. *Cont.*

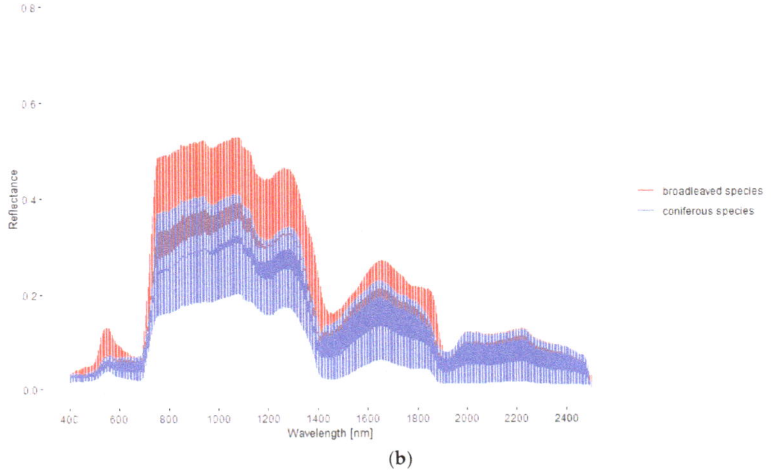

(b)

Figure 2. Impact of leaf/needle optical properties on canopy signatures. (**a**) Mean spectral signatures of leaves/needles from seven Central European tree species measured in the laboratory with an ASD FieldSpec Pro FR spectrometer following the method presented in Einzmann et al. [45]. Species and colors are identical to Figure 1. (**b**) Simulations of forest canopy reflectance using the INFORM radiative transfer model. For the simulations (n = 5000), leaf optical properties typical for broadleaved species (red) and coniferous species (blue) were used. Other INFORM parameters were randomly selected and kept identical for the two classes, so that solely differences resulting from the leaf optical properties are shown.

Together, it is not surprising that the sometimes very strong and persistent overlap of spectral-temporal signatures between two or more tree species cannot be untangled—independent of which classifier is used [26,46–48].

With respect to the sole use of spectral vegetation indices (VI), the situation is even worse, as spectral indices are only simple (equivariant) arithmetic combinations of 2–3 spectral bands (e.g., with an infinite number of band combinations leading to the same VI value). VIs should, therefore, never replace the spectral signatures, but should instead be added to the predictive feature set, for example to provide some "correction" for undesired factors of spectral variation [23].

In the same line of thinking, it is also unsurprising that accuracies obtained from a limited band setting are usually not able to perform as well as using the full spectral resolution [49]. Similar observations have been reported (Figure 3) when comparing single-frequency SAR data in two polarizations to multi-spectral optical sensors [22].

As expected, spectral observations over multiple time steps facilitate species identification. As we will show later, multi-temporal (optical) time series inherently capture and analyze the temporal (co)variation of the canopies' structural characteristics and biochemical composition. Analyzing such time series offer, therefore, strong improvements compared to single date analyses. Similar findings were reported in other studies [50–53].

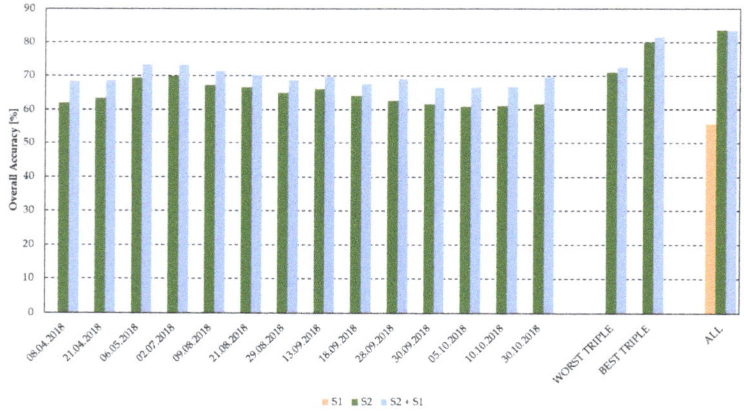

Figure 3. Overall classification accuracy for 12 tree species based on Sentinel-2 (S2) and Sentinel-1 (S1) data: mono-temporal S2 data, two seasonal selections using the three worst and the three best-performing S2 triples, and all available S2 data—alone and in combination with S1 phenological data (adapted from Lechner et al. [22]).

As active sensor modalities, such as LiDAR and SAR, react primarily to canopy properties not/less strongly impacting the spectral signatures recorded by optical sensors (e.g., structural canopy properties other than LAI), the combination of optical data with the active devices is generally beneficial. Similar findings have been reported in [27,54]. However, the improvement is usually not overly strong, as the active data is often very noisy (in particular, SAR) and/or focusing on a single vegetation trait (e.g., 3D structure/clumping in the case of LiDAR). For the same reason, neither SAR nor LiDAR alone were able to achieve the accuracies which can be achieved with optical (time series) data, confirming the conclusion of the review paper by Fassnacht et al. [46]. The research studies in our special issue also confirmed the importance of representative and comprehensive field samples [46]. When reference data are collected, foremost attention should be put on samples near the decision boundaries in the respective feature space, as pointed out by Foody et al. [55]. Indeed, the "difficult cases" matter most when dividing the feature space into thematic classes [56]. Compared to overly cleaned and curated data for training and evaluation, these data generate more reliable error statistics. Several studies [47,50,57] highlighted the fact that auxiliary variables such as altitude can help to improve the performance of tree species classification models. Consequently, a representative distribution of reference data for each class has to be ensured. Unlike spectral measurements, variables such as altitude are not a true explanatory variable for actual occurrence of a species. A comprehensive summary of the main findings and conclusions of our special issue, major limits, and possible amendments, is provided in Figure 4.

Important findings and main conclusions

- The quality of input data is key for successful vegetation analysis and classifications. Hence, a proper preprocessing (cloud detection, atmospheric, radiometric and topographic corrections etc.) is very important.
- Coniferous and broadleaved tree species can be easily distinguished due to their spectral differences throughout the electromagnetic spectrum and, in particular, the NIR.
- Several major tree species can usually be distinguished relatively well, while other species—often rare species—are more difficult to separate, which is due to different number of reference data.
- Optical data are the most useful EO data source for tree species classification.
- Combination of optical data with structural information derived from LiDAR and SAR can improve the models.
- To separate different species, the full spectral signature is most useful—ideally, the entirety of available spectral information should be used, complemented by some spectral indices to compensate for shadow effects and other biases.
- Multi-temporal data (several data sets over the year) or times series data (continuous data set used to derive phenological indicators) offer clear advantages compared to single observations, but still do not yet permit a model transfer to other seasons.
- Reference data quality and quantity are very important. One should aim for a balanced dataset, avoiding spatial autocorrelation (and leakage)—avoiding, for example, the use of several samples from the same forest stand.
- Research papers should always specify which (tree) species were not included in the classification and their approximative relevance, as species which were not considered in the training will obviously not be mapped. Usually, such thematic simplifications will not be reflected in error statistics.
- No consensus has been reached on a statistically sound validation strategy—often species with more samples obtain better results.
- Classification approaches do not have a clear impact. Good results can be obtained with different classifiers; quality of the EO data matters most.

Limits of current approaches

- Mis-classifications are mainly due to similar spectral behavior (e.g., class-overlap) resulting from high intra-class variabilities compared to the inter-class distances;
- Only a limited number of species/age-classes are addressed—problem related to the thematic simplification of many studies;
- To date, no generalization of spectral-temporal signatures has been achieved—problem to identify permanent properties which would permit a model transfer to other regions and/or seasons;
- No roll-out over large areas has been attempted—problem of data gaps, which increase with the size of the area of interest;
- Mixtures at crown levels are inevitable with deca-metric EO data—problem related to the use of hard instead of soft classifications.

Possible amendments

- Leveraging multi-year spectral datasets while extracting temporal/phenological features from the time series instead of simply stacking layers;
- Leveraging bi-directional properties of tree canopies (e.g., multi-angular EO data);
- Integrating high(er) spatial resolution data to minimize spectral mixtures and to leverage information content in color textures;
- 2-step approaches: expansion of reference data with very high-resolution data (e.g., drone);
- Combination of spectral datasets with LiDAR or other modalities (e.g., microwaves);
- Conversion of spectral properties into biophysical variables and use of species-inherent co-variations in those biophysical variables for tree species mapping.

Figure 4. Summary of the special issue "Tree species diversity mapping": main findings and conclusions, limits, and possible amendments.

4. A More Fundamental View Going Forward: Use of Physically-Based Approaches

The studies published in this special issue have in common the proposal of data-driven machine-learning based approaches, in which the classification is performed in the spectral-temporal feature space (e.g., based on spectral signatures and other spectral/temporal metrics). We argue that this is possibly contributing to the reported difficulties, as spectral-temporal features are not easily transferable across space and time, depend on observation conditions, and are not necessarily species-inherent. As an alternative approach, we propose instead, working in the feature space of biophysical variables retrieved using appropriate forest RTMs. In our opinion, this has not yet been investigated enough.

To illustrate our statement, let us assume deca-metric, multi/hyper-spectral optical EO data such as Landsat, S2, or EnMAP, where individual trees are not resolved. For simplicity, we also ignore possible atmospheric perturbations and calibration issues in the EO data. In this case, and under a more fundamental physical view, spectral signatures of trees under a given illumination/observation geometry and at a given acquisition date (t), are the expression of the biochemical composition and structural settings of the trees within the IFOV of the sensor (see also discussion of Baret & Buis [58] with respect to agricultural crops). If we denote the biochemical properties of the trees as Θbio, the canopy structural properties as Θstruc, background properties as Θbg, and observation and illumination geometry as Θview, this can be expressed as:

$$R(\lambda,t) = f(\Theta bio_t, \Theta struc_t, \Theta bg_t, \Theta view_t). \quad (1)$$

Important biochemical and structural variables are listed in Table 2. Note that the tree-related variables themselves (short: Θtree = Θbio + Θstruc), and their temporal (co)evolution within and across seasons, are the result of the growth and development of the respective genotype under the historical/current environmental setting and management practices (time-dependent notation here omitted for sake of clarity):

$$\Theta tree = f(gene, environment). \quad (2)$$

In this context, it is also worth noting that the "growth form or habitus", i.e., the phenotypical expression of the genotype leading to the trees we literally see with our eyes (Figure 5 right—examples of bark, branching, leaf form, growth form) is fundamentally different compared to EO measurements. When we observe tree species with our eyes, we tend to characterize and differentiate them based on leaf form, fruits and reproductive organs, bark structure, habitus, tree growth form etc. (Equation (3)). The spectral signatures of trees are predominantly influenced by factors other than the mentioned plant traits (Equation (2)), unless specifically examining high-resolution data from centimeter-level UAV and LiDAR analysis.

$$\text{Human discernible plant traits (pheno-type)} = f(\text{gene, environment}) \quad (3)$$

Table 2. Important biochemical and structural properties of trees. These biochemical and structural variables determine the spectral signature of a tree canopy. In **bold**, the spectrally most important variables. Note that understory/background variables are not included here.

Biochemical Properties	Structural Properties		
	Micro-Scale	Macro-Scale	Meso-Scale
Chlorophyll A + B content	**Leaf structure**	**Leaf Area Index**	**Crown size**
Leaf water thickness	Leaf hairing	**Leaf angle distribution**	**Gap size distribution/gap fraction**
Protein content	Waxing	**Leaf clumping/ arrangement**	Crown shape
Dry matter content/specific leaf area		Tree branching	Stem density
Other photosynthetic pigment content		Leaf size	
Non-photosynthetic pigment content		Leaf form	

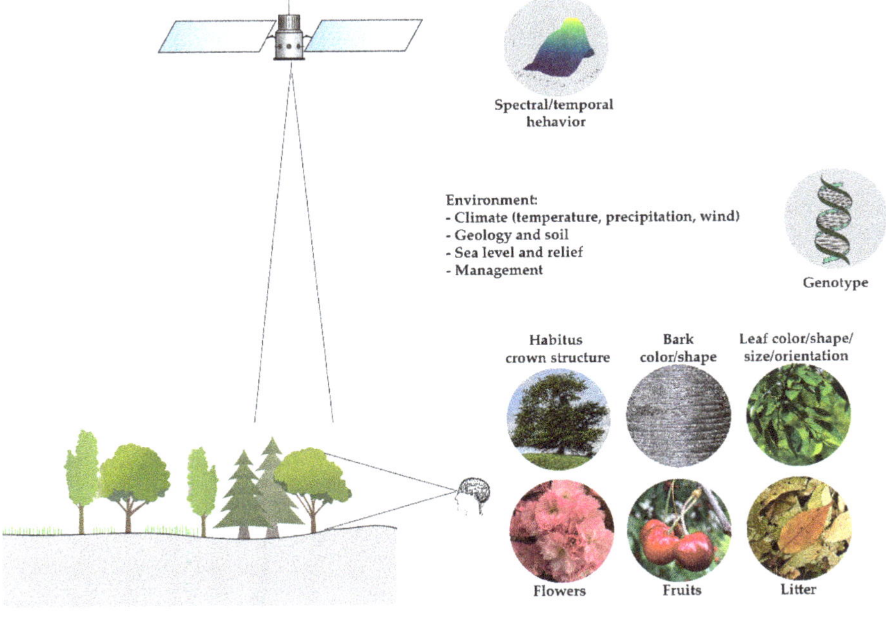

Figure 5. Tree species identification using (space-borne) EO measurements and human vision. Note the fundamental differences resulting from the different perspectives: features that we literally see from the ground (and partly from UAV and terrestrial sensors) (bottom right) and the spectral-temporal signatures that we measure from space (top right).

Using remote sensing data at deca-metric spatial resolution, we thus capture a completely different expression of the tree's geno-type, compared to computer/human vision and centi-metric resolution UAV/LiDAR systems (Table 3).

Table 3. Simplified enumeration of main traits and approaches used for the identification of tree species using different sensor modalities from laboratory (genetics) to satellite-based Earth Observation. In the right-most column, our recommendation for future research.

Genetics	Botany & Human Vision	Computer Vision	LiDAR	EO Current	EO Recommended
DNS analysis	Leaf shape & size	Crown structure	Point-cloud derived metrics	Spectral-temporal features	Biophysical variables (Table 2) extracted using RTM
	Bark structure & color	Crown shape	Crown shape	Supervised classification in spatial-spectral-temporal feature space	Learning of the temporal co-evolution of derived biochemical & structural variables
	Fruits and flowers	Branching	Branching pattern		
	Branching pattern				
	Habitus/growth form				

From the above it is also evident that the link of tree species to the biochemical and structural properties (Θtree) is inherently closer than the link to direct use of spectral signatures, as the latter relation is shaped by external factors (Θbg and Θview) and subjected to a strong time factor that limits generalization over large geographic regions. In botany and human vision, on the other hand, one recognizes (more or less) permanent and species-inherent properties (Table 4) that allow a trained observer to correctly classify many different tree species as the observed traits are relatively invariant. Unfortunately, the remote sensing literature is very short on attempts to retrieve the species-inherent co-evolution of biochemical and structural properties (Θtree), even though this would potentially lead to a more generalizable pattern.

Table 4. Important plant traits for human visual recognition of tree species.

Individual Tree	Tree Components
Habitus/crown form/shape Crown structure/branching	Size, shape, color, orientation of leaves/needles Structure and color of bark Color, shape, size, orientation of flowers Color, shape, size, orientation of fruits

In our opinion, this idea amounts to more than a simple hypothesis. Indeed, the fact that simple classifications of time series of spectral data are successful (at least to some degree), provides strong evidence that such species-inherent structural/biochemical pattern do indeed exist.

The relationships between the fundamental properties and the spectral signatures have already been formalized and modeled using various kinds of RTMs (e.g., [31,59]). Existing formalizations are mainly based on analytical expressions (e.g., SCOPE [60]), 3D geometric-optical models (e.g., FLIM [61]), and mixtures of both (e.g., INFORM [32]; PARAS [62,63]). In addition, ray-tracing models such as DART track the path of individual photons as they interact with the canopy elements, where the canopy is discretized into voxels with specific optical properties [64]. RTMs can be readily inverted using a panoply of methods such as look-up-tables, numerical optimization, predictive equations, Bayesian approaches and neural networks (e.g., [58,62,65]).

Existing radiative transfer models have definitively ample room for improvement, as highlighted by systematic but irregular evaluations against each other, and some natural and synthetic benchmark datasets (e.g., RAMI [31]). However, with some effort the models can certainly be improved so that the biophysical and biochemical traits can be derived with sufficient accuracy, using remotely sensed data. This seems to be a more plausible path towards robust and generalized models, compared to the current data-driven approaches. Indeed, as trees are living organisms embedded in their respective biotic/abiotic environment, the spectral signatures will change over the course of the season and evolve as trees age, as both the biochemical composition and the structural setting change/evolve. Hence, instead of "viewing" the classification task in the wavelength and reflectance space, (Figure 6a) one should, in our opinion, view species "living" in the space of Θbio and Θstruc (Figure 6b).

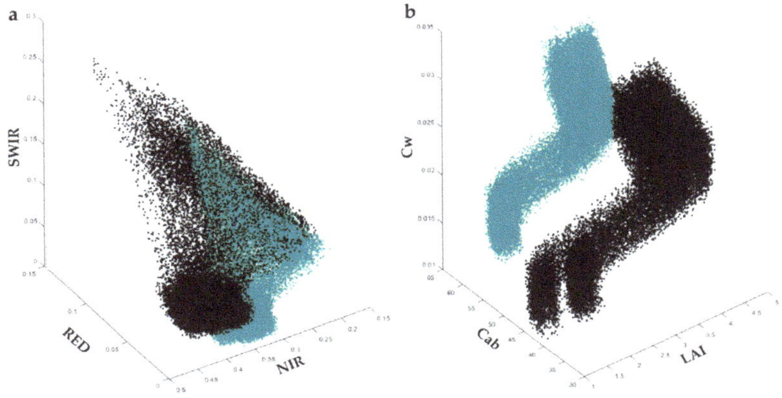

Figure 6. Illustration of two hypothetical species (black and green) "living" in two different feature spaces over the course of a growing season (t1 to t12): (**a**) shown in the 3-dimensional Red, NIR and SWIR feature space; (**b**) temporal co-evolution of species in the 3-dimensional feature space of Leaf area index (LAI), Chlorophyll A + B (Cab) and Water content (Cw).

Hence, from a physics-based deduction, as well as from an empirical inference, we strongly believe that this under-researched area needs to be exploited in order to make EO data analysis more generalizable across large geographic regions and more transferable across time and space. This is needed not only for the classical applications of remote sensing and biodiversity research, but also for more efficient implementations of financial instruments such as nature-based solutions (NBS). The latter rely heavily on robust and cost-efficient monitoring-reporting-verification (MRV) for supporting the development of (voluntary) carbon markets, which progressively integrate attempts to protect/enhance biodiversity.

Author Contributions: The entire paper development (conceptualization, visualization and writing) was done by both authors together, M.I. and C.A. All authors have read and agreed to the published version of the manuscript.

Funding: This research received no external funding.

Data Availability Statement: Not applicable.

Acknowledgments: The authors thank Katja Berger for valuable inputs.

Conflicts of Interest: The authors declare no conflict of interest.

References

1. *IPBES Global Assessment Report on Biodiversity and Ecosystem Services of the Intergovernmental Science-Policy Platform on Biodiversity and Ecosystem Services*; IPBES Secretariat: Bonn, Germany, 2019.
2. UNEP UN Biodiversity Conference (COP 15). Available online: http://www.unep.org/un-biodiversity-conference-cop-15 (accessed on 26 March 2023).

3. Pettorelli, N.; Safi, K.; Turner, W. Satellite Remote Sensing, Biodiversity Research and Conservation of the Future. *Phil. Trans. R Soc. B* **2014**, *369*, 20130190. [CrossRef] [PubMed]
4. Skidmore, A.K.; Pettorelli, N.; Coops, N.C.; Geller, G.N.; Hansen, M.; Lucas, R.; Mücher, C.A.; O'Connor, B.; Paganini, M.; Pereira, H.M.; et al. Agree on Biodiversity Metrics to Track from Space. *Nature* **2015**, *523*, 403–405. [CrossRef] [PubMed]
5. Dash, J.; Ogutu, B.O. Recent Advances in Space-Borne Optical Remote Sensing Systems for Monitoring Global Terrestrial Ecosystems. *Prog. Phys. Geogr. Earth Environ.* **2016**, *40*, 322–351. [CrossRef]
6. Pettorelli, N.; Wegmann, M.; Skidmore, A.; Mücher, S.; Dawson, T.P.; Fernandez, M.; Lucas, R.; Schaepman, M.E.; Wang, T.; O'Connor, B.; et al. Framing the Concept of Satellite Remote Sensing Essential Biodiversity Variables: Challenges and Future Directions. *Remote Sens. Ecol. Conserv.* **2016**, *2*, 122–131. [CrossRef]
7. Vihervaara, P.; Auvinen, A.-P.; Mononen, L.; Törmä, M.; Ahlroth, P.; Anttila, S.; Böttcher, K.; Forsius, M.; Heino, J.; Heliölä, J.; et al. How Essential Biodiversity Variables and Remote Sensing Can Help National Biodiversity Monitoring. *Glob. Ecol. Conserv.* **2017**, *10*, 43–59. [CrossRef]
8. Michałowska, M.; Rapiński, J. A Review of Tree Species Classification Based on Airborne LiDAR Data and Applied Classifiers. *Remote Sens.* **2021**, *13*, 353. [CrossRef]
9. Lu, T.; Brandt, M.; Tong, X.; Hiernaux, P.; Leroux, L.; Ndao, B.; Fensholt, R. Mapping the Abundance of Multipurpose Agroforestry Faidherbia Albida Trees in Senegal. *Remote Sens.* **2022**, *14*, 662. [CrossRef]
10. Pearse, G.D.; Watt, M.S.; Soewarto, J.; Tan, A.Y.S. Deep Learning and Phenology Enhance Large-Scale Tree Species Classification in Aerial Imagery during a Biosecurity Response. *Remote Sens.* **2021**, *13*, 1789. [CrossRef]
11. Tian, L.; Fu, W. Bi-Temporal Analysis of Spatial Changes of Boreal Forest Cover and Species in Siberia for the Years 1985 and 2015. *Remote Sens.* **2020**, *12*, 4116. [CrossRef]
12. Krzystek, P.; Serebryanyk, A.; Schnörr, C.; Červenka, J.; Heurich, M. Large-Scale Mapping of Tree Species and Dead Trees in Šumava National Park and Bavarian Forest National Park Using Lidar and Multispectral Imagery. *Remote Sens.* **2020**, *12*, 661. [CrossRef]
13. Dong, C.; Zhao, G.; Meng, Y.; Li, B.; Peng, B. The Effect of Topographic Correction on Forest Tree Species Classification Accuracy. *Remote Sens.* **2020**, *12*, 787. [CrossRef]
14. Egli, S.; Höpke, M. CNN-Based Tree Species Classification Using High Resolution RGB Image Data from Automated UAV Observations. *Remote Sens.* **2020**, *12*, 3892. [CrossRef]
15. Xu, K.; Zhang, Z.; Yu, W.; Zhao, P.; Yue, J.; Deng, Y.; Geng, J. How Spatial Resolution Affects Forest Phenology and Tree-Species Classification Based on Satellite and Up-Scaled Time-Series Images. *Remote Sens.* **2021**, *13*, 2716. [CrossRef]
16. Yan, S.; Jing, L.; Wang, H. A New Individual Tree Species Recognition Method Based on a Convolutional Neural Network and High-Spatial Resolution Remote Sensing Imagery. *Remote Sens.* **2021**, *13*, 479. [CrossRef]
17. Kovačević, J.; Cvijetinović, Ž.; Lakušić, D.; Kuzmanović, N.; Šinžar-Sekulić, J.; Mitrović, M.; Stančić, N.; Brodić, N.; Mihajlović, D. Spatio-Temporal Classification Framework for Mapping Woody Vegetation from Multi-Temporal Sentinel-2 Imagery. *Remote Sens.* **2020**, *12*, 2845. [CrossRef]
18. Takahashi Miyoshi, G.; Imai, N.N.; Garcia Tommaselli, A.M.; Antunes de Moraes, M.V.; Honkavaara, E. Evaluation of Hyperspectral Multitemporal Information to Improve Tree Species Identification in the Highly Diverse Atlantic Forest. *Remote Sens.* **2020**, *12*, 244. [CrossRef]
19. Jackson, C.M.; Adam, E. Machine Learning Classification of Endangered Tree Species in a Tropical Submontane Forest Using WorldView-2 Multispectral Satellite Imagery and Imbalanced Dataset. *Remote Sens.* **2021**, *13*, 4970. [CrossRef]
20. Varin, M.; Chalghaf, B.; Joanisse, G. Object-Based Approach Using Very High Spatial Resolution 16-Band WorldView-3 and LiDAR Data for Tree Species Classification in a Broadleaf Forest in Quebec, Canada. *Remote Sens.* **2020**, *12*, 3092. [CrossRef]
21. Wan, H.; Tang, Y.; Jing, L.; Li, H.; Qiu, F.; Wu, W. Tree Species Classification of Forest Stands Using Multisource Remote Sensing Data. *Remote Sens.* **2021**, *13*, 144. [CrossRef]
22. Lechner, M.; Dostálová, A.; Hollaus, M.; Atzberger, C.; Immitzer, M. Combination of Sentinel-1 and Sentinel-2 Data for Tree Species Classification in a Central European Biosphere Reserve. *Remote Sens.* **2022**, *14*, 2687. [CrossRef]
23. Immitzer, M.; Neuwirth, M.; Böck, S.; Brenner, H.; Vuolo, F.; Atzberger, C. Optimal Input Features for Tree Species Classification in Central Europe Based on Multi-Temporal Sentinel-2 Data. *Remote Sens.* **2019**, *11*, 2599. [CrossRef]
24. Karasiak, N.; Dejoux, J.-F.; Fauvel, M.; Willm, J.; Monteil, C.; Sheeren, D. Statistical Stability and Spatial Instability in Mapping Forest Tree Species by Comparing 9 Years of Satellite Image Time Series. *Remote Sens.* **2019**, *11*, 2512. [CrossRef]
25. Chaurasia, A.N.; Dave, M.G.; Parmar, R.M.; Bhattacharya, B.; Marpu, P.R.; Singh, A.; Krishnayya, N.S.R. Inferring Species Diversity and Variability over Climatic Gradient with Spectral Diversity Metrics. *Remote Sens.* **2020**, *12*, 2130. [CrossRef]
26. Immitzer, M.; Atzberger, C.; Koukal, T. Tree Species Classification with Random Forest Using Very High Spatial Resolution 8-Band WorldView-2 Satellite Data. *Remote Sens.* **2012**, *4*, 2661–2693. [CrossRef]
27. Waser, L.T.; Rüetschi, M.; Psomas, A.; Small, D.; Rehush, N. Mapping Dominant Leaf Type Based on Combined Sentinel-1/-2 Data—Challenges for Mountainous Countries. *ISPRS J. Photogramm. Remote Sens.* **2021**, *180*, 209–226. [CrossRef]
28. Maschler, J.; Atzberger, C.; Immitzer, M. Individual Tree Crown Segmentation and Classification of 13 Tree Species Using Airborne Hyperspectral Data. *Remote Sens.* **2018**, *10*, 1218. [CrossRef]
29. Lukeš, P.; Stenberg, P.; Rautiainen, M.; Mõttus, M.; Vanhatalo, K.M. Optical Properties of Leaves and Needles for Boreal Tree Species in Europe. *Remote Sens. Lett.* **2013**, *4*, 667–676. [CrossRef]

30. Hosgood, B.; Jacquemoud, S.; Andreoli, G.; Verdebout, J.; Pedrini, G.; Schmuck, G. *Leaf Optical Properties Experiment 93 (LOPEX93)*; Report EUR 16095 EN; European Commission Joint Research Center: Brussels, Belgium, 1994; pp. 1–46.
31. Widlowski, J.-L.; Mio, C.; Disney, M.; Adams, J.; Andredakis, I.; Atzberger, C.; Brennan, J.; Busetto, L.; Chelle, M.; Ceccherini, G.; et al. The Fourth Phase of the Radiative Transfer Model Intercomparison (RAMI) Exercise: Actual Canopy Scenarios and Conformity Testing. *Remote Sens. Environ.* **2015**, *169*, 418–437. [CrossRef]
32. Atzberger, C. Development of an Invertible Forest Reflectance Model: The INFOR-Model. *Decade Trans-Eur. Remote Sens. Coop.* **2000**, *14*, 39–44.
33. Schlerf, M.; Atzberger, C. Inversion of a Forest Reflectance Model to Estimate Structural Canopy Variables from Hyperspectral Remote Sensing Data. *Remote Sens. Environ.* **2006**, *100*, 281–294. [CrossRef]
34. Modzelewska, A.; Kamińska, A.; Fassnacht, F.E.; Stereńczak, K. Multitemporal Hyperspectral Tree Species Classification in the Białowieża Forest World Heritage Site. *For. Int. J. For. Res.* **2021**, *94*, 464–476. [CrossRef]
35. Immitzer, M.; Vuolo, F.; Atzberger, C. First Experience with Sentinel-2 Data for Crop and Tree Species Classifications in Central Europe. *Remote Sens.* **2016**, *8*, 166. [CrossRef]
36. Hovi, A.; Raitio, P.; Rautiainen, M. A Spectral Analysis of 25 Boreal Tree Species. *Silva Fenn.* **2017**, *51*, 7753. [CrossRef]
37. Leckie, D.G.; Walsworth, N.; Gougeon, F.A. Identifying Tree Crown Delineation Shapes and Need for Remediation on High Resolution Imagery Using an Evidence Based Approach. *ISPRS J. Photogramm. Remote Sens.* **2016**, *114*, 206–227. [CrossRef]
38. Rautiainen, M.; Lukeš, P.; Homolová, L.; Hovi, A.; Pisek, J.; Mõttus, M. Spectral Properties of Coniferous Forests: A Review of In Situ and Laboratory Measurements. *Remote Sens.* **2018**, *10*, 207. [CrossRef]
39. Rautiainen, M.; Stenberg, P.; Nilson, T.; Kuusk, A. The Effect of Crown Shape on the Reflectance of Coniferous Stands. *Remote Sens. Environ.* **2004**, *89*, 41–52. [CrossRef]
40. Jensen, R.R.; Hardin, P.J.; Hardin, A.J. Classification of Urban Tree Species Using Hyperspectral Imagery. *Geocarto Int.* **2012**, *27*, 443–458. [CrossRef]
41. Shang, X.; Chisholm, L.A. Classification of Australian Native Forest Species Using Hyperspectral Remote Sensing and Machine-Learning Classification Algorithms. *IEEE J. Sel. Top. Appl. Earth Obs. Remote Sens.* **2014**, *7*, 2481–2489. [CrossRef]
42. Gao, L.; Darvishzadeh, R.; Somers, B.; Johnson, B.A.; Wang, Y.; Verrelst, J.; Wang, X.; Atzberger, C. Hyperspectral Response of Agronomic Variables to Background Optical Variability: Results of a Numerical Experiment. *Agric. For. Meteorol.* **2022**, *326*, 109178. [CrossRef]
43. Rautiainen, M.; Mõttus, M.; Heiskanen, J.; Akujärvi, A.; Majasalmi, T.; Stenberg, P. Seasonal Reflectance Dynamics of Common Understory Types in a Northern European Boreal Forest. *Remote Sens. Environ.* **2011**, *115*, 3020–3028. [CrossRef]
44. Eriksson, H.M.; Eklundh, L.; Kuusk, A.; Nilson, T. Impact of Understory Vegetation on Forest Canopy Reflectance and Remotely Sensed LAI Estimates. *Remote Sens. Environ.* **2006**, *103*, 408–418. [CrossRef]
45. Einzmann, K.; Ng, W.; Immitzer, M.; Bachmann, M.; Pinnel, N.; Atzberger, C. Method Analysis for Collecting and Processing In-Situ Hyperspectral Needle Reflectance Data for Monitoring Norway Spruce. *Photogramm.-Fernerkund.-Geoinf.* **2014**, *2014*, 351–367. [CrossRef]
46. Fassnacht, F.E.; Latifi, H.; Stereńczak, K.; Modzelewska, A.; Lefsky, M.; Waser, L.T.; Straub, C.; Ghosh, A. Review of Studies on Tree Species Classification from Remotely Sensed Data. *Remote Sens. Environ.* **2016**, *186*, 64–87. [CrossRef]
47. Nasiri, V.; Beloiu, M.; Asghar Darvishsefat, A.; Griess, V.C.; Maftei, C.; Waser, L.T. Mapping Tree Species Composition in a Caspian Temperate Mixed Forest Based on Spectral-Temporal Metrics and Machine Learning. *Int. J. Appl. Earth Obs. Geoinf.* **2023**, *116*, 103154. [CrossRef]
48. Dalponte, M.; Ørka, H.O.; Gobakken, T.; Gianelle, D.; Næsset, E. Tree Species Classification in Boreal Forests with Hyperspectral Data. *IEEE Trans. Geosci. Remote Sens.* **2013**, *51*, 2632–2645. [CrossRef]
49. Ghosh, A.; Fassnacht, F.E.; Joshi, P.K.; Koch, B. A Framework for Mapping Tree Species Combining Hyperspectral and LiDAR Data: Role of Selected Classifiers and Sensor across Three Spatial Scales. *Int. J. Appl. Earth Obs. Geoinf.* **2014**, *26*, 49–63. [CrossRef]
50. Grabska, E.; Hostert, P.; Pflugmacher, D.; Ostapowicz, K. Forest Stand Species Mapping Using the Sentinel-2 Time Series. *Remote Sens.* **2019**, *11*, 1197. [CrossRef]
51. Pflugmacher, D.; Rabe, A.; Peters, M.; Hostert, P. Mapping Pan-European Land Cover Using Landsat Spectral-Temporal Metrics and the European LUCAS Survey. *Remote Sens. Environ.* **2019**, *221*, 583–595. [CrossRef]
52. Xi, Y.; Ren, C.; Tian, Q.; Ren, Y.; Dong, X.; Zhang, Z. Exploitation of Time Series Sentinel-2 Data and Different Machine Learning Algorithms for Detailed Tree Species Classification. *IEEE J. Sel. Top. Appl. Earth Obs. Remote Sens.* **2021**, *14*, 7589–7603. [CrossRef]
53. Hemmerling, J.; Pflugmacher, D.; Hostert, P. Mapping Temperate Forest Tree Species Using Dense Sentinel-2 Time Series. *Remote Sens. Environ.* **2021**, *267*, 112743. [CrossRef]
54. Dalponte, M.; Bruzzone, L.; Gianelle, D. Tree Species Classification in the Southern Alps Based on the Fusion of Very High Geometrical Resolution Multispectral/Hyperspectral Images and LiDAR Data. *Remote Sens. Environ.* **2012**, *123*, 258–270. [CrossRef]
55. Foody, G.M.; Pal, M.; Rocchini, D.; Garzon-Lopez, C.X.; Bastin, L. The Sensitivity of Mapping Methods to Reference Data Quality: Training Supervised Image Classifications with Imperfect Reference Data. *ISPRS Int. J. Geo-Inf.* **2016**, *5*, 199. [CrossRef]
56. Fowler, J.; Waldner, F.; Hochman, Z. All Pixels Are Useful, but Some Are More Useful: Efficient in Situ Data Collection for Crop-Type Mapping Using Sequential Exploration Methods. *Int. J. Appl. Earth Obs. Geoinf.* **2020**, *91*, 102114. [CrossRef]

57. Hościło, A.; Lewandowska, A. Mapping Forest Type and Tree Species on a Regional Scale Using Multi-Temporal Sentinel-2 Data. *Remote Sens.* **2019**, *11*, 929. [CrossRef]
58. Baret, F.; Buis, S. Estimating Canopy Characteristics from Remote Sensing Observations: Review of Methods and Associated Problems. In *Advances in Land Remote Sensing*; Liang, S., Ed.; Springer: Dordrecht, The Netherlands, 2008; pp. 173–201. ISBN 978-1-4020-6449-4.
59. Ross, J. *The Radiation Regime and Architecture of Plant Stands*; Springer Science & Business Media: Berlin/Heidelberg, Germany, 1981; ISBN 978-90-6193-607-7.
60. Van der Tol, C.; Verhoef, W.; Timmermans, J.; Verhoef, A.; Su, Z. An Integrated Model of Soil-Canopy Spectral Radiances, Photosynthesis, Fluorescence, Temperature and Energy Balance. *Biogeosciences* **2009**, *6*, 3109–3129. [CrossRef]
61. Rosema, A.; Verhoef, W.; Noorbergen, H.; Borgesius, J.J. A New Forest Light Interaction Model in Support of Forest Monitoring. *Remote Sens. Environ.* **1992**, *42*, 23–41. [CrossRef]
62. Schraik, D.; Varvia, P.; Korhonen, L.; Rautiainen, M. Bayesian Inversion of a Forest Reflectance Model Using Sentinel-2 and Landsat 8 Satellite Images. *J. Quant. Spectrosc. Radiat. Transf.* **2019**, *233*, 1–12. [CrossRef]
63. Rautiainen, M. The Spectral Signature of Coniferous Forests: The Role of Stand Structure and Leaf Area Index. Doctroal thesis, University of Helsinki, Helsinki, Finland, 2005.
64. Gastellu-Etchegorry, J.P.; Martin, E.; Gascon, F. DART: A 3D Model for Simulating Satellite Images and Studying Surface Radiation Budget. *Int. J. Remote Sens.* **2004**, *25*, 73–96. [CrossRef]
65. Kimes, D.S.; Knyazikhin, Y.; Privette, J.L.; Abuelgasim, A.A.; Gao, F. Inversion Methods for Physically-Based Models. *Remote Sens. Rev.* **2000**, *18*, 381–439. [CrossRef]

remote sensing

MDPI

Review

A Review of Tree Species Classification Based on Airborne LiDAR Data and Applied Classifiers

Maja Michałowska *,†,‡ and Jacek Rapiński †,‡

Institute of Geodesy and Civil Engineering, Faculty of Geoengineering, University of Warmia and Mazury in Olsztyn, 10-724 Olsztyn, Poland; jacek.rapinski@uwm.edu.pl
* Correspondence: maja.michalowska@uwm.edu.pl
† Current address: Prawocheńskiego 15, 10-724 Olsztyn, Poland.
‡ These authors contributed equally to this work.

Abstract: Remote sensing techniques, developed over the past four decades, have enabled large-scale forest inventory. Light Detection and Ranging (LiDAR), as an active remote sensing technology, allows for the acquisition of three-dimensional point clouds of scanned areas, as well as a range of features allowing for increased performance of object extraction and classification approaches. As many publications have shown, multiple LiDAR-derived metrics, with the assistance of classification algorithms, contribute to the high accuracy of tree species discrimination based on data obtained by laser scanning. The aim of this article is to review studies in the species classification literature which used data collected by Airborne Laser Scanning. We analyzed these studies to figure out the most efficient group of LiDAR-derived features in species discrimination. We also identified the most powerful classification algorithm, which maximizes the advantages of the derived metrics to increase species discrimination performance. We conclude that features extracted from full-waveform data lead to the highest overall accuracy. Radiometric features with height information are also promising, generating high species classification accuracies. Using random forest and support vector machine as classifiers gave the best species discrimination results in the reviewed publications.

Keywords: LiDAR; ALS; forestry; tree species; classification

Citation: Michałowska, M.; Rapiński, J. A Review of Tree Species Classification Based on Airborne LiDAR Data and Applied Classifiers. *Remote Sens.* **2021**, *13*, 353. https://doi.org/10.3390/rs13030353

Academic Editor: Lin Cao and Alfonso Fernández-Manso

Received: 21 November 2020
Accepted: 19 January 2021
Published: 20 January 2021

1. Introduction

Forest inventories, consisting of the systematic collection of information about forests, covers various spatial scales, from the stand level to the woodlot level, and from the regional to national and global levels. Depending on the purpose of the inventory, different scales require different levels of detail [1]. Getting complete information about a single tree in traditional forest inventory is a challenging task. Inspecting large tracts of forest areas is labor-, time-, and cost-intensive. Moreover, there exists a risk of not carrying out a comprehensive assessment of all trees due to potentially limiting factors, such as the area's topography preventing entry to specific territories, a lack of access to protected areas, or areas with dangerous wildlife [2]. Obtaining the spatial composition of tree species is crucial for economic, ecological, and technical reasons, as well as being essential for forest inventory. The importance of tree species maps has been highlighted in publications, as standalone products for forest management and planning [3,4], as input for species-specific growth and yields model, or for species-specific allometric models [5]. During the last four decades, remote sensing (RS) techniques have developed into a standard in large-scale forest inventory [1,6–14]. The acquisition of high-resolution data using RS techniques with a relatively small set of field sample plots allows for the efficient and automated forest inventory of large areas. Analysis on remote sensing data dedicated to forestry have become an alternative to traditional inventory methods. Unfortunately, tree species identification constitutes a bottleneck in remote sensing-based forest inventory. In optical remote-sensed imagery, the differentiating features overlap, and bidirectional reflectance makes the analysis difficult. Recent developments in active remote sensing, particularly Light Detection and Ranging (LiDAR) technology, have demonstrated great potential in forest mapping, due to their capability to gather a three-dimensional point cloud of a forest environment in a short time [15,16]. Laser scanning may be provided at:

1. Terrestrial level (Terrestrial Laser Scanning, TLS) for, among others, architectural heritage [17,18], buildings [19–21], roads [22], and bridges [23,24]; or
2. airborne level (Airborne Laser Scanning, ALS), using a flying platform (aircraft, helicopter, or Unmanned Aerial Vehicle; UAV), for aerial objects such as forests [25–27] or urban environments [28,29].

Airborne Laser Scanning systems consist of, among other things, a laser scanner, a Global Positioning System (GPS) receiver, and an Inertial Measurement Unit (IMU). The laser scanner uses laser light as an emission energy source and detects extremely minute energy returns from distant surfaces. The laser scanner's principle is to generate, release, and capture reflected light radiation, as well as recording the associated data. The characteristics of distant objects can be determined based on changes in the return energy from object backscatter, while knowing the characteristics of the emitted laser light. The 3D positions of all measured points can be calculated when the laser scanner's position and orientation are known. This is why a GPS receiver is used to register the flying platform's position, while the IMU simultaneously records attitude parameters at every measurement time [16,30]. Airborne laser scanning has gained acceptance as a rapid and accurate method for collecting georeferenced data as a three-dimensional point clouds for forest and urban environment surveying. Many publications have presented promising results for determining tree parameters based on LiDAR data. The studies performed so far have concerned the determination of parameters such as the location of the tree trunk, height, crown size, crown base height [31–33], species [27,34,35], type of the tree [36], stem volume [37–41], diameter at breast height [37], and biomass [42]. LiDAR data has been found to be advantageous for tree detection, being less affected by occlusions and shading than passive optical data [43–45]. Tree detection corresponds to the extraction of data that represents a tree, or a group of trees, in an entire geospatial data set. There are two commonly used approaches to relate source data to ground-measured response variables: Area-based approach (ABA) and individual tree crown (ITC) methods [46–48]. ABA methods provide statistically calibrated maps of forest stand parameters, which are useful for large-area inventories. ITC is referred to as single tree extraction, the smallest unit in the forest inventory, to present the spatial distribution of a single tree and its basic features (e.g., height, species, crown size) [48,49]. Latifi et al. (2015) emphasized that both ABA and ITC methods are appropriate for forest inventory. Naturally, these methods have some advantages over the others: ABA does not for allow the recognition of individual tree characteristics, while ITC does not detect trees in the understory and young, dense forest stands. A considerable group of raster-based approaches is available for single tree extraction which uses a canopy height model (CHM), a raster surface interpolated from LiDAR points hitting the tree canopy surface, as basic information. CHM methods have been often used in commercial and business environments, due to the speed of processing and format accessibility of the software. Unfortunately, these methods introduce errors and uncertainties due to the interpolation methods, grid spacing [50], or CHM smoothing filters [51,52]. These errors affect the subsequent estimation of tree metrics, leading to underestimation or overestimation of tree height. Over the years, more sophisticated methods for tree delineation on LiDAR data have been explored and described, with normalized-cut segmentation, supervoxel algorithms, graph-based approaches, and others [46,53–57]. Tree segmentation is an essential step in obtaining the characteristics of the specified object used in the classification. Accurate and proper extraction of the object is a key to achieving high accuracy in determining the object's features, which have an impact on species classification results [48,58,59]. To ensure and improve the correctness of tree extraction, manual or semi-manual approaches for tree identification have been used in many studies [43,60,61].

Unfortunately, manual and semi-manual tree extraction is time-consuming and challenging to implement in large-scale data sets.

The motivation for writing this review was an interest in the commercial market for species recognition based solely on LiDAR data. We reviewed research on tree species classification based on LiDAR data acquired by ALS. The objective of this article answering the following questions:

1. What LiDAR features are the most effective for tree species classification?
2. What is the most effective classification algorithm to take maximum advantage of the information extracted from LiDAR data to potentially increase the accuracy of tree species classification?

The remainder of this article is structured as follows: Section 2 describes the review procedure, quantification of species classification efficiency, and LiDAR-derived features and classifiers that might be used for species discrimination. Section 3 presents the results of an analysis of the effectiveness of LiDAR metrics and classifiers in species classification. Finally, in Section 4, our conclusions and discussions are drawn.

2. Materials and Methods

2.1. Review Approach

In this paper, we review recent publications about tree species classification performed on ALS data, with LiDAR-derived features used for the classification. In total, 44 publications were reviewed. From these publications, we dismantled comparative studies into sub-studies and treated them as individual cases. In this way, we obtained 97 study cases in total. We would like to point out that some reviewed publications described experiments based on LiDAR data combined with diverse sensor data (e.g., multispectral, hyperspectral, satellite imagery). We have included them in this review, but only within the scope that corresponds to LiDAR data [42,59,62–68].

The reviewed tree species classification studies presented many approaches to species classification using LiDAR-derived metrics. Scanning data was acquired under various conditions and using different sensors. There were different referenced data, very often different species to be recognized, and different methods of tree segmentation. In this paper, we focus only on verifying the effectiveness of groups of features and algorithms in species discrimination. We analyzed their performance by comparing the accuracies developed in every experiment.

2.2. Accuracy of Species Classification

The overall accuracy (OA) and, usually, the κ coefficient were calculated to quantify the species classification capability in the studies. The overall accuracy is calculated from the number of trees correctly classified relative to the total number of trees, while the κ coefficient is the proportion of chance-expected disagreements that do not occur, or the proportion of agreement after the chance agreement is removed from consideration. κ is calculated by Equation (1) [69]:

$$\kappa = \frac{N * \sum_{i=1}^{m} X_{ii} - \sum_{i=1}^{m} (X_{i+} * X_{+i})}{N^2 - \sum_{i=1}^{m} (X_{i+} * X_{+i})}, \tag{1}$$

where:

N—the total number of samples,
X_{ii}—the value on the diagonal of the confusion matrix,
m—the number of classes/species, and
X_{i+}, X_{+i}—the sums of values on the ith row and ith column, respectively.

There is an agreed explanation for the κ coefficient, the value of which ranges from −1 to 1. When κ is less than 0, it indicates less than chance agreement; κ from 0.01 to 0.20 indicates slight agreement; 0.21–0.40: fair agreement; 0.41–0.60: moderate agreement;

0.61–0.80: substantial agreement; and κ from 0.81 to 0.99 indicates almost perfect agreement [70]. Following Wang and Liu (2018), classification approaches that develop high overall accuracy also produce a high κ coefficient. κ values can be significantly lower than the overall accuracy. This might be observed in tree classification results, when the number of delineated species is too small or when there is a dominant species with a vast sample size [71].

Additionally, we analyzed the Number of Categories Adapted Index (NOCAI) to compare the accuracy of species classification. We calculated NOCAI due to suggestions that it is a more reasonable comparator than overall accuracy [71,72]. NOCAI is calculated by Equation (2):

$$NOCAI = OA * \frac{1}{k} * 100\%, \quad (2)$$

where:

OA—overall accuracy, and
k—the number of identified species ($\frac{1}{k}$ is expected to be achieved if trees were randomly assigned to a species).

NOCAI provides a more legitimate basis for comparison when different numbers of species are identified, by excluding the influence of species number in a defined approach. Higher NOCAI values suggest that the developed approaches for tree species classification are more efficient [72].

2.3. *Lidar Features for Tree Species Classification*

LiDAR data represent scanned objects in point cloud form. Every measured point stores additional information that might be used during the classification process. LiDAR-derived features can be classified into three major groups: geometric, radiometric, and full-waveform features. Geometric metrics describe the point cloud structure representing an object (or part of it) in 3D space. Radiometric parameters relate to the point's radiometric characteristics, while full-waveform metrics are characteristics assigned to the backscatter waveform [73].

All LiDAR-derived metrics can be determined based on a series of features derived from laser scanning, referring to vertical and horizontal structure of the scanned objects point cloud, the intensity of the backscattered signal, number of returns and return number, and parameters from the full-waveform decomposition. Structure parameters with the assistance of other LiDAR features, such as discrete-return intensity, return or echo types, and full-waveform intensity response can help to obtain complex information for forest inventory, especially for tree classification.

2.3.1. Geometric—Point Cloud

Laser scanning registers different tree foliage distribution and canopy branching structures, which are often typical for the specified species, thus helping with their recognition during species classification [74,75]. Geometric features used for tree species classification represent the tree's shape and geometry, or those of its parts [73]. Apart from height information, more complex LiDAR-derived geometric metrics that represent tree spatial neighborhoods in 3D or 2D (two-dimensional), such as grid, raster, voxel, or height layers of neighboring points, are valuable. Geometric features are regularly expressed in statistics, such as the arithmetic mean of the normalized height, standard deviation of normalized height, range of normalized height of all points, ratios of crown base height to tree height, ratios of crown volume to the crown area, and more. As well as a vast range of height statistics, there are many features describing the point distribution, such as kurtosis, skewness, penetration, percentiles of height distribution of points in the cluster, density, and number of points in the 10th level layer of the tree height [34,35,59,76]. Geometric tree features that might be obtained from LiDAR data, along with detailed definitions, are given in Table 1.

Table 1. Geometric features extracted from normalized LiDAR data.

Feature	Definition
H_{max}	Maximum of the normalized height of all points
H_{min}	Arithmetic mean of normalized height of all points above stated in meters threshold
H_{std}	Standard deviation of normalized height of all points above stated in meters threshold
H_{med}	Median of the normalized height of all points
H_{range}	Range of normalized height of all points above stated in meters threshold
HP_{10} to HP_{90}	10% to 90% percentiles of normalized height of all points above stated in meters threshold with a 10% increment
P	Penetration as a ratio between number of returns below stated meter and total returns
CA	Crown area as the area of 2D convex hull
CV	Crown volume as the 3D convex hull
CD	Crown equivalent diameter calculated from crown area considering crown as a circle
As	Asymmetry as a difference between two crown diameters taken in two directions
dns	Canopy density
D_1 to D_{10}	Density at a fixed height. $D_i = N_i/N_{total}$, where i = 1 to 10, N_i is the number of points within ith layer when tree height was divided into 10 intervals starting from specified meter, N_{total} is the number of all points

The shape of the tree crown might be advantageous for species recognition, as it is species-dependent. This has been confirmed, by Ørka et al. (2009), who pointed out that the major difference between spruce and birch is the rounder crown of birch and the conical shape of a spruce crown. Holmgren et al. (2008) indicated that Scots pine can be separated from spruce and deciduous trees based on the tree's relative crown base height, as Scots pine has a higher crown base than other species. However, Holmgren et al. (2008) emphasized that the height of the base crown varies in the group of species and may depend on factors not included in the classification process. Axelsson et al. (2018), for tree species classification, used geometric features to identify variations between species with large leaves and a dense crown and other species with thinner leaves and a sparser crown. Spheroid-type ellipsoid layers fitted to the tree's crown points defined differences in density and reflectance inside and outside the crown. This approach allowed the authors to find structural and spectral combinations for species recognition. Their experiment results indicated that the outer ellipsoid layers, as the highest horizontal layer of the tree crown, were highly ranked as one of the most effective features for species discrimination [76]. The same conclusions have been reported in previous investigations by Heinzel and Koch (2011) and Shi et al. (2018).

There have been many investigations in which the features of the crown's internal structure were identified and used for species classification. Li et al. 2013 developed algorithms to describe the internal structure of an individual tree crown and to test the effectiveness of those features for species recognition. LiDAR-derived features characterized the 3D texture, foliage clustering degree relative to tree envelope, foliage clustering scale, and gap distribution of an individual tree in horizontal and vertical directions. The approach developed an overall accuracy of 77.5% (κ = 0.7) for the classification of four species. Li et al. 2013 emphasized that the extraction of trees performed in the experiment needed improvement, such that the overall accuracy of classification could be even higher. Harikumar et al. (2017) presented a method of conifer species classification based on internal and external crown structure features. They defined new internal crown features using a novel internal branch structure model. This model uses a 3D region growing and principal component analysis to precisely delineate the branch structure. The proposed approach developed an overall accuracy of 93.7% for species classification; its effectiveness was also confirmed in conifer species recognition. Further experiments with the recommended method are needed, to consider the effects on the modeling process of crown overlap and understory vegetation, damaged trees with missing branches, or trees with non-symmetrical crown shapes. Features that represent the internal crown structure are useful in tree species classification [77–79]. Extraction of such features requires highly dense LiDAR data. The reviewed studies confirmed that the species classification effectiveness depends, among others, on the point cloud density [77,80,81]. Höfle et al. (2012) [82] investigated the effect of data density on classification and showed that classification accuracies remain relatively stable for point densities higher than 5 points/m^2. Li et al. 2013 investigated a positive linear correlation (R^2 = 0.88) between the LiDAR data point density and species classification accuracy. Their examination showed that higher overall accuracy of species classification was developed on the dense point cloud: For a point density of 2–5 points/m^2 the overall accuracy was approximately 50% while for over 50 points/m^2, the overall accuracy increased to 70%.

Some reviewed studies have shown that using geometric features solely for species classification generates low overall accuracy, compared to the classification accuracy obtained while using radiometric features, a combination of geometric and radiometric metrics, or full-waveform metrics. Observed misclassifications while using only geometric features might be due to many factors, such as similarities of morphology among different species, structural variations within the same species, the point cloud density, and so on. LiDAR height distribution might be useful for species discrimination for the forest tree species with different heights, but ineffective for multi-layered forest stands with smaller trees below the top-most canopy layer [83].

Therefore, species classification based on LiDAR-derived geometric features remains a demanding task in different types of forests [59].

2.3.2. Radiometric—Intensity and Reflection Type

The maximum energy of the backscattered echo, which represents the reflectivity of every point measured by the laser scanner as an intensity [16], has been demonstrated to be a useful and very promising feature in the tree species classification process [36,74,75,83,84]. Intensity features that might be derived from single-channel ALS data, along with explanations, are given in Table 2.

Table 2. Single-channel intensity features derived from LiDAR data.

Feature	Definition
I_{min}	Minimum of intensity
I_{max}	Maximum of intensity
I_m	Mean of intensity
I_{std}	Standard deviation of intensity
I_{cv}	Coefficient of variation of intensity
I_{med}	Median of intensity
I_{sk}	Skewness of intensity
I_{range}	Range of intensity
I_{kut}	Kurtosis of intensity
I_{10to90}	Intensity value at percentiles from 10% to 90% with 10% increment

The backscattered signal intensity is related to the foliage type, leaf size, leaf orientation, and foliage density [43,80,85]. Many studies have presented that the combination of LiDAR height information with radiometric features could achieve the complementary advantages of different details and generate better results for species classification [34,76,79,80,83,86]. Intensity features generate a higher accuracy of tree species discrimination when used together with geometric features, compared to the results generated with geometric or radiometric features alone [34,76,79,83,86,87]. Zhang and Liu (2012), while researching the effectiveness of classifiers for tree species classification, analyzed three sets of LiDAR features—structural features alone, intensity features alone, and a combination of structural and intensity features—for the classification of three species. Their results showed that intensity features alone generated lower overall accuracy (OA = 75.4%) of species classification than structural features alone (OA = 88.6%), while a combination of both variables yielded better overall accuracy (OA = 92.8%). You et al. (2020) also studied species classification based on LiDAR height, intensity, and a combination of height and intensity parameters. Their results showed that structural features generated the lowest overall accuracy (OA = 87.54%), not the intensity features, as in the study of Zhang and Liu (2012). The combination of both features gave the best overall accuracy (OA = 92.35%), in agreement with the results of Zhang and Liu (2012). Lin and Hyyppä (2016) proposed a modular framework for species classification based on deriving LiDAR data feature parameters of point-distribution, laser pulse intensity, crown (internal), and tree (external) structures. The accuracy of species classification based on point-distribution features generated the lowest overall accuracy (65.0%), as in other studies [83]. Intensity features generated 80.0% overall accuracy, crown-internal categorized 85.0%, while the integration of all individual optimal parameters generated the highest classification overall accuracy of 92.5%. The approach proposed by Lin and Hyyppä (2016) was presented as an effective tree species classification framework that, among others, reduces the complexity of determining which combination of feature parameters gives the best classification result.

In many studies, the type of reflection has been analyzed for species classification. Ørka et al. (2010, 2012) analyzed LiDAR-derived intensity metrics and echo categories to delineate tree species. Their results showed that first-returns provided the most information for species classification, single-returns the least information, and last-returns offered additional information. Ørka et al. (2010), for two sets of LiDAR data, emphasized that the most important classification features seemed to be normalized heights from the last echoes, density features from the last echoes, and intensity features from the first and last echoes. Different echo categories were affected differently by canopy conditions. Reitberger et al. (2008) found that the mean intensity of laser echoes inside the tree led to a higher classification of species accuracy than the mean intensity of the upper 10% of the tree crown. Korpela et al. (2010) compared tree species classification based on data from two laser sensors (data sets) with a one-year temporal mismatch using the first-or-only echo intensity features. Their results suggested that leaf size, orientation, and foliage density affect the intensity, therefore making it not affected only by reflectance. In their experiment, intensity normalization improved the classification performance in the data from each sensor (ALTM3100 sensor OA = 83.5%, improved by 1.8%; ALS50 OA = 88.5%, improved by 5.9%), as well as in the combined data set (OA = 90.8%, improved by 5.7%). Hovi et al. (2016) and Shi et al. (2018) confirmed the results of Korpela et al. (2010) that only returns were informative of tree species and less sensitive to structural differences within the species, compared to the first or single returns. Shi et al. (2018) emphasized that the mean intensity of first or single returns and the mean value of echo width are the most powerful LiDAR metrics for tree species classification, in agreement with the previous studies of Yao et al. (2012), in which the results showed that the mean intensity in the entire tree is the most crucial feature in the leaf-on case and the mean pulse width of single and first reflections in an entire tree segment works better in the leaf-off condition. Vauhkonen et al. (2010), in a study on spruce and pine classification based on laboratory measurements with active hyperspectral LiDAR instrument data, tested whether these species could be separated using combined range and reflectance measurements. The analysis focused on pulses that penetrated through the foliage, improving species classification accuracy with very similar reflectance properties. Vauhkonen et al. (2010) found out that species with highly similar reflectance properties can be better distinguished by analyzing the reflectance values of the types of returns separately.

Most ALS systems use one wavelength scanner, either near-infrared or shortwave infrared. The recently developed technique of Multispectral Laser Scanning (MLS) allows for the acquisition of a dense point cloud with additional spectral information of three different wavelength channels: 1550 nm, 1064 nm, and 532 nm. As a single-sensor solution, MLS is currently becoming an innovative tool for providing tree-specific valuable information required for forest mapping (i.e., structure and reflectance). Having broader spectral information could simplify the processing of data and facilitate its interpretation. However, the cost of data acquisition with MLS is higher than acquisition with aerial spectral sensors and ALS on the same flight. Nevertheless, it is expected that MLS acquisition costs will decrease as MLS techniques advance [34]. MLS has a great potential for the species classification process, which has been emphasized in many publications [34,35,76,88]. Features derived from multispectral ALS data that might be used for species classification are given in Table 3.

Table 3. Multi-channel intensity features derived from LiDAR data.

Feature	Definition
I^i_{min}	Minimum of intensity of *i*th channel
I^i_{max}	Maximum of intensity of *i*th channel
I^i_m	Mean of intensity of *i*th channel
I^i_{std}	Standard deviation of intensity of *i*th channel
I^i_{cv}	Coefficient of variation of intensity of *i*th channel
I^i_{med}	Median of intensity of *i*th channel
I^i_{sk}	Skewness of intensity of *i*th channel
I^i_{range}	Range of intensity of *i*th channel
I^i_{kut}	Kurtosis of intensity of *i*th channel
I^i_{10to90}	Percentiles of intensity of *i*th channel from 10% to 90% with 10% increment
I^{1+2+3}, I^{1+2}, I^{1+3}, I^{3+2}	Intensity computed from combined set of echoes from different channels
$I^i_F/(I^1 + I^2 + I^3)$	Ratios of intensity features, F refers to different single-channel intensity features
I^3/I^2, I^3/I^1, I^2/I^1	Ratios of intensity features
$(I^2 - I^3)/(I^2 + I^3)$	Index of intensity features

Yu et al. (2017) performed a study assessing tree classification using a different combinations of features gathered by MLS: point cloud features alone, single-channel intensity features, multi-channel intensity features, point cloud with single-channel intensity features, and all features combined. The highest overall accuracy of 85.9% for three species classification was generated using point cloud and single-channel intensity features. In contrast, the lowest overall accuracy (OA = 76.0%) was produced by point cloud features alone. The results showed that there is no significant difference in classification accuracy while using single-channel features (OA = 85.42%), single-channel and point cloud features (OA = 85.59%), or all features generated from the three channels (OA = 85.6%). This means that the point cloud features did not improve the accuracy of the species classification. Moreover, using multi-channel intensity features (OA = 81.60%) for species classification did not provide more information than single-channel intensity features. Yu et al. (2017) emphasized that additional information provided by multispectral laser scanning might be a valuable source of information. Still, as the results above showed, it cannot guarantee higher accuracy of tree classification. In a recent study, Kukkonen et al. (2019) compared the results of three main species classification using MLS with unispectral LiDAR (1064 nm wavelength channel) and showed that multispectral LiDAR obtained better results (OA = 88.2%, κ = 0.79) than unispectral LiDAR (OA = 85%, κ = 0.72). It should be noted that in the study of Kukkonen et al. (2019), multispectral LiDAR predicted broadleaved trees much better (PA = 51.1%) than unispectral LiDAR (PA = 16.8%). Yu et al. (2017) also confirmed that multi-channel intensity features are powerful while classifying birch (point cloud features PA = 45.8%, MLS features in the best scenario PA = 70.99%). In the experiment of Kukkonen et al. (2019), pine and spruce prediction for both multispectral and unispectral LiDAR was comparable. Kukkonen et al. (2019) also compared the results of three main species classification using multispectral LiDAR and unispectral LiDAR with 4-band aerial images collected at a different times. Their results showed that the overall accuracy of the classification performed on fused unispectral LiDAR data with aerial imaging is comparable to the overall accuracy of species classification performed on multispectral LiDAR data. Krüger geb. Amiri et al. (2018) simulated an MLS sensor by combining two different LiDAR sensors providing three wavelength channels and marked that the third spectral channel (532 nm) did not have any meaningful effect for tree species classification. Budei et al. (2017) examined discrete-return MLS data for identifying species and concluded that species identification accuracy could be improved by using MLS, mostly when the number of species classes is relatively high (seven or more species).

LiDAR intensity data contain plenty of information about different forest tree species. Unfortunately, it still has some limitations for tree species with similar canopy reflectances. Some of the reviewed publications have presented that the combination of intensity and height metrics could lead to complementary advantages in tree species classification, thus generating better results [34,76,79,80,83,86].

2.3.3. Full-Waveform Metrics

Some studies have demonstrated the high potential of full-waveform data for tree species classification by applying a set of LiDAR full-waveform features [4,36,37,60,84]. In general, full-waveform LiDAR systems sample and record the entire backscattered signal intensity at regular time intervals, unlike discrete-return LiDAR systems. To normalize information and decrease sources of errors in further processing steps, pre-processing of the signal is used. A cross-section profile can be directly derived from the recorded waveform and mapped into 3D data using deconvolution techniques. Extraction of detection peaks from the waveform can be also performed with a decomposition algorithm [89,90]. The most commonly used method is Gaussian decomposition, which is used to fit a generalized Gaussian function, according to the Levenberg–Marquardt algorithm [90]. Metrics that can be derived from the full waveform include, among others, the height of median energy, waveform extent, waveform distance, number of peaks, roughness of the outermost canopy, front slope angle, and return waveform energy [91–93]. All these metrics have physical explanations; for instance, waveform extent relates to maximum tree height, while return waveform energy describes the canopy surface characteristics. Features that might be derived from full-waveform decomposition for species classification are given in Table 4.

Table 4. Single waveform LiDAR metrics.

Feature	Definition
WD	Waveform distance–distance between waveform begin and the ground (last peak)
WE	Waveform extent–distance between waveform begin and end
HOME	Height of median energy–distance from waveform centroid to the ground (last peak)
HTRT	Height of median energy/canopy height ratio
FS	Front slope angle–vertical angle from waveform begin to the first peak of canopy return energy
NP	Number of peaks–number of detected peaks within each normalized composite waveform
H_x	x% quartile height from 25% to 100% with 25% increment–elevation at x% of the returned energy subtracted by the ground elevation
ROUGH	Roughness of outer canopy–distance from the waveform beginning to the first peak

Table 4. *Cont.*

Feature	Definition
RWE	Return waveform energy–the area below the waveform between beginning and end
GE	Ground energy–total intensity of the last mode
CE	Canopy energy–difference between return waveform energy and ground return energy
GRND	Ground return ratio–ground return energy divided by canopy return energy

Hovi et al. (2016) studied four LiDAR data sets acquired within 4 years for the classification of three species and concluded that waveform features outperformed the discrete-return intensity data (OA = 74–85%, κ = 0.57–0.75 vs. OA = 84–91%, κ = 0.74–0.86 for manually extracted trees; OA = 78–89%, κ = 0.64–0.82 vs. OA = 88–95% , κ = 0.79–0.91 for trees extracted by watershed segmentation). Improvements in this experiment were ascribed to the total backscattered energy feature. Reitberger et al. (2008) presented an approach for the decomposition of full-waveform data, using full-waveform features for the classification of deciduous and coniferous trees. The full-waveform decomposition provides additional attributes for the detected points, such as the intensity and pulse width, which improved the classification results in general. Additionally, full-waveform decomposition provided increased spatial point density, which allowed the internal tree structure to be highly resolved. With the proposed approach, an OA = 85.4% of classification was generated in leaf-on condition, while OA = 95.7% was obtained in leaf-off condition. In the study of Reitberger et al. (2008), the pulse width turned out to be the feature that provided the best result in the leaf-on case. Yao et al. (2012) demonstrated the usefulness of full-waveform features for tree segmentation and species classification of deciduous and coniferous trees, achieving OA = 95%. Heinzel and Koch (2011) found that the most important full-waveform parameter for species delineation was the mean intensity (singular reflections have the highest impact), the median of the width, and the mean total number of targets within a beam (all reflections). Heinzel and Koch (2012) investigated five multiple data sources (full-waveform, height, texture, hyperspectral, and CIR metrics) and showed that features derived from full-waveform data generated the highest classification accuracy (OA = 79.22%) for four species. In contrast, LiDAR features generated one of the lowest overall accuracies (OA = 47.27%). Their experiment showed that using combined features from all groups for species classification generated the greatest overall accuracy (OA = 88.03%). The combination of full-waveform backscattering properties with other metrics has a high potential for tree species classification, which has also been emphasized in other studies [81,94]. Yu et al. (2014) compared the results of three tree species classification using low-density LiDAR data (average 1 pulse/m^2) obtained with full-waveform features for the classification, discrete-return features, and a combination of those features. The best overall classification accuracy (OA = 73.4%) was generated when full-waveform features were included in the classification with features of discrete-return. Classification with solely full-waveform features developed OA = 71.5%, while only discrete-return features developed OA = 62.1%. The results suggested that full-waveform LiDAR data contains more information about structural and physical properties. Zhou et al. (2018) [95] investigated the capacity of full-waveform LiDAR data for species identification through the integration of waveform metrics with random forest and conditional inference forest methods and Bayesian multinomial logistic regression. Their experiment showed that composite waveforms are more informative than raw waveforms and waveform-based point clouds for characterizing tree species in the study area. Cao et al. (2016) indicated that full-waveform LIDAR data have significant potential for classifying tree species in subtropical forests. In their study, full-waveform metrics were extracted using a voxel-based composite waveform approach and tested to distinguish species at three levels of discrimination. The applied approach ensured a relatively high accuracy of species classification: OA = 68.6% for six species, OA = 75.8% for the four main species, and OA = 86.2% for conifers and deciduous trees. Full-waveform metrics, based on the height of median energy, waveform distance, and several waveform peaks, showed high classification importance and were stable across various voxel sizes. The obtained results indicated that voxel-based approaches can alleviate some of the problems associated with large scanning angles. Blomley et al. (2017) presented a novel methodology to derive features for tree species classification by capturing the geometric distribution of laser scanning returns or their waveform attributes. The spin image algorithm used in this approach relies on spinning the image plane around a specific axis, collecting the number of LiDAR returns or mean values of returns attributes per pixel as respective values. The results showed that the used approach improved the classification of three species in almost all test cases for two waveform LiDAR data sets that differed in footprint size, pulse width, density, and laser wavelength. Although the basic classification accuracy given by the non-spatial metrics was very high (OA = 90.1% for the LMS-Q680i data set, OA = 90.5% for the ALS60 data set), the overall improvement in the combined case of a spin image with non-spatial metrics was equal to 1.1% for LMS-Q680i data and 2.0% for ALS60 data.

All studies presented above demonstrated the great potential of tree species classification based on data from full-waveform LiDAR. It has been confirmed that attributes and metrics derived from full-waveform data can improve tree species classification performance, compared to the species classification based on geometric or radiometric features, or their combination.

2.4. Seasonal Tree Condition

Tree species classification in all reviewed studies was carried out based on LiDAR data acquired under leaf-on condition [60,81], leaf-off condition [94,96], or under both conditions [37,61,84,86]. Seasonal changes in canopy leaf conditions influence the physical structure of the vegetation and impact the penetration and concentration of ALS pulses near the top of the canopy [97]. In some of reviewed studies, the authors concluded that the leaf-off data set produced higher species classification accuracy than the leaf-on data set [84–86]. Reitberger et al. (2008), for broadleaved and coniferous species classification, obtained OA = 95.7% for leaf-off condition, while the leaf-on condition LiDAR data generated OA = 85.4%. In this study, the classification of deciduous trees worked better in leaf-off condition. Kim et al. (2009), for eight broadleaved species and seven coniferous species, obtained better results on the leaf-off data set, similar to Reitberger et al. (2008), with leaf-off OA = 83.4%, while on the leaf-on data set gave OA = 73.1%. Shi (2018) reported only a slight improvement in classification accuracy when using leaf-off data, compared to leaf-on data. However, in the study of Shi et al. (2018), there was no statistically significant difference in species classification accuracy between leaf-on and leaf-off LiDAR features. The combination of features of both data sets significantly increased the overall accuracy of species classification from OA = 58.2% (leaf-on) and OA = 62.0% (leaf-off) to OA = 66.5% [86]. Other studies have confirmed that the integration of LiDAR data acquired in two seasons under leaf-on and leaf-off conditions improves tree classification [61,85]. Sumnall et al. (2015) [98] concluded that both leaf-on and leaf-off data benefit greater complementary information about the forest canopy profile than discrete-return or full-waveform LiDAR data.

In summary, data acquisition in the leaf-off condition offers a higher potential for representing the tree crown structure using radiometric features. All internal structure metrics are more efficiently used for species classification when there is no foliage cover—the dense canopy under leaf-on conditions fully intercepts the laser beam in the crown. It causes a lower proportion of returns interacting with understory vegetation. The leaf-off condition produces more intermediate returns, which provides more information of the vertical tree structure, driving an improvement in the accuracy of species discrimination [61,81,86,97]. The combination of the two data sets of leaf-on and leaf-off conditions can increase the overall accuracy of tree classification.

2.5. Classification Algorithms

In recent years, machine learning techniques, due to the accuracy of the produced classifications, have received increasing interest and gained popularity in many applications. Machine learning classifiers are algorithms that allow for the prediction of a class of input data based on specific methods built into the algorithms based on the use of training data. Combining machine learning techniques with remote sensing data creates possibilities for solving complex problems, such as tree species classification [61].

All machine learning algorithms can be divided into three classes: classification tree methods (e.g., decision trees, Random Forest—RF), grouping and separability methods (e.g., Support Vector Machine—SVM, k-nearest neighbors—kNN), and rule creation and application methods (Convolutional Neural Network—CNN) [61]. Species classification based on remote sensing data has evolved along with the methodological developments of statistical learning. In the reviewed studies, the following species classifiers have been applied: logistic regression [35,43,66], linear discriminant analysis [43,75,80,87], and kNN [43,66], with the most popular methods recently being random forest [3,34,43,61,64,81,88,99–101] and support vector machines [3,5,61,66,87,101–104].

2.5.1. Decision Trees and Random Forest

Decision trees are effective tools for prediction and are the most widely used methods for supervised learning. Decision tree algorithms identify ways to split a set of data based on every feature of the data set. The model obtains nodes—locations where the tree splits, according to the value of some features—and edges—the outcome of a split to the next node. Building a decision tree and specifying features that split the data might be challenging. This is why many decision trees use a randomly selected, independently sampled feature at every single split; this is called the Random Forest technique. It has been noted that random inputs and random features produce good results in classification with the random forest method [105].

2.5.2. Support Vector Machine

Support vector machine is a non-linear classification supervised machine learning model, which analyzes data and recognizes patterns. An SVM uses geometry to make a categorical prediction: it classifies input variables by finding an optimal separating hyperplane that maximizes the margin width between categories in a data set. The training points near the hyperplane are called support vectors. Support vector machines can be used for regression and classification analysis [106].

Nguyen et al. (2019) performed a weighted SVM (WSVM) classifier for tree species recognition using data from ALS as a strategy to correct the loss of SVM performance-dependent distribution of tree species and the presence of unreliable data due to inaccurate positions of field data and errors in tree extraction. The WSVM classifier extracts and uses species weights and sample weights (i.e., high weights and reduced weights), as importance is defined based on a strategy that relies on the distribution of unlabeled samples in the feature space. Their experimental results showed an improved classification accuracy of minority classes: WSVM provided 2.4% higher OA and 3.9% higher κ coefficient, compared to traditional SVM.

3. Results

In this paper, we review publications investigating tree species classification performed on ALS data with particular emphasis on LiDAR-derived features used for classification. The reviewed studies were categorized into three LiDAR-derived metric groups, which were used for species classification:

- *geometry* (G)—geometric features: statistics, point distribution, vertical profile, shape;
- *intensity+* (I+)—radiometric features, intensity metrics combined with geometric features, intensity features from three Multispectral Laser Scanning channels;

- *waveform* (WF)—features extracted from the full-waveform decomposition, full-waveform metrics combined with other metrics.

All generated results from the reviewed publications, along with information about the number of discriminated species and the calculated NOCAI, are listed in Table A1 in Appendix A, for studies where *geometry* features were assessed for species classification; Table A2 for the *intensity+* features group; and Table A3 for the *waveform* features group.

Most of the studies were conducted for the classification of three species (39 cases), followed by studies for two species (21 cases), four species (18 cases), six species (10 cases), and five species (5 cases). The classification results of seven, nine, and 15 species were conducted only in one publication each. The number of publications and studies grouped by the number of discriminated species is presented in Figure 1.

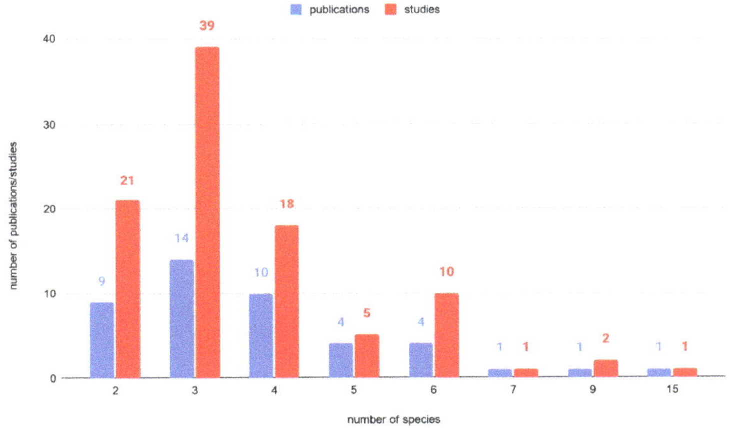

Figure 1. Number of reviewed publications and studies for specified numbers of classified species.

3.1. Effectiveness of the Features Groups in Species Classification

Of the reviewed publications, 39 studies were conducted equally for species classification using *intensity+* features and *waveform* features, while only 19 studies were conducted on *geometry* features solely. The number of conducted studies, with their division into feature groups used to classify a specified number of discriminated species, is presented in Figure 2. The most common studies were processed to classify three species using *intensity+* features (18 cases) and *waveform* features (15 cases).

It should be noted that in general, the overall accuracy of species classification can be influenced by many factors, such as LiDAR data density, season of acquisition, number of discriminated species, approaches for tree delineation from the point cloud, sampling situations, features taken for the classification process, and so on. In this paper, the overall accuracies of all reviewed studies were compared only based on the number of discriminated species and the feature group that was taken for the species classification. Figure 3 presents the relationship between the overall accuracy and the number of classified species for studies where two, three, four, or six species were classified. Results for five, seven, nine, and 15 species were excluded, due to not being representative for statistical analysis. The overall accuracy of species classification declined with the number of classified species. This was similar to the relationship between NOCAI and the number of species, which is presented in Figure 4.

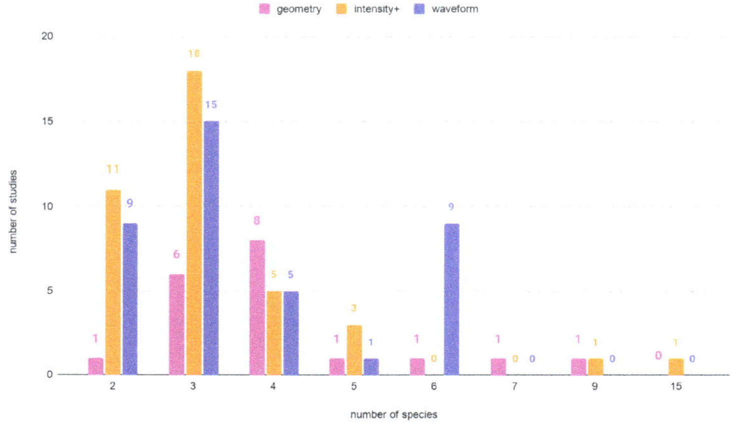

Figure 2. The number of conducted studies with division into feature groups (*geometry*, *intensity+*, *waveform*) used to classify a specified number of species.

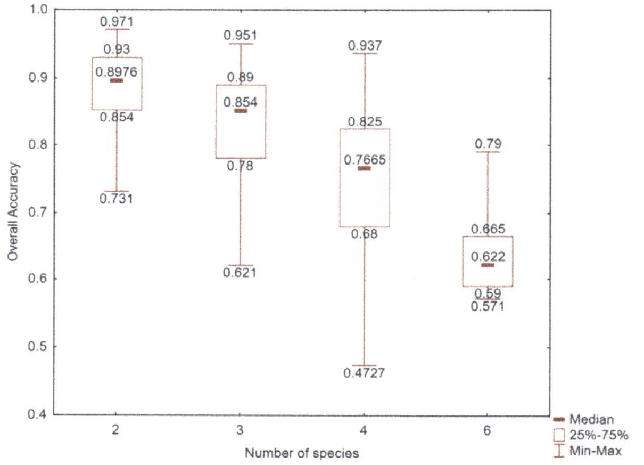

Figure 3. Relationship between the overall accuracy and number of discriminated species in the reviewed studies.

Studies with overall accuracy more significant than the median overall accuracy of all studies for a discriminated number of species (2, 3, 4, and 6 species) are listed in Table A4. Table A4 includes a description of which quartile the result of the study belonged to: QM-U represents the range from the median to upper quartile (75%), while QU-UW represents the range from upper quartile (75%) to the upper whisker. By the analysis of the number of these studies, it was observed that the most numerous studies were those in which the *waveform* feature group was used for species discrimination (see Figure 5). Investigations proceeded with metrics from the full-waveform decomposition were useful for tree species classification (10 cases in QM-U, 12 cases in QU-UW). The *waveform* group metrics were followed by the *intensity+* group (7 cases in QM-U, 6 cases in QU-UW) and *geometry* group (4 cases in QM-U, 6 cases in QU-UW), respectively.

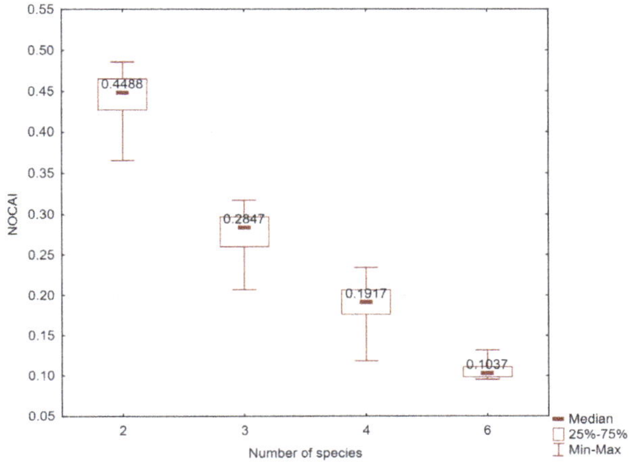

Figure 4. Relationship between NOCAI and the number of discriminated species in the reviewed studies.

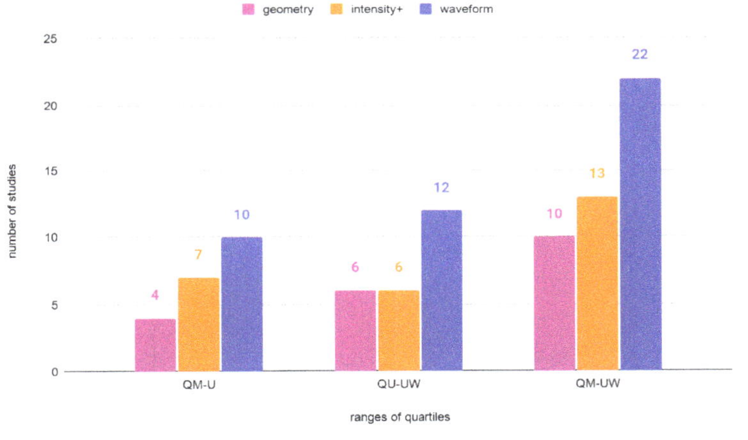

Figure 5. The number of studies with OA higher than median OA, grouped by feature groups (*geometry*, *intensity+*, *waveform*) used for classification with a division into the range of quartiles; QM-U—the range from the median to upper quartile, QU-U—the range from the upper quartile to the upper whisker, and QM-UW—the range from the median to the upper whisker.

To compare the overall accuracy of classification generated when using the different feature groups, the most relevant studies of three species classification with calculated κ coefficient were analyzed. Figure 6 presents a comparison of the classification accuracy using the overall accuracy, κ coefficient, and NOCAI for the two most relevant studies of every feature group. Species classification using full-waveform features seemed to generate better results in general (OA = 95.0–95.1%, κ = 0.91–0.912, NOCAI = 31.67–31.70%) than geometry features (OA = 76.04–86.3%, κ = 0.84–0.82, NOCAI=25.35–28.77%) and intensity features or their combination (OA = 90.80–89.00%, κ = 0.57–0.759, NOCAI = 29.67–30.27%), corresponding to the previous conclusion. It should be emphasized that the *intensity+* group features also produced high accuracy in species discrimination.

Figure 6. Classification accuracy using overall accuracy, κ coefficient, and NOCAI value of the studies where three tree species were classified with feature group; G—*geometry* group, I+—*intensity+* group, WF—*waveform* group; the studies of classification results are presented in the square bracket.

3.2. Effectiveness of the Classifiers in Species Classification

In several of the reviewed studies, tree species classification was carried out using more than one classifier under each scheme to examine the performance of classifiers. Table 5 presents a list of publications with overall accuracy and κ coefficient generated by the following classifiers: k-NN, LDA, LR (Logistic Regression), MLC (Maximum-Likelihood Classifier), RF, SVM, and DL (Deep Learning methods).

Based on the reviewed publication results, random forest and support vector machine classifiers generated the highest overall accuracy for species classification, compared to other classifiers, while the differences in RF and SVM results can be treated as marginal [61,67]. Nevertheless, in some of the studies, the SVM classifier has been proved to be more effective in tree species discrimination than other classifiers [3,66,87,101]. Despite the high accuracy that random forest and support vector machine methods might generate, it should be emphasized that using a particular classifier always has advantages and disadvantages. Followed by Fassnacht et al. (2016), with a non-parametric SVM approach, no distribution assumption is required and, so, it is appropriate when incorporating non-remote sensing variables into the classification for robustness to noise and rapid prediction of high-dimensional data, where relatively few training data are needed; meanwhile, the optimal design of multi-class SVM is demanding. There is a relatively high computational cost for training due to its algorithmic complexity. For random forest, no distribution assumption is required, the classifier is less sensitive to the number of input variables and overfitting, and there is an intuitive derivation of the accuracy and importance of the variable. At the same time, it may be overfitted in the presence of noisy data and may be biased in the case of response classes that have a different number of levels [107]. Fassnacht et al. (2016) emphasized that from the experience of the authors of their review, the choice of the classifier itself is often of low importance if the data is adequately preprocessed to match the requirements of the classifier.

Table 5. List of reviewed studies where more than one classifier was used for tree species classification; results are presented as overall accuracy and κ coefficient (if calculated); k-NN—k-Nearest Neighbors, LDA—Linear Discriminant Analysis, LR—Logistic Regression, MLC—Maximum-Likelihood classifier, RF—Random Forest, SVM—Support Machine Vectors, DL—Deep Learning methods.

Article	kNN	LDA	LR	MLC	RF	SVM	DL
[61]					83.1% N/A	82.0% N/A	
[43]	73.4% 0.639	77.1% 0.628			81.7% 0.690		
	76.0% 0.681	78.9% 0.656			83.5% 0.720		
	78.4% 0.644	79.6% 0.664			82.6% 0.706		
	85.1% 0.774	87.4% 0.785			88.5% 0.804		
	83.1% 0.715	82.2% 0.703			85.1% 0.750		
	89.0% 0.816	88.5% 0.805			89.3% 0.819		
	89.4% 0.840	90.0% 0.830			90.8% 0.844		
[87]		83.2% N/A				88.6% N/A	
		74.6% N/A				75.4% N/A	
		86.4% N/A				892.8% N/A	
[65]						60.3% 0.476	62.3% 0.513
[66]	83.6% 0.780		86.7% 0.822	84.7% 0.79		89.7% 0.863	
	49.0% 0.377		47.3% 0.364	46.3% 0.351		50.5% 0.393	
[67]					79% 0.75	79% 0.74	
					78% 0.73	78% 0.74	

4. Conclusions and Discussion

In this paper, we reviewed publications investigating tree species classification performed on airborne laser scanning data, with particular emphasis on studies where LiDAR-derived features were used for the classification. In total, 44 publications including tree species classification were reviewed, for a total of 97 performed studies. All reviewed studies were categorized into the three metric groups used for species discrimination: *geometry* (geometric features), *intensity+* (radiometric features, intensity metrics combined with geometric features, intensity features from three Multispectral Laser Scanning channels), and *waveform* (features extracted from the full-waveform decomposition, full-waveform metrics combined with other metrics). It was observed that the most numerous studies with high overall accuracy (greater than the median overall accuracy of the studies) were those in which the waveform feature group was used for species discrimination. This corresponds to the conclusion that full-waveform data have great potential in the tree species classification process [4,36,37,60,84]. The benefit of full-waveform features is derived from the available waveform information, which can be directly accessed or derived by further processing. Additionally, it has a higher point density and a theoretically unlimited number of recorded returns for each emitted pulse [73]. It should be emphasized that radiometric features (*intensity+* group) also generated high accuracy. Radiometric features seem to be useful and very promising features in the tree species classification process, as also demonstrated in many studies [36,74,75,83,84]. Moreover, the combination of radiometric features with LiDAR height information could lead to achieving the complementary advantages of the different information needed in the species classification process to improve the overall accuracy [34,76,79,80,83,86]. In this paper, studies where species classification was carried out by more than one classifier under each scheme were reviewed, to examine the performance of particular classifiers. Based on the gathered results, the random forest and support vector machine classifiers generated the highest overall accuracy for species classification. In some studies, the SVM classifier has been proved to be more effective in tree species discrimination than other classifiers [3,66,87,101]; however, in other studies, the differences between RF and SVM results were marginal [61,67]. Unfortunately, a specific group of features dedicated to species classification and the particular classifiers used for the classification do not guarantee a high overall accuracy. Employing multiple features for species classification is required to increase the discrimination accuracy. It is recommended to exploit more features, especially using data fusion of active and passive optical systems, which is in common with the observed trend of an increasing number of studies with the usage of more than one sensor type data for species classification [107]. LiDAR technology, which is valued for its unique capabilities for detecting 3D structural information, is a great supplement to discrimination based on any optical remote-sensed imagery. It has been proven, in many publications, that the combination of LiDAR data with multispectral or hyperspectral imagery generates greater species classification accuracy than LiDAR data alone [42,59,62–68]. LiDAR data also complements imagery data used for other purposes, such as mapping dead trees with crowns and snags (dead trees without crowns) [108,109], analyzing forest structural complexity [110,111], and determining bird species habitats [112], where passive imagery alone would not be as efficient as a set fused with LiDAR data. Fusion of data from different sensors deserves a great attention, as it is a valuable data set to provide sophisticated analysis and get satisfactory results. Although multispectral, hyperspectral, and LiDAR data provide complementary information, the integration of these data into an information extraction procedure can be challenging. Nevertheless, combining remote sensing data-derived features and machine learning techniques creates possibilities for solving complex problems, such as tree classification [61]. Although there are multiple features, different classifiers could be considered simultaneously by exploiting the best processing chain for each specific tree class. Improvement of accuracy of individual tree extraction and the optimization of feature selection for discrimination could lead to higher classification accuracy.

Apart from all the algorithms and methods used in tree classification procedures, the technological development of sensors plays an important role. As the technical parameters of the equipment increase, it will become possible to obtain more accurate data and more powerful sensor-derived metrics; thus, the classification results could be more accurately determined. In terms of classification algorithms, the fastest development has been predicted for artificial intelligence-related algorithms; we have already seen their application in tree classification. Future research should put more effort into the investigation of species-specific feature characteristics and detailed analyses of the impacts of technical factors, tree extraction methods, and various growing conditions on the feature characteristics. Additionally, for different types of forests, simplified data sets combining the most powerful significant tree species characteristics from different sensors could be examined and developed. Approaches and algorithms that fuse different remote sensing data sets in a more profound way should be developed, as multiple sources of remote sensing data should be included in the classification attempts.

Author Contributions: Conceptualization, J.R. and M.M.; investigation, M.M.; Writing—Original draft preparation, M.M.; Writing—Review and editing, M.M. and J.R.; visualization, M.M.; supervision, J.R.; project administration, J.R. All authors have read and agreed to the published version of the manuscript.

Funding: This research was funded by Ministerstwo Nauki i Szkolnictwa Wyższego (Ministry of Science and Higher Education), Republic of Poland, under the program entilted "Doktorat wdrożeniowy" (Industrial Doctoral Program) [DWD/3/19/2019].

Institutional Review Board Statement: Not applicable.

Informed Consent Statement: Not applicable.

Data Availability Statement: Data sharing not applicable.

Conflicts of Interest: The authors declare no conflict of interest.

Abbreviations

ABA	Area-Based Approach
ALS	Airborne Laser Scanning
CHM	Canopy Height Model

CNN	Convolutional Neural Network
DL	Deep learning
G	geometry features group
GPS	Global Positioning System
I+	radiometric features group
IMU	Inertial Measurement Unit
ITC	Individual Tree Crown
kNN	k-nearest neighbors
LiDAR	Light Detection and Ranging
LR	Logistic Regression
MLC	Maximum-Likelihood Classifier
MLS	Multispectral Laser Scanning
NOCAI	Number of Categories Adapted Index
OA	Overall Accuracy
QM-U	quartile range from the median to upper quartile (75%)
QM-UW	quartile range from the median to the upper whisker
QU-UW	quartile range from upper quartile (75%) to the upper whisker
PA	Producer's Accuracy
RF	Random Forest
RS	Remote Sensing
SVM	Support Vector Machine
TLS	Terrestrial Laser Scanning
UAV	Unmanned Aerial Vehicle
WF	full-waveform features group
WSVM	Weighted Support Vector Machine
2D	two-dimensional
3D	three-dimensional

Appendix A

Table A1. List of results of reviewed studies where species classification was performed with a *geometry* feature group; list sorted by species number and overall accuracy; DM—additional information for studies where trees were delineated manually.

Number of Species	OA	κ	NOCAI	Article
2	95.0%	N/A	47.50%	[75]
3	88.6%	N/A	29.53%	[87]
3	86.3%	0.759	28.77%	[63]
3	85.4% (DM)	0.739	28.47%	[63]
3	83.15% (DM)	N/A	27.72%	[61]
3	76.04%	0.57	25.35%	[34]
3	62.1%	N/A	20.70%	[81]
4	93.7%	N/A	23.43%	[78]
4	89.7%	0.86	22.43%	[66]
4	85%	N/A	21.25%	[79]
4	82.5%	N/A	20.63%	[79]
4	77.5%	0.7	19.38%	[77]
4	68%	N/A	17.00%	[35]
4	65%	N/A	16.25%	[79]
4	47.27%	N/A	11.82%	[4]
5	87.54%	0.81	17.51%	[83]
6	79%	0.75	13.17%	[67]
7	50.5%	0.39	7.21%	[66]
9	41.3%	N/A	4.59%	[76]

Table A2. List of results of reviewed studies where species classification was performed an *intensity+* feature group; list sorted by species number and overall accuracy; DM—additional information for studies where trees were delineated manually.

Number of Species	OA	κ	NOCAI	Article
2	97.1%	0.94	48.55%	[40]
2	97%	N/A	48.50%	[113]
2	90.6%	N/A	45.30%	[85]
2	90.0%	0.80	45.00%	[40]
2	89%	N/A	44.50%	[113]
2	86.9%	0.74	43.42%	[40]
2	86%	N/A	43.00%	[113]
2	83.4%	N/A	41.70%	[85]
2	83%	N/A	41.50%	[113]
2	78%	N/A	39.00%	[113]
2	73.1%	N/A	36.55%	[85]
3	92.8%	N/A	30.93%	[87]
3	90.8% (DM)	0.84	30.27%	[43]
3	89%	0.82	29.67%	[60]
3	88.2%	0.79	29.40%	[27]
3	88%	0.82	29.33%	[62]
3	88%	0.79	29.33%	[60]
3	85.42%	0.75	28.47%	[34]
3	85% (DM)	0.75	28.33%	[60]
3	85.0%	0.72	28.33%	[27]
3	83%	0.71	27.67%	[60]
3	82% (DM)	0.71	27.33%	[60]
3	81.60%	0.68	27.20%	[34]
3	78.1%	N/A	26.03%	[100]
3	78% (DM)	0.64	26.00%	[60]
3	78%	0.64	26.00%	[60]
3	75.4%	N/A	25.13%	[87]
3	74.38%	0.56	24.79%	[5]
3	74% (DM)	0.57	24.67%	[60]
4	92.5%	N/A	23.13%	[79]
4	80%	N/A	20.00%	[79]
4	75%	N/A	18.75%	[35]
4	73%	N/A	18.25%	[35]
4	68.0%	0.56	17.60%	[80]
5	92.35%	0.88	18.47%	[83]
5	69.3%	0.58	13.86%	[68]
5	65.1%	0.57	13.02%	[59]
9	76.5%	N/A	8.50%	[76]
15	61%	0.58	4.07%	[64]

Table A3. List of results of reviewed studies where species classification was performed with a *waveform* feature group; list sorted by species number and overall accuracy; DM—additional information for studies where trees were delineated manually.

Number of Species	OA	κ	NOCAI	Article
2	95.7%	0.92	47.85%	[84]
2	93.45%	0.61	46.73%	[114]
2	93.0%	0.72	46.5%	[37]
2	92.96%	0.570	46.48%	[114]
2	91.7%	N/A	45.85%	[36]

Table A3. *Cont.*

Number of Species	OA	κ	NOCAI	Article
2	89.76%	0.389	44.88%	[114]
2	86.2%	0.72	43.10%	[91]
2	85.4%	0.71	42.70%	[84]
2	84.4%	0.69	42.20%	[91]
3	95.1%	0.912	31.70%	[99]
3	95%	0.91	31.67%	[60]
3	94.8%	0.905	31.60%	[99]
3	94.5%	0.901	31.50%	[99]
3	92%	0.86	30.67%	[60]
3	92%	0.86	30.67%	[60]
3	91% (DM)	0.86	30.33%	[60]
3	88% (DM)	0.8	29.33%	[60]
3	88%	0.79	29.33%	[60]
3	87% (DM)	0.79	29.00%	[60]
3	84% (DM)	0.74	28.00%	[60]
3	85.0%	0.62	28.33%	[94]
3	75.0%	N/A	25.00%	[115]
3	73.4%	N/A	24.47%	[81]
3	71.5%	N/A	23.83%	[81]
4	80.4%	N/A	20.10%	[36]
4	79.22%	N/A	19.81%	[4]
4	75.8%	0.68	18.95%	[91]
4	73.2%	0.64	18.30%	[91]
4	53%	0.44	13.25%	[42]
5	85.4%	0.817	17.08%	[103]
6	68.6%	0.62	11.43%	[91]
6	66.5%	0.58	11.08%	[86]
6	66.1%	0.59	11.02%	[91]
6	62.4%	0.54	10.40%	[86]
6	62.0%	0.51	10.33%	[86]
6	60.0%	0.49	10.00%	[86]
6	59.0%	N/A	9.83%	[36]
6	58.2%	0.47	9.70%	[86]
6	57.1%	0.46	9.52%	[86]

Table A4. List of studies with overall accuracy greater than the median overall accuracy of all studies for each number of species with information about the quartiles; QM-U—the range from the median to upper quartile, QU-UW—the range from the upper quartile to the upper whisker; list sorted by species number and OA.

Number of Species	Quartile	Feature Group	OA	κ	Article
2	QU-UW	intensity+	97.1%	0.94	[40]
2	QU-UW	intensity+	97%	N/A	[113]
2	QU-UW	waveform	95.7%	0.92	[84]
2	QU-UW	geometry	95.0%	N/A	[75]
2	QU-UW	waveform	93.45%	0.61	[114]
2	QU-UW	waveform	93.0%	0.72	[37]
2	QM-U	waveform	92.96%	0.570	[114]
2	QM-U	waveform	91.7%	N/A	[36]
2	QM-U	intensity+	90.6%	N/A	[85]
2	QM-U	intensity+	90.0%	0.80	[40]
2	QM-U	waveform	89.76%	0.389	[114]

Table A4. *Cont.*

Number of Species	Quartile	Feature Group	OA	κ	Article
3	QU-UW	waveform	95.1%	0.912	[99]
3	QU-UW	waveform	95%	0.91	[60]
3	QU-UW	waveform	94.8%	0.905	[99]
3	QU-UW	waveform	94.5%	0.901	[99]
3	QU-UW	intensity+	92.8%	N/A	[87]
3	QU-UW	waveform	92%	0.86	[60]
3	QU-UW	waveform	92%	0.86	[60]
3	QU-UW	waveform	91%	0.86	[60]
3	QU-UW	intensity+	90.8%	0.84	[43]
3	QU-UW	intensity+	89%	0.82	[60]
3	QM-U	geometry	88.6%	N/A	[87]
3	QM-U	intensity+	88.2%	0.79	[27]
3	QM-U	intensity+	88%	0.82	[62]
3	QM-U	waveform	88%	0.8	[60]
3	QM-U	intensity+	88%	0.79	[60]
3	QM-U	waveform	88%	0.79	[60]
3	QM-U	waveform	87%	0.79	[60]
3	QM-U	geometry	86.3%	0.759	[63]
3	QM-U	intensity+	85.42%	0.75	[34]
3	QM-U	geometry	85.4%	0.739	[63]
4	QU-UW	geometry	93.7%	N/A	[78]
4	QU-UW	intensity+	92.5%	N/A	[79]
4	QU-UW	geometry	89.7%	0.86	[66]
4	QU-UW	geometry	85%	N/A	[79]
4	QU-UW	geometry	82.5%	N/A	[79]
4	QM-U	waveform	80.4%	N/A	[36]
4	QM-U	intensity+	80%	N/A	[79]
4	QM-U	waveform	79.22%	N/A	[4]
4	QM-U	geometry	77.5%	0.7	[77]
6	QU-UW	geometry	79%	0.75	[67]
6	QU-UW	waveform	68.6%	0.62	[91]
6	QU-UW	waveform	66.5%	0.58	[86]
6	QM-U	waveform	66.1%	0.59	[91]
6	QM-U	waveform	62.4%	0.54	[86]

References

1. White, J.C.; Coops, N.C.; Wulder, M.A.; Vastaranta, M.; Hilker, T.; Tompalski, P. Remote sensing technologies for enhancing forest inventories: A review. *Can. J. Remote Sens.* **2016**, *42*, 619–641. [CrossRef]
2. Kangas, A.; Maltamo, M. *Forest Inventory: Methodology and Applications*; Springer Science & Business Media: Dordrecht, The Netherlands, 2006; Volume 10. [CrossRef]
3. Dalponte, M.; Bruzzone, L.; Gianelle, D. Tree species classification in the Southern Alps based on the fusion of very high geometrical resolution multispectral/hyperspectral images and LiDAR data. *Remote Sens. Environ.* **2012**, *123*, 258–270. [CrossRef]
4. Heinzel, J.; Koch, B. Investigating multiple data sources for tree species classification in temperate forest and use for single tree delineation. *Int. J. Appl. Earth Obs. Geoinf.* **2012**, *18*, 101–110. [CrossRef]
5. Ørka, H.; Dalponte, M.; Gobakken, T.; Næsset, E.; Ene, L. Characterizing forest species composition using multiple remote sensing data sources and inventory approaches. *Scand. J. For. Res.* **2013**, *28*. [CrossRef]
6. Peterson, D.L.; Aber, J.D.; Matson, P.A.; Card, D.H.; Swanberg, N.; Wessman, C.; Spanner, M. Remote sensing of forest canopy and leaf biochemical contents. *Remote Sens. Environ.* **1988**, *24*, 85–108. [CrossRef]
7. Baret, F.; Vanderbilt, V.C.; Steven, M.D.; Jacquemoud, S. Use of spectral analogy to evaluate canopy reflectance sensitivity to leaf optical properties, *Remote Sens. Environ.* **1994**, *48*, 253–260. [CrossRef]
8. Asner, G.P.; Martin, R.E. Airborne spectranomics: Mapping canopy chemical and taxonomic diversity in tropical forests. *Front. Ecol. Environ.* **2009**, *7*, 269–276. [CrossRef]

9. Schimel, D.S.; Asner, G.P.; Moorcroft, P. Observing changing ecological diversity in the Anthropocene. *Front. Ecol. Environ.* **2013**, *11*, 129–137. [CrossRef]
10. Baldeck, C.A.; Asner, G.P.; Martin, R.E.; Anderson, C.B.; Knapp, D.E.; Kellner, J.R.; Wright, S.J. Operational tree species mapping in a diverse tropical forest with airborne imaging spectroscopy. *PLoS ONE* **2015**, *10*, e0118403. [CrossRef]
11. Brosofske, K.D.; Froese, R.E.; Falkowski, M.J.; Banskota, A. A review of methods for mapping and prediction of inventory attributes for operational forest management. *For. Sci.* **2014**, *60*, 733–756. [CrossRef]
12. Maltamo, M.; Næsset, E.; Vauhkonen, J. Forestry applications of airborne laser scanning. *Concepts Case Stud. Manag. Ecosys* **2014**, *27*, 460. [CrossRef]
13. Kelly, M.; Di Tommaso, S. Mapping forests with Lidar provides flexible, accurate data with many uses. *Calif. Agric.* **2015**, *69*, 14–20. [CrossRef]
14. Krzystek, P.; Serebryanyk, A.; Schnörr, C.; Červenka, J.; Heurich, M. Large-Scale Mapping of Tree Species and Dead Trees in Šumava National Park and Bavarian Forest National Park Using Lidar and Multispectral Imagery. *Remote Sens.* **2020**, *12*, 661. [CrossRef]
15. Bachman, C.G. *Laser Radar Systems and Techniques*; Artech House: Dedham, MA, USA, 1979.
16. Wehr, A.; Lohr, U. Airborne laser scanning—An introduction and overview. *ISPRS J. Photogramm. Remote Sens.* **1999**, *54*, 68–82. [CrossRef]
17. Pritchard, D.; Sperner, J.; Hoepner, S.; Tenschert, R. Terrestrial laser scanning for heritage conservation: The Cologne Cathedral documentation project. *ISPRS Ann. Photogramm. Remote Sens. Spat. Inf. Sci.* **2017**, *4*. [CrossRef]
18. Pérez-Álvarez, R.; de Luis, J.; Pereda-García, R.; Fernández-Maroto, G.; Malagón-Picón, B. 3D Documentation with TLS of Caliphal Gate (Ceuta, Spain). *Appl. Sci.* **2020**, *10*, 5377. [CrossRef]
19. Pu, S. Generating building outlines from terrestrial laser scanning. *ISPRS08 B* **2008**, *5*, 451–455.
20. Nowak, R.; Orłowicz, R.; Rutkowski, R. Use of TLS (LiDAR) for building diagnostics with the example of a historic building in Karlino. *Buildings* **2020**, *10*, 24. [CrossRef]
21. Suchocki, C.; Damięcka-Suchocka, M.; Katzer, J.; Janicka, J.; Rapinski, J.; Stałowska, P. Remote Detection of Moisture and Bio-Deterioration of Building Walls by Time-of-Flight and Phase-Shift Terrestrial Laser Scanners. *Remote Sens.* **2020**, *12*, 1708. [CrossRef]
22. Guan, H.; Li, J.; Cao, S.; Yu, Y. Use of mobile LiDAR in road information inventory: A review. *Int. J. Image Data Fusion* **2016**, *7*, 219–242. [CrossRef]
23. Truong-Hong, L.; Laefer, D.F. Documentation of bridges by terrestrial laser scanner. In Proceedings of the IABSE Geneva Conference 2015, Geneva, Switzerland, 23–25 September 2015. [CrossRef]
24. Artese, S.; Zinno, R. TLS for Dynamic Measurement of the Elastic Line of Bridges. *Appl. Sci.* **2020**, *10*, 1182. [CrossRef]
25. Hyyppä, J.; Yu, X.; Hyyppä, H.; Vastaranta, M.; Holopainen, M.; Kukko, A.; Kaartinen, H.; Jaakkola, A.; Vaaja, M.; Koskinen, J.; et al. Advances in forest inventory using airborne laser scanning. *Remote Sens.* **2012**, *4*, 1190–1207. [CrossRef]
26. Latifi, H.; Fassnacht, F.E.; Müller, J.; Tharani, A.; Dech, S.; Heurich, M. Forest inventories by LiDAR data: A comparison of single tree segmentation and metric-based methods for inventories of a heterogeneous temperate forest. *Int. J. Appl. Earth Obs. Geoinf.* **2015**, *42*, 162–174. [CrossRef]
27. Kukkonen, M.; Maltamo, M.; Korhonen, L.; Packalen, P. Multispectral airborne LiDAR data in the prediction of boreal tree species composition. *IEEE Trans. Geosci. Remote Sens.* **2019**, *57*, 3462–3471. [CrossRef]
28. Prieto, I.; Izkara, J.L.; Usobiaga, E. The application of lidar data for the solar potential analysis based on urban 3D model. *Remote Sens.* **2019**, *11*, 2348. [CrossRef]
29. Wang, X.; Chan, T.O.; Liu, K.; Pan, J.; Luo, M.; Li, W.; Wei, C. A robust segmentation framework for closely packed buildings from airborne LiDAR point clouds. *Int. J. Remote Sens.* **2020**, *41*, 5147–5165. [CrossRef]
30. Heritage, G.; Large, A. *Laser Scanning for the Environmental Sciences*; John Wiley & Sons: Hoboken, NJ, USA, 2009. [CrossRef]
31. Pyysalo, U.; Hyyppa, H. Reconstructing tree crowns from laser scanner data for feature extraction. *Int. Arch. Photogramm. Remote Sens. Spat. Inf. Sci.* **2002**, *34*, 218–221.
32. Solberg, S.; Naesset, E.; Bollandsas, O.M. Single tree segmentation using airborne laser scanner data in a structurally heterogeneous spruce forest. *Photogramm. Eng. Remote Sens.* **2006**, *72*, 1369–1378. [CrossRef]
33. Chamberlain, C.P.; Meador, A.J.S.; Thode, A.E. Airborne lidar provides reliable estimates of canopy base height and canopy bulk density in southwestern ponderosa pine forests. *For. Ecol. Manag.* **2020**, *481*, 118695. [CrossRef]
34. Yu, X.; Hyyppä, J.; Litkey, P.; Kaartinen, H.; Vastaranta, M.; Holopainen, M. Single-sensor solution to tree species classification using multispectral airborne laser scanning, *Remote Sens.* **2017**, *9*, 108. [CrossRef]
35. Krüger geb. Amiri, N.; Heurich, M.; Krzystek, P.; Skidmore, A. Feature Relevance Assessment of Multispectral Airborne LiDAR Data for Tree Species Classification. *ISPRS Int. Arch. Photogramm. Remote Sens. Spat. Inf. Sci.* **2018**, *XLII-3*, 31–34. [CrossRef]
36. Heinzel, J.; Koch, B. Exploring full-waveform LiDAR parameters for tree species classification. *Int. J. Appl. Earth Obs. Geoinf.* **2011**, *13*, 152–160. [CrossRef]
37. Yao, W.; Krzystek, P.; Heurich, M. Tree species classification and estimation of stem volume and DBH based on single tree extraction by exploiting airborne full-waveform LiDAR data. *Remote Sens. Environ.* **2012**, *123*, 368–380. [CrossRef]
38. Naesset, E. Estimating timber volume of forest stands using airborne laser scanner data. *Remote Sens. Environ.* **1997**, *61*, 246–253. [CrossRef]

39. Hyyppa, J.; Kelle, O.; Lehikoinen, M.; Inkinen, M. A segmentation-based method to retrieve stem volume estimates from 3-D tree height models produced by laser scanners. *IEEE Trans. Geosci. Remote Sens.* **2001**, *39*, 969–975. [CrossRef]
40. Ørka, H.O.; Næsset, E.; Bollandsås, O.M. Effects of different sensors and leaf-on and leaf-off canopy conditions on echo distributions and individual tree properties derived from airborne laser scanning. *Remote Sens. Environ.* **2010**, *114*, 1445–1461. [CrossRef]
41. Bont, L.; Hill, A.; Waser, L.; Bürgi, A.; Ginzler, C.; Blattert, C. Airborne-laser-scanning-derived auxiliary information discriminating between broadleaf and conifer trees improves the accuracy of models for predicting timber volume in mixed and heterogeneously structured forests. *For. Ecol. Manag.* **2020**, *459*, 117856. [CrossRef]
42. Sarrazin, D.; van Aardt, J.; Asner, G.; Mcglinchy, J.; Messinger, D.; Wu, J. Fusing small-footprint waveform LiDAR and hyperspectral data for canopy-level species classification and herbaceous biomass modeling in savanna ecosystems. *Can. J. Remote Sens.* **2011**, *37*, 653–665. [CrossRef]
43. Korpela, I.; Ørka, H.; Maltamo, M.; Tokola, T.; Hyyppä, J.; Tokola, M.; Maltamo, T. Tree Species Classification Using Airborne LiDAR—Effects of Stand and Tree Parameters, Downsizing of Training Set, Intensity Normalization, and Sensor Type. *Silva Fenn.* **2010**, *44*. [CrossRef]
44. Culvenor, D. *A Spatial Clustering Approach to Automated Tree Crown Delineation*; Pacific Forestry Centre: Victoria, BC, Canada, 1998; pp. 67–88.
45. Voss, M.; Sugumaran, R. Seasonal effect on tree species classification in an urban environment using hyperspectral data, LiDAR, and an object-oriented approach. *Sensors* **2008**, *8*, 3020–3036. [CrossRef]
46. Reitberger, J.; Schnörr, C.; Krzystek, P.; Stilla, U. 3D segmentation of single trees exploiting full waveform LIDAR data. *ISPRS J. Photogramm. Remote Sens.* **2009**, *64*, 561–574. [CrossRef]
47. Yu, X.; Hyyppä, J.; Holopainen, M.; Vastaranta, M. Comparison of Area-Based and Individual Tree-Based Methods for Predicting Plot-Level Forest Attributes. *Remote Sens.* **2010**, *2*. [CrossRef]
48. Yao, W. *A Sensitivity Analysis for a Novel Individual Tree Segmentation Algorithm Using 3D Lidar Point Cloud Data*; Silvilaser: Vancouver, BC, Canada. 2012.
49. Eysn, L.; Hollaus, M.; Lindberg, E.; Berger, F.; Monnet, J.M.; Dalponte, M.; Kobal, M.; Pellegrini, M.; Lingua, E.; Mongus, D.; Pfeifer, N. A Benchmark of Lidar-Based Single Tree Detection Methods Using Heterogeneous Forest Data from the Alpine Space. *Forests* **2015**, *6*, 1721–1747. [CrossRef]
50. Smith, S.; Holland, D.; Longley, P. The importance of understanding error in lidar digital elevation models. *Int. Arch. Photogramm. Remote Sens. Spat. Inf. Sci.* **2004**, *35*, 996–1001.
51. Tiede, D.; Hochleitner, G.; Blaschke, T. A full GIS-based workflow for tree identification and tree crown delineation using laser scanning. *ISPRS Workshop CMRT* **2005**, *5*, 2005.
52. Zhao, K.; Popescu, S. Hierarchical watershed segmentation of canopy height model for multi-scale forest inventory. *Proc. ISPRS Work. Group* **2007**, *442*, 436.
53. Zhong, L.; Cheng, L.; Xu, H.; Wu, Y.; Chen, Y.; Li, M. Segmentation of Individual Trees From TLS and MLS Data. *IEEE J. Sel. Top. Appl. Earth Obs. Remote Sens.* **2016**, 1–14. [CrossRef]
54. Xu, S.; Ye, N.; Xu, S.; Zhu, F. A supervoxel approach to the segmentation of individual trees from LiDAR point clouds. *Remote Sens. Lett.* **2018**, *9*, 515–523. [CrossRef]
55. Xu, Y.; Sun, Z.; Hoegner, L.; Stilla, U.; Yao, W. Instance Segmentation of Trees in Urban Areas from MLS Point Clouds Using Supervoxel Contexts and Graph-Based Optimization. In Proceedings of the 10th IAPR Workshop on Pattern Recognition in Remote Sensing (PRRS), Beijing, China, 19–20 August 2018; pp. 1–5. [CrossRef]
56. Anandakumar, R.; Nidamanuri, R.; Krishnan, R. Assessment of various parameters on 3D semantic object-based point cloud labelling on urban LiDAR dataset. *Geocarto Int.* **2018**, *34*, 1–29. [CrossRef]
57. Anandakumar, R.; Nidamanuri, R.; Krishnan, R. Individual tree detection from airborne laser scanning data based on supervoxels and local convexity. *Remote Sens. Appl. Soc. Environ.* **2019**, *15*, 100242. [CrossRef]
58. Ørka, H.; Næsset, E.; Bollandsås, O. Classifying species of individual trees by intensity and structure features derived from airborne laser scanner data. *Remote Sens. Environ.* **2009**, *113*, 1163–1174. [CrossRef]
59. Shi, Y.; Skidmore, A.; Holzwarth, S.; Heiden, U.; Pinnel, N.; Zhu, X.; Heurich, M. Tree species classification using plant functional traits from LiDAR and hyperspectral data. *Int. J. Appl. Earth Obs. Geoinf.* **2018**, *73*. [CrossRef]
60. Hovi, A.; Korhonen, L.; Vauhkonen, J.; Korpela, I. LiDAR waveform features for tree species classification and their sensitivity to tree- and acquisition related parameters. *Remote Sens. Environ.* **2016**, *173*, 224–237. [CrossRef]
61. Ba, A.; Laslier, M.; Dufour, S.; Hubert-Moy, L. Riparian Trees Genera Identification Based on Leaf-on/Leaf-off Airborne Laser Acanner Data and Machine Learning Classifiers in Northern France. *Int. J. Remote Sens.* **2019**. [CrossRef]
62. Holmgren, J.; Persson, A.; Söderman, U. Species identification of individual trees by combining high resolution LiDAR data with multi-spectral images. *Int. J. Remote Sens. Int. Remote Sens.* **2008**, *29*, 1537–1552. [CrossRef]
63. Dalponte, M.; Ørka, H.; Ene, L.; Gobakken, T.; Næsset, E. Tree crown delineation and tree species classification in boreal forests using hyperspectral and ALS data. *Remote Sens. Environ.* **2014**, *140*, 306–317. [CrossRef]
64. Liu, L.; Coops, N.; Aven, N.; Pang, Y. Mapping urban tree species using integrated airborne hyperspectral and LiDAR remote sensing data. *Remote Sens. Environ.* **2017**, *200*, 170–182. [CrossRef]

65. Liao, W.; Van Coillie, F.; Gao, L.; Li, L.; Chanussot, J. Deep Learning for Fusion of APEX Hyperspectral and Full-waveform LiDAR Remote Sensing Data for Tree Species Mapping. *IEEE Access* **2018**, *6*, 68716–68729. [CrossRef]
66. Yang, G.; Zhao, Y.; Li, B.; Ma, Y.; Li, R.; Jing, J.; Dian, Y. Tree Species Classification by Employing Multiple Features Acquired from Integrated Sensors. *J. Sens.* **2019**, *2019*, 1–12. [CrossRef]
67. Li, Q.; Wong, F.; Fung, T. Classification of Mangrove Species Using Combined WordView-3 and LiDAR Data in Mai Po Nature Reserve, Hong Kong. *Remote Sens.* **2019**, *11*, 2114. [CrossRef]
68. Shi, Y.; Skidmore, A.; Heurich, M. Improving LiDAR-based tree species mapping in Central European mixed forests using multi-temporal digital aerial colour-infrared photographs. *Int. J. Appl. Earth Obs. Geoinf.* **2019**, *84*. [CrossRef]
69. Cohen, J. A Coefficient of Agreement for Nominal Scales. *Psychol. Bull.* **1960**, *20*, 37. [CrossRef]
70. Landis, J.; Koch, G. The Measurement of Observer Agreement For Categorical Data. *Biometrics* **1977**, *33*, 159–174. [CrossRef] [PubMed]
71. Wang, K.; Liu, X. A Review: Individual Tree Species Classification Using Integrated Airborne LiDAR and Optical Imagery with a Focus on the Urban Environment. *Forests* **2018**, *10*, 1. [CrossRef]
72. Zhang, C.; Qiu, F. Mapping individual tree species in an urban forest using airborne lidar data and hyperspectral imagery. *Photogramm. Eng. Remote Sens.* **2012**, *78*, 1079–1087. [CrossRef]
73. Koenig, K.; Höfle, B. Full-Waveform Airborne Laser Scanning in Vegetation Studies—A Review of Point Cloud and Waveform Features for Tree Species Classification. *Forests* **2016**, *7*, 198. [CrossRef]
74. Ørka, H.; Næsset, E.; Bollandsås, O. Utilizing Airborne Laser Intensity for Tree Species Classification. *Int. Arch. Photogramm. Remote Sens. Spat. Inf. Sci.* **2012**, *36*, 300–304.
75. Holmgren, J.; Persson, A. Identifying species of individual trees using airborne laser scanner. *Remote Sens. Environ.* **2004**, *90*, 415–423. [CrossRef]
76. Axelsson, A.; Lindberg, E.; Olsson, H. Exploring Multispectral ALS Data for Tree Species Classification. *Remote Sens.* **2018**, *10*, 183. [CrossRef]
77. Li, J.; Hu, B.; Noland, T. Classification of tree species based on structural features derived from high density LiDAR data. *Agric. For. Meteorol.* **2013**, *171–172*, 104–114. [CrossRef]
78. Harikumar, A.; Bovolo, F.; Bruzzone, L. An Internal Crown Geometric Model for Conifer Species Classification With High-Density LiDAR Data. *IEEE Trans. Geosci. Remote Sens.* **2017**, *55*, 2924–2940. [CrossRef]
79. Lin, Y.; Hyyppä, J. A comprehensive but efficient framework of proposing and validating feature parameters from airborne LiDAR data for tree species classification. *Int. J. Appl. Earth Obs. Geoinf.* **2016**, *46*, 45–55. [CrossRef]
80. Suratno, A.; Seielstad, C.; Queen, L. Tree species identification in mixed coniferous forest using airborne laser scanning. *ISPRS J. Photogramm. Remote Sens. ISPRS Photogramm.* **2009**, *64*, 683–693. [CrossRef]
81. Yu, X.; Litkey, P.; Hyyppä, J.; Holopainen, M.; Vastaranta, M. Assessment of Low Density Full-Waveform Airborne Laser Scanning for Individual Tree Detection and Tree Species Classification. *Forests* **2014**, *5*, 1011–1031. [CrossRef]
82. Höfle, B.; Hollaus, M.; Hagenauer, J. Urban vegetation detection using radiometrically calibrated small-footprint full-waveform airborne LiDAR data. *ISPRS J. Photogramm. Remote Sens.* **2012**, *67*, 134–147. [CrossRef]
83. You, H.; Lei, P.; Li, M.; Ruan, F. Forest Species Classification Based on Three-dimensional Coordinate and Intensity Information of Airborne LiDAR Data with Random Forest Method. *ISPRS Int. Arch. Photogramm. Remote Sens. Spat. Inf. Sci.* **2020**, *XLII-3/W10*, 117–123. [CrossRef]
84. Reitberger, J.; Krzystek, P.; Stilla, U. Analysis of full waveform LIDAR data for the classification of deciduous and coniferous trees. *Int. J. Remote Sens.* **2008**, *29*, 1407–1431. [CrossRef]
85. Kim, S.; Mcgaughey, R.; Andersen, H.E.; Schreuder, G. Tree species differentiation using intensity data derived from leaf-on and leaf-off airborne laser scanner data. *Remote Sens. Environ.* **2009**, *113*, 1575–1586. [CrossRef]
86. Shi, Y.; Skidmore, A.; Heurich, M. Important LiDAR metrics for discriminating forest tree species in Central Europe. *ISPRS J. Photogramm. Remote Sens.* **2018**, *137*, 163–174. [CrossRef]
87. Zhang, Z.; Liu, X. Support vector machines for tree species identification using LiDAR-derived structure and intensity variables. *Geocarto Int.* **2012**, *28*, 1–15. [CrossRef]
88. Budei, B.; St-Onge, B.; Hopkinson, C.; Audet, F.A. Identifying the genus or species of individual trees using a three-wavelength airborne lidar system. *Remote Sens. Environ.* **2017**, *204*. [CrossRef]
89. Pirotti, F. Analysis of full-waveform LiDAR data for forestry applications: A review of investigations and methods. *Iforest Biogeosci. For.* **2011**, 100–106. [CrossRef]
90. Wagner, W.; Ullrich, A.; Ducic, V.; Melzer, T.; Studnicka, N. Gaussian decomposition and calibration of a novel small-footprint full-waveform digitising airborne laser scanner. *ISPRS J. Photogramm. Remote Sens.* **2006**, *60*, 100–112. [CrossRef]
91. Cao, L.; Coops, N.; Innes, J.; Dai, J.; Ruan, H.; She, G. Tree species classification in subtropical forests using small-footprint full-waveform LiDAR data. *Int. J. Appl. Earth Obs. Geoinf.* **2016**, *49*, 39–51. [CrossRef]
92. Drake, J.B.; Dubayah, R.O.; Clark, D.B.; Knox, R.G.; Blair, J.; Hofton, M.A.; Chazdon, R.L.; Weishampel, J.F.; Prince, S. Estimation of tropical forest structural characteristics using large-footprint LiDAR. *Remote Sens. Environ.* **2002**, *79*, 305–319. [CrossRef]
93. Duong, V. Processing and Application of ICESat Large Footprint Full Waveform Laser Range Data. Ph.D. Thesis, Delft University of Technology, Delft, The Netherlands, 2010.

94. Hollaus, M.; Mücke, W.; Höfle, B.; Dorigo, W.; Pfeifer, N.; Wagner, W.; Bauerhansl, C.; Regner, B. Tree species classification based on full-waveform airborne laser scanning data. *SilviLaser* **2009**, *54*, 14–16.
95. Zhou, T.; Popescu, S.; Lawing, A.M.; Eriksson, M.; Strimbu, B.; Burkner, P. Bayesian and Classical Machine Learning Methods: A Comparison for Tree Species Classification with LiDAR Waveform Signatures. *Remote Sens.* **2018**, *10*, 39. [CrossRef]
96. Brandtberg, T.; Warner, T.; Landenberger, R.; Mcgraw, J. Detection and analysis of individual leaf-off tree crowns in small footprint, high sampling density lidar data from the eastern deciduous forest in North America. *Remote Sens. Environ.* **2003**, *85*, 290–303. [CrossRef]
97. Wasser, L.; Day, R.; Chasmer, L.; Taylor, A. Influence of Vegetation Structure on Lidar-derived Canopy Height and Fractional Cover in Forested Riparian Buffers During Leaf-Off and Leaf-On Conditions. *PLoS ONE* **2013**, *8*, e54776. [CrossRef]
98. Sumnall, M.; Hill, R.; Hinsley, S. Comparison of small-footprint discrete return and full waveform airborne lidar data for estimating multiple forest variables. *Remote Sens. Environ.* **2015**, *173*. [CrossRef]
99. Blomley, R.; Hovi, A.; Weinmann, M.; Hinz, S.; Korpela, I.; Jutzi, B. Tree species classification using within crown localization of waveform LiDAR attributes. *ISPRS J. Photogramm. Remote Sens.* **2017**, *133*, 142–156. [CrossRef]
100. Vauhkonen, J.; Korpela, I.; Maltamo, M.; Tokola, T. Imputation of single-tree attributes using airborne laser scanning-based height, intensity, and alpha shape metrics. *Remote Sens. Environ.* **2010**, *114*, 1263–1276. [CrossRef]
101. Ballanti, L.; Blesius, L.; Hines, E.; Kruse, B. Tree Species Classification Using Hyperspectral Imagery: A Comparison of Two Classifiers. *Remote Sens.* **2016**, *8*, 445. [CrossRef]
102. Jones, T.; Coops, N.; Sharma, T. Assessing the utility of airborne hyperspectral and LiDAR data for species distribution mapping in the coastal Pacific Northwest, Canada. *Remote Sens. Environ.* **2010**, *114*, 2841–2852. [CrossRef]
103. Vaughn, N.; Moskal, L.; Turnblom, E. Tree Species Detection Accuracies Using Discrete Point Lidar and Airborne Waveform Lidar. *Remote Sens.* **2012**, *4*, 377–403. [CrossRef]
104. Nguyen, H.; Demir, B.; Dalponte, M. Weighted Support Vector Machines for Tree Species Classification Using Lidar Data. In Proceedings of the IGARSS 2019-2019 IEEE International Geoscience and Remote Sensing Symposium, Yokohama, Japan, 28 July–2 August 2019; pp. 6740–6743. [CrossRef]
105. Breiman, L. Random Forests. *Mach. Learn. Vol.* **2001**, *45*, 5–32. [CrossRef]
106. Cortes, C.; Vapnik, V. Support-vector networks. *Chem. Biol. Drug Des.* **2009**, *297*, 273–297. [CrossRef]
107. Fassnacht, F.; Latifi, H.; Stereńczak, K.; Modzelewska, A.; Lefsky, M.; Waser, L.; Straub, C.; Ghosh, A. Review of studies on tree species classification from remotely sensed data. *Remote Sens. Environ.* **2016**, *186*, 64–87. [CrossRef]
108. Polewski, P.P. Reconstruction of Standing and Fallen Single Dead Trees in Forested Areas from LiDAR Data and Aerial Imagery. Ph.D. Thesis, Technische Universität München, München, Germany, 2017.
109. Polewski, P.; Yao, W.; Heurich, M.; Krzystek, P.; Stilla, U. Free Shape Context descriptors optimized with genetic algorithm for the detection of dead tree trunks in ALS point clouds. *ISPRS Geospat. Week* **2015**, *W5*, 41–48. [CrossRef]
110. Jayathunga, S.; Owari, T.; Tsuyuki, S. Analysis of forest structural complexity using airborne LiDAR data and aerial photography in a mixed conifer–broadleaf forest in northern Japan. *J. For. Res.* **2018**, *29*, 479–493. [CrossRef]
111. Latifi, H.; Heurich, M.; Hartig, F.; Müller, J.; Krzystek, P.; Jehl, H.; Dech, S. Estimating over-and understorey canopy density of temperate mixed stands by airborne LiDAR data. *For. Int. J. For. Res.* **2016**, *89*, 69–81. [CrossRef]
112. Zielewska-Büttner, K.; Heurich, M.; Müller, J.; Braunisch, V. Remotely sensed single tree data enable the determination of habitat thresholds for the three-toed woodpecker (Picoides tridactylus). *Remote Sens.* **2018**, *10*, 1972. [CrossRef]
113. Vauhkonen, J.; Hakala, T.; Suomalainen, J.; Kaasalainen, S.; Nevalainen, O.; Vastaranta, M.; Holopainen, M.; Hyyppä, J. Classification of Spruce and Pine Trees Using Active Hyperspectral LiDAR. *IEEE Geosci. Remote Sens. Lett.* **2013**, *10*, 1138–1141. [CrossRef]
114. Bruggisser, M.; Roncat, A.; Schaepman, M.; Morsdorf, F. Retrieval of higher order statistical moments from full-waveform LiDAR data for tree species classification. *Remote Sens. Environ.* **2017**, *196*, 28–41. [CrossRef]
115. Vaughn, N.; Moskal, L.; Turnblom, E. Fourier transformation of waveform Lidar for species recognition. *Remote Sens. Lett.* **2011**, *2*, 347–356. [CrossRef]

remote sensing

MDPI

Article

Mapping the Abundance of Multipurpose Agroforestry *Faidherbia albida* Trees in Senegal

Tingting Lu [1,*], Martin Brandt [1], Xiaoye Tong [1], Pierre Hiernaux [2], Louise Leroux [3,4,5], Babacar Ndao [5] and Rasmus Fensholt [1]

1 Department of GeoSciences and Natural Resource Management, University of Copenhagen, 1350 Copenhagen, Denmark; mabr@ign.ku.dk (M.B.); xito@ign.ku.dk (X.T.); rf@ign.ku.dk (R.F.)
2 Pastoralisme Conseil, 30 Chemin de Jouanal, 82160 Caylus, France; pierre.hiernaux2@orange.fr
3 CIRAD, UPR AÏDA, Dakar, Sénégal and AÏDA, University of Montpellier, CIRAD, 34398 Montpellier, France; louise.leroux@cirad.fr
4 Agroecology and Sustainable Intensification for Annual Crop, University of Montpellier, CIRAD, 34090 Montpellier, France
5 Centre de Suivi Ecologique, Dakar, Senegal, Institute of Environmental Sciences, Faculty of Sciences and Technics, Cheikh Anta Diop University of Dakar, Dakar 15532, Senegal; babacar.ndao@cse.sn
* Correspondence: tlu@ign.ku.dk; Tel.: +45-71625774

Abstract: Multi-purpose *Faidherbia albida* trees represent a vital component of agroforestry parklands in West Africa as they provide resources (fodder for livestock, fruits and firewood) and support water lifting and nutrient recycling for cropping. *Faidherbia albida* trees are characterized by their inverse phenology, growing leaf flowers and pods during the dry season, thereby providing fodder and shedding leaves during the wet season, which minimizes competition with pastures and crops for resources. Multi-spectral and multi-temporal satellite systems and novel computational methods open new doors for classifying single trees and identifying species. This study used a Multi-Layer Perception feedforward artificial neural network to classify pixels covered by *Faidherbia albida* canopies from Sentinel-2 time series in Senegal, West Africa. To better discriminate the *Faidherbia albida* signal from the background, monthly images from vegetation indices were used to form relevant variables for the model. We found that NDI54/NDVI from the period covering onset of leaf senescence (February) until end of senescence (leaf-off in June) to be the most important, resulting in a high precision and recall rate of 0.91 and 0.85. We compared our result with a potential *Faidherbia albida* occurrence map derived by empirical modelling of the species ecology, which deviates notably from the actual species occurrence mapped by this study. We have shown that even small differences in dry season leaf phenology can be used to distinguish tree species. The *Faidherbia albida* distribution maps, as provided here, will be key in managing farmlands in drylands, helping to optimize economic and ecological services from both tree and crop products.

Keywords: multi-layer perception; savanna; species distribution model

Citation: Lu, T.; Brandt, M.; Tong, X.; Hiernaux, P.; Leroux, L.; Ndao, B.; Fensholt, R. Mapping the Abundance of Multipurpose Agroforestry *Faidherbia albida* Trees in Senegal. *Remote Sens.* **2022**, *14*, 662. https://doi.org/10.3390/rs14030662

Academic Editors: Markus Immitzer, Onisimo Mutanga and Clement Atzberger

Received: 24 November 2021
Accepted: 25 January 2022
Published: 29 January 2022

1. Introduction

Faidherbia albida (Del.) A. Chev. (synonym *Acacia albida* Del.) trees belong to the *mimosoideae* sub family of the *Fabaceae* family and are a key component of West African agroforestry parklands, where crops are grown under a discontinuous cover of scattered trees. *Faidherbia albida* trees provide resources such as fodder for livestock, fruits, firewood, wood products for construction and traditional medicine [1–6]. The deep-rooted trees have access to deep water layers and mineral resources. In addition, *Faidherbia albida* trees contribute to symbiotic nitrogen fixation, and thus serve multiple ecosystem services, such as water lifting, nutrients recycling and carbon sequestration [7]. What makes *Faidherbia albida* trees distinct is their unique leaf phenology, which is characterized by leafing out during the dry season and shedding leaves in the rainy season. Due to this "inverse phenology" [8], the trees do not intercept rainfall and sunlight from crops growing below the tree canopy during the rainy season, and the litter dropped by trees provides organic matter and nutrients to the soil. *Faidherbia albida* trees are therefore kept within fields while other woody species are often removed [9].

Faidherbia albida trees are found in parklands throughout the Sudano-Sahelian zone of West Africa, as well as in eastern and southern Africa [10–14]. Several scholars have demonstrated that *Faidherbia albida* trees improve crop yields as well as agricultural resilience and sustainability [15–19]. Roupsard [20] used unmanned aerial vehicles (UAV) to derive detailed field data to quantify the radius where *Faidherbia albida* trees influence crop yields in an agroforestry parkland in Senegal. Accurate mapping of *Faidherbia albida* trees is therefore important for a better understanding and management of the interaction of trees with crops within the agricultural system (species, density, planting design, etc.). This has become even more important in recent times with increases in human population driving agricultural intensification, and climate change threatening established agroforestry systems. The effectiveness and optimal implementation of restoration plans in relation to Farmer-Managed Natural Regeneration projects in Sub-Saharan Africa [21–24], also calls for accurate mapping of the spatial distribution of *Faidherbia albida* as a prerequisite.

The intriguing question of how specific tree species are spatially distributed has inspired many scholars within plant ecology. In addition to detailed and laborious field inventories, species distribution modelling (SDM) has been the main method to predict the spatial distribution of plant species. Since woody species are known to be highly dependent on biotic and abiotic conditions, most SDM approaches use computer algorithms to predict the spatiotemporal distribution of a species using geo-spatial data of the physical environment combined with observed species distribution points [25]. Data used for modelling are related to climatic, geomorphologic and pedologic variables like temperature, precipitation, elevation and soil type as well as human management, like land use [26]. However, such empirical models generate maps of the probability or potential occurrence rather than the actual distribution of particular species. The coarse resolution of gridded input data, as well as difficulties regarding how to adequately account for anthropogenic drivers, limits the prediction capacity of these models [27,28]. Species distribution points, serving as input to the SDM model, still rely on field records of tree species data, and accurate remotely sensed tree species mapping across large scales would provide valuable complementary information. Large-scale actual distribution of *Faidherbia albida* generated from remote sensing would thereby open new research avenues for data-driven SDM approaches that allow studying the importance of input environmental variables on species distribution for an improved variable parameterization of SDMs.

As an alternative to modelling the potential occurrence of a given species, recent advances in remote sensing enable the mapping of actual species distributions, yet previous studies remain at a local scale [29,30]. The principles behind the majority of the methods are linked to an understanding that woody species have unique individual biochemical and biophysical characteristics such as spectral reflectance, canopy structure, branch and foliage density, as well as phenological cycles. Mapping attempts grounded in these characteristics have been conducted along three separate pathways in relation to the use of satellite and aerial data: the first pathway utilizes optical remote sensing data to capture the biochemical character of tree species [31–34]. The second relies on the vertical structure of vegetation, and here LiDAR [35,36] is an effective data source. A third approach integrates synthetic aperture radar (SAR) with optical data [37–40]. The recent availability of high temporal resolution optical data enables the capture of the seasonal phenology of tree leaves with the timing of the images being the key for a successful classification [41]. Schriever et al. [42] compared images from different periods and found that the period of leaf senescence is most important for classifying tree species. Hesketh et al. [43] also found that inter-seasonal spectral variation allowed the accurate discernment of species of trees using automated classification methods. Moreover, several studies showed that the combination of images from the leaf-out (May/June) and senescence (October) phases provide a higher accuracy as compared to the use of single image [44–46].

So far, most attempts to conduct tree species mapping from remote sensing have focused on forest environments, and mapping tree species of scattered woody populations in drylands on a large-scale remains a challenge. The scattered distribution of agroforestry parkland trees in the Sahel makes the mapping different from trees in forest stands, as the signal from a pixel can be influenced by the soil or herbaceous background, especially when the pixel is at the edge of the canopy. Despite the improvement in spatial, temporal and spectral resolution of sensor systems, remote sensing has not yet been widely applied for the mapping of *Faidherbia albida* on a country scale. For instance, Triboulet et al. [47] pioneered this area by using pixel value thresholding, but due to the lack of field data the final result was not validated. Lelong [48] utilized WorldView-3 imagery to classify 7 tree species including *Faidherbia albida* in Senegal. However, these studies were limited to a local scale and a snapshot in time. It is still a challenging endeavour to achieve species-level detection across large geographic areas due to restrictions related to data costs and the workload associated with the data analysis.

Previous studies have demonstrated that large savanna trees can be identified with the 10-m resolution of Sentinel-2 [49]. Classification algorithms are rooted in computer science and statistical learning [29], for example, machine learning algorithms like Random Forest and neural networks [50–52]. Considering trade-offs between data cost and spatiotemporal resolution requirements, this study investigates the possibility of mapping the distribution of *Faidherbia albida* over Senegal based on publicly available Sentinel-2 time series data. The constellation of Sentinel-2A and 2B multi-spectral satellite systems were launched in 2015 and 2017 as a part of European Commission's Copernicus Programme. This system, with a 5-day revisiting time under the same viewing conditions [53], provides an unprecedented opportunity for *mapping Faidherbia albida* trees. The high temporal resolution may capture the inverse phenology of *Faidherbia albida*, contrasting with other woody and herbaceous vegetation types in the semi-arid Sahel. Our specific objectives are (1) to select dates and bands from Sentinel-2 images suited to map *Faidherbia albida* trees, (2) to use a Multi-Layer Perception (MLP) artificial neural network fed with a number of features to generate a 10-m resolution *Faidherbia albida* distribution map over Senegal and (3) to compare this distribution map with a potential occurrence map derived from an SDM model.

2. Materials and Methods

2.1. Study Area and Sample Data

Most of Senegal extends in the Sahelian and Sudanian regions with annual rainfall ranging between 100 to 1000 mm (Figure 1). The western part of Senegal comprises a vast area of subsistence rain-fed cultivation, called the peanut basin. Livelihoods are based on traditional farming systems rotating millet grown for on-farm consumption and groundnuts as cash crops. The peanut basin is the most important crop production area in the country and is characterized by intensively cultivated parklands [54] comprising randomly scattered trees and woody shrubs within cropped fields. The two major soil types are arenosols and calcisols, according to the FAO classification [55]. They are characterized by a high sand content and low organic matter. Among tree species forming the parklands, *Faidherbia albida* is widely present across the peanut basin. Generally, the natural habitat of *Faidherbia albida* species occurs throughout most areas in West, East and Southern Africa where annual rainfall ranges between 400 and 900 mm [56]. As a leguminous nitrogen-fixing tree species, *Faidherbia albida* adds nitrogen and organic matter through leaf fall, and hence improves soil fertility and boosts crop productivity [57]. Consequently, *Faidherbia albida* occurs widely in croplands as a result of human management promoting the presence of this species on deep light sands or sandy clays [58], which is also the case for the peanut basin in Senegal [11]. *Faidherbia albida* rarely occurs in pristine savannas but is mainly disseminated by livestock in croplands.

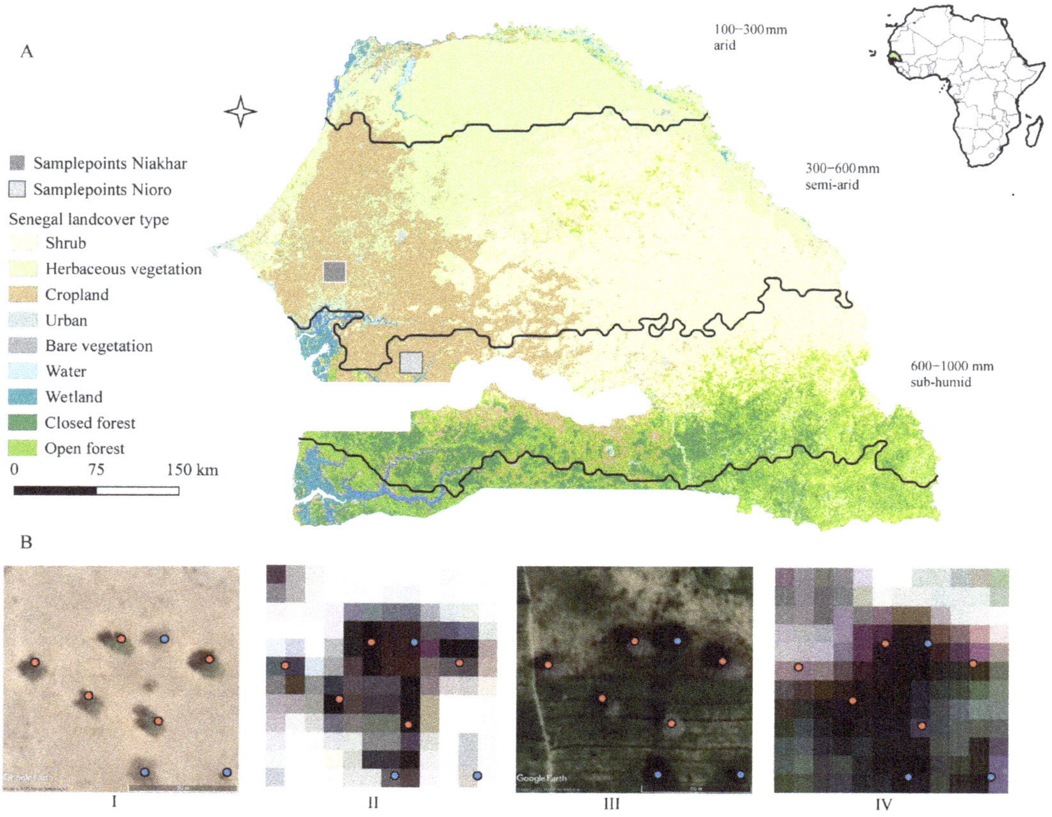

Figure 1. (**A**) Location of Senegal and land use/cover map of 2019 (Copernicus Global Land Service at 100-m resolution). The precipitation data are derived from the CHIRPS long-term (1981–2019) yearly average. The rainfall zones of 100–300 mm, 300–600 mm and 600–1000 mm are categorized as arid area, semi-arid area and sub-humid area. The two grey boxes represent locations of data sampling in Niakhar and Nioro. (**B**) Examples of *Faidherbia albida* trees (red dots) during the (I) dry and (III) rainy season from Google Earth. Image (II) and (IV) represent the corresponding stretched true colour images from Sentinel-2 at 10-m resolution. Blue points represent other tree species.

Field samples were collected in two different semi-arid regions of Senegal (Niakhar and Nioro; Figure 1) in 2018 and 2019 using an optimized sampling strategy based on a landscape heterogeneity classification [59,60]. At each sampling site, an exhaustive inventory of all trees was conducted in a 1-hectare plot. The coordinates of each individual tree were recorded with a GPS. The collection of tree species contains over 70 common species, including 6612 sample points in the Nioro area where *Faidherbia albida* accounted for 3.07% (203) of the samples. A total of 9258 trees were collected in Niakhar among which *Faidherbia albida* accounted for 41.82% (3872). Since the spatial resolution of Sentinel-2 is 10 m, we only selected *Faidherbia albida* samples with a canopy diameter greater or equal than the Sentinel-2 pixel size as the reference data used for the building the MLP model. The individual tree canopy diameter information was derived from a previous study [61]. We further selected and added samples of other tree species, cropland, water, urban, herbaceous vegetation, bare soil and shrubs. In total, 2310 samples were used as reference points for a binary classification of two classes encompassing 516 points representing *Faidherbia albida*, and 1794 point for the other class. The Sentinel-2 pixels that contain the reference points were sampled and used for building the model. Since the sampled pixels are often a mixture of the tree canopy and the immediate surroundings of the tree, the *Faidherbia albida* mapping is thus conducted at a canopy level with a tree canopy minimum mapping unit of one pixel.

2.2. Sentinel-2 Data Pre-Processing

Sentinel-2 top of atmosphere data are available in the Google Earth Engine (GEE) cloud computing platform. All available Sentinel-2 images from January 2017 to December 2019 with cloud cover less than 50% were fused to monthly composites. Cloud masking was done according to the flags of the Sentinel-2 QA band. The spectral noise caused by scene-specific illumination conditions and residual cloud effects is expected to be reduced by using 3-year average values from the period 2017–2019. Vegetation indices and bands (next section) in specific months and their dynamic changes were selected as feature collection for the following image classification based on Recursive Feature Elimination. The detailed Sentinel-2 data processing is described in the following sections.

2.2.1. Vegetation Indices

The Sentinel-2A and B instruments sample 13 spectral bands: visible and near-infrared at 10-m, red edge and SWIR at 20-m and atmospheric bands at 60-m spatial resolution. The bands with a resolution >20 m have been discarded from the current study due to the mismatch between pixel resolution and tree size, and the six bands of 20-m resolution were resampled to 10-m by the nearest neighbour, leaving 10 bands for the monthly composites. To better isolate the signal inherent to photosynthetic activity from other signals within spatially mixed pixels, 10 vegetation indices, which are widely applied to Sentinel-2 imagery [62], were selected (Table S1). As some of those vegetation indices were derived from the same spectral bands, considerable redundancy in information exists. To determine the similarity of indices, the pair-wise correlation coefficient was calculated (Table S2). Based on the high co-variability, those indices were categorized into 3 classes. The first class is composed of NDVI, NDVI74, $PSSR_a$, EVI, MCARI, S2REP and EVI2. The second class is composed of NDI54 and GNDVI and the third class includes only MTCI. To select only one of each group for feature collection, the correlation coefficient between vegetation index values of *Faidherbia albida* and other species (Table S3) was taken into consideration, with a lower coefficient value indicating a higher discrepancy between *Faidherbia albida* and other tree species. Hence only NDVI and NDI54 were retained, and the MTCI was omitted because its monthly values showed a high standard deviation during the growing season (Figure S1H). The initial feature bands used to construct the model are the monthly values of each band and 2 vegetation index bands, yielding in total $12 \times (2 + 10) = 144$ bands.

The foliage production of *Faidherbia albida* trees is high in the dry season and low in the rainy season, and their foliage production reaches a peak in December (Figure 2). As biochemical and structure properties of woody plants can influence the spectral signature, the inverse phenology of *Faidherbia albida* potentially makes the tree distinguishable from other deciduous or evergreen tree species. The green herbaceous layer and understory impact the *Faidherbia albida* canopy pixels in Sentinel-2 images, so this study discriminates *Faidherbia albida* trees not only based on monthly band/vegetation index values but also uses seasonal dynamics. This was done by calculating all possible monthly image combinations (by subtraction of images) producing 792 dynamic images, leading to a total of 936 feature bands.

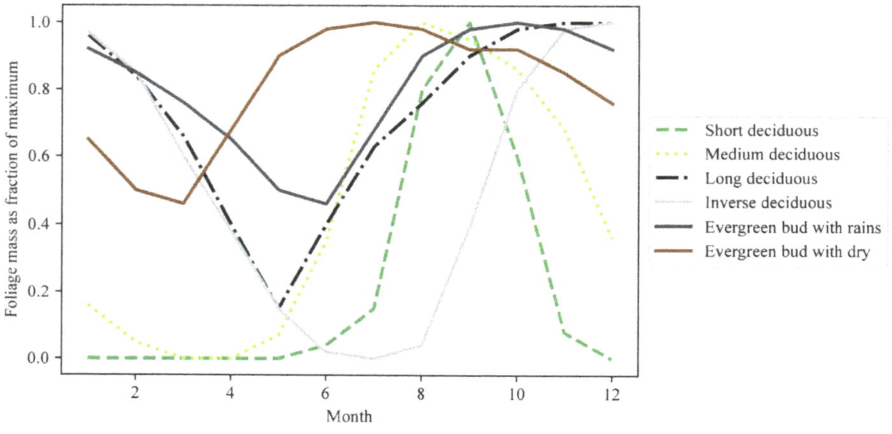

Figure 2. Seasonal distribution of leaf mass depending on the phenological type of woody plants in the Sahel [8]. The inverse deciduous type represents *Faidherbia albida* trees.

2.2.2. Feature Selection

A recursive feature elimination was then applied to reduce redundant features. The Random Forest classifier ranks the importance of features through their impact on the model accuracy. Recursive feature elimination can be used to rank the feature by looping through Random Forest classifiers and removing the least important feature until the accuracy remains stable. Here we generated collections of features and a 4-fold cross-validation was utilized to test the performance of the various feature collections on the MLP model. Cross-validation is a universal method for assessing the accuracy of a classifier based on a limited data sample. The dataset was generated from the field-observed data in combination with the Sentinel-2 imagery, and the whole dataset was split into 5 individual sets. A test set was held out for final evaluation. Then, three of the remaining datasets were used as training datasets, initially used to fit the model. The last dataset is used as a validation dataset, which provides an unbiased evaluation of a model fit on the training dataset while tuning the parameters. The process is repeated 4 times. The feature collection including the fewest number of features achieving the same performance as using all features was selected for the final model.

The performance metrics of recall and precision were used to evaluate models using different feature collections [63]. Precision is defined as the fraction of relevant instances among the retrieved instances, i.e., the number of true positives divided by the total number of elements labelled as belonging to the positive class. Recall is defined as the fraction of relevant instances that were retrieved, i.e., the number of true positives divided by the total number of elements that actually belong to the positive class. The true positives refer to the elements correctly labelled as belonging to the positive class, here, *Faidherbia albida* trees. The elements labelled as belonging to the positive class is the sum of true positives and false positives, the latter which are instances incorrectly labelled as belonging to the class. The elements that actually belong to the positive class are the sum of true positives and false negatives, which are the elements that were not labelled as belonging to the positive class but should have been. Recall and precision scores are often inversely related. Taking the harmonic average of precision and recall leads to the F-measure (a measure that combines precision and recall):

$$\text{F-measure} = 2 \times \text{recall} \times \text{precision}/(\text{recall} + \text{precision}) \quad (1)$$

Precision and recall weigh equally in F-measure and a higher F-measure represents a superior model.

2.3. Mapping *Faidherbia albida* Trees Using Artificial Neural Networks

The Multi-Layer Perception (MLP) model was adopted for the binary classification of *Faidherbia albida* trees. MLP is an artificial neural network with multiple densely connected layers between the input and output layers. The densely connected layers in MLP models combine all the features of the previous layer, which is suitable for complex classification tasks. We used the artificial neural network instead of other non-parametric classifiers, such as Random Forests or Support Vector Machines, which have been applied in previous tree species classifications, because of the irregular spectral signatures of the dataset and the uneven composition of species in the data structure. Considering the image resolution and the size of tree canopy, any given *Faidherbia albida* pixel will usually be a mix of reflectance from *Faidherbia albida* canopies and background vegetation or bare soil. Therefore, even the same tree species would show considerable intra-species spectral variability among training and testing data. Neural networks as a general-purpose technique for classification hold the potential to map *Faidherbia albida* in such a challenging case. The MLP model is created with three layers, and the learning rate, batch size and epochs are fine-tuned through grid search. The input of the MLP is the feature collection selected by cross-validation. The output of the MLP is the probability that a pixel includes a *Faidherbia albida* canopy, with values ranging from 0 to 1. Here, pixel values > 0.5 were mapped as *Faidherbia albida* canopies.

2.4. Post Processing

To comply with the nature of parklands and savanna ecosystems where *Faidherbia albida* trees are predominantly scattered without forming a closed canopy, all clusters of predicted *Faidherbia albida* pixels covering more than 1000 m^2 were masked. This threshold was selected based on Brandt et al. [61], where very high spatial resolution image analysis revealed that <0.0001 of all single tree canopies surpass this size in the study area.

We combined the *Faidherbia albida* canopy map derived from MLP with the woody canopy map produced by Zhang et al. [49], which is also based on time-series Sentinel-2 data. A map representing the relative cover of *Faidherbia albida* related to all woody plant cover was generated by calculating the fraction of the *Faidherbia albida* cover (per hectare) and the woody canopy cover (per hectare) map. To explore the agreement between our remote sensing based *Faidherbia albida* distribution map and the potential occurrence based on ecological modelling, we obtained a *Faidherbia albida* potential occurrence map that was provided by Kindt [64] at a 5-km resolution. The occurrence map was generated from 9 bioclimatic layers, 2 topographic layers and 3 soil maps by the R package BiodiversityR. Details about the data sources and resolution of the layers can be found in Table S4. This map consists of a unitless *Faidherbia albida* habitat suitability index varying from 0 (lowest) to 910 (highest). The absence–presence threshold for suitability of *Faidherbia albida* was provided by the SDM model as 534, meaning that suitability values above 534 correspond to the likely presence of the species with increasing values associated with higher likelihood of species presence.

3. Results

3.1. Intra-Annual Difference in Vegetation Indices

Monthly values of NDVI and NDI54 for pixels covering *Faidherbia albida* and other tree species were analysed for Nioro and Niakhar. It is noticeable that even pixels covering the same group of tree species show varying index values in the two study regions. This is likely because the majority of the sampled pixels in Sentinel-2 images are mixed pixels, and the index value is the result of the reflectance from the tree canopy and a complex set of other covers underneath such as crops and herbaceous vegetation or bare soil, especially in the wet season when the *Faidherbia albida* trees are defoliated. However, the values of pixels with *Faidherbia albida* show the same seasonality in both regions different to other tree species (Figure 3). Specifically, pixels with *Faidherbia albida* have a lower NDVI value in the rainy season and a higher NDVI value in the dry season as compared to the pixels representing other tree species (Figure 3A), which means that the relative difference of *Faidherbia albida* NDVI values between the dry and wet seasons is smaller as compared to other tree species. While all tree species reach their NDVI peak in October, the NDVI value of pixels with *Faidherbia albida* decrease much slower after the peak month as compared to other tree species. As for NDI54 (Figure 3B), *Faidherbia albida* pixels have the highest values in November while the pixels of other tree species show a peak in October. Consequently, the mapping of *Faidherbia albida* ideally should not only depend on the absolute index values of individual months, but also take the dynamics between different months into consideration.

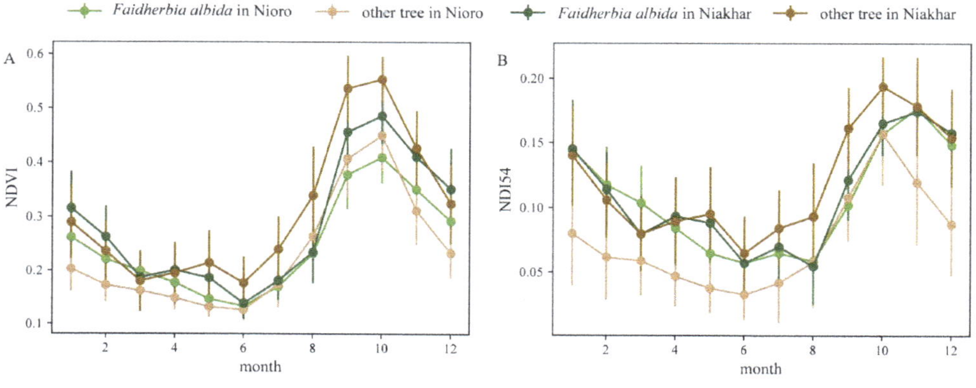

Figure 3. Averaged monthly (**A**) NDVI and (**B**) NDI54 values for sample pixels of *Faidherbia albida* and other species in the regions of Nioro and Niakhar. Vertical lines represent the standard deviation of the sample population. Overall, 1418 pixels were sampled and averaged over the study period, including 57 *Faidherbia albida* pixels and 112 pixels with other tree species in the Nioro region, and 459 *Faidherbia albida* pixels and 790 pixels with other species in the Niakhar region.

3.2. Mapping Faidherbia albida Using the MLP Model

By conducting a recursive feature elimination on the original 936 features (Section 2.2) we tested subsets of feature combinations as a function of the model mean accuracy using a 4-fold cross-validation approach (Figure 4). The evaluation of the *Faidherbia albida* mapping was performed by mean accuracy of all iterations using a test set. Using only the seasonal dynamic feature of NDI54 between November and August, the classifier achieves an average F-measure of 60%. When increasing the features to 40, the total variability in the feature layers reaches the same level of accuracy (precision, recall and the F-measure) as when using all features, indicating that 40 features are sufficient. Using 40 features, the model achieves an overall F-measure of 89%, with a fairly well-balanced rate of 91% precision and 85% recall, showing that the pixels of *Faidherbia albida* canopies are captured well (Table S5). The final selected features for the MLP model (Table 1) show that the seasonal dynamic features account for more than 50% of all selected features, and the typical span of the two months forming the difference feature is around 5 months, usually representing one month in the start of the dry season and the other month at the end of the dry season. The acquired times of the selected data by recursive feature elimination are February (start of *Faidherbia albida* senescence), July (leaf-off) and October (leaf-out). In terms of feature type (vegetation index or single band reflectance), the NDVI and NDI54 dominate the selected features.

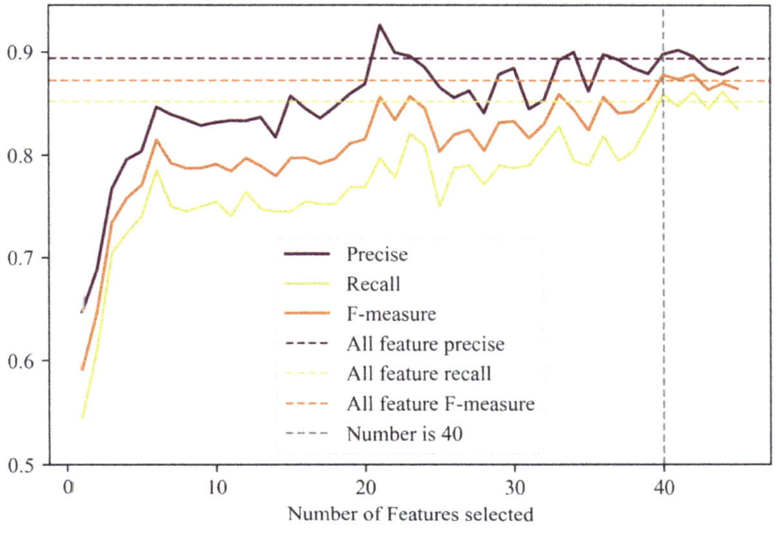

Figure 4. Relationship between mean accuracy and number of features fed to the MLP classifier. The flat horizontal dashed line represents the accuracy from a model using all features.

Table 1. Feature information of the 40 features finally retained to feed the MLP model including monthly features (band/spectral index and specific month) and seasonal dynamics features (band/spectral index and related months).

Monthly Feature			Seasonal Dynamic Feature		
Band/Index	Month	Feature Importance	Band/Index	Month	Feature Importance
NDI54	2	0.040	NDVI	2–6	0.063
Band12	6	0.035	NDI54	2–6	0.043
Band12	3	0.032	NDVI	2–5	0.041
Band11	6	0.031	NDI54	8–12	0.038
Band12	5	0.030	NDVI	4–6	0.031
NDI54	12	0.029	NDVI	7–12	0.031
NDI54	3	0.029	NDI54	2–5	0.029
Band2	10	0.026	NDI54	2–8	0.024
Band12	4	0.025	Band12	3–11	0.023
NDI54	4	0.025	NDVI	2–7	0.023
NDI54	6	0.025	NDVI	6–12	0.022
NDVI	6	0.023	NDI54	10–11	0.020
Band3	10	0.022	NDI54	3–8	0.020
Band8	10	0.019	NDI54	2–10	0.018
NDVI	10	0.017	Band12	3–10	0.018
NDVI	7	0.017	Band12	4–11	0.017
			Band3	10–12	0.017
			NDVI	4–7	0.015
			Band4	10–11	0.014
			NDI54	10–12	0.014
			NDI54	1–8	0.014
			Band4	6–12	0.014
			NDVI	3–7	0.010

3.3. Comparing *Faidherbia albida* Canopy Cover Maps with Potential Occurrence Maps

A *Faidherbia albida* canopy map at 10-m resolution covering Senegal was generated based on 3 years of Sentinel-2 data. To illustrate the mapping at the national scale, the per-pixel mapping was aggregated to 1 hectare showing the percentage of *Faidherbia albida* cover

(Figure 5A). *Faidherbia albida* trees are mainly distributed in the western cropland areas (Figure 5A) where annual precipitation ranges from 300 to 600 mm. *Faidherbia albida* trees are to a less extent also found in the croplands of the northern part of the country, where rainfall is less than 300 mm/year. For most of the cropland areas, the percentage of *Faidherbia albida* cover is in the range between 2–15%. Our remote sensing-based distribution agrees to some extent with the potential occurrence map from ecological modelling (Figure 5B) in which the potential occurrence is shown by SDM suitability values. While similar patterns are shown in western Senegal, the southeast, the Cap Vert peninsula and the limited cropland areas along the north-western coast do not match with our map.

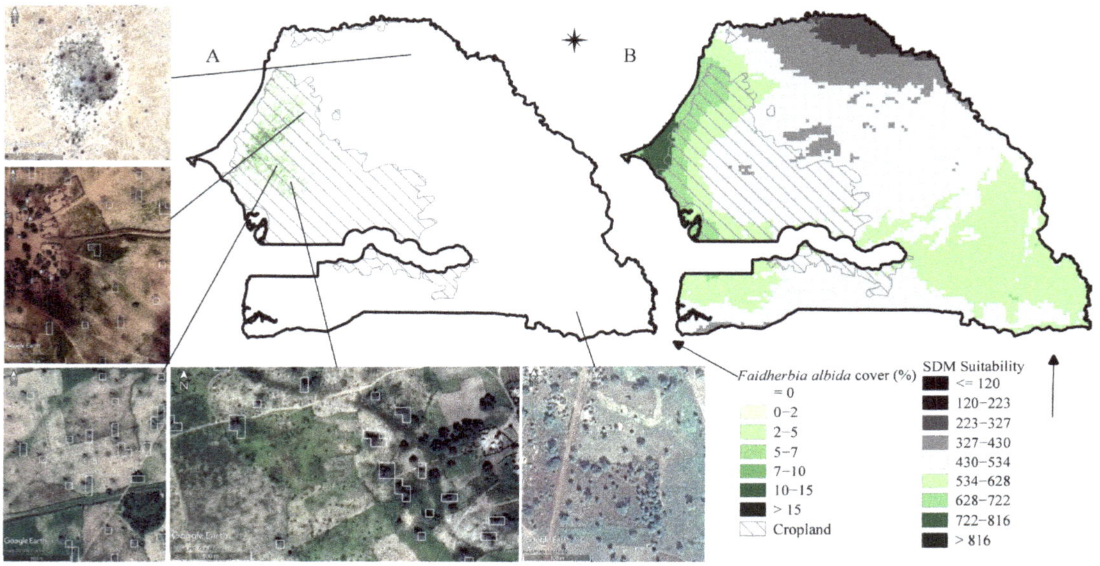

Figure 5. (**A**) *Faidherbia albida* cover map (%) at 1-hectare resolution from our classification. Included are close-up maps showing detailed *Faidherbia albida* classification maps of different regions as white polygons superimposed on images from Google Earth. (**B**) *Faidherbia albida* potential occurrence map provided by Kindt [64].

The remote sensing-based map of *Faidherbia albida* cover was directly compared with the SDM suitability score within different land use/cover types (Figure 6). The SDM map at a 5-km resolution was resampled to 1 ha using the nearest neighbour. *Faidherbia albida* trees mainly occur in areas with suitability values above 560 and are almost absent below this threshold except in croplands where they occur even with SDM values of 300. *Faidherbia albida* trees were barely found in the north/north-eastern shrub and open forest areas of Senegal (Figure 5), even though the SDM suitability was around 600. When suitability values exceed 680, we observe that the high suitability score relates to low *Faidherbia albida* cover (<0.25%). This discrepancy is linked to the Senegal land cover, which is related to the near coastal areas around the Dakar peninsula (Figures 1 and 5).

By combing the woody cover map from Zhang et al. [49] and our results, we calculated the relative contribution of *Faidherbia albida* trees to the overall woody plant cover per hectare (Figure 7). The map of this relative contribution shows the degree of dominance of *Faidherbia albida* among woody plant cover. The higher the contribution of *Faidherbia albida* to the woody plant cover, the lower is the woody plant diversity. Furthermore, in areas where woody cover is nonzero, the contribution of *Faidherbia albida* to cover as a function of all woody species cover is illustrated in Figure 7B. The relative contribution of

Faidherbia albida in croplands is on average 7.23%, with several areas being dominated by *Faidherbia albida*.

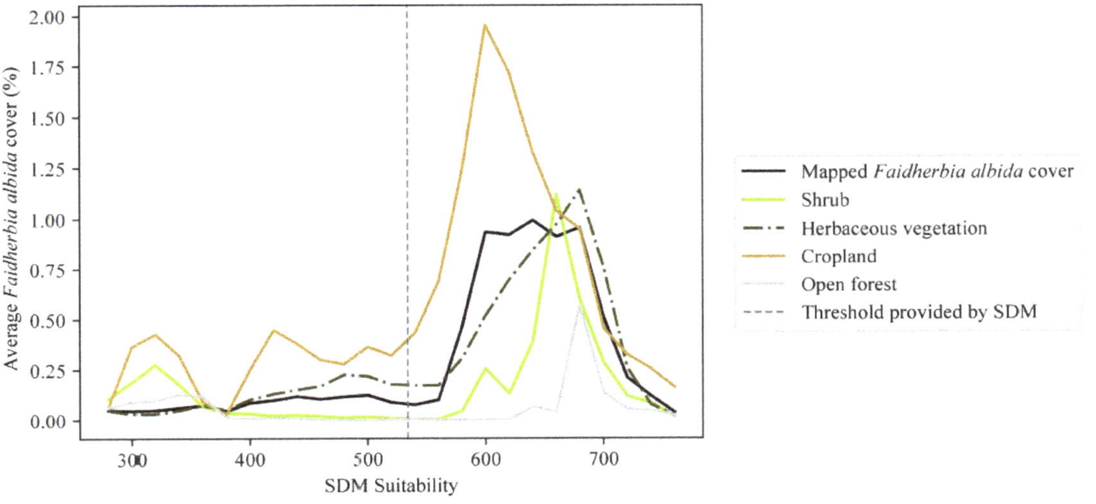

Figure 6. Average *Faidherbia albida* cover (%) in relation to modelled suitability scores (intervals of 20 units) for different land use/cover types. The land use/cover map of 2019 is provided by Copernicus Global Land Service at 100-m resolution. Wetland, urban, water, bare vegetation and closed forest types were not included in this comparison due to the limited number of observations.

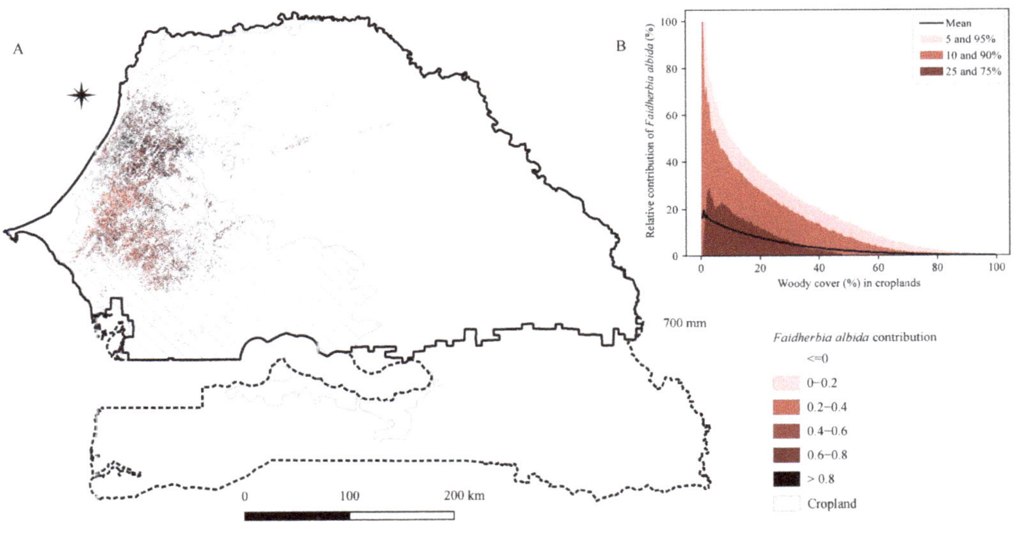

Figure 7. (**A**) Map of relative contribution of *Faidherbia albida* canopy cover to the cover of all woody plants shown as a percentage at the hectare scale. The rainfall zones are derived from the CHIRPS long-term (1981–2019) yearly average. (**B**) Relation between the relative contribution of *Faidherbia albida* to canopy cover in croplands. The land use/cover map of 2019 is provided by Copernicus Global Land Service at 100-m resolution.

4. Discussion

4.1. Mapping Faidherbia albida

This study presents a method to estimate *Faidherbia albida* cover across Senegal from Sentinel-2 spatiotemporal signatures at 10-m resolution. Previously, only Lelong et al. [48] used very high resolution Worldview-3 imagery to map 7 tree species including *Faidherbia albida*. Our approach was based on the assumption that the classification could take advantage of the distinct and inverse phenological behaviour of *Faidherbia albida* trees as compared to other tree species. However, we found that this phenological behaviour was not seen in satellite time series from Sentinel-2. A possible explanation is that the herbaceous layer growing below the canopy of *Faidherbia albida* trees during the non-leaf season dominates the signal. It impedes the capacity of satellite time series to capture the inverse phenology during the leaf-off period. As for the leaf-on period, the difference is also unexpectedly small and varies with location and time. Since most remote sensing applications in mapping forest types or tree species rely on the phenological differences in satellite time series between the target and the background [45], the mapping of *Faidherbia albida* trees is more challenging than initially assumed.

Given this situation, we maximized the input capacity by generating a total of 936 features, which include monthly and seasonal features of both multiple vegetative indices and bands. Instead of directly using sparse temporal features and single vegetation indices [43,65], we made use of recursive feature elimination to reduce data redundancy. The accuracy stabilized when 40 features were included, mainly from the senescence month of *Faidherbia albida* and including both spectral bands and vegetation indices [44,45]. Features related to seasonal dynamics were most important, which aligns with the results of [43,66]. NDVI, NDI54 and the second SWIR band of Sentinel-2 are often picked as high-ranking features, indicating the importance of the red, red edge and SWIR bands for tree species mapping [29]. Additionally, the background reflectance causes different values for different regions with *Faidherbia albida* trees (Figure 3A). This effect can be mitigated by using images of seasonal changes in addition to monthly composites (Table 1).

We did not use a convolutional neural network (CNN) because CNNs rely on features which are extracted by sliding windows within an image, and the widow size N is usually larger than the tree canopy pixel (usually one pixel). Thus, the importance from a given tree canopy pixel might be diluted on the extracted window's feature. Contrastingly, the input for the MLP is the labelled feature collection, which is derived from single pixels in an image, which is considered as more suitable when using Sentinel-2 imagery.

In the past decades, mapping forest cover and type have achieved satisfactory outcomes globally and operationally using remote sensing techniques [37,39]. Mapping tree species at large scale has however only progressed slowly, mainly because the required remote sensing datasets are expensive to acquire, such as LiDAR and other very high-resolution data [35,40]. Furthermore, the *Faidherbia albida* agroforestry multipurpose trees studied here grow in drylands where trees grow as individuals, and such trees outside of forest have been underestimated [61]. As such, our study represents a first attempt to conduct a national scale mapping of scattered non-forest tree species. It would be interesting to further map the distribution of *Faidherbia albida* trees using higher spatial resolution images like PlanetScope, which are now provided as a pre-processed product in an analysis-ready data format of monthly composites of 4.7-m spatial resolution free of charge from the Norwegian International Climate & Forests Initiative (NICFI). However, PlanetScope data has a lower spectral resolution (only four classical bands of R, G, B and NIR) as compared to Sentinel-2. The degraded spectral information may thus decrease the mapping capacity. With a monthly temporal resolution and spatial resolution of 4.7 m, recurrent CNN techniques could boost the mapping by interpreting temporal and textural information of individual tree canopies simultaneously [67].

4.2. Comparison with Ecological Niche Modelling Products

Our results show that Sentinel-2 image time series are able to map the actual distribution of *Faidherbia albida* trees, which is a significant advance as compared to SDM-based potential species occurrence maps [64]. Our results indicate that *Faidherbia albida* trees are mainly found in croplands where they are the dominating tree species. The relation in Figure 6 showed that when the land suitability exceeds the threshold defined by the SDM model, there is indeed often an observed occurrence of *Faidherbia albida* trees. The average *Faidherbia albida* cover is always higher in croplands than for other land use/cover types because farmers are well aware of the benefits of *Faidherbia albida* trees being an integral part of agroforestry systems [68–70]. In the sparsely populated south-eastern part of Senegal where land cover is mainly open woodland and shrublands, we did not predict to find any *Faidherbia albida*, while the SDM model predicts a high suitability, likely because human management was not carefully considered in the applied model.

5. Conclusions

A 10-m resolution *Faidherbia albida* canopy map was generated for Senegal based on a time series of Sentinel-2 images from the period of 2017 to 2019, used to form 12 monthly composites covering the phenological phases of the *Faidherbia albida* tree. Ground observations of tree species from two different regions in Senegal were used to generate a robust model mapping the occurrence of *Faidherbia albida* trees. Compared to an existing map showing the potential occurrence of *Faidherbia albida* trees from a suitability modelling approach (species distribution modelling; SDM), our study took advantage of the unique phenology of *Faidherbia albida* trees to separate this tree species from other trees. We showed that the actual distribution does not always agree with the potential distribution. We also showed that *Faidherbia albida* dominates tree cover in some cropland areas, supporting the well-known benefits of the tree in agroforestry systems of West Africa (termed parklands). *Faidherbia albida* is managed and utilised by farmers in different ways as this multi-purpose tree does not only provide fuel wood and fodder like other species, but also affects crop traits for improved productivity and yield by shedding leaves in the crop growing season. The derived 10-m map of *Faidherbia albida* occurrence could be used as a base-map in support of an empirically based understanding of the role of *Faidherbia albida* in parkland ecosystems in relation to crop yields. Ultimately, this could serve as a vantage point for spatially explicit analyses of the extent to which the use/management of this tree resource is optimally implemented.

Supplementary Materials: The following supporting information can be downloaded at: https://www.mdpi.com/article/10.3390/rs14030662/s1, Figure S1: Averaged monthly (**A**) NDVI74, (**B**) MCARI, (**C**)S2REP, (**D**) PSSRa, (**E**) EVI, (**F**) EVI2, (**G**) GNDVI, (**H**) MTCI values for sample pixels of *Faidherbia albida* and other species in the regions of Nioro and Niakhar. Vertical lines represent the standard deviation of the sample population. Overall, 1418 pixels were sampled and averaged over the study period, including 57 *Faidherbia albida* pixels and 112 pixels with other tree species in the Nioro region, and 459 *Faidherbia albida* pixels and 790 pixels with other species in the Niakhar region; Table S1: Vegetation indices used in the study. Their formulation and the related spectral bands (band numbering is referring to the Sentinel-2 sensor system); Table S2: Correlation coefficient of monthly vegetation index of *Faidherbia albida*; Table S3: Correlation coefficient between vegetation index values of *Faidherbia albida* and other species; Table S4: Climatic, geomorphologic, and pedologic layers used for the Species Distribution Modelling [71]; Table S5: Results from the 4-fold cross-validation using a test set of 40 features.

Author Contributions: T.L.: Conceptualization, Methodology, Writing—original draft. M.B.: Supervision, Writing—review and editing. X.T.: Methodology. P.H.: Validation, Writing—review and editing. L.L.: Validation, Writing—review and editing. B.N.: Resource. R.F.: Supervision, Writing—review and editing. All authors have read and agreed to the published version of the manuscript.

Funding: This research was funded by the China Scholarship Council. R.F (grant No. 201906400012). Acknowledge support by the Villum Foundation through the project 'Deep Learning and Remote

Sensing for Unlocking Global Ecosystem Resource Dynamics' (DeReEco), grant number 34306. This work was also supported by the European Research Council (ERC) under the European Union's Horizon 2020 research and innovation programme (grant No. 947757 TOFDRY) and a DFF Sapere Aude (grant No. 9064–00049B).

Institutional Review Board Statement: Not applicable.

Informed Consent Statement: Not applicable.

Data Availability Statement: All the Sentinel-2 images used in this study can be downloaded from Google Earth Engine (https://code.earthengine.google.com/, accessed on 1 December 2021). Please contact Louise Leroux for sharing of the species data.

Acknowledgments: We thank Roeland Kindt who provided the ecologically *Faidherbia albida* occurrence map.

Conflicts of Interest: The authors declare no conflict of interest.

References

1. Gonzalez, P.; Tucker, C.J.; Sy, H. Tree density and species decline in the African Sahel attributable to climate. *J. Arid. Environ.* **2012**, *78*, 55–64. [CrossRef]
2. Karlson, M.; Reese, H.; Ostwald, M. Tree crown mapping in managed woodlands (parklands) of semi-arid West Africa using WorldView-2 imagery and geographic object based image analysis. *Sensors* **2014**, *14*, 22643–22669. [CrossRef] [PubMed]
3. Sinare, H.; Gordon, L.J. Ecosystem services from woody vegetation on agricultural lands in Sudano-Sahelian West Africa. *Agr. Ecosyst. Environ.* **2015**, *200*, 186–199. [CrossRef]
4. Kuyah, S.; Öborn, I.; Jonsson, M.; Dahlin, A.S.; Barrios, E.; Muthuri, C.; Malmer, A.; Nyaga, J.; Magaju, C.; Namirembe, S. Trees in agricultural landscapes enhance provision of ecosystem services in Sub-Saharan Africa. *Int. J. Biodivers. Sci. Ecosyst. Serv. Manag.* **2016**, *12*, 255–273. [CrossRef]
5. Brandt, M.; Rasmussen, K.; Hiernaux, P.; Herrmann, S.; Tucker, C.J.; Tong, X.; Tian, F.; Mertz, O.; Kergoat, L.; Mbow, C. Reduction of tree cover in West African woodlands and promotion in semi-arid farmlands. *Nat. Geosci.* **2018**, *11*, 328–333. [CrossRef]
6. Koffi, C.; Lourme-Ruiz, A.; Djoudi, H.; Bouquet, E.; Dury, S.; Gautier, D. The contributions of wild tree resources to food and nutrition security in sub-Saharan African drylands: A review of the pathways and beneficiaries. *Int. Forest. Rev.* **2020**, *22*, 64–82. [CrossRef]
7. Tschakert, P.; Khouma, M.; Sene, M. Biophysical potential for soil carbon sequestration in agricultural systems of the Old Peanut Basin of Senegal. *J. Arid. Environ.* **2004**, *59*, 511–533. [CrossRef]
8. Hiernaux, P.H.; Cissé, M.; Diarra, L.; De Leeuw, P. Fluctuations saisonnières de la feuillaison des arbres et des buissons sahéliens. Conséquences pour la quantification des ressources fourragères. *Rev. d'Elevage. Méd. Vét. Pays. Trop* **1994**, *47*, 117–125. [CrossRef]
9. Garrity, D.P.; Akinnifesi, F.K.; Ajayi, O.C.; Weldesemayat, S.G.; Mowo, J.G.; Kalinganire, A.; Larwanou, M.; Bayala, J. Evergreen Agriculture: A robust approach to sustainable food security in Africa. *Food Secur.* **2010**, *2*, 197–214. [CrossRef]
10. Wood, P.J. *Faidherbia albida (Del.) A Chev. (Synonym: Acacia albida Del.): A Monograph*; CTFT: Nogent-sur-Marne, France, 1989.
11. Boffa, J.M. *Agroforestry Parklands in Sub-Saharan Africa;* Food and Agriculture Organization of the United Nations: Rome, Italy, 1999.
12. Barnes, R.; Fagg, C.W. *Faidherbia albida* Monograph and Annotated Bibliography. 2003. Available online: https://ora.ox.ac.uk/objects/uuid:fe18e8c9-1a92-435f-94c2-7c5827cbea57 (accessed on 23 November 2021).
13. Glover, J.D.; Reganold, J.P.; Cox, C.M. Plant perennials to save Africa's soils. *Nature* **2012**, *489*, 359–361. [CrossRef]
14. Noulekoun, F.; Birhane, E.; Chude, S.; Zenebe, A. Characterization of *Faidherbia albida* (Del.) A. Chev. population in agroforestry parklands in the highlands of Northern Ethiopia: Impact of conservation, environmental factors and human disturbances. *Agroforest. Syst.* **2017**, *91*, 123–135. [CrossRef]
15. Vandenbeldt, R.; Williams, J. The effect of soil surface temperature on the growth of millet in relation to the effect of *Faidherbia albida* trees. *Agr. Forest Meteorol.* **1992**, *60*, 93–100. [CrossRef]
16. Saka, A.; Bunderson, W.; Itimu, O.; Phombeya, H.; Mbekeani, Y. The effects of Acacia albida on soils and maize grain yields under smallholder farm conditions in Malawi. *Forest. Ecol. Manag.* **1994**, *64*, 217–230. [CrossRef]
17. Hadgu, K.M.; Kooistra, L.; Rossing, W.A.; van Bruggen, A.H. Assessing the effect of *Faidherbia albida* based land use systems on barley yield at field and regional scale in the highlands of Tigray, Northern Ethiopia. *Food Secur.* **2009**, *1*, 337–350. [CrossRef]
18. Sida, T.S.; Baudron, F.; Kim, H.; Giller, K.E. Climate-smart agroforestry: *Faidherbia albida* trees buffer wheat against climatic extremes in the Central Rift Valley of Ethiopia. *Agric. Forest Meteorol.* **2018**, *248*, 339–347. [CrossRef]
19. Leroux, L.; Falconnier, G.N.; Diouf, A.A.; Ndao, B.; Gbodjo, J.; Tall, L.; Balde, A.; Clermont-Dauphin, C.; Bégué, A.; Affholder, F. Using remote sensing to assess the effect of trees on millet yield in complex parklands of Central Senegal. *Agric. Syst.* **2020**, *184*, 102918. [CrossRef]

20. Roupsard, O.; Audebert, A.; Ndour, A.P.; Clermont-Dauphin, C.; Agbohessou, Y.; Sanou, J.; Koala, J.; Faye, E.; Sambakhe, D.; Jourdan, C. How far does the tree affect the crop in agroforestry? New spatial analysis methods in a Faidherbia parkland. *Agric. Ecosyst. Environ.* **2020**, *296*, 106928. [CrossRef]
21. Rinaudo, T. The development of farmer managed natural regeneration. *LEISA Mag.* **2007**, *23*, 32–34.
22. Haglund, E.; Ndjeunga, J.; Snook, L.; Pasternak, D. Dry land tree management for improved household livelihoods: Farmer managed natural regeneration in Niger. *J. Environ. Manag.* **2011**, *92*, 1696–1705. [CrossRef]
23. Weston, P.; Hong, R.; Kaboré, C.; Kull, C.A. Farmer-managed natural regeneration enhances rural livelihoods in dryland West Africa. *Environ. Manag.* **2015**, *55*, 1402–1417. [CrossRef]
24. Chomba, S.; Sinclair, F.; Savadogo, P.; Bourne, M.; Lohbeck, M. Opportunities and constraints for using farmer managed natural regeneration for land restoration in sub-Saharan Africa. *Front. For. Glob. Chang.* **2020**, *3*, 122. [CrossRef]
25. Elith, J.; Leathwick, J.R. Species distribution models: Ecological explanation and prediction across space and time. *Ann. Rev. Ecol. Evol. Syst.* **2009**, *40*, 677–697. [CrossRef]
26. Chakraborty, A.; Joshi, P.; Sachdeva, K. Predicting distribution of major forest tree species to potential impacts of climate change in the central Himalayan region. *Ecol. Eng.* **2016**, *97*, 593–609. [CrossRef]
27. Guisan, A.; Thuiller, W. Predicting species distribution: Offering more than simple habitat models. *Ecol. Lett.* **2005**, *8*, 993–1009. [CrossRef] [PubMed]
28. He, K.S.; Bradley, B.A.; Cord, A.F.; Rocchini, D.; Tuanmu, M.N.; Schmidtlein, S.; Turner, W.; Wegmann, M.; Pettorelli, N. Will remote sensing shape the next generation of species distribution models? *Remote Sens. Ecol. Conserv.* **2015**, *1*, 4–18. [CrossRef]
29. Fassnacht, F.E.; Latifi, H.; Stereńczak, K.; Modzelewska, A.; Lefsky, M.; Waser, L.T.; Straub, C.; Ghosh, A. Review of studies on tree species classification from remotely sensed data. *Remote Sens. Environ.* **2016**, *186*, 64–87. [CrossRef]
30. Karlson, M.; Ostwald, M.; Reese, H.; Bazié, H.R.; Tankoano, B. Assessing the potential of multi-seasonal WorldView-2 imagery for mapping West African agroforestry tree species. *Int. J. Appl. Earth Observ. Geoinform.* **2016**, *50*, 80–88. [CrossRef]
31. Wang, L.; Sousa, W.; Gong, P. Integration of object-based and pixel-based classification for mapping mangroves with IKONOS imagery. *Int. J. Remote Sens.* **2004**, *25*, 5655–5668. [CrossRef]
32. Tigges, J.; Lakes, T.; Hostert, P. Urban vegetation classification: Benefits of multitemporal RapidEye satellite data. *Remote Sens. Environ.* **2013**, *136*, 66–75. [CrossRef]
33. Adelabu, S.; Dube, T. Employing ground and satellite-based QuickBird data and random forest to discriminate five tree species in a Southern African Woodland. *Geocarto. Int.* **2015**, *30*, 457–471. [CrossRef]
34. Cho, M.A.; Malahlela, O.; Ramoelo, A. Assessing the utility WorldView-2 imagery for tree species mapping in South African subtropical humid forest and the conservation implications: Dukuduku forest patch as case study. *Int. J. Appl. Earth Observ. Geoinform.* **2015**, *38*, 349–357. [CrossRef]
35. Korpela, I.; Ørka, H.O.; Maltamo, M.; Tokola, T.; Hyyppä, J. Tree species classification using airborne LiDAR–effects of stand and tree parameters, downsizing of training set, intensity normalization, and sensor type. *Silva. Fenn.* **2010**, *44*, 319–339. [CrossRef]
36. Yao, W.; Krzystek, P.; Heurich, M. Tree species classification and estimation of stem volume and DBH based on single tree extraction by exploiting airborne full-waveform LiDAR data. *Remote Sens. Environ.* **2012**, *123*, 368–380. [CrossRef]
37. Heinzel, J.; Koch, B. Exploring full-waveform LiDAR parameters for tree species classification. *Int. J. Appl. Earth Observ. Geoinform.* **2011**, *13*, 152–160. [CrossRef]
38. Cho, M.A.; Mathieu, R.; Asner, G.P.; Naidoo, L.; Van Aardt, J.; Ramoelo, A.; Debba, P.; Wessels, K.; Main, R.; Smit, I.P. Mapping tree species composition in South African savannas using an integrated airborne spectral and LiDAR system. *Remote Sens. Environ.* **2012**, *125*, 214–226. [CrossRef]
39. Dalponte, M.; Bruzzone, L.; Gianelle, D. Tree species classification in the Southern Alps based on the fusion of very high geometrical resolution multispectral/hyperspectral images and LiDAR data. *Remote Sens. Environ.* **2012**, *123*, 258–270. [CrossRef]
40. Kukkonen, M.; Maltamo, M.; Korhonen, L.; Packalen, P. Multispectral Airborne Lidar Data in the Prediction of Boreal Tree Species Composition. *IEEE Trans. Geosci. Remote Sens.* **2019**, *57*, 3462–3471. [CrossRef]
41. Gärtner, P.; Förster, M.; Kleinschmit, B. The Benefit of Synthetically Generated Rapideye and Landsat 8 Data Fusion Time Series for Riparian Forest Disturbance Monitoring. *Remote Sens. Environ.* **2016**, *177*, 237–247. [CrossRef]
42. Schriever, J.R.; Congalton, R.G. Evaluating Seasonal Variability as an Aid to Cover-Type Mapping from Landsat Thematic Mapper Data in the Northeast. *Photogramm. Eng. Remote Sens.* **1995**, *61*, 321–327.
43. Hesketh, M.; Sánchez-Azofeifa, G.A. The Effect of Seasonal Spectral Variation on Species Classification in the Panamanian Tropical Forest. *Remote Sens. Environ.* **2012**, *118*, 73–82. [CrossRef]
44. Key, T.; Warner, T.A.; McGraw, J.B.; Fajvan, M.A. A Comparison of Multispectral and Multitemporal Information in High Spatial Resolution Imagery for Classification of Individual Tree Species in a Temperate Hardwood Forest. *Remote Sens. Environ.* **2001**, *75*, 100–112. [CrossRef]
45. Sheeren, D.; Fauvel, M.; Josipović, V.; Lopes, M.; Planque, C.; Willm, J.; Dejoux, J.-F. Tree Species Classification in Temperate Forests Using Formosat-2 Satellite Image Time Series. *Remote Sens.* **2016**, *8*, 734. [CrossRef]
46. Mickelson, J.G.; Civco, D.L.; Silander, J.A. Delineating Forest Canopy Species in the Northeastern United States Using Multi-Temporal Tm Imagery. *Photogramm. Eng. Remote Sens.* **1998**, *64*, 891–904.
47. Triboulet, C. Identification des parcs à *Faidherbia albida* par télédétection. *Cirad-Forêt Cahiers Sci.* **1996**, *12*, 203–216.

48. Lelong, C.C.; Tshingomba, U.K.; Soti, V. Assessing Worldview-3 multispectral imaging abilities to map the tree diversity in semi-arid parklands. *Int. J. Appl. Earth Observ. Geoinform.* **2020**, *93*, 102211. [CrossRef]
49. Zhang, W.; Brandt, M.; Wang, Q.; Prishchepov, A.V.; Tucker, C.J.; Li, Y.; Lyu, H.; Fensholt, R. From woody cover to woody canopies: How Sentinel-1 and Sentinel-2 data advance the mapping of woody plants in savannas. *Remote Sens. Environ.* **2019**, *234*, 111465. [CrossRef]
50. Chan, J.C.-W.; Paelinckx, D. Evaluation of Random Forest and Adaboost tree-based ensemble classification and spectral band selection for ecotope mapping using airborne hyperspectral imagery. *Remote Sens. Environ.* **2008**, *112*, 2999–3011. [CrossRef]
51. Dalponte, M.; Ørka, H.O.; Gobakken, T.; Gianelle, D.; Næsset, E. Tree species classification in boreal forests with hyperspectral data. *IEEE Transac. Geosci. Remote Sens.* **2012**, *51*, 2632–2645. [CrossRef]
52. Immitzer, M.; Atzberger, C.; Koukal, T. Tree species classification with random forest using very high spatial resolution 8-band WorldView-2 satellite data. *Remote Sens.* **2012**, *4*, 2661–2693. [CrossRef]
53. Drusch, M.; Del Bello, U.; Carlier, S.; Colin, O.; Fernandez, V.; Gascon, F.; Hoersch, B.; Isola, C.; Laberinti, P.; Martimort, P. Sentinel-2: ESA's optical high-resolution mission for GMES operational services. *Remote Sens. Environ.* **2012**, *120*, 25–36. [CrossRef]
54. Tottrup, C.; Rasmussen, M.S. Mapping long-term changes in savannah crop productivity in Senegal through trend analysis of time series of remote sensing data. *Agr. Ecosyst. Environ.* **2004**, *103*, 545–560. [CrossRef]
55. Lufafa, A.; Diédhiou, I.; Samba, S.; Séné, M.; Khouma, M.; Kizito, F.; Dick, R.; Dossa, E.; Noller, J. Carbon stocks and patterns in native shrub communities of Senegal's Peanut Basin. *Geoderma* **2008**, *146*, 75–82. [CrossRef]
56. Lamprecht, H. *Silviculture in the Tropics*; GTZ: Eschborn, Germany, 1989; p. 296.
57. Sileshi, G.W. The magnitude and spatial extent of influence of *Faidherbia albida* trees on soil properties and primary productivity in drylands. *J. Arid. Environ.* **2016**, *132*, 1–14. [CrossRef]
58. Tschakert, P. Carbon for farmers: Assessing the potential for soil carbon sequestration in the old peanut basin of Senegal. *Clim. Chang.* **2004**, *67*, 273–290. [CrossRef]
59. Ndao, B.; Leroux, L.; Gaetano, R.; Diouf, A.A.; Soti, V.; Bégué, A.; Mbow, C.; Sambou, B. Landscape heterogeneity analysis using geospatial techniques and a priori knowledge in Sahelian agroforestry systems of Senegal. *Ecol. Indic.* **2021**, *125*, 107481. [CrossRef]
60. Ndao, B.; Leroux, L.; Hema, A.; Diouf, A.A.; Bégué, A.; Sambou, B. Tree Species Diversity Mapping Using Species Distribution Models: A *Faidherbia albida* Parkland Case Study in Senegal. *Ecol. Model* **2021**. *submitted*.
61. Brandt, M.; Tucker, C.J.; Kariryaa, A.; Rasmussen, K.; Abel, C.; Small, J.; Chave, J.; Rasmussen, L.V.; Hiernaux, P.; Diouf, A.A. An unexpectedly large count of trees in the West African Sahara and Sahel. *Nature* **2020**, *587*, 78–82. [CrossRef]
62. Frampton, W.J.; Dash, J.; Watmough, G.; Milton, E.J. Evaluating the capabilities of Sentinel-2 for quantitative estimation of biophysical variables in vegetation. *ISPRS J. Photogramm. Remote Sens.* **2013**, *82*, 83–92. [CrossRef]
63. Hripcsak, G.; Rothschild, A.S. Agreement, the f-measure, and reliability in information retrieval. *J. Am. Med. Inform. Assoc.* **2005**, *12*, 296–298. [CrossRef]
64. Kindt, R. Ensemble species distribution modelling with transformed suitability values. *Environ. Model Softw.* **2018**, *100*, 136–145. [CrossRef]
65. Yan, E.; Wang, G.; Lin, H.; Xia, C.; Sun, H. Phenology-Based Classification of Vegetation Cover Types in Northeast China Using Modis Ndvi and Evi Time Series. *Int. J. Remote Sens.* **2015**, *36*, 489–512. [CrossRef]
66. Bradley, B.A. Remote detection of invasive plants: A review of spectral, textural and phenological approaches. *Biol. Invas.* **2014**, *16*, 1411–1425. [CrossRef]
67. Franklin, S.E.; Hall, R.J.; Moskal, L.M.; Maudie, A.J.; Lavigne, M.B. Incorporating Texture into Classification of Forest Species Composition from Airborne Multispectral Images. *Int. J. Remote Sens.* **2000**, *21*, 61–79. [CrossRef]
68. Lykke, A.M. Local perceptions of vegetation change and priorities for conservation of woody-savanna vegetation in Senegal. *J. Environ. Manag.* **2000**, *59*, 107–120. [CrossRef]
69. Lykke, A.; Kristensen, M.; Ganaba, S. Valuation of local use and dynamics of 56 woody species in the Sahel. *Biodiv. Conserv.* **2004**, *13*, 1961–1990. [CrossRef]
70. Sambou, A.; Sambou, B.; Ræbild, A. Farmers' contributions to the conservation of tree diversity in the Groundnut Basin, Senegal. *J. Forest. Res.* **2017**, *28*, 1083–1096. [CrossRef]
71. Kindt, R. A climate change atlas for Africa. In *Presentation Made during a Workshop on the Estimation of the Potential of Agroforestry to Mitigate Climate Change in Sub-Saharan Africa*; CIRAD: Montpellier, France, 2021. [CrossRef]

remote sensing

MDPI

Article

Deep Learning and Phenology Enhance Large-Scale Tree Species Classification in Aerial Imagery during a Biosecurity Response

Grant D. Pearse [1,*], Michael S. Watt [2], Julia Soewarto [1] and Alan Y. S. Tan [1]

1 Scion, Private Bag 3020, Rotorua 3046, New Zealand; julia.soewarto@scionresearch.com (J.S.); alan.tan@scionresearch.com (A.Y.S.T.)
2 Scion, 10 Kyle Street, Christchurch 8011, New Zealand; Michael.Watt@scionresearch.com
* Correspondence: grant.pearse@scionresearch.com; Tel.: +64-7-343-5899

Abstract: The ability of deep convolutional neural networks (deep learning) to learn complex visual characteristics offers a new method to classify tree species using lower-cost data such as regional aerial RGB imagery. In this study, we use 10 cm resolution imagery and 4600 trees to develop a deep learning model to identify *Metrosideros excelsa* (pōhutukawa)—a culturally important New Zealand tree that displays distinctive red flowers during summer and is under threat from the invasive pathogen *Austropuccinia psidii* (myrtle rust). Our objectives were to compare the accuracy of deep learning models that could learn the distinctive visual characteristics of the canopies with tree-based models (XGBoost) that used spectral and textural metrics. We tested whether the phenology of pōhutukawa could be used to enhance classification by using multitemporal aerial imagery that showed the same trees with and without widespread flowering. The XGBoost model achieved an accuracy of 86.7% on the dataset with strong phenology (flowering). Without phenology, the accuracy fell to 79.4% and the model relied on the blueish hue and texture of the canopies. The deep learning model achieved 97.4% accuracy with 96.5% sensitivity and 98.3% specificity when leveraging phenology—even though the intensity of flowering varied substantially. Without strong phenology, the accuracy of the deep learning model remained high at 92.7% with sensitivity of 91.2% and specificity of 94.3% despite significant variation in the appearance of non-flowering pōhutukawa. Pooling time-series imagery did not enhance either approach. The accuracy of XGBoost and deep learning models were, respectively, 83.2% and 95.2%, which were of intermediate precision between the separate models.

Keywords: tree species; classification; deep learning; convolutional networks; biosecurity; forest pathology; myrtle rust; urban forestry; machine learning; aerial imagery

Citation: Pearse, G.D.; Watt, M.S.; Soewarto, J.; Tan, A.Y.S. Deep Learning and Phenology Enhance Large-Scale Tree Species Classification in Aerial Imagery during a Biosecurity Response. *Remote Sens.* **2021**, *13*, 1789. https://doi.org/10.3390/rs13091789

Academic Editor: Markus Immitzer

Received: 17 April 2021
Accepted: 29 April 2021
Published: 4 May 2021

1. Introduction

The early stages of a biosecurity response to a newly arrived plant pathogen can have a significant bearing on the final outcome and cost [1,2]. Once an unwanted pathogen has been positively identified, mapping and identification of potential host species become essential for managing the incursion [3]. Identification of host plants must be carried out by trained personnel and the hosts may be located across a mixture of public and private property or in hard to access areas. For these reasons, carrying out large-scale searches for host plants can be very costly and challenging to resource.

The level of host detection and surveillance required in the face of an incursion is usually defined by the response objective. Eradication of a pathogen necessitates exhaustive detection of host species to monitor spread and enable the destruction of infected plants or even hosts showing no signs of infection to limit future spread. A monitoring objective may require only the identification of key indicator species to define the infection front and monitor impacts and host range. Finally, long-term management strategies may require large-scale but inexhaustive host identification to locate resistant individuals within a population for breeding programmes or other approaches to biological control [4,5].

Remote sensing can complement all of these objectives by offering an efficient and scalable means of identifying host species [6,7]. Imagery acquired from UAVs, aircraft or even space-borne optical sensors can be used to identify both potential hosts as well as the symptoms of pathogen infection on susceptible host species [6,8]. However, the detection and classification of species from remotely sensed data comprise a complex sub-discipline. Fassnacht et al. [9] carried out a comprehensive review of methods for tree species classification using remotely sensed data and highlighted clear themes in the literature. Multispectral and hyperspectral data were identified as being the most useful data sources for accurate species classification with LiDAR data being highly complementary. Through capturing reflected light outside the visible spectrum, the use of multi/hyperspectral data sources increases the chance of observing patterns of reflectance related to structural or biochemical traits that may be unique or distinctive to species or groups.

Multispectral data (4–12 bands) are relatively easy to capture and have been widely used in combination with machine learning methods to perform species classification [10–12]. However, accurate classification is often limited to broad groups such as conifer vs. deciduous forest types [13]. Hyperspectral data contain many more (>12) narrow spectral bands—enhancing the ability to observe small differences that may be present between the spectra of tree species and has been well studied for fine-grained species classification tasks [14,15]. The idea of unique spectral 'signatures' for species has been present in the literature for several decades; however, [9] concluded that these signatures appear to be rare in practice and, when present, require observation of a wide portion of the spectrum using sophisticated sensors [16].

Although hyperspectral data have been successfully used to classify as many as 42 species [6,9,17], large-scale applications of hyperspectral-based species classification face challenges related to practicality and cost. The increased spectral resolution usually demands careful acquisition from expensive sensors and is constrained by illumination and atmospheric requirements. The post-processing of these data can also be complex and requires careful correction of atmospheric impacts and noise reduction. Finally, the substantial volumes of data must often be subjected to dimensionality reduction before analysis can proceed [13,18]. Classification is based on patterns in the calibrated reflectance spectra from the canopy and differences in data sources and quality can reduce the transferability of the classifiers [19]. Other information content, such as the structure, shape, texture and other distinctive but hard to quantify characteristics are often neglected or partially utilised. Efforts to characterise the texture or the shape of the crown or other attributes typically rely on a small number of engineered features to summarise complex attributes [11,13].

In contrast, the human visual system allows experienced individuals to distinguish many species by visual inspection alone. Some cryptic species remain hard to tell apart visually, but trained experts (and even non-experts) can discriminate a surprising number of species [20]. This has led to the development of sites such as iNaturalist, where members of the public can upload images of species for experts to identify [21]. Recently, the advent of deep learning models based on convolutional neural networks (hereafter referred to as deep learning) has transformed the capability of machines to perform fine-grained classification of images, often reaching or exceeding human-level accuracy [22,23]. The architecture of these networks allows these networks to effectively learn the features important for classification. This is an important contrast with other approaches as the features are not engineered or pre-selected but rather learned by the network from labelled training examples with little requirement for image pre-processing.

Deep learning has been used for tree species classification from various combinations of LiDAR, hyperspectral and multispectral imagery [24–27]. Many studies have also successfully used simpler RGB imagery for species detection and classification. Importantly, these approaches have demonstrated a remarkable capacity to perform fine-grained species classification from consumer-grade camera imagery that is poorly suited to traditional remote sensing [28,29]. However, these studies have mostly used RGB data collected from UAV [30–32] and to a lesser extent high-resolution satellites [33,34], which constrains the ability to scale predictions in the former case or limits the spatial resolution of predictions in the latter case.

Although RGB imagery is routinely captured at regional levels by fixed-wing aircraft in many countries, few studies have undertaken large-scale host species identification using this ubiquitous data source. These data often include only RGB colour channels in uncalibrated radiance values rather than reflectance. The simplicity of these data means that large areas can be captured at high-resolution (<10 cm) for lower unit cost. Successful application of deep learning for large-scale host species identification using aerial imagery offers a scalable method to support biosecurity responses that bypasses many issues facing ground-based surveillance such as permissions and safe accessibility.

Classification of tree species is generally enhanced when there is low spectral variability within a species and high spectral variability between the target and other species [35]. Often there are times during the year when interspecies spectral variability is greater because of variation in phenological attributes such as leaf flush, senescence, or flowering. Little research has examined how phenological variation can be used by deep learning to improve species classification in trees, although we are aware of one such study for an invasive weed [36]. Collection of data from a species during a period of distinctive phenology could assist the use of deep learning through both enhancing predictive precision and providing a means to rapidly generate large training datasets.

Myrtle rust, caused by the fungal plant pathogen *Austropuccinia psidii* (G. Winter) Beenken (syn. *Puccinia psidii*), affects a broad range of hosts in the *Myrtaceae* family, causing lesions, dieback and, in some cases, mortality [37,38]. The pathogen is airborne and has spread rapidly around the globe [39–42]. New Zealand is home to at least 37 native myrtaceous species [43]. Of these, *Metrosideros excelsa* Sol. Ex Gaertn (pōhutukawa) has very high cultural value and has been widely planted for amenity purposes. This coastal evergreen tree has a sprawling habit of up to 20 m and produces dense masses of red flowers over the Christmas period [44], earning it the name 'the New Zealand Christmas tree'. Observations from pōhutukawa growing in other countries where myrtle rust is present indicate that the species is susceptible to myrtle rust [45,46].

In May 2017, myrtle rust was detected on the New Zealand mainland for the first time [47]. The disease has spread rapidly and has established on numerous native and exotic host species [48].

The overarching goal of this research was to test novel methods suitable for large-scale identification of key *Metrosideros* host species focussing on pōhutukawa as a test case. Specifically, the objectives of the research were to (1) test two state-of-the-art classification methods (XGBoost and deep convolutional neural networks) applied to three-band aerial imagery leveraging the strong phenology of pōhutukawa, i.e., distinctive flowering in summer, (2) test classification of the same trees without the assistance of phenology by using historical aerial imagery (3) test how practical and generally applicable these techniques are in real-world conditions by creating a combined dataset from objectives 1 and 2 that contained imagery captured using different sensors in different years and that showed a mixture of flowering and non-flowering trees.

2. Materials and Methods

2.1. Ground Truth Data

New Zealand maintains an extensive biosecurity surveillance system and an established incursion response protocol. During the first months after the incursion of myrtle rust, sites that were confirmed to contain infected hosts received intensive ground-based surveys to identify and inspect all potential host species within a fixed radius from the infected site. New, confirmed infections triggered additional searches around the new site. A mobile app used by trained inspectors was used to record the genus, GPS location and infection status for every host inspected during the response. These efforts produced a substantial volume of ground surveillance data including GPS locations and positive identification of *Metrosideros* spp. by trained inspectors. Many of the trees inspected were present within the coastal city of Tauranga (Figure 1), and nearly all the records for *Metrosideros* spp. in this region were pōhutukawa.

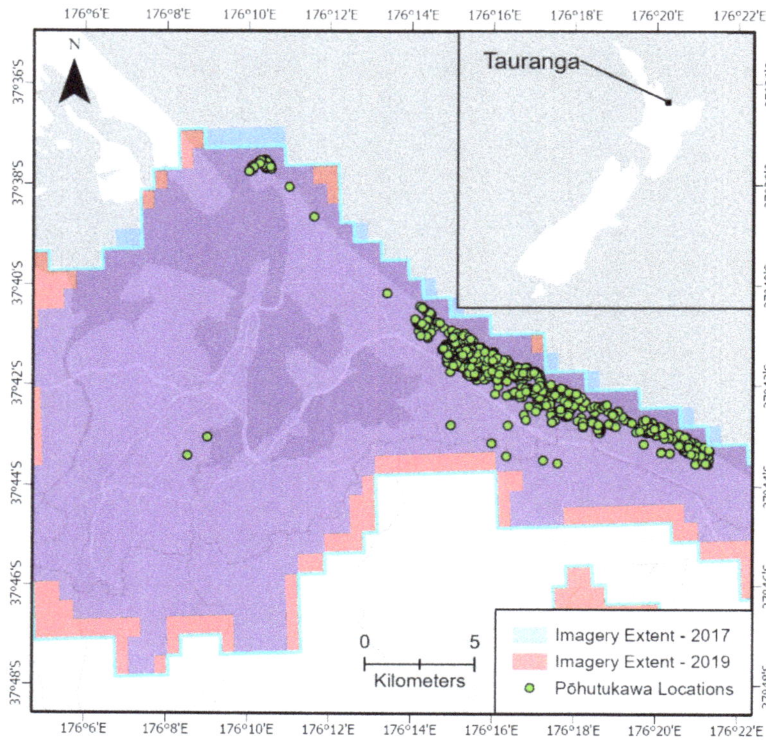

Figure 1. The extent of aerial imagery datasets and location of sample trees around Tauranga in the Bay of Plenty, New Zealand.

The extensive and distinctive red flowers of pōhutukawa are easily identifiable from above in the summer which made this species an ideal candidate to test the potential to utilise phenology to enhance species identification in RGB aerial imagery. For much of the rest of the year, some degree of buds, flowers, or seed capsules are present but less distinctive. However, the multi-leader crown shape and blueish hue of the large, waxy and elliptical leaves are also distinctive and present all year round (Figure 2).

Aerial imagery captured over Tauranga during the 2018–2019 summer period (Table 1) was overlaid with the ground surveillance locations in a GIS. Locations were collected using consumer-grade GPS and could only be considered approximate.

Figure 2. Images of pōhutukawa trees illustrating the distinctive features such as multi-stem form, leaf colour and texture and extensive buds and flowers usually present in summer. The bottom row shows aerial views from very high-resolution UAV imagery.

Table 1. Summary of multitemporal imagery used to develop classification models.

Imagery Dataset	Phenology	Resolution, Colour Channels
Tauranga—summer 2018–2019	Wide-spread flowering	10 cm/pixel, 3-band RGB
Tauranga—March 2017	Limited flowering	10 cm/pixel, 3-band RGB

For each inspection record, a trained analyst examined the GPS point and identified the corresponding tree in the aerial imagery. If the tree showed at least some evidence of flowering, then the imagery was annotated by delineating a bounding box around the canopy extent. The distinctive features of the canopy and strong flowering observed in the imagery greatly assisted identification and annotation; however, inspection records were only at the genus level and other species with similar phenological traits such as *Metrosideros robusta* (rātā) may occasionally be found within this region. In addition, some cultivated *Metrosideros excelsa* 'Aurea' ('yellow' pōhutukawa) appeared to be present within the dataset but these were removed due to the small number of samples available.

We assessed the purity of the training dataset by inspecting the majority of trees using publicly available, street-level imagery followed by on-site inspections for a smaller subset of trees. The results show that all study trees identified through combining the aerial imagery and surveillance records were pōhutukawa. After completing this process, we considered that the assembled training dataset consisted of only pōhutukawa and any misclassifications would have been very small in number.

Development of the classifiers also required negative examples. The candidate negative examples were any tree other than *Metrosideros* spp., hereafter referred to as other species. We once again leveraged the ground inspection efforts to develop this dataset. The intensity of the initial surveillance efforts meant that within inspected areas, such as streets or parks, the locations for nearly every pōhutukawa were recorded. We used these areas to select negative examples and cross-referenced a substantial portion of the dataset against other imagery and field inspections. This approach reduced the chances of accidentally including pōhutukawa or biasing the training set by excluding species that were visually similar to pōhutukawa due to uncertainty. In addition, this provided a realistic set of non-target tree canopies that the classifier might encounter in the areas surveyed for the biosecurity response. Bounding boxes around the canopies were defined against the aerial imagery and annotation proceeded until the dataset was balanced. Figure 3 shows examples of typical and atypical pōhutukawa and other tree species as seen in the aerial imagery.

Figure 3. Examples of canopy images used to train classification models. Pōhutukawa canopies from the 2019 imagery are shown in panel (**a**) with strong phenology (flowering and buds) visible. Panel (**b**) shows the same canopies in the 2017 imagery with less visible phenology. Examples of non-*Metrosideros* spp. seen in the 2019 imagery, including some harder examples, are shown in panel (**c**). The same canopies seen in the 2017 imagery are shown in panel (**d**).

2.2. *Imagery Datasets*

The aerial imagery datasets consisted of large orthomosaics generated from campaigns carried out in 2017 and 2019 using different aerial cameras (Table 1). The imagery from 2017 showed lower levels of detail, probably due to poorer image matching, and the trees had less visual detail (Figure 3). The bounding boxes were used to extract sub-images from the larger orthomosaics and each image 'chip' showing a tree canopy was labelled with the dataset year and class (pōhutukawa or other spp.). Very small trees (canopy radius $< \sim 1.5$ m) were excluded as these canopies contained too few pixels.

The final datasets included 2300 images of tree canopies evenly split between pōhutukawa and other spp. with images available for both 2017 and 2019 (Table 2). Images of pōhutukawa from 2019 and 2017 were used, respectively, to test the classification with and without the assistance of strong phenological features (Table 2). The imagery from the 2017 and 2019 datasets was combined to assess how well the model would generalise under real-world conditions (Table 2). The images of tree canopies were randomly split into training data (70%) used to fit the models. Validation data (15%) were used to select hyperparameters and evaluate model performance during training and a test set (15%)

was used to assess final model performance on completely withheld data (Table 2). Trees were assigned to the same splits in the 2017 and 2019 datasets for a fair comparison of the models. For the combined dataset, data were re-shuffled at the tree level and the imagery from both years was included in the assigned split to prevent data leakage.

Table 2. Summary of dataset splits used to train and validate classification models before testing on withheld data.

Dataset	Purpose	Tree Counts (Pōhutukawa/Other spp.)	Data Splits Training/Validation/Test
Tauranga 2019	Classification using phenology	2300 (1150/1150)	1610/345/345 (70/15/15%)
Tauranga 2017	Classification without phenology	2300 (1150/1150)	1610/345/345 (70/15/15%)
Tauranga 2017 and 2019 combined	Combined classification with and without phenology	4600 (2300/2300)	3220/690/690 (70/15/15%)

2.3. Deep Learning Models

We selected the ResNet model architecture [49] for classification of the tree canopies. The ResNet model is made up of small building blocks called the residual block. Each residual block is primarily made up of two to three convolution layers (this is dependent on the depth of the network) stacked together. The convolution layers are designed to learn and fit against the residual of the target function. The learned residual is then mapped back to the learned function through a skip connection that connects the input of the residual block to the output of the stacked convolution layers. By designing the neural network to learn and optimise on the residual instead of the original function, ResNet can learn the unknown original function more easily, thereby improving accuracy. We used the ResNet-50 architecture, which comprises 49 convolution layers organised into residual blocks and a fully connected layer for classification (Figure 4).

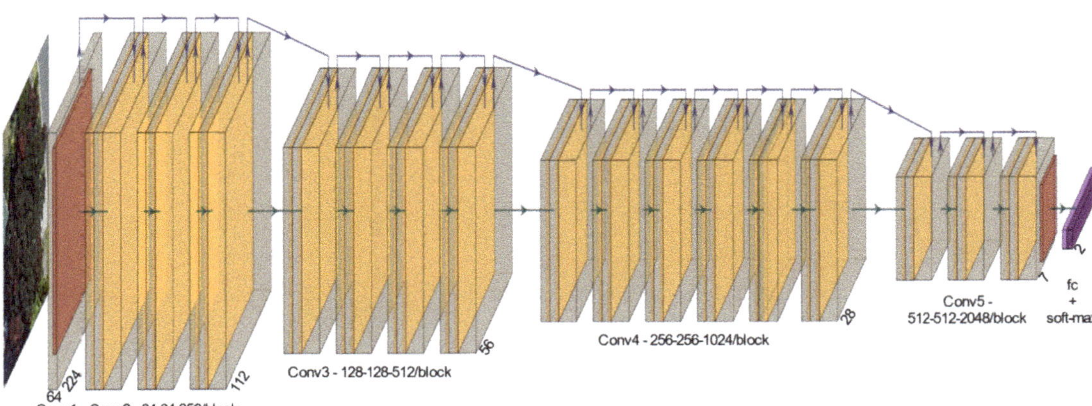

Figure 4. Overview of the architecture of ResNet-50. Adapted from [49].

A randomly initialized, fully connected layer was trained for 2 epochs to adapt a model pre-trained on the ImageNet [50] task to the binary classification task in this study. Thereafter, differential learning rates in the range 1×10^{-3}–1×10^{-6} were used to adapt deeper layers of the network at linearly decreasing learning rates for another 30 epochs. At this point, the performance validation metrics showed no further benefits from additional training. All deep learning models and metrics were implemented using the PyTorch 1.4 deep learning library [51] and the Scikit-Learn Python package [52]. Model training was carried out using a Nvidia Tesla K80 GPU with 12 GB of memory.

2.4. XGBoost Models

Approaches to species classification frequently use imagery to generate variables (metrics), such as vegetation indices, to capture features or characteristics useful for discriminating different species [9,13]. This may be done using rule-based methods [53] or machine learning methods such as decision trees [54]. We chose variables that target the distinctive properties of pōhutukawa canopies. These included spectral metrics aimed at capturing the blueish hue of the leathery, elliptical leaves and the strong and distinctive sprays of red flowers present in summer. The canopies also exhibit distinctive textural properties arising from the multi-stem structure and leaf and bud arrangements independent of the presence or absence of flowers (Figures 2 and 3). Texture analysis using grey-level co-occurrence matrices (GLCMs) [55] was used to try and capture these characteristics. Computation of the texture images was done using the 'glcm' package [56] in R [57]. The GLCM metric classes and parameters were selected based on the analysis and recommendations of [58]. The raw digital numbers (pixel radiance values) within each canopy bounding box were used to generate patch-level mean values for the predictive variables (Table 3). This was necessary because this type of imagery is optimised for visual appearance and lacks the information required to calculate reflectance.

We selected the XGBoost algorithm to perform binary classification using the 'xgboost' package for R [59]. XGBoost is a tree-based machine learning algorithm that is scalable, fast and has produced benchmark results on classification tasks [60]. The spectral and textural variables were used to train the XGBoost classifier for a maximum of 400 iterations, with early stopping based on validation set metrics used to prevent over-fitting. Subsampling of variables and observations for individual tree learners was also implemented alongside fine-tuning of the *gamma* hyperparameter to further guard against over-fitting.

Table 3. Vegetation indices and metrics computed from 3-band RGB aerial imagery of tree canopies for use in the XGBoost classification model. All metrics used the raw image digital numbers (DNs) (0–255) from the input pixels.

Variable Name	Description	Definition	Source
Mean red	Mean of red channel DNs	$\frac{\sum Red}{Num\ Red}$	NA
Mean green	Mean of green channel DNs	$\frac{\sum Green}{Num\ green}$	NA
Mean blue	Mean of blue channel DNs	$\frac{\sum Blue}{Num\ blue}$	NA
SD red	Standard deviation of red channel DNs	$\sqrt{\frac{\sum(Red-Mean\ red)^2}{Num\ red-1}}$	NA
SD green	Standard deviation of green channel DNs	$\sqrt{\frac{\sum(Green-Mean\ green)^2}{Num\ green-1}}$	NA
SD blue	Standard deviation of blue channel DNs	$\sqrt{\frac{\sum(Blue-Mean\ blue)^2}{Num\ blue-1}}$	NA
RG ratio	Red green ratio index	$\frac{Red}{Green}$	[61]
Normdiff RG	Normalised difference red/green ratio	$\frac{Red-Green}{Red+Green}$	NA
Scaled red	Scaled red ratio	$\frac{Red}{(Red+Green+Blue)}$	NA
Scaled green (SG)	Scaled green ratio	$\frac{Green}{(Red+Green+Blue)}$	NA
Scaled blue	Scaled blue ratio	$\frac{Blue}{(Red+Green+Blue)}$	NA
SD GI	Standard deviation of the scaled green index	$\sqrt{\frac{\sum(SG-Mean\ SG)^2}{Num\ SG-1}}$	NA
GLCM correlation	Textural metric computed on RGB channels	Grey-level co-occurrence correlation	[55]
GLCM homogeneity	Textural metric computed on RGB channels	Grey-level co-occurrence homogeneity	[55]
GLCM mean	Textural metric computed on RGB channels	Grey-level co-occurrence mean	[55]
GLCM entropy	Textural metric computed on RGB channels	Grey-level co-occurrence entropy	[55]

2.5. Performance Metrics

The predictions made by the final models on the withheld test splits of the three imagery datasets were used to compute the number of correct classifications (true positives) and incorrect classifications (false positives) for the pōhutukawa and 'other spp.' classes. These values were used to compute the classification performance metrics shown in Table 4.

Table 4. Performance metrics used to assess classification models. TP = true positive, FP = false positive, TN = true negative, FN = false negative.

Metric	Description	Definition
Accuracy	A measure of how often the classifier's predictions were correct.	$\frac{TP+TN}{TP+FP+TN+FN}$
Error	A measure of how often the classifier's predictions were wrong.	$1 - Accuracy$
Cohen's kappa	A measure of a classifier's prediction accuracy that accounts for chance agreement.	$\frac{observedAgreement-chanceAgreemeent}{1-chanceAgreement}$
Precision (Positive predictive value)	A measure of the proportion of positive predictions that were correct.	$\frac{TP}{TP+FP}$
Sensitivity (Recall)	The proportion of actual positives (*Metrosideros*) that were correctly identified by the classifier.	$\frac{TP}{TP+FN}$
Specificity	The proportion of actual negatives (other species) that were correctly identified by the classifier.	$\frac{TN}{TN+FP}$

3. Results

3.1. XGBoost Models

The results from the XGBoost and deep learning models applied to the withheld portions of the datasets used to test classification with phenology (2019 imagery), without phenology (2017 imagery) and classification of the combined datasets are shown in Table 5. The XGBoost classifiers showed moderately high accuracy on all three datasets (Table 5). The strong phenological traits of the pōhutukawa captured in the 2019 summer imagery produced the model with the highest accuracy (86.7%). The sensitivity and specificity were similar, reflecting nearly equal rates of false negatives and false positives. The variable importance scores extracted from XGBoost are shown in Figure 5. The scaled green pixel values and the *RG ratio* metric capturing the ratio of red to green pixels had the highest importance in the 2019 model utilising phenology. These two metrics most likely captured differences between the mostly green canopies of other spp. and the extensive showers of red flowers present on many pōhutukawa. The misclassified pōhutukawa often showed lower levels of flowering or were very small (Figure 6a). There were several hundred non-pōhutukawa trees in the dataset with canopies that appeared red in colour. These trees were often falsely classified as pōhutukawa (Figure 6b). This suggests that variables capturing the strong flowering patterns drove the high performance of the XGBoost model but struggled to separate other species with reddish or darker canopies.

Table 5. Classification results obtained by applying the trained models to the test split of the respective dataset.

	Classification with Strong Phenology (2019)	Classification without Strong Phenology (2017)	Classification of Combined 2017 & 2019 Datasets with and without Phenology
	XGBoost		
Accuracy	86.7%	79.4%	83.2%
Error	13.3%	20.6%	16.8%
kappa	0.733	0.588	0.664
Precision (PPV)	0.861	0.793	0.831
Sensitivity (recall)	0.871	0.788	0.827
Specificity	0.863	0.800	0.837
	Deep Learning		
Accuracy	97.4%	92.7%	95.2%
Error	3.6%	7.3%	4.8%
kappa	0.948	0.855	0.904
Precision (PPV)	0.982	0.939	0.973
Sensitivity (recall)	0.965	0.912	0.932
Specificity	0.983	0.943	0.973

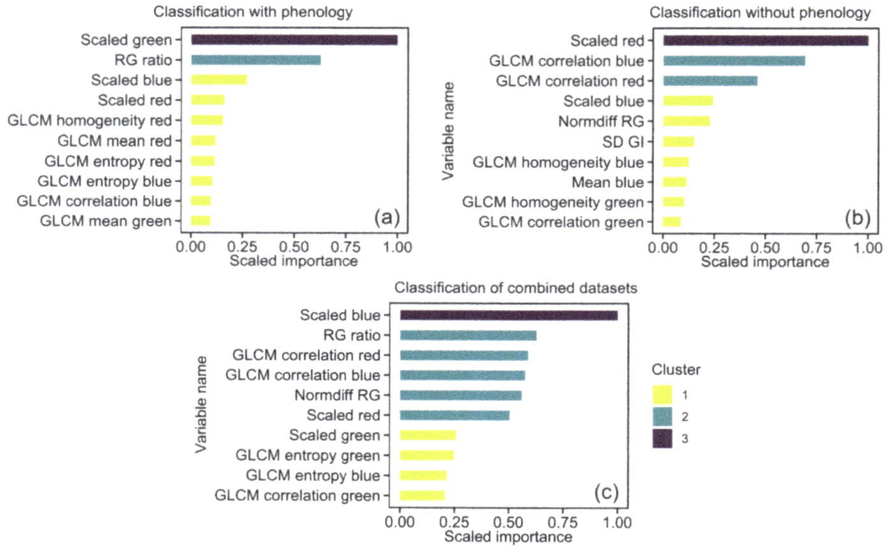

Figure 5. Plots of scaled variable importance metrics from the XGBoost species classification models using 2019 imagery with strong phenology (**a**), imagery from 2017 without strong phenology (**b**), and combined imagery from both years (**c**). Variables are clustered into groups with similar importance scores.

Figure 6. Examples of errors from XGBoost models. (**a**) Pōhutukawa canopies incorrectly classified as other species (false negatives) and (**b**) examples of other species canopies incorrectly classified as pōhutukawa (false positives) using the 2019 imagery with strong phenology. False negatives (**c**) and false positives (**d**) from the 2017 imagery with limited phenology. False negatives (**e**) and false positives (**f**) from the XGBoost classifier trained on imagery from both years.

3.2. Deep Learning Models

The deep learning models performed substantially better than the XGBoost models on all three datasets. The classifier developed using the 2019 imagery with strong phenology achieved an accuracy of 97.4%. The specificity of the model (98.3%) was slightly higher than the sensitivity (96.5%). The false negatives often showed similarities to pōhutukawa with a few exceptions (Figure 7a). The few false positives (Figure 7b) included a single relatively obvious error and some examples with limited or irregular flowering. The classification performance indicated that the model was highly effective at discriminating flowering trees from most other species with reddish canopies or other flowering species present in the data.

Figure 7. Examples of errors from deep learning models. (**a**) Pōhutukawa canopies incorrectly classified as other species (false negatives) and (**b**) examples of other species canopies incorrectly classified as pōhutukawa (false positives) using the 2019 imagery with strong phenology. False negatives (**c**) and false positives (**d**) from the 2017 imagery with limited phenology. False negatives (**e**) and false positives (**f**) from the deep learning classifier trained on imagery from both years.

Without using the strong flowering, the accuracy of the deep learning classifier dropped to 92.7% (Table 5). The model appeared to struggle more with the canopies affected by the lower quality of the imagery—small, blurry canopies without the characteristic appearance visible in other images frequently appeared in the misclassified images and the false negatives and false positives were visually similar to each other (Figure 7c,d).

As with the XGBoost models, the deep learning model trained on the combined imagery from both 2017 and 2019 (with and without strong phenology) did not show improved performance with a larger dataset. The model achieved 95.2% accuracy and showed the largest difference between sensitivity (93.2%) and specificity (97.3%), reflecting additional false negatives. Most of the misclassified canopies were from the 2017 dataset, and once again these images often showed blurry and indistinct features relative to other correctly classified examples (Figure 7e,f).

4. Discussion

This study demonstrated that deep learning algorithms could classify pōhutukawa in the study area with a very high level of accuracy using only three-band RGB aerial imagery, with or without the use of phenology to enhance detection. Existing remote sensing approaches to tree species classification rely extensively on calibrated multi or hyperspectral data that can be expensive and complex to capture over larger areas [9,13,18,24,26,27]. In contrast, RGB aerial imagery is routinely captured over large areas. Our results suggest that combining deep learning with this type of imagery enables large-scale mapping of visually distinctive species.

Significant gains in deep learning model accuracy were realised through leveraging the visual distinctiveness of pōhutukawa flowering that was clearly identifiable in 2019 aerial imagery. This distinctive phenological attribute also greatly assisted the collation of a robust number of samples (1150) that was large in comparison to many other tree classification studies [9,24,62]. Although we had access to ground-truth data, the characteristic flowering would have allowed most trees to be readily identified without the ground inspections. Through linking these clearly visible tree locations to previously collected imagery of pōhutukawa that were not flowering, it was possible to rapidly assemble data and train deep learning models that could accurately classify pōhutukawa without the strong phenology. Through combining these two sets of phenologically contrasting images we were able to assemble and train a model from a dataset that more closely approximated a real-world scenario where pōhutukawa exhibited variation in phenological expression. This workflow highlights how imagery with clear phenological traits can be used to rapidly assemble a more general dataset and through this approach mitigate a common bottleneck for training deep learning models.

The phenology of tree species has previously been used to enhance remote sensing classification [63,64]. However, attempting classification using only three-band imagery—with or without phenology—is less common [9]. This imagery lacks the spectral bandwidth required by most traditional methods to discriminate species. The few indices that can be derived are not widely generalisable, as the imagery represents sensor radiance rather than reflectance from the canopy and the imagery is manipulated to enhance visual appearance. To overcome this limitation, we derived features such as textural metrics and simple band ratios aimed at capturing the bright-red, extensive flowering of these species and the characteristic blueish hue and textural properties of the canopies.

This approach was successfully used by the XGBoost classifier for classification in the presence of phenology, and although not as accurate, the model without phenology was still reasonably robust. The performance of both models was high compared to other examples in the literature. For example, [62] achieved 68.3% classification accuracy of pōhutukawa using multispectral satellite data from the Coromandel region in New Zealand. The addition of LiDAR-derived features improved this result to 81.7% but pōhutukawa were noted to be more difficult to detect than several other species targeted in that study. It is likely that having multispectral and LiDAR data would have further improved the XGBoost results in our study, but this would come with higher costs for data acquisition, storage and processing.

The deep learning approach differed in fundamental ways to traditional remote sensing methods. While the models will utilise the colour of the canopies, as demonstrated by the accurate classification of flowering trees, the deep learning approach is also capable of learning harder-to-quantify features. For example, the characteristic appearance of the multi-stem canopy, distinctive canopy texture and extensive budding are relatively easy for knowledgeable analysts to identify in the aerial imagery and the deep learning models can 'learn' that these or similar features are important. This makes the models harder to interpret but powerful for complex classification tasks [65].

One key methodological limitation of our approach was the need to manually delineate individual tree canopies before training and inference could be carried out. This requirement is present in many traditional remote sensing approaches to species classification. A common workflow is to use LiDAR-derived elevation data alone [66] or in combination with multispectral data (especially the vegetation-sensitive NIR band) to delineate tree canopies [15,67]. While effective, this method introduces the need for costly LiDAR data and substantial analysis to extract canopies. More complex deep learning frameworks may offer an alternative option to perform both segmentation and classification, although the training data are more expensive to collect [31,68].

The high classification accuracies observed in this study are likely subject to some caveats. The models were exposed to the unique characteristics and properties of both aerial imagery datasets. Deep learning approaches do not expect or require calibrated or corrected imagery, but it is possible that subtle differences in resolution or other dataset characteristics may reduce transferability to new, unseen aerial imagery. The level of flowering seen in the 2019 dataset varied widely and many trees showed limited flowering. However, the imagery was also sharper and many of the other characteristic features of pōhutukawa were more easily visible in the imagery (e.g., buds, canopy form and hue, e.g., Figure 3). This provided additional features for the deep learning models and likely contributed to the high accuracy above and beyond the distinctiveness of the flowering. The 2017 imagery had the same nominal resolution (10 cm) but had markedly lower quality and detail (Figure 3). The pōhutukawa all exhibited a blueish hue in this imagery and some of the textural attributes were still discernible—both of which are likely to have contributed to the performance of the deep learning model. For predictions to work in new areas, the features learned from these datasets would need to be discernible in the new imagery. A brief test conducted by reducing the resolution of some of the imagery (bilinear resampling) showed that the accuracy of the combined classifier declined rapidly as the distinctive features were lost, with simulated 15 cm imagery showing only a 70% accuracy rate.

The resolution of the imagery also placed a limit on the size of the trees that could be classified. Many canopies fell between 30 and 60 pixels in size. At this size, the characteristic traits were difficult for a human observer to discern and the models would also have had limited information to learn from. This problem was reduced when phenology could be utilised, but smaller canopies were more frequently misclassified. It is very likely that higher-resolution imagery would have improved the classification accuracy still further and may enhance the transferability of the models. Outside of this domain, for example, where only moderate to low resolution imagery is available, traditional multispectral or hyperspectral methods may be more appropriate as they attempt to recover and utilise the spectral attributes of the canopy that can persist at coarser resolutions or be retrieved through unmixing.

Future work should explore expanding these methods to a greater number of species and validate the transferability of deep learning models across multiple, regional datasets. An extremely promising area of research is the potential for combined deep learning architectures that offer localisation and segmentation as well as classification [31,33]. This work could enable large-scale and repeatable mapping of tree species across a range of environments from lower-cost RGB datasets. This would be useful for biosecurity as well as many other applications.

5. Conclusions

In this study, we combined distinctive phenological traits and biosecurity surveillance records to develop a high-quality dataset to train and test novel algorithms to detect pōhutukawa from simple three-band (RGB) aerial imagery. Both modelling approaches performed well when the dataset included distinctive phenological traits (extensive, bright red flowers). However, the deep learning algorithm was able to achieve very high accuracies even in the absence of some key traits such as the distinctive flowers. The results of this study suggest that deep learning-based approaches could be used to rapidly and accurately map certain species over large areas using only RGB aerial imagery. Candidate species include those where classification is achievable by an experienced analyst using the same input data. The deep learning approach did appear sensitive to image resolution and quality and higher resolution imagery would likely expand the range of species suitable for classification using this method.

Author Contributions: Conceptualization, G.D.P.; methodology, G.D.P.; software, G.D.P. and A.Y.S.T.; validation, M.S.W. and A.Y.S.T.; formal analysis, G.D.P.; investigation, M.S.W. and G.D.P.; resources, G.D.P. and J.S.; data curation, G.D.P.; writing—original draft preparation, G.D.P., M.S.W. and J.S.; writing—review and editing, M.S.W. and A.Y.S.T.; visualization, G.D.P., A.Y.S.T. and J.S.; supervision, M.S.W.; project administration, G.D.P. and M.S.W.; funding acquisition, G.D.P. and M.S.W. All authors have read and agreed to the published version of the manuscript.

Funding: This research was funded by the Ministry for Primary Industries (Contract 18607). The study received additional support from the Strategic Science Investment Funding provided by the Ministry of Business, Innovation and Employment to Scion Research.

Data Availability Statement: The imagery used in this study is freely available under a Creative Commons License from Land Information New Zealand. https://data.linz.govt.nz/ (accessed on 26 October 2019).

Acknowledgments: We gratefully acknowledge the provision of data and research input from Deirdre Nagle and Quenten Higgan of AsureQuality Ltd. We would like to thank the many field staff who carried out the ground inspections and Honey Estarija for her tireless work annotating and checking data. We would also like to thank Tauranga District Council and the Bay of Plenty Regional council who funded data capture and provided early access to the 2019 imagery (BOPLAS2019).

Conflicts of Interest: The authors declare no conflict of interest.

References

1. Goldson, S.; Bourdôt, G.; Brockerhoff, E.; Byrom, A.; Clout, M.; McGlone, M.; Nelson, W.; Popay, A.; Suckling, D.; Templeton, M. New Zealand pest management: Current and future challenges. *J. R. Soc. N. Z.* **2015**, *45*, 31–58. [CrossRef]
2. Kriticos, D.; Phillips, C.; Suckling, D. Improving border biosecurity: Potential economic benefits to New Zealand. *N. Z. Plant Prot.* **2005**, *58*, 1–6. [CrossRef]
3. Kalaris, T.; Fieselmann, D.; Magarey, R.; Colunga-Garcia, M.; Roda, A.; Hardie, D.; Cogger, N.; Hammond, N.; Martin, P.T.; Whittle, P. The role of surveillance methods and technologies in plant biosecurity. In *The Handbook of Plant Biosecurity*; Springer: Dordrecht, The Netherlands, 2014; pp. 309–337.
4. DiTomaso, J.M.; Van Steenwyk, R.A.; Nowierski, R.M.; Vollmer, J.L.; Lane, E.; Chilton, E.; Burch, P.L.; Cowan, P.E.; Zimmerman, K.; Dionigi, C.P. Enhancing the effectiveness of biological control programs of invasive species through a more comprehensive pest management approach. *Pest. Manag. Sci.* **2017**, *73*, 9–13. [CrossRef] [PubMed]
5. Mundt, C.C. Durable resistance: A key to sustainable management of pathogens and pests. *Infect. Genet. Evol.* **2014**, *27*, 446–455. [CrossRef] [PubMed]
6. Asner, G.P.; Martin, R.E.; Keith, L.M.; Heller, W.P.; Hughes, M.A.; Vaughn, N.R.; Hughes, R.F.; Balzotti, C. A Spectral Mapping Signature for the Rapid Ohia Death (ROD) Pathogen in Hawaiian Forests. *Remote Sens.* **2018**, *10*, 404. [CrossRef]
7. Huang, C.; Asner, G.P. Applications of remote sensing to alien invasive plant studies. *Sensors* **2009**, *9*, 4869–4889. [CrossRef] [PubMed]
8. He, Y.; Chen, G.; Potter, C.; Meentemeyer, R.K. Integrating multi-sensor remote sensing and species distribution modeling to map the spread of emerging forest disease and tree mortality. *Remote Sens. Environ.* **2019**, *231*, 111238. [CrossRef]
9. Fassnacht, F.E.; Latifi, H.; Stereńczak, K.; Modzelewska, A.; Lefsky, M.; Waser, L.T.; Straub, C.; Ghosh, A. Review of studies on tree species classification from remotely sensed data. *Remote Sens. Environ.* **2016**, *186*, 64–87. [CrossRef]
10. Dash, J.P.; Watt, M.S.; Pearse, G.D.; Dungey, H.S. UAV Based Monitoring of Physiological Stress in Trees is Affected by Image Resolution and Choice of Spectral Index. *ISPRS J. Photogramm. Remote Sens.* **2017**, *131*, 1–14. [CrossRef]
11. Ferreira, M.P.; Wagner, F.H.; Aragão, L.E.O.C.; Shimabukuro, Y.E.; de Souza Filho, C.R. Tree species classification in tropical forests using visible to shortwave infrared WorldView-3 images and texture analysis. *ISPRS J. Photogramm. Remote Sens.* **2019**, *149*, 119–131. [CrossRef]
12. Krzystek, P.; Serebryanyk, A.; Schnörr, C.; Červenka, J.; Heurich, M. Large-scale mapping of tree species and dead trees in šumava national park and bavarian forest national park using lidar and multispectral imagery. *Remote Sens.* **2020**, *12*, 661. [CrossRef]
13. Ballanti, L.; Blesius, L.; Hines, E.; Kruse, B. Tree species classification using hyperspectral imagery: A comparison of two classifiers. *Remote Sens.* **2016**, *8*, 445. [CrossRef]

14. Clark, M.L.; Roberts, D.A.; Clark, D.B. Hyperspectral discrimination of tropical rain forest tree species at leaf to crown scales. *Remote Sens. Environ.* **2005**, *96*, 375–398. [CrossRef]
15. Dalponte, M.; Orka, H.O.; Gobakken, T.; Gianelle, D.; Naesset, E. Tree Species Classification in Boreal Forests With Hyperspectral Data. *IEEE Trans. Geosci. Remote Sens.* **2012**, *51*, 2632–2645. [CrossRef]
16. Hesketh, M.; Sánchez-Azofeifa, G.A. The effect of seasonal spectral variation on species classification in the Panamanian tropical forest. *Remote Sens. Environ.* **2012**, *118*, 73–82. [CrossRef]
17. Maschler, J.; Atzberger, C.; Immitzer, M. Individual tree crown segmentation and classification of 13 tree species using airborne hyperspectral data. *Remote Sens.* **2018**, *10*, 1218. [CrossRef]
18. Bioucas-Dias, J.M.; Plaza, A.; Camps-Valls, G.; Scheunders, P.; Nasrabadi, N.; Chanussot, J. Hyperspectral remote sensing data analysis and future challenges. *IEEE Geosci. Remote Sens. Mag.* **2013**, *1*, 6–36. [CrossRef]
19. Bannari, A.; Morin, D.; Bonn, F.; Huete, A.R. A review of vegetation indices. *Remote Sens. Rev.* **1995**, *13*, 95–120. [CrossRef]
20. De Lacerda, A.E.B.; Nimmo, E.R. Can we really manage tropical forests without knowing the species within? Getting back to the basics of forest management through taxonomy. *For. Ecol. Manag.* **2010**, *259*, 995–1002. [CrossRef]
21. Van Horn, G.; Mac Aodha, O.; Song, Y.; Cui, Y.; Sun, C.; Shepard, A.; Adam, H.; Perona, P.; Belongie, S. The inaturalist species classification and detection dataset. In Proceedings of the IEEE Conference on Computer Vision and Pattern Recognition, Salt Lake, UT, USA, 18–23 June 2018; pp. 8769–8778.
22. De Fauw, J.; Ledsam, J.R.; Romera-Paredes, B.; Nikolov, S.; Tomasev, N.; Blackwell, S.; Askham, H.; Glorot, X.; O'Donoghue, B.; Visentin, D.; et al. Clinically applicable deep learning for diagnosis and referral in retinal disease. *Nat. Med.* **2018**, *24*, 1342–1350. [CrossRef]
23. Krizhevsky, A.; Sutskever, I.; Hinton, G.E. Imagenet classification with deep convolutional neural networks. In Proceedings of the Advances in Neural Information Processing Systems, Lake Tahoe, NV, USA, 3–8 December 2012; pp. 1097–1105.
24. Fricker, G.A.; Ventura, J.D.; Wolf, J.A.; North, M.P.; Davis, F.W.; Franklin, J. A convolutional neural network classifier identifies tree species in mixed-conifer forest from hyperspectral imagery. *Remote Sens.* **2019**, *11*, 2326. [CrossRef]
25. Hartling, S.; Sagan, V.; Sidike, P.; Maimaitijiang, M.; Carron, J. Urban tree species classification using a WorldView-2/3 and LiDAR data fusion approach and deep learning. *Sensors* **2019**, *19*, 1284. [CrossRef]
26. Mäyrä, J.; Keski-Saari, S.; Kivinen, S.; Tanhuanpää, T.; Hurskainen, P.; Kullberg, P.; Poikolainen, L.; Viinikka, A.; Tuominen, S.; Kumpula, T.; et al. Tree species classification from airborne hyperspectral and LiDAR data using 3D convolutional neural networks. *Remote. Sens. Environ.* **2021**, *256*, 112322. [CrossRef]
27. Trier, Ø.D.; Salberg, A.-B.; Kermit, M.; Rudjord, Ø.; Gobakken, T.; Næsset, E.; Aarsten, D. Tree species classification in Norway from airborne hyperspectral and airborne laser scanning data. *Eur. J. Remote. Sens.* **2018**, *51*, 336–351. [CrossRef]
28. Cui, Y.; Song, Y.; Sun, C.; Howard, A.; Belongie, S. Large scale fine-grained categorization and domain-specific transfer learning. In Proceedings of the IEEE Conference on Computer Vision and Pattern Recognition, Salt Lake, UT, USA, 18–23 June 2018; pp. 4109–4118.
29. Wäldchen, J.; Rzanny, M.; Seeland, M.; Mäder, P. Automated plant species identification—Trends and future directions. *PLoS Comput. Biol.* **2018**, *14*, e1005993. [CrossRef]
30. Onishi, M.; Ise, T. Explainable identification and mapping of trees using UAV RGB image and deep learning. *Sci. Rep.* **2021**, *11*, 903. [CrossRef]
31. Schiefer, F.; Kattenborn, T.; Frick, A.; Frey, J.; Schall, P.; Koch, B.; Schmidtlein, S. Mapping forest tree species in high resolution UAV-based RGB-imagery by means of convolutional neural networks. *ISPRS J. Photogramm. Remote Sens.* **2020**, *170*, 205–215. [CrossRef]
32. Egli, S.; Höpke, M. CNN-Based Tree Species Classification Using High Resolution RGB Image Data from Automated UAV Observations. *Remote Sens.* **2020**, *12*, 3892. [CrossRef]
33. Wagner, F.H.; Sanchez, A.; Tarabalka, Y.; Lotte, R.G.; Ferreira, M.P.; Aidar, M.P.M.; Gloor, E.; Phillips, O.L.; Aragão, L.E.O.C. Using the U-net convolutional network to map forest types and disturbance in the Atlantic rainforest with very high resolution images. *Remote Sens. Ecol. Conserv.* **2019**, *5*, 360–375. [CrossRef]
34. Omer, G.; Mutanga, O.; Abdel-Rahman, E.M.; Adam, E. Performance of support vector machines and artificial neural network for mapping endangered tree species using WorldView-2 data in dukuduku forest, South Africa. *IEEE J. Sel. Top. Appl. Earth Obs. Remote Sens.* **2015**, *8*, 4825–4840. [CrossRef]
35. Castro-Esau, K.L.; Sánchez-Azofeifa, G.A.; Rivard, B.; Wright, S.J.; Quesada, M. Variability in leaf optical properties of Mesoamerican trees and the potential for species classification. *Am. J. Bot.* **2006**, *93*, 517–530. [CrossRef] [PubMed]
36. Tian, J.; Wang, L.; Yin, D.; Li, X.; Diao, C.; Gong, H.; Shi, C.; Menenti, M.; Ge, Y.; Nie, S.; et al. Development of spectral-phenological features for deep learning to understand Spartina alterniflora invasion. *Remote Sens. Environ.* **2020**, *242*, 111745. [CrossRef]
37. Carnegie, A.J.; Kathuria, A.; Pegg, G.S.; Entwistle, P.; Nagel, M.; Giblin, F.R. Impact of the invasive rust Puccinia psidii (myrtle rust) on native Myrtaceae in natural ecosystems in Australia. *Biol. Invasions* **2016**, *18*, 127–144. [CrossRef]
38. Glen, M.; Alfenas, A.C.; Zauza, E.A.V.; Wingfield, M.J.; Mohammed, C. Puccinia psidii: A threat to the Australian environment and economy—A review. *Australas. Plant Pathol.* **2007**, *36*, 1–16. [CrossRef]
39. Carnegie, A.J.; Cooper, K. Emergency response to the incursion of an exotic myrtaceous rust in Australia. *Australas. Plant Pathol.* **2001**, *40*, 346. [CrossRef]

40. Coutinho, T.A.; Wingfield, M.J.; Alfenas, A.C.; Crous, P.W. Eucalyptus Rust: A Disease with the Potential for Serious International Implications. *Plant Dis.* **1998**, *82*, 819–825. [CrossRef]
41. McTaggart, A.R.; Roux, J.; Granados, G.M.; Gafur, A.; Tarrigan, M.; Santhakumar, P.; Wingfield, M.J. Rust (Puccinia psidii) recorded in Indonesia poses a threat to forests and forestry in South-East Asia. *Australas. Plant Pathol.* **2015**, *45*, 83–89. [CrossRef]
42. Roux, J.; Greyling, I.; Coutinho, T.A.; Verleur, M.; Wingfield, M.J. The Myrtle rust pathogen, Puccinia psidii, discovered in Africa. *IMA Fungus* **2013**, *4*, 155–159. [CrossRef]
43. De Lange, P.J.; Rolfe, J.R.; Barkla, J.W.; Courtney, S.P.; Champion, P.D.; Perrie, L.R.; Beadel, S.M.; Ford, K.A.; Breitwieser, I.; Schoenberger, I.; et al. *Conservation Status of New Zealand Indigenous Vascular Plants, 2017*; Department of Conservation: Wellington, New Zealand, 2018; ISBN 978-1-98-85146147-1.
44. Allan, H.H. *Flora of New Zealand Volume I Indigenous Tracheophyta-Psilopsida, Lycopsida, Filicopsida, Gymnospermae, Dicotyledones*; Flora of New Zealand-Manaaki Whenua Online Reprint Series; Government Printer Publication: Wellington, New Zealand, 1982; Volume 1, ISBN 0-477-01056-3.
45. Loope, L. A summary of information on the rust Puccinia psidii Winter (guava rust) with emphasis on means to prevent introduction of additional strains to Hawaii. In *Open-File Report*; US Geological Survey: Reston, VA, USA, 2010; pp. 1–31. Available online: https://pubs.usgs.gov/of/2010/1082/of2010-1082.pdf (accessed on 17 June 2019).
46. Sandhu, K.S.; Park, R.F. *Genetic Basis of Pathogenicity in Uredo Rangelii*; University of Sydney: Camperdown, Sydney, 2013.
47. Ho, W.H.; Baskarathevan, J.; Griffin, R.L.; Quinn, B.D.; Alexander, B.J.R.; Havell, D.; Ward, N.A.; Pathan, A.K. First Report of Myrtle Rust Caused by Austropuccinia psidii on Metrosideros kermadecensis on Raoul Island and on M. excelsa in Kerikeri, New Zealand. *Plant Dis.* **2019**, *103*, 2128. [CrossRef]
48. Beresford, R.M.; Turner, R.; Tait, A.; Paul, V.; Macara, G.; Yu, Z.D.; Lima, L.; Martin, R. Predicting the climatic risk of myrtle rust during its first year in New Zealand. *N. Z. Plant Prot.* **2018**, *71*, 332–347. [CrossRef]
49. He, K.; Zhang, X.; Ren, S.; Sun, J. Deep Residual Learning for Image Recognition. *arXiv* **2015**, arXiv:1512.03385.
50. Russakovsky, O.; Deng, J.; Su, H.; Krause, J.; Satheesh, S.; Ma, S.; Huang, Z.; Karpathy, A.; Khosla, A.; Bernstein, M.; et al. ImageNet Large Scale Visual Recognition Challenge. *Int. J. Comput. Vis.* **2015**, *115*, 211–252. [CrossRef]
51. Paszke, A.; Gross, S.; Massa, F.; Lerer, A.; Bradbury, J.; Chanan, G.; Killeen, T.; Lin, Z.; Gimelshein, N.; Antiga, L.; et al. PyTorch: An Imperative Style, High-Performance Deep Learning Library. In Proceedings of the Advances in Neural Information Processing Systems, Vancouver, BC, Canada, 8–14 December 2019; pp. 8024–8035.
52. Pedregosa, F.; Varoquaux, G.; Gramfort, A.; Michel, V.; Thirion, B.; Grisel, O.; Blondel, M.; Prettenhofer, P.; Weiss, R.; Dubourg, V.; et al. Scikit-learn: Machine Learning in Python. *J. Mach. Learn. Res.* **2011**, *12*, 2825–2830.
53. Blaschke, T. Object based image analysis for remote sensing. *ISPRS J. Photogramm. Remote Sens.* **2010**, *65*, 2–16. [CrossRef]
54. Belgiu, M.; Drăguţ, L. Random forest in remote sensing: A review of applications and future directions. *ISPRS J. Photogramm. Remote Sens.* **2016**, *114*, 24–31. [CrossRef]
55. Haralick, R.M.; Shanmugam, K. Textural Features for Image Classification. *IEEE Trans. Syst. Man Cybern.* **1973**, *SMC-3*, 610–621. [CrossRef]
56. Zvoleff, A. *Glcm: Calculate Textures from Grey-Level Co-Occurrence Matrices (GLCMs)*. R-CRAN Project. 2019. Available online: https://cran.r-project.org/web/packages/glcm/index.html (accessed on 14 June 2019).
57. *R Core Team R: A Language and Environment for Statistical Computing*; R Foundation for Statistical Computing: Vienna, Austria, 2019.
58. Hall-Beyer, M. Practical guidelines for choosing GLCM textures to use in landscape classification tasks over a range of moderate spatial scales. *Int. J. Remote Sens.* **2017**, *38*, 1312–1338. [CrossRef]
59. Chen, T.; He, T.; Benesty, M.; Khotilovich, V.; Tang, Y.; Cho, H.; Chen, K.; Mitchell, R.; Cano, I.; Zhou, T.; et al. Xgboost: Extreme gradient boosting. *R Package Version* **2019**, *1*, 0.4–2.
60. Chen, T.; Guestrin, C. XGBoost: A Scalable Tree Boosting System. In *22nd ACM SIGKDD International Conference on Knowledge Discovery and Data Mining*; ACM: New York, NY, USA, 2016; pp. 785–794.
61. Gamon, J.A.; Surfus, J.S. Assessing leaf pigment content and activity with a reflectometer. *New Phytol.* **1999**, *143*, 105–117. [CrossRef]
62. Pham, L.T.; Brabyn, L.; Ashraf, S. Combining QuickBird, LiDAR, and GIS topography indices to identify a single native tree species in a complex landscape using an object-based classification approach. *Int. J. Appl. Earth Obs. Geoinf.* **2016**, *50*, 187–197. [CrossRef]
63. Dymond, C.C.; Mladenoff, D.J.; Radeloff, V.C. Phenological differences in Tasseled Cap indices improve deciduous forest classification. *Remote Sens. Environ.* **2002**, *80*, 460–472. [CrossRef]
64. Wolter, P.T.; Mladenoff, D.J.; Host, G.E.; Crow, T.R. Improved forest classification in the Northern Lake States using multi-temporal Landsat imagery. *Photogramm. Eng. Remote Sens.* **1995**, *61*, 1129–1144.
65. LeCun, Y.; Bengio, Y.; Hinton, G. Deep learning. *Nature* **2015**, *521*, 436–444. [CrossRef] [PubMed]
66. Zörner, J.; Dymond, J.R.; Shepherd, J.D.; Wiser, S.K.; Jolly, B. LiDAR-Based Regional Inventory of Tall Trees—Wellington, New Zealand. *Forests* **2018**, *9*, 702. [CrossRef]
67. MacFaden, S.W.; O'Neil-Dunne, J.P.; Royar, A.R.; Lu, J.W.; Rundle, A.G. High-resolution tree canopy mapping for New York City using LIDAR and object-based image analysis. *J. Appl. Remote Sens.* **2012**, *6*, 063567. [CrossRef]
68. Kattenborn, T.; Eichel, J.; Fassnacht, F.E. Convolutional Neural Networks enable efficient, accurate and fine-grained segmentation of plant species and communities from high-resolution UAV imagery. *Sci. Rep.* **2019**, *9*, 17656. [CrossRef]

remote sensing

MDPI

Letter

Bi-Temporal Analysis of Spatial Changes of Boreal Forest Cover and Species in Siberia for the Years 1985 and 2015

Lei Tian and Wenxue Fu *

Key Lab. of Digital Earth Science, Aerospace Information Research Institute, Chinese Academy of Sciences, Beijing 100094, China; bstian@ceode.ac.cn

* Correspondence: fuwx@aircas.ac.cn

Received: 2 November 2020; Accepted: 14 December 2020; Published: 16 December 2020

Abstract: Boreal forest is a sensitive indicator of the influence of climate change. It can quantify the level and spatial divergence of forest change for forest resources and carbon cycle research. This study selected a typical boreal forest affected by few human activities as a research area, in Siberia, with a latitude span of 51°N–69°N. A total of 150 Landsat images of this area acquired in 1985 and 2015 were collected. A hierarchical classification approach was first established to retrieve the information of forest cover and species. The forested and nonforested lands were discriminated by the decision tree method and, furthermore, the forested land was classified to broad-leaved and coniferous forests by a random forest algorithm. The overall accuracy was 90.37%, which indicates the validity of the approach. Finally, the quantitative information of the forest cover and species changes in each latitude zone of every 2° was analyzed. The results show that the overall boreal forest cover increased by 5.11% over the past three decades, with broad-leaved forest increasing by 3.54% and coniferous forest increasing by 1.57%. In addition, boreal forest increased in every latitude zone, and the spatial divergence of the changes of the boreal forest cover and species in different latitude zones were significant. Finally, broad-leaved forest increased more rapidly than coniferous forest, and the greatest increase, of up to 5.77%, occurred in the zone of 55°N–57°N.

Keywords: boreal forest; forest cover and species; Siberia; Landsat; spatial divergence

1. Introduction

Forests cover more than one-third of the earth's land surface. They are one of the critical biomes globally and provide ecosystem services that benefit society at levels ranging from local to global [1,2]. Forests have been shown to strongly interact with climate change [3]. Furthermore, forests can modify the global carbon budget by altering the sequestration and release of carbon to influence climate change. Conversely, climate change has a profound impact on global forestry, and continues to accelerate with increasing anthropogenic greenhouse gas (GHG) emissions [4]. Due to rapid climate change, forest cover and species ranges continue to shift [5]. To a degree, the above coupling process balances climate change, and these influences also cause changes in forest cover and abundance of species. Therefore, quantifying the spatial changes of forest cover and species is important for improving carbon accounting, climate change modeling, and biodiversity monitoring.

It is an indisputable fact that the global climate is now warming [6–10]. Recent years (2015–2018) have been the warmest since 1850 [11,12]. In addition, higher latitudinal areas, such as boreal forest areas, are expected to undergo the largest increases in temperature [13–16] and experience variable shifts in precipitation regimes [17–19]. Global warming has influenced the suitable climate conditions for different forest species, and studies have shown that climate zones in boreal forests are moving northwards ten times faster than trees' ability to migrate [20,21].

Boreal forests, which span the northern hemisphere across three continents and ten different countries, comprise about 30% of the total forest area globally [20,22] and are considered among the most vulnerable to climate change [13]. The boreal forest is floristically simple and consists of hardy genera of larch (*Larix*), pine (*Pinus*), spruce (*Picea*), and fir (*Abies*), interspersed with deciduous hardwoods of birch (*Betula*), aspen (*Populus*), willow (*Salix*), and alder (*Alnus*). With temperatures in the arctic and boreal domains recently warming, boreal forests have been shown as one of the ecosystems most affected by climate change [23]. Furthermore, quantitative rates of the spatial changes of boreal forest cover and species are important indicators for research of the influence of climate change and are helpful for improving carbon accounting and biodiversity monitoring, which are not yet fully understood. A practical solution to examining trends in forest cover change at a large scale is to employ remotely sensed data [24]. Satellite-based monitoring can be implemented consistently across large regions at annual and interannual intervals [25–27].

Numerous studies have investigated the effect of climate change on boreal forest by remote sensing techniques. It has been observed that the growth of boreal forest was influenced by global warming over the past decade using Moderate-resolution Imaging Spectroradiometer (MODIS) images; however, there is obvious spatial divergence [28,29]. In practice, it is a challenge to quantify the effects of climate change using remote sensing data with low spatial resolution because, due to the low resolution, it is difficult to distinguish the subtle spatial distribution of different forest species. Furthermore, significant uncertainties are attached to possible interactions among forests and climate change, in addition to other land-use pressures [3].

Landsat images provide an ideal option for remote sensing with a medium spatial resolution of 30 m and temporal resolution of 16 days, thus allowing for monitoring of natural and human-induced landscape changes at local or regional scales over years and decades [30,31]. Landsat satellite data have been most widely adopted for studying vegetation phenology, agriculture and forestry, surface temperature monitoring, and air pollution monitoring due to their high spatial resolution and global coverage [32–35]. The long-term data record of nearly five decades allows it to be used for forest investigations for any number of time periods. However, longer time series will require more data processing and storage capacity [36].

In this work, we selected a typical study area in Siberia in which the boreal forests have been affected primarily by natural factors to quantify the spatial changes and divergences of boreal forest cover and species over the past three decades using the Landsat series dataset. The aim was to reveal the latitude zones in which the forest cover and species were more sensitive to climate change, and to provide quantitative information for climate change and ecosystem research.

2. Study Area and Datasets

2.1. Study Area

The boreal forest in Krasnoyarskiy Kray in Central Russia, located in the middle of Siberia (Figure 1), was selected as the study area. The approximate latitude range of the area is 51°N to 69°N, and the longitude is 84°E to 110°E. This work concerns the spatial changes of forest cover and forest species related to latitude in the context of climate change. We divided the study area into nine latitude zones by every 2° from south to north. The information for every latitude zone is shown in Table 1. The latitude range of this area is considered sufficiently large for sensitivity analysis of the forest change.

The climate in the area is strongly continental, with large temperature variations from south to north. The territory of the study area experiences the conditions of three climate belts: Arctic, Subarctic, and moderate. For the central and southern regions, long winters and short hot summers are characteristic. In the north, there are fewer than 40 days with temperatures above 10 °C, whereas, in the south, there are 110–120 such days. The average temperature in January is −36 °C in the north and −18 °C in the south, and in July it is 10 °C in the north and 20 °C in the south. The annual precipitation is 200–300 mm in the north and about 1000 mm in the south. Permafrost is widespread in this area, particularly in the north. The south of the permafrost zone is covered by primeval forest, and the forest species are mainly broad-leaved and coniferous. This area is sparsely populated, with a small number of towns and villages spread throughout the south, surrounded by areas of farmland.

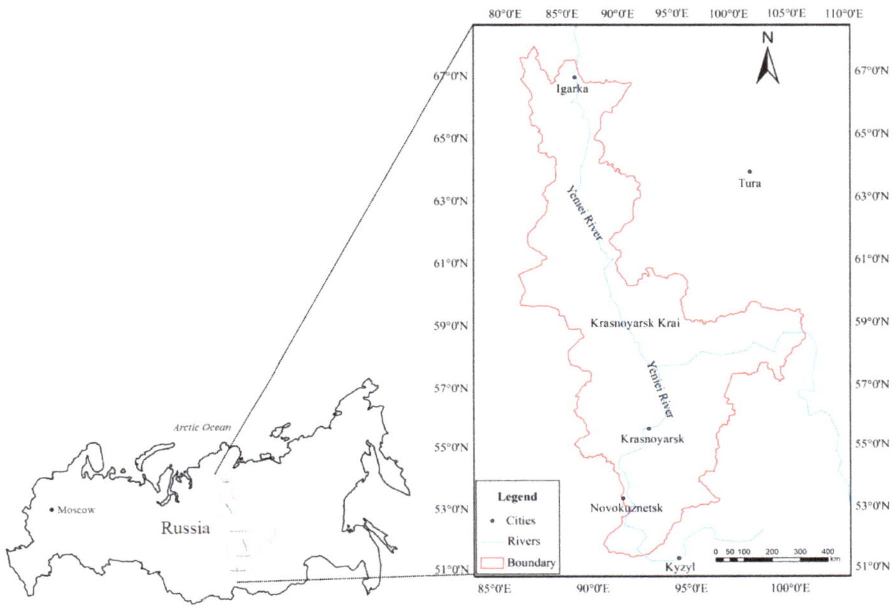

Figure 1. Location of the study area.

Table 1. The areas of each latitude zone.

Latitude Zones (°N)	Area (km^2)
67–69	19,596.45
65–67	58,813.18
63–65	58,364.09
61–63	66,364.15
59–61	130,507.24
57–59	155,232.73
55–57	114,456.67
53–55	77,579.58
51–53	33,564.27
Total	714,478.36

2.2. Datasets

A total of 150 Landsat Thematic Mapper (TM) and Operational Land Imager (OLI) series scenes with little to no cloud cover in the study area, collected from the United States Geological Survey (USGS) (http://glovis.usgs.gov/), were used in this work. These images were acquired during 1985 and 2015, respectively. Most of the images were acquired during summer, mainly between June and September. To account for lacking data, three images with no snow over the south area acquired in October were also used. Some data from adjacent years were used to make up the images due to data deficiencies in the current years. The data product is L1T level with a spatial resolution of 30 m.

3. Data Process

3.1. Data Preprocess

The dataset was preprocessed to eliminate radiometric and geometric distortions. Image preprocessing includes radiometric and atmospheric corrections (Figure 2). To minimize the influence of cloud noise, a haze optimized transformation (HOT) algorithm was used to identify and remove thin clouds [37–39]. Firstly, the clear line was determined according to the high correlation between the blue and red bands in the clear region, and then the HOT value was calculated using the transform equation:

$$\mathrm{HOT} = \rho_1 \sin\theta - \rho_3 \cos\theta - |I| \cos\theta \quad (1)$$

where ρ_1 and ρ_3 are the reflectance of the blue and red bands of TM and OLI images, respectively. I represents the intercept of the clear line, and θ is the inclination of the clear line.

The cloud images were classified by the HOT value which represented the cloud thickness, and then the Landsat image in the cloud region and clear region was automatically classified using just one near-infrared band and two shortwave infrared bands. The image in the cloud region of visible bands was matched to the image in the clear region according to the cloud class and object classification to remove the effect of the cloud.

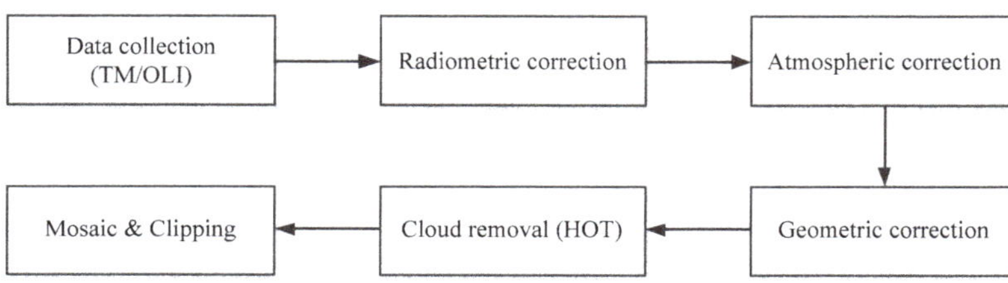

Figure 2. Preprocessing of Landsat Thematic Mapper (TM)/Operational Land Imager (OLI) scenes.

3.2. Forest Classification Approach

This study focused on the changes of boreal forest coverage and species in Siberia, and a hierarchical information extraction approach was established and used for the forest classification. Firstly, a simple decision tree method was adopted to discriminate between forest and nonforest land in the Landsat images, and then a random forest (RF) algorithm was employed to distinguish the broad-leaved and coniferous species from forest land.

3.2.1. Forest and Nonforest Land Classification by Decision Tree Method

The decision tree algorithm is intuitive and flexible, and has high computational efficiency in processing a large quantity of data. In this classification, two vegetation indices, normalized vegetation index (NDVI) and ratio vegetation index (RVI), were used for the discrimination between forest and nonforest land. NDVI can effectively weaken the effects of complex terrain in image information extraction, and enhance the distinction between vegetation and other land types, which is helpful for improving the accuracy and credibility of forest information extraction. The RVI can better reflect the difference of vegetation growth and coverage, and is suitable for vegetation monitoring in areas with vigorous vegetation growth and high coverage. The annual maximum NDVI (NDVImax) values of built-up areas, barren lands, and sparsely vegetated lands are usually lower than 0.30, whereas forest NDVImax values are usually higher than 0.50 [40]. Subsequently, we determined the decision tree classification rules based on sample training: NDVI values greater than 0.62 and RVI values greater than 6.0 were selected as vegetation land, otherwise land was regarded as nonvegetation land.

Next, we distinguished the vegetation land between forest and other vegetation. We found that different plants have different spectral reflectance peaks in the near-infrared band, which is highly sensitive to the difference in reflectance caused by the internal structure of different leaves. Therefore, the reflectance of the near-infrared band can be used to distinguish forest and other vegetation. The vegetation objects classified in the upper layer with reflectance values in the near-infrared band of less than 0.38 were selected as forest land, otherwise they were categorized as nonforest land. The first layer of nonvegetation land and the second layer of nonforest land were combined as the final nonforest land classification result, and the second layer of forest land classification as the final forested land classification result (Figure 3).

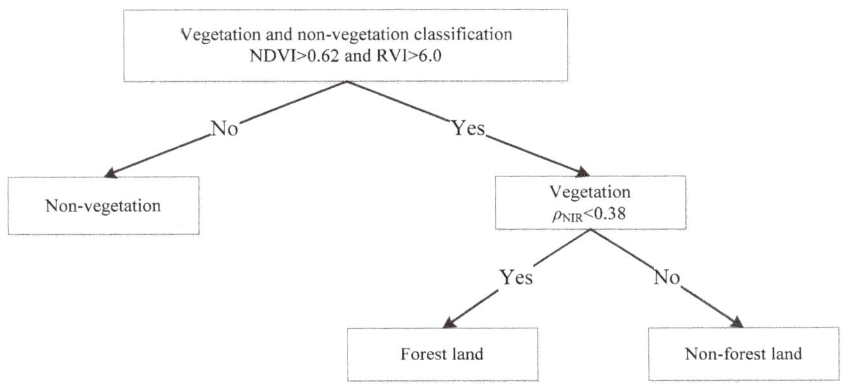

Figure 3. TM/OLI scene classification based on the decision tree classification model.

3.2.2. Broad-Leaved and Coniferous Forest Classification by Random Forest Algorithm

The distinction of natural attributes within forest species is limited by the spectral resolution. It is difficult to detect the subtle spectral differences between coniferous forest and broad-leaved forest with a single image spectral feature. Therefore, it is necessary to use a combination of various spectral features and vegetation indices to classify coniferous and broad-leaved forests. To avoid the impact of other ground features, the nonforested lands were masked.

The random forest algorithm is an extension of the concept of the decision tree [41–43]. As a common algorithm applied in the field of machine learning, RF is often used as a classification method [44–47]. Representative training samples are one of the most critical components in the RF algorithm. The quality and distribution of these samples are also important. In this work, we selected the broad-leaved and coniferous forest sample points from Landsat images following the following principles: (1) the training samples must avoid land-cover mixtures and heterogeneous areas; (2) the sample points should be distributed in a scene that is as representative as possible; (3) the forest-cover category must be interpretable using TM images. Additionally, Google Earth images of the same areas were also used as a reference [48]. Six bands, Landsat TM bands 1–5 and 7, and Landsat OLI bands 2–7, were selected as spectral characteristic variables; NDVI, normalized difference index (NDI) [49] and RVI were selected as index characteristic variables for classification in RF. Additionally, in the RF algorithm, the parameter *ntree* controls the growth number of decision trees, and another parameter *mtry* represents the maximum number of features randomly selected from the above nine variables in the individual tree. In this study, considering the computational efficiency and classification accuracy, the values of *ntree* and *mtry* were set to 1000 and 3, respectively. Finally, the broad-leaved and coniferous forests were classified using the above random sample points using the RF algorithm.

3.2.3. Accuracy Assessment

In this study, four simultaneous Gaofen-2 (GF-2) satellite panchromatic band images acquired in 2015 were used to validate the classification results. The spatial resolution of GF-2 images was 0.81 m. We selected 987 randomly distributed sampling points from GF-2 images for the validation, and the overall accuracy (OA) and F1-score were calculated to evaluate the reliability of the classification results [50], as shown in Table 2. The overall accuracy was 90.37%, and the F1-scores of broad-leaved, coniferous and nonforest land were 0.85, 0.93, and 0.91, respectively. Considering the consistency of the Landsat series images, the above validation result is still considered to be valid for 1985 because it is difficult to obtain the measured data or the high-resolution satellite images of the study area for 1985.

Moreover, the tree-cover-2010 data from Global Forest Watch (GFW) was also used to validate the classification results by spatial overlay analysis [51]. The GFW project is focused on developing global tree cover change data products based on Landsat satellite imagery, and provides global tree cover data with a resolution of 30 m [52]. The spatial consistency of the tree-cover-2010 data and our classification results of forest land was 85.26%. Additionally, Figure 4 presents examples of classified scenes from various areas.

Table 2. Accuracy assessment using Gaofen-2 (GF-2) images.

	Land Classification	GF-2 Results			
		Broad-Leaved Forest	**Coniferous Forest**	**Nonforest Land**	**Total**
Landsat Results	Broad-leaved forest	202	12	22	236
	Coniferous forest	29	421	24	474
	Nonforest land	8	0	269	277
	Total	239	433	315	987
	F1-score	0.85	0.93	0.91	
	Overall accuracy = 90.37%				

Figure 4. Typical classified scenes in various areas. (**a**) Images of forest increase; (**b**) classification results of the images in (**a**); (**c**,**d**) forest classification results in different areas.

4. Result and Analysis

4.1. Spatial Change of Boreal Forest Cover

Firstly, the forest and nonforest lands in the two periods of 1985 and 2015 were classified, and the results are shown in Figure 5a,b. The spatial change of the boreal forest cover between the two years is shown in Figure 5c, and the quantitative information is displayed in Table 3 and Figure 6.

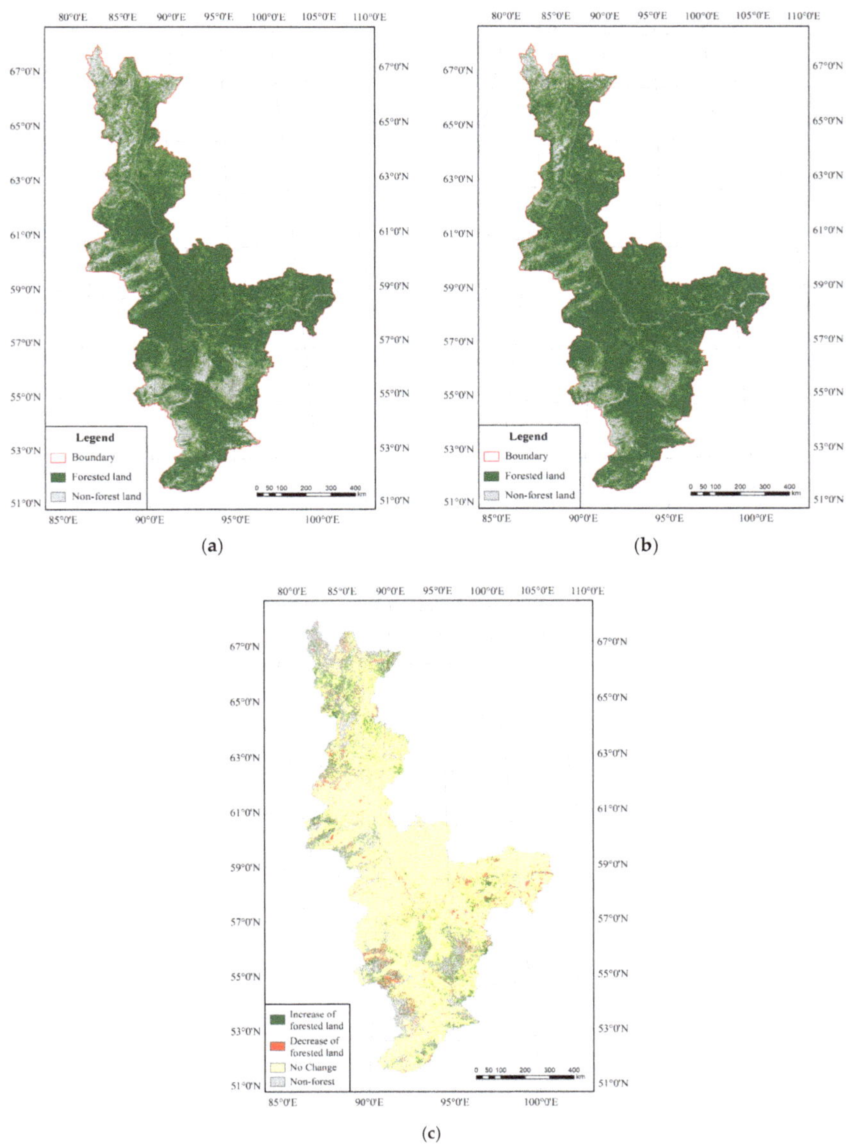

Figure 5. Forest and nonforest land classification: (**a**) in 1985; (**b**) in 2015; (**c**) the spatial change of the forest and nonforest lands.

From Figure 5 and Table 3, it can be seen that substantial change has occurred in the boreal forest cover in the research area over the past three decades. Overall, the rate of forest cover R_{fl}, expressed as the ratio between the area of forest land and the total land area, increased by 5.11% from 75.42% to 80.52%, and R_{fl} was high in the south and low in the north. In 53°N–57°N, human activities occurred that slightly reduced the forest cover. In the region 57°N–69°N, the forest cover and change were mainly influenced by natural factors, with little human activity taking place. The largest rate of forest cover was in 57°N–59°N, of about 91.00%, in 2015. In the latitude zone 67°N–69°N, the north border of the boreal forest, R_{fl} was as high as 34.35% in 2015.

In this research, we were particularly concerned about the quantitative spatial divergence of the forest change in different latitude zones. Thus, we calculated the change of forest cover rate ΔR_{fl} between 1985 and 2015 for every latitude zone to reflect the absolute change velocity of forest cover (Table 3). It can be seen that the rates of forest cover increased in all latitude zones from 1985 to 2015, although with a clear divergence. Compared to other zones, the increase in forest cover was greatest in the north region from 63°N to 69°N, and the greatest increase was up to 13.50% in 65°N–67°N, indicating that the forest cover was more sensitive to climate change in this latitude zone. A relatively slow increase of 1.00% occurred in the latitude zone from 57°N to 59°N, in which the forest cover was up to about 91.00% (Figure 6).

Additionally, we also calculated ΔR_{fo}, expressed as the ratio between the area of changed forest cover and the original forest land, which can be used to reflect the relative increase rate of the forest area itself. The result shows that the greatest ΔR_{fo} was up to 24.61% in the latitude zone 65°N–67°N (Figure 6), which indicates that the forest has the strongest regeneration ability in this region against the background of global warming.

Table 3. Forest land area and ratio.

Latitude Zone (°N)	1985	2015	1985–2015	
	R_{fl}	R_{fl}	ΔR_{fl}	ΔR_{fo}
67–69	29.80%	34.35%	4.55%	15.28%
65–67	54.84%	68.34%	13.50%	24.61%
63–65	68.68%	77.52%	8.84%	12.87%
61–63	84.25%	85.82%	1.57%	1.86%
59–61	81.73%	84.97%	3.24%	3.96%
57–59	89.99%	90.99%	1.00%	1.11%
55–57	69.07%	77.03%	7.96%	11.52%
53–55	68.65%	74.93%	6.28%	9.14%
51–53	77.64%	82.66%	5.02%	6.47%
Total	75.42%	80.52%	5.11%	6.77%

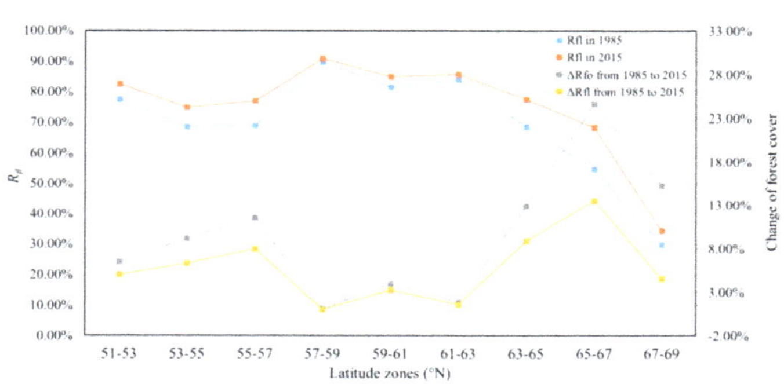

Figure 6. Line chart of the rates and the changes of forest cover in different latitude zones.

4.2. Spatial Changes of Boreal Forest Species

The broad-leaved and coniferous forests were distinguished from forest land, as shown in Figure 7a,b, and the quantitative information of the two forest species is shown Table 4. The total broad-leaved forest area in the research region increased by 39,341.30 km^2 over the past three decades. Overall, the broad-leaved and coniferous forest cover increased in the research area. First, we calculated the ratio R_{bl} between the areas of broad-leaved forest and the total land, and R_{cl} for the coniferous forest. These two ratios can be used to reflect the absolute cover of the two forest species of the total land. The line charts of the cover and change of the forest species are shown in Figure 8. In the study area, the broad-leaved forest cover increased by 3.54%, from 23.83% to 27.37%, and coniferous forest increased by 1.57%, from 51.58% to 53.15%. It also can be seen that overall coniferous forest was more dominant than broad-leaved forest during the study period; however, the broad-leaved forest increased more rapidly than the coniferous forest.

We also analyzed the spatial divergence of the cover of each forest species in different latitude zones. The largest R_{bl} was about 45.00%, and occurred in the latitude zone 55°N–57°N in 2015. It can also be seen that, in 51°N–65°N, the change trend of R_{bl} and R_{cl} were almost contrary with the latitude zones (Figure 8a). R_{bl} and R_{cl} were both lowest in the 67°N–69°N zone, which was the north border of boreal forest in Siberia, with values of about 5.35% and 29.00% in 2015, respectively.

The ratio R_{bo} between the area of broad-leaved forest and the area of forested land, and R_{co} for coniferous forest, can be used to reflect the abundances of the two forest species in the forested land. The broad-leaved forest was the dominant species in 53°N–57°N (Figure 8b). However, in the 51°N–53°N zone, the R_{bo} of broad-leaved forest was about 32.79% in 1985 and 35.59% in 2015, and significantly less than that in 53°N–57°N. This is because the average altitude is over 1000 m in the 51°N–53°N zone, which is significantly higher than that of about 450 m in 53°N–57°N. In the latitude zones from 57°N to 69°N, coniferous forests were more abundant than broad-leaved forests, mainly due to the relatively cold weather in this region. However, an abnormal phenomenon is that R_{bl} in the 59°N–61°N zone was almost the lowest in the research area, and surprisingly was substantially lower than that in 63°N–65°N, which may be due to the influence of multiple factors, such as climate and artificial logging.

It is important for carbon cycle research that the spatial divergence of the boreal forest species changes in different latitude zones (Figure 8c,d). Here, we firstly calculated ΔR_{bl} and ΔR_{bo}, which represent the changes of R_{bl} and R_{bo} over the past three decades, respectively (Table 4 and Figure 8c). It can be seen that the broad-leaved forest cover R_{bl} increased in all latitude zones. The greatest increase was up to 5.77% in 55°N–57°N. The lowest was about 0.45% in 67°N–69°N, which can be considered to represent no change in the broad-leaved forest cover in this zone considering the accuracy of the classification. We also note that the change of R_{bl} was about 1.33% in 59°N–61°N, which is relatively low compared with other latitude zones, with the exception of 67°N–69°N. The explanation for this phenomenon requires further study. ΔR_{bo} in 67°N–69°N was −0.89%, which indicates that the proportion of broad-leaved forest decreased slightly in this zone, although R_{bl} increased.

In addition, ΔR_{cl} was also calculated, and represents the change of R_{cl} over the past three decades (Figure 8d). It can be seen that the coniferous forest cover decreased in the zones 57°N–59°N and 61°N–63°N; however, coniferous forest increased significantly in the latitude from 65°N to 69°N, where broad-leaved forest increased relatively slowly.

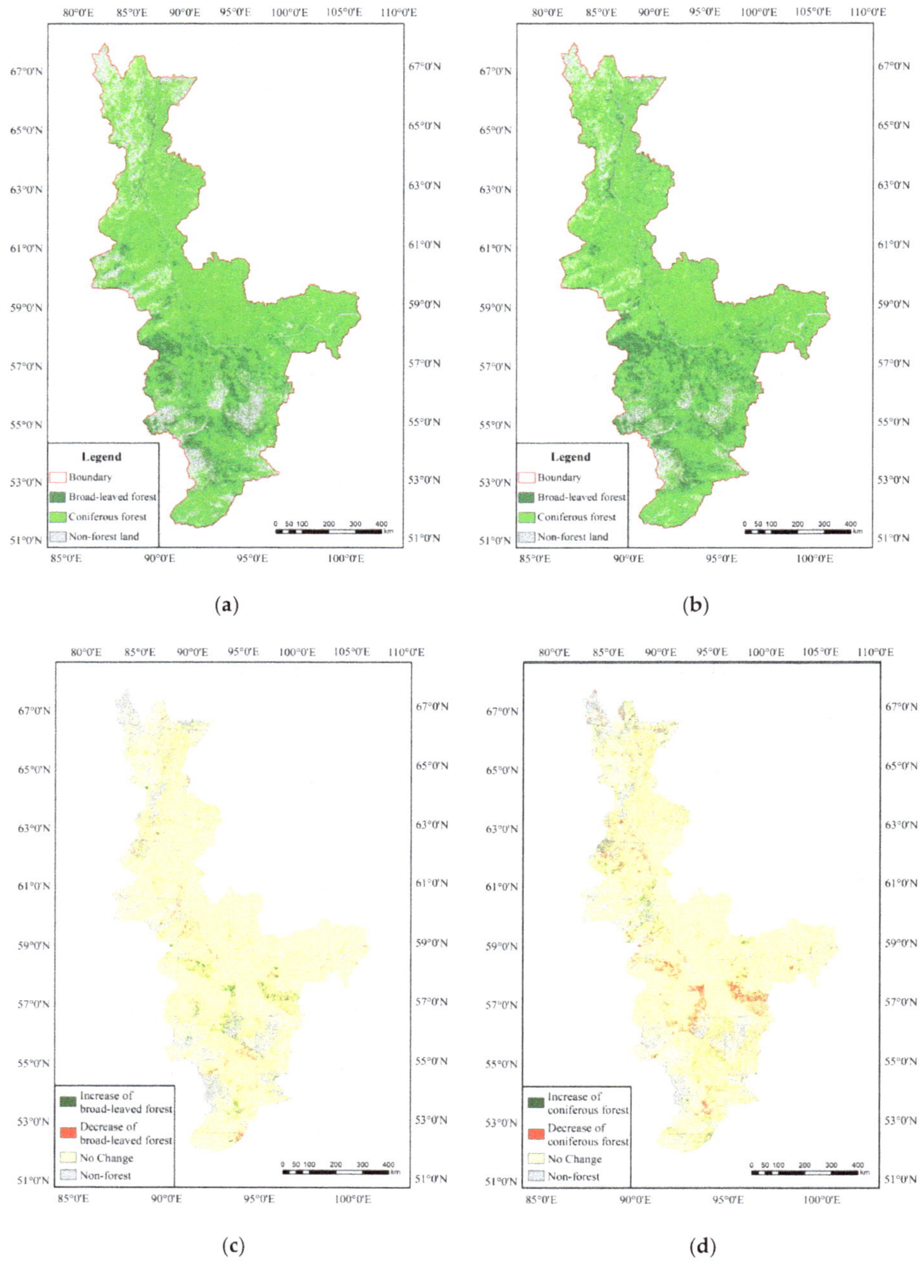

Figure 7. Broad-leaf forest and coniferous forest classification: (**a**) 1985; (**b**) 2015; (**c**) change in the broad-leaved forest; (**d**) change in the coniferous forest.

Table 4. Broad-leaved forest area.

Latitude Zone (°N)	Broad-Leaved Forest						Coniferous Forest				
	1985		2015		1985–2015		1985		2015		1985–2015
	R_{bl}	R_{bo}	R_{bl}	R_{bo}	ΔR_{bl}	ΔR_{bo}	R_{cl}	R_{co}	R_{cl}	R_{co}	ΔR_{cl}
67–69	4.91%	16.47%	5.35%	15.59%	0.45%	−0.89%	24.89%	83.53%	29.00%	84.41%	4.11%
65–67	14.08%	25.68%	17.97%	26.29%	3.89%	0.62%	40.76%	74.32%	50.37%	73.71%	9.61%
63–65	17.51%	25.50%	21.41%	27.61%	3.89%	2.11%	51.17%	74.50%	56.11%	72.39%	4.95%
61–63	14.47%	17.18%	18.22%	21.23%	3.74%	4.05%	69.77%	82.82%	67.60%	78.77%	−2.18%
59–61	12.03%	14.71%	13.36%	15.72%	1.33%	1.01%	69.71%	85.29%	71.62%	84.28%	1.91%
57–59	28.20%	31.34%	31.92%	35.08%	3.72%	3.74%	61.79%	68.66%	59.07%	64.92%	−2.72%
55–57	39.22%	56.78%	44.99%	58.41%	5.77%	1.63%	29.85%	43.22%	32.04%	41.59%	2.19%
53–55	36.50%	53.17%	39.98%	53.35%	3.47%	0.18%	32.15%	46.83%	34.96%	46.65%	2.80%
51–53	25.46%	32.79%	29.42%	35.59%	3.96%	2.79%	52.18%	67.21%	53.25%	64.41%	1.06%
Total	23.83%	31.60%	27.37%	33.99%	3.54%	2.39%	51.58%	68.40%	53.15%	66.01%	1.57%

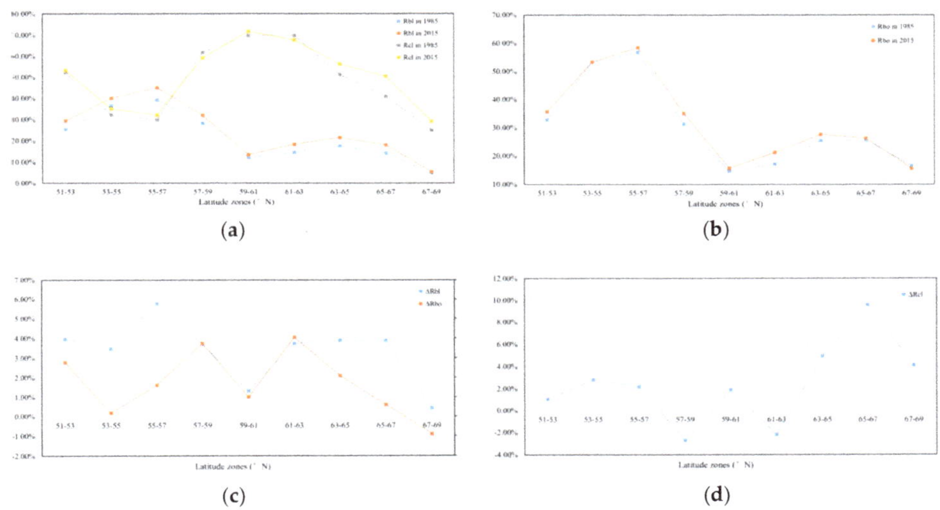

Figure 8. Spatial divergence of the cover and change of forest species with different latitude zones: (**a**) R_{bl} and R_{cl} in 1985 and 2015; (**b**) R_{bo} in 1985 and 2015; (**c**) ΔR_{bl} and ΔR_{bo}; (**d**) ΔR_{cl}.

5. Conclusions

Boreal forests have been affected by climate change over the past three decades, which is predominantly reflected in the changes of forest cover and species. This work selected a typical area in central Siberia, and collected 150 Landsat TM/OLI images covering this area acquired in 1985 and 2015. A hierarchical classification process was established for the retrieval of boreal forest cover and species. Finally, quantitative spatial information of the forest cover and species was compiled for latitude zones for every 2°. The results show that the impact of climate change on the boreal forest has a significant spatial divergence. Compared with other zones, the increase in forest cover was greatest in the region from 63°N to 69°N, indicating that the forest cover was more sensitive to climate change in this region. Furthermore, the broad-leaved forest cover increased by 3.54% and coniferous forest increased by 1.57%. Overall, coniferous forest was also more dominant than broad-leaved forest during the study period, although broad-leaved forest increased more rapidly than coniferous forest. In 67°N–69°N, the proportion of broad-leaved forest decreased slightly.

This study considered only two periods, i.e., 1985 and 2015. However, it will be necessary in future work to study forest change during other periods. In addition, more research is required to determine which climate factors, such as temperature or precipitation, are the main driving forces of forest change.

Author Contributions: Conceptualization, W.F.; methodology, W.F. and L.T.; investigation, L.T. Both wrote the original draft and performed the final review. All authors have read and agreed to the published version of the manuscript.

Funding: This work was supported by the Strategic Priority Research Program of the Chinese Academy of Sciences (XDA19070102) and supported by the National Natural Science Foundation of China (No. 61971417).

Conflicts of Interest: The authors declare no conflict of interest.

References

1. Trumbore, S.; Brando, P.; Hartmann, H. Forest health and global change. *Science* **2015**, *349*, 814–818. [CrossRef] [PubMed]
2. Bos, A.B.; Sy, V.D.; Duchelle, A.E.; Herold, M.; Martius, C.; Tsendbazar, N.E. Global data and tools for local forest cover loss and REDD+ performance assessment: Accuracy, uncertainty, complementarity and impact. *Int. J. Appl. Earth Obs. Geoinf.* **2019**, *80*, 295–311. [CrossRef]
3. Price, D.T.; Alfaro, R.I.; Brown, K.J.; Flannigan, M.D.; Fleming, R.A.; Hogg, E.H. Anticipating the consequences of climate change for Canada's boreal forest ecosystems1. *Environ. Rev.* **2013**, *21*, 322–365. [CrossRef]
4. IPCC. Summary for policy makers. In *Climate Change 2014: Synthesis Report. Contibution of Working Groups I, II, and III to the Fifth Assessment Report if the Intergovernmental Panel on Climate Change*; IPCC: Geneva, Switzerland, 2014.
5. Pecl, G.T.; Araújo, M.B.; Bell, J.D.; Blanchard, J.; Bonebrake, T.C.; Chen, I.C. Biodiversity redistribution under climate change: Impacts on ecosystems and human well-being. *Science* **2017**, *355*, 1–9. [CrossRef] [PubMed]
6. Folland, C.K.; Karl, T.R.; Christy, J.R.; Clarke, R.A.; Gruza, G.V.; Jouzel, J.; Mann, M.E.; Oerlemands, J.; Salinger, M.J.; Wang, S.W. Observed climate variability and change. In *The Scientific Basis: Contribution of Working Group I to the Third Assessment Report of the Intergovernmental Panel on Climate Change*; Cambridge University Press: New York, NY, USA, 2001; pp. 99–181.
7. Thuiller, W. Biodiversity: Climate change and the ecologist. *Nature* **2007**, *448*, 550–552. [CrossRef] [PubMed]
8. Gauthier, S.; Bernier, P.; Burton, P.J.; Edwards, J.; Isaac, K.; Isabel, N.; Jayen, K.; Le Goff, H.; Nelson, E.A. Climate change vulnerability and adaptation in the managed. *Environ. Rev.* **2014**, *22*, 256–285. [CrossRef]
9. Lawal, S.; Lennard, C.; Hewitson, B. Response of southern African vegetation to climate change at 1.5 and 2.0° global warming above the pre-industrial level. *Clim. Serv.* **2019**, *16*, 1–16. [CrossRef]
10. Engelbrecht, C.J.; Engelbrecht, F.A. Shifts in Köppen-Geiger climate zones over southern Africa in relation to key global temperature goals. *Theor. Appl. Climatol.* **2015**, *123*, 247–261. [CrossRef]
11. Jones, P.D.; Osborn, T.J. Climate record: Surface temperature trends. In *Reference Module in Earth Systems and Environmental Sciences*; 2018; pp. 110–120. Available online: https://doi.org/10.1016/B978-0-12-382225-3.00005-0 (accessed on 16 December 2020).
12. Kumar, A.; Yadav, J.; Mohan, R. Global warming leading to alarming recession of the Arctic sea-ice cover: Insights from remote sensing observations and model reanalysis. *Heliyon* **2020**, *6*, 1–10. [CrossRef]
13. Diffenbaugh, N.S.; Field, C.B. Changes in ecologically critical terrestrial climate conditions. *Science* **2013**, *341*, 486–492. [CrossRef]
14. Ouyang, W.; Gao, X.; Hao, Z.; Liu, H.; Hao, F. Farmland shift due to climate warming and impacts on temporal-spatial distributions of water resources in a middle-high latitude agricultural watershed. *J. Hydrol.* **2017**, *547*, 156–167. [CrossRef]
15. Sun, Q.; Miao, C.; Duan, Q. Projected changes in temperature and precipitation in ten river basins over China in 21st century. *Int. J. Climatol.* **2015**, *35*, 1125–1141. [CrossRef]
16. Huang, J.; Zhang, X.; Zhang, Q.; Lin, Y.; Hao, M.; Luo, Y.; Zhao, Z.; Yao, Y.; Chen, X.; Wang, L.; et al. Recently amplified arctic warming has contributed to a continual global warming trend. *Nat. Clim. Chang.* **2017**, *7*, 875–879. [CrossRef]
17. Gauthier, S.; Bernier, P.; Kuuluvainen, T.; Shvidenko, A.Z.; Schepaschenko, D.G. Boreal forest health and global change. *Science* **2015**, *349*, 819–822. [CrossRef]
18. Reyer, C.P.O.; Brouwers, N.; Rammig, A.; Brook, B.W.; Epila, J.; Grant, R.F. Forest resilience and tipping points at different spatio-temporal scales: Approaches and challenges. *J. Ecol.* **2015**, *103*, 5–15. [CrossRef]

19. Donnelly, C.; Wouter, G.; Andersson, J.; Gerten, D.; Pisacane, G.; Roudier, P.; Ludwig, F. Impacts of climate change on European hydrology at 1.5, 2 and 3 degree mean global warming above preindustrial level. *Clim. Chang.* **2017**, *143*, 13–26. [CrossRef]
20. Soja, A.J.; Tchebakova, N.M.; French, N.H.F.; Flannigan, M.D.; Shugart, H.H.; Stocks, B.J.; Sukhinin, A.I.; Parfenova, E.I.; Chapin, F.S.; Stackhouse, P.W. Climate-induced boreal forest change: Predictions versus current observations. *Glob. Planet. Chang.* **2007**, *56*, 274–296. [CrossRef]
21. Toshichika, I.; Ramankutty, N. How do weather and climate influence cropping area and intensity? *Glob. Food Secur.* **2015**, *4*, 46–50.
22. Brecka, A.F.J.; Shahi, C.; Chen, H.Y.H. Climate change impacts on boreal forest timber supply. *For. Policy Econ.* **2018**, *92*, 11–21. [CrossRef]
23. Oboite, F.O.; Comeau, P.G. Competition and climate influence growth of black spruce in western boreal forests. *For. Ecol. Manag.* **2019**, *443*, 84–94. [CrossRef]
24. Kivinen, S.; Koivisto, E.; Keski-Saari, S.; Poikolainen, L.; Tanhuanpää, T.; Kuzmin, A.; Viinikka, A.; Heikkinen, R.K.; Pykälä, J.; Virkkala, R.; et al. A keystone species, European aspen (*Populus tremula* L.), in boreal forests: Ecological role, knowledge needs and mapping using remote sensing. *For. Ecol. Manag.* **2020**, *462*, 118008. [CrossRef]
25. Ols, C.; Kålås, I.H.; Drobyshev, I.; Söderström, L.; Hofgaard, A. Spatiotemporal variation in the relationship between boreal forest productivity proxies and climate data. *Dendrochronologia* **2019**, *58*, 125648. [CrossRef]
26. Kato, A.; Thau, D.; Hudak, A.T.; Meigs, G.W.; Moskal, L.M. Quantifying fire trends in boreal forests with Landsat time series and self-organized criticality. *Remote Sens. Environ.* **2020**, *237*, 111525. [CrossRef]
27. Moreno-Martínez, Á.; Camps-Valls, G.; Kattge, J.; Robinsona, N.; Reichstein, M.; van Bodegom, P.; Kramer, K.; Cornelissen, J.H.C.; Reich, P.; Bahn, M.; et al. A methodology to derive global maps of leaf traits using remote sensing and climate data. *Remote Sens. Environ.* **2018**, *218*, 69–88. [CrossRef]
28. Hou, M.; Venäläinen, A.K.; Wang, L.; Pirinen, I.P.; Gao, Y.; Jin, S.; Zhu, Y.; Qin, F.; Hu, Y. Spatio-temporal divergence in the responses of Finland's boreal forests to climate variables. *Int. J. Appl. Earth Obs. Geoinf.* **2020**, *92*, 1–9. [CrossRef]
29. Alibakhshi, S.; Naimi, B.; Hovi, A.; Crowther, T.W.; Rautiainen, M. Quantitative analysis of the links between forest structure and land surface albedo on a global scale. *Remote Sens. Environ.* **2020**, *246*, 1–12. [CrossRef]
30. Rendenieks, Z.; Nita, M.D.; Nikodemus, O.; Radeloff, V.C. Half a century of forest cover change along the Latvian-Russian border captured by object-based image analysis of Corona and Landsat TM/OLI data. *Remote Sens. Environ.* **2020**, *249*, 1–14. [CrossRef]
31. Senf, C.; Laštovička, J.; Okujeni, A.; Heurich, M.; Linden, S. A generalized regression-based unmixing model for mapping forest cover fractions throughout three decades of Landsat data. *Remote Sens. Environ.* **2020**, *240*, 1–10. [CrossRef]
32. Wulder, M.; Loveland, T.; Roy, D.; Crawford, C.; Masek, J.; Woodcock, C. Current status of Landsat program, science, and applications. *Remote Sens. Environ.* **2019**, *225*, 127–147. [CrossRef]
33. Wu, Z.; Snyder, G.; Vadnais, C.; Arora, R.; Babcock, M.; Stensaas, G.; Doucette, P.; Newman, T. User needs for future Landsat missions. *Remote Sens. Environ.* **2019**, *231*, 1–13. [CrossRef]
34. Jönsson, P.; Cai, Z.; Melaas, E.; Friedl, M.A.; Eklundh, L. A method for robust estimation of vegetation seasonality from Landsat and sentinel-2 time series data. *Remote Sens.* **2018**, *10*, 635. [CrossRef]
35. Restrepo, A.M.C.; Yang, Y.R.; Hamm, N.A.; Gray, D.J.; Barnes, T.S.; Williams, G.M.; Clements, A.C. Land cover change during a period of extensive landscape restoration in Ningxia Hui autonomous region. *China Sci. Total Environ.* **2017**, *598*, 669–679. [CrossRef]
36. Bolton, D.K.; Tompalski, P.; Coops, N.C.; White, J.C.; Wulder, M.A.; Hermosilla, T.; Queinnec, M.; Luther, J.E.; Lier, O.R.; Fournier, R.A.; et al. Optimizing Landsat time series length for regional mapping of lidar-derived forest structure. *Remote Sens. Environ.* **2020**, *239*, 111645. [CrossRef]
37. Zhang, Y.; Guindon, B.; Cihlar, J. An image transform to characterize and compensate for spatial variations in thin cloud contamination of Landsat images. *Remote Sens. Environ.* **2002**, *82*, 173–187. [CrossRef]
38. Li, W.; Li, Y.; Chen, D.; Chan, J.C. Thin cloud removal with residual symmetrical concatenation network. *ISPRS J. Photogramm. Remote Sens.* **2019**, *153*, 137–150. [CrossRef]
39. Liu, Q.; Gao, X.; He, L.; Lu, W. Haze removal for a single visible remote sensing image. *Signal Process.* **2017**, *137*, 33–43. [CrossRef]

40. Qin, Y.; Xiao, X.; Dong, J.; Zhang, G.; Shimada, M.; Liu, J. Forest cover maps of china in 2010 from multiple approaches and data sources: PALSAR, Landsat, MODIS, FRA, and NFI. *ISPRS J. Photogramm. Remote Sens.* **2015**, *109*, 1–16. [CrossRef]
41. Breiman, L. Random forests. *Mach. Learn.* **2001**, *45*, 5–32. [CrossRef]
42. Strobl, C.; Boulesteix, A.L.; Zeileis, A.; Hothorn, T. Bias in random forest variable importance measures: Illustrations, sources and a solution. *BMC Bioinform.* **2007**, *8*, 25. [CrossRef]
43. Cutler, D.R.; Edwards, J.T.C.; Beard, K.H.; Cutler, A.; Hess, K.T.; Gibson, J.; Lawler, J.J. Random forest for classification in ecology. *ESA Online J.* **2008**. [CrossRef]
44. Svetnik, V.; Liaw, A.; Tong, C.; Culberson, J.C.; Sheridan, R.P.; Feuston, B.P. Random forest: A classification and regression tool for compound classification and QSAR modeling. *J. Chem. Inf. Comput. Sci.* **2003**, *43*, 1947–1958. [CrossRef]
45. Rodriguez-Galiano, V.F.; Ghimire, B.; Rogan, J.; Chica-Olmo, M.; Rigol-Sanchez, J.P. An assessment of the effectiveness of a random forest classifier for land-cover classification. *ISPRS J. Photogramm. Remote Sens.* **2012**, *67*, 93–104. [CrossRef]
46. Assiri, A. Anomaly classification using genetic algorithm-based random forest model for network attack detection. *Comput. Mater. Contin.* **2021**, *66*, 769–778. [CrossRef]
47. Climent, F.; Momparler, A.; Carmona, P. Anticipating bank distress in the Eurozone: An Extreme Gradient Boosting approach. *J. Bus. Res.* **2019**, *101*, 885–896. [CrossRef]
48. Gong, P.; Wang, J.; Yu, L.; Zhao, Y.C.; Zhao, Y.Y.; Liang, L.; Niu, Z.G.; Huang, X.M.; Fu, H.H.; Liu, S.; et al. Fine resolution observation and monitoring of global land cover: First mapping results with Landsat TM and ETM+ data. *Int. J. Remote Sens.* **2013**, *34*, 2607–2654. [CrossRef]
49. Rodríguez-Moreno, V.M.; Bullock, S.H. Vegetation response to hydrologic and geomorphic factors in an arid region of the Baja California Peninsula. *Environ. Monit. Assess.* **2014**, *186*, 1009–1021. [CrossRef]
50. Pontius, R.G., Jr.; Millones, M. Death to Kappa: Birth of quantity disagreement and allocation disagreement for accuracy assessment. *Int. J. Remote Sens.* **2011**, *32*, 4407–4429. [CrossRef]
51. Zhao, K.; Jin, B.; Fan, H.; Song, W.; Zhou, S.; Jiang, Y. High-performance overlay analysis of massive geographic polygons that considers shape complexity in a cloud environment. *ISPRS Int. J. Geo-Inf.* **2019**, *8*, 290. [CrossRef]
52. Hansen, M.C.; Potapov, P.V.; Moore, R.; Hancher, M.; Turubanova, S.A.; Thau, D.; Stehman, S.V.; Goetz, S.J.; Loveland, T.R.; Kommareddy, A.; et al. High-resolution global maps of 21st-century forest cover change. *Science* **2013**, *342*, 850–853. [CrossRef]

Publisher's Note: MDPI stays neutral with regard to jurisdictional claims in published maps and institutional affiliations.

remote sensing

MDPI

Article

Large-Scale Mapping of Tree Species and Dead Trees in Šumava National Park and Bavarian Forest National Park Using Lidar and Multispectral Imagery

Peter Krzystek [1,*], Alla Serebryanyk [2], Claudius Schnörr [3], Jaroslav Červenka [4] and Marco Heurich [5,6]

1 Department of Geoinformatics, Munich University of Applied Sciences, Karlstr. 6, D-80333 Munich, Germany
2 Jambit GmbH, Erika-Mann-Straße 63, D-80636 Munich, Germany; alla.serebryanyk@jambit.com
3 Department of Computer Science and Mathematics, Munich University of Applied Sciences, Lothstr. 64, D-80335 Munich, Germany; schnoerr@cs.hm.edu
4 Šumava National Park, Sušická 339, CZ-34192 Kašperské Hory, Czech Republic; jaroslav.cervenka@npsumava.cz
5 Faculty of Environment and Natural Resources, University of Freiburg, 79085 Freiburg im Breisgau, Germany; marco.heurich@wildlife.uni-freiburg.de or marco.heurich@npv-bw.bayern.de
6 Bavarian Forest National Park, Department of Visitor Management and National Park Monitoring, Freyungerstr. 2, D-94481 Grafenau, Germany
* Correspondence: krzystek@hm.edu

Received: 22 December 2019; Accepted: 7 February 2020; Published: 17 February 2020

Abstract: Knowledge of forest structures—and of dead wood in particular—is fundamental to understanding, managing, and preserving the biodiversity of our forests. Lidar is a valuable technology for the area-wide mapping of trees in 3D because of its capability to penetrate vegetation. In essence, this technique enables the detection of single trees and their properties in all forest layers. This paper highlights a successful mapping of tree species—subdivided into conifers and broadleaf trees—and standing dead wood in a large forest 924 km^2 in size. As a novelty, we calibrate the critical stopping criterion of the tree segmentation based on a normalized cut with regard to coniferous and broadleaf trees. The experiments were conducted in Šumava National Park and Bavarian Forest National Park. For both parks, lidar data were acquired at a point density of 55 points/m^2. Aerial multispectral imagery was captured for Šumava National Park at a ground sample distance (GSD) of 17 cm and for Bavarian Forest National Park at 9.5 cm GSD. Classification of the two tree groups and standing dead wood—located in areas of pest infestation—is based on a diverse set of features (geometric, intensity-based, 3D shape contexts, multispectral-based) and well-known classifiers (Random forest and logistic regression). We show that the effect of under- and oversegmentation can be reduced by the modified normalized cut segmentation, thereby improving the precision by 13%. Conifers, broadleaf trees, and standing dead trees are classified with overall accuracies better than 90%. All in all, this experiment demonstrates the feasibility of large-scale and high-accuracy mapping of single conifers, broadleaf trees, and standing dead trees using lidar and aerial imagery.

Keywords: classification; segmentation; single trees; forest structure analysis; dead wood

1. Introduction

Forest inventories based on remote sensing data vary in terms of scale, sensors, and calculated forest structural attributes. For example, Latifi and Heurich [1] reported that lidar point clouds advantageously fused with optical imagery are the most prominent choices for inventory of forest structural variables at the landscape and regional scales. Spatial distribution of trees and standing or fallen dead trees are key forest variables for estimation of forest attributes such as aboveground biomass and growing stock. Aside from area-based methods, tree-level approaches estimate forest inventory parameters utilizing segmented single trees. The study of Latifi et al. [2] compared both approaches for a temperate forest area of around 300 km^2 in a large-scale experiment and showed that the cost of data collection could be reduced by 90% compared to a conventional forest inventory.

For single tree detection, a huge set of raster-based approaches are available that use the canopy height model (CHM) as basic information [3]. Aside from these low-level methods, more sophisticated approaches operate on the point cloud instead of using the CHM. Strîmbu and Strîmbu [4] present a three-dimensional approach in which a weighted graph is built from hierarchical topologies of lidar data, thereby generating three-dimensional tree segments. The authors indicate an average detection accuracy of 98% for simple, regular tree structures and 85% for complex tree structures. The approach of Dai et al. [5] merges mean-shift clusters initially generated from different lidar-based feature spaces to individual trees using both the spatial domain and the multispectral domain. In the case of multispectral lidar data, this study reports a detection rate of 88%. Reitberger et al. [6] proposes a novel segmentation method based on normalized cut [7] that detects single trees using a graph cut approach. The study successfully shows a significant improvement of the single tree detection (up to 20%) in the lower forest layers.

Graph-cut based segmentation methods, such as normalized cut, partition the underlying weighted graph that is being finished with a dependence on a static stopping criterion. The stopping criterion controls the iterative segmentation and has no physical or object specific meaning. It must be either defined based on practical experience or optimized by means of a sensitivity analysis using reference data [8]. However, the stopping criterion is applied statically, without accounting for the actual local tree species; therefore over- and under-segmentation occurs when using the stopping criterion averaged over the reference data. Recently, Amiri et al. [9] improved the normalized cut based tree segmentation by replacing the static stopping criterion with an adaptive classifier that analyzes the point cloud of the cluster to be partitioned. However, this extension of the basic method only works for conifers and fails for broadleaf trees.

Optical imagery has become a standard source to discriminate tree species [10,11]. Recently, dense matching has emerged as a mature technique used for reconstructing objects from a series of highly overlapping images at the pixel level [12]. When applied in a forest area, the dense point cloud of a canopy surface can be used for tree species classification either at the tree level, or using an area-based approach [13]. Moreover, the combination of multispectral images with lidar advantageously enriches the limited radiometric range of lidar. Ullah et al. [14] demonstrated an improvement in the estimation of forest structural parameters at the stand and forest compartment levels using point clouds generated from aerial imagery.

Over the past decade, lidar has emerged as an excellent data source for classifying tree species. Several studies have demonstrated the feasibility of tree species classification using structural features, such as crown shapes, height distribution percentiles, and proportions of first/single returns, obtained from lidar point clouds [15–18]. The application of the newly introduced full waveform lidar systems has also improved accuracy by applying waveform features that use detailed backscattered pulse information, such as the intensity and pulse width [19,20]. Reitberger et al. [6] also demonstrated the advantages of full waveform lidar for tree species mapping in the Bavarian Forest National Park by utilizing the intensity and pulse width. For the same forest area, Yao et al. [21] showed that broadleaf trees and conifers could be classified with an overall accuracy of 78%. However, Shi et al. [22] found that the classification accuracy decreases by 30% if the detailed tree species mapping (six different tree species) is attempted for the same study area. The study in Hovi et al. [20] reported an overall accuracy of 75% for the classification of three main tree species in Finland using lidar waveform features.

The advent of high resolution lidar point clouds made possible the successful detection of individual standing dead trees located in areas of pest infestation. Yao et al. [21] was able to detect standing dead trees with crowns with an accuracy between 71% and 73%. The approach only used the lidar intensity (single wavelength 1550 nm) and geometric features such as the crown shape and point height distribution. Recently, Polewski [23] presented a method that combines single 3D tree segments with multispectral aerial imagery for mapping of standing dead trees with crowns. A two-class classification (dead tree and non-dead tree) led to an accuracy of around 88% using features generated from the covariance matrix of three image channels. The detection of standing dead trees without crowns (i.e., snags) was tackled in Polewski et al. [24]. The approach generates salient features by free shape contexts suitable to describe the single snags in a lidar point cloud. The study reports a classification accuracy between 71% to 73%. At present, these techniques for mapping dead trees have only been applied to small control plots.

The structure of a forest is one of the most important variables for the future development of forest stands [25] and has considerable impact on the presence of protected species [26]. For adequate management, the description of species composition, spatial structure, natural dynamics, and various tree characteristics is mandatory. These forest characteristics are also needed as explanatory variables for many research issues. Šumava National Park and Bavarian Forest National Park are collecting forest structure data in the field on a permanent research plot, whose total area is relatively small when compared with the size of the entire forest area [27]. So far, slightly different measurement methodologies have been applied in both parks, such as update periods and plot size, hence complicating forest data comparisons and cross-border mapping of forest structures. Lidar offers the unique opportunity to describe the forest structure of both neighboring parks in a consistent data set generated at the same time and based on identical methods.

In summary, tree segmentation based on normalized cut using a static stopping criterion generates over- and undersegmentation effects depending on the distribution of tree species. Therefore, it appears advantageous to apply a stopping criterion that is based on a tree species's information obtained from a preliminary classification of conifers and broadleaf trees. Moreover, the complete mapping of conifers, broadleaf trees, as well as standing dead trees, has not yet been successfully demonstrated in a large temperate forest area using lidar and multispectral imagery. Contrary to the experiments normally conducted in small test areas, the results of a large-scale experiment are statistically more relevant and more reliable. Finally, the area-wide mapping of individual trees—especially of dead trees and snags—across the border allows for a seamless representation of the ecosystem that is important for forest biodiversity conservation.

The main objectives and the novelties of the current study were (i) to optimise tree segmentation based on normalized cut using two different stopping criteria for conifers and broadleaf trees and (ii) to map conifers, broadleaf trees, standing dead trees, and snags in a large temperate forest 924 km^2 in size situated in Šumava National Park and Bavarian Forest National Park. Thereby, we verified the statistical performance of the methods in a large-scale experiment and provided spatial information about the distribution of trees usable for area-wide information on the abundance and distribution of key habitat structures [28].

2. Materials and Methods

2.1. Study Area

Established in the Bohemian Forest in 1991 and 1970, respectively, Šumava National Park (49°12′N, 13°30′E) and Bavarian Forest National Park (49°12′N, 12°58′E) protect 92,464 ha of mountain spruce forests. The forest cover more than 82% of this area, one of the largest complexes in central Europe. Both parks are part of the Natura 2000 network, which was established to protect the most endangered habitats and species in Europe, as defined in both the Habitats Directive (1992) and Birds Directive (1979). This forest area comprises natural and secondary habitats with numerous rare and protected plant and animal species of exceptional natural value of European-wide significance [27]. Although divided by the state border between Czech Republic and Germany, the forest nature is consistent for the area of both parks.

Both parks are dominated by Norway spruce (*Picea abies*) with European beech (*Fagus sylvatica*), silver fir (*Abies alba*) and larch (*Larix*). The parks also contain some other tree species, such as white birch (*Betula pendula*), sycamore maple (*Acer pseudoplatanus*), and common rowan (*Sorbus aucuparia*) [29]. Due to bark beetle infestation, large areas are now characterized by dead wood, comprising fallen dead trees, standing dead trees, and standing dead trees without crowns (=snags). Note that we did not focus on the detection of fallen dead trees. Moreover, we excluded dead trees located in windthrow areas. Figure 1 shows a map of the two parks and the reference plots (see Section 2.4.1).

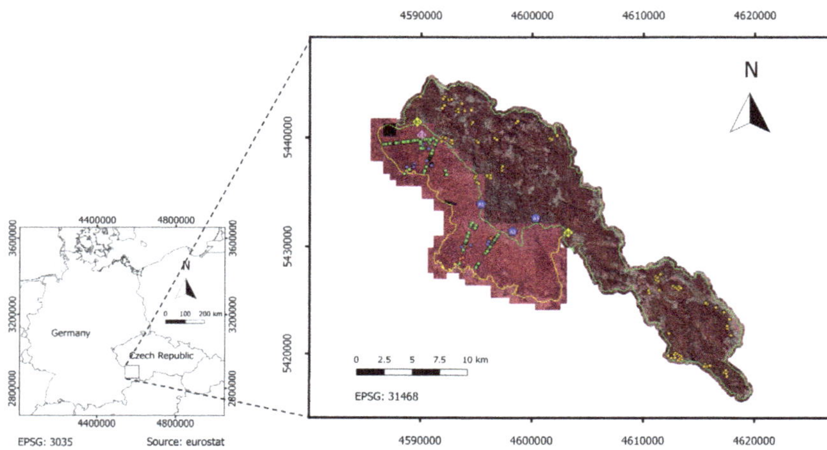

Figure 1. Map of Šumava National Park (green border) and Bavarian Forest National Park (yellow border) superimposed with orthophotos. Reference areas measured by field measurements: *Bioklim CZ* (orange circle), *HTO* (violet circle), *Bioklim* (green circle), *LAI* (red). Additional reference areas were manually labelled: *A1*, *A2*, and *A3* (blue pentagon) for dead trees and snags; *B1*, *B2* (yellow diamond) and *C1* (purple diamond) for broadleaf trees and conifers.

2.2. Lidar Data

The airborne full waveform data were acquired in June 2017 (leaf-on condition) using a RIEGL LMSQ 680i instrument, which was carried by a helicopter at a flying altitude of 550 m and at a speed of 60 kts. Parallel flight strips were flown with side lap of 60%. Therefore, the resulting average point density was 55 points/m^2. In total, the area of the lidar campaign was 943 km^2. The lidar data were provided as LAS-files containing an amplitude value for each laser point. Table 1 summarizes the parameters of the RIEGL LMSQ 680i scanner and the lidar flight campaign.

The raw amplitude values I_{raw} were corrected with regard to travelling distance s of the laser beam assuming a mean flight height s_0 = 400 m with

$$I_{corr} = I_{raw} \times (\frac{s}{s_0})^n. \tag{1}$$

The parameter n was calibrated using the lidar data of a calibration flight that was conducted on a near-by airfield. Following the procedure suggested by [6], we calculated for parameter $n = 2.03$.

Finally, the entire lidar set was geometrically checked in planimetry and height using enclosing polygons for flat buildings and the mean height of small horizontal areas. On average, the mean horizontal and vertical displacements were 5 cm and 6 cm, respectively.

Table 1. Flight parameters of lidar campaign.

Scanner Type	RIEGL LMS-680i
Platform	Helicopter D-HFCE/AS350
Spectral wavelength (*nm*)	1550
Beam divergence (*mrad*)	0.5
Flight speed (*kts*)	60
Flight height (*m*)	550
Side lap (%)	60
Point density (*pts/m*^{2})	55
Footprint size (*mm*)	275
Area (*km*^{2})	943
Number of raw laser points (*Mio*)	52.400

2.3. Aerial Multispectral Data

For both parks, the multispectral aerial imagery was acquired on October 2017 and June 2017 in two separate aerial flight campaigns. In Šumava National Park, the fall foliage change had already started for most of the broadleaf trees (see Figure 2a). Therefore, the spectral appearance was similar for beeches and dead trees. Moreover, because of low solar altitude, the shadows of the trees were fairly long. In Bavarian Forest National Park, the color infrared imagery was optimal because the data were acquired in June 2017. Note that the acquisition time for the lidar campaign is almost the same. Figure 2a,b show typical areas containing conifers, broadleaf trees, and dead wood. The ground sample distance (GSD) was not identical for the two park areas and had values of 17 cm and 9.5 cm for the Šumava and Bavarian Forest National Parks, respectively. In total, 669 images were captured for Šumava and 2553 for Bavarian Forest. Additionally, Global Navigation Satellite System (GNSS) data and Inertial Navigation System (INS) data were also acquired to provide initial information about the exterior orientation. The mean above-ground flight heights were 3900 m and 2880 m for the Šumava and Bavarian Forest National Parks, respectively. The images had end-laps of 80% and side-laps of 60%. They contained four spectral bands (blue, green, red, and near-infrared). The aerotriangulation (AT) was calculated using the following input: the aerial images, the camera calibration model, and the image measurements for the control points. As expected, the sigma naught of the AT was excellent, with 0.22 GSD for Šumava and 0.3 GSD for Bavarian Forest. For the software packages, ISAT [30] and MATCH-AT [31] were used for Šumava National Park and Bavarian Forest National Park, respectively. The software packages OrthoMaster/OrthoVista [32] generated true orthophotos utilizing the digital surface model of the total forest area. Table 2 shows the details of the aerial flight campaigns and the subsequent photogrammetric processing of the aerial imagery.

Table 2. Flight parameters of aerial image acquisition, results of subsequent aerial triangulation, and software packages used.

Park Area	Šumava	Bavarian Forest
Acquisition time	October 2017	June 2017
Aerial camera	Xp-w/a	DMCIII
End lap/Side lap (%)	80/60	80/60
Focal length (*mm*)	100	92
Number of images	669	2553
MSL flight height (*m*)	3900	2880
GSD (*cm*)	17	9.5
Sigma naught AT (*GSD*)	0.22	0.3
Software for AT	ISAT	MATCH-AT
Software for Orthophotos	OrthoMaster	OrthoMaster

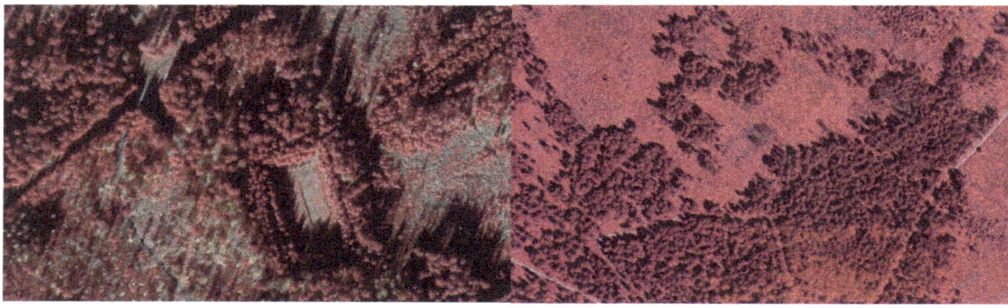

(a) (b)

Figure 2. Cut-outs of two aerial images. (**a**) Šumava National Park. Approximate location is 49°7.26′N, 13°19.27′E. Approximate size is 514 m × 243 m. Note the spectral appearance of broadleaf trees due to fall foliage. (**b**) Bavarian Forest National Park. Approximate location is 49°6′N, 13°18.61′E. Approximate size is 624 m × 377 m.

2.4. Reference Data

2.4.1. Field Measurements

Ground truth data for sample plots were acquired by field measurements in four reference data set *Bioklim CZ* [33], *HTO* [34], *Bioklim* [33], and *LAI Bioklim* [35]. In both parks, tree height measurements were carried out using the "Vertex" III system [36] considering trees with a diameter at breast height (DBH) > 7 cm. In rectangular *HTO* plots, each stem position was precisely surveyed by means of tacheometry and double-checked with differential GPS (DGPS) using the Leica GPS System 500 instrument. The coordinate differences indicated an excellent standard deviation of 5 cm [34]. The centers of all remaining circular plots were measured by means of differential GNSS (DGNNS). In the case of Czech plots *Bioklim CZ*, the Leica ATX1230 GG instrument provided plot center locations, reporting a position accuracy between 3 cm and 96 cm. For the *Bioklim* and *LAI* plots, a Leica GS 14 Professional GNSS receiver was used [33]. The range of the internal accuracy estimation was between a few centimeters and several decimeters. In Czech plots *Bioklim CZ*, tree positions were collected with a laser range-finder and a magnetic compass IFER-MMS from Field-Map Technology (see Zenáhlíková et al. [27]). In Bavarian plots *Bioklim* and *LAI*, the distance was surveyed using a "Vertex" III system and a compass for the azimuth (see Bässler et al. [33]). Table 3 gives an overview of the tree species distribution in the four reference data sets.

The reference trees were further subdivided into three layers with respect to the top tree height *htop* in the plot. The parameter *htop* is defined as the average height of the 100 highest trees per ha [37]. The lower layer contains all trees below 50% of *htop*, the intermediate layer corresponds to all trees between 50% and 80% of *htop*, and the upper layer contains the remainder of the trees.

2.4.2. Reference Trees by Manual Labelling

Additional reference data were collected by visual inspection using polygons of segmented trees. In total, six different rectangular plots $A1, A2, A3$, $B1, B2$, and $C1$ (see Figure 1) were selected. These reference plots are dominated by broadleaf trees, conifers, dead trees and snags, thereby providing a varied composition of trees for classification. Table 4 gives an overview on the total number of trees manually labelled. We used an interactive tool that displays polygons of tree segments in the aerial image, as well as the segmented point cloud of tree segments. The point cloud can be rotated in 3D, thereby aiding the expert in verifying the class of investigated tree objects. In detail, we labelled samples for the classes 'conifer' and 'broadleaf', 'dead tree (with crown)', 'non-dead tree', 'snag', and 'non-snag'. The identification of living and dead conifers in both parks was fairly easy (see Figure 3a,e,g). The broadleaf trees in Bavarian Forest National Park could also be classified without problems (see Figure 3f). However, broadleaf trees in Šumava National Park were difficult to label because many were already at the leaf-off stage and were therefore visually similar to dead trees (see Figure 3b). For the same reason, the labelling of dead trees in Šumava National Park was also a fairly difficult task (see Figure 3c,d). Instead, single dead trees were in many cases perfectly silhouetted against the background in Bavarian Forest National Park (see Figure 3g). Some problems occurred when they were overlapped with nearby spruces (see Figure 3h). In both parks, snags could only be found using the point cloud of the tree segments because the snags were practically invisible in an aerial image (see Figure 3i,j).

Table 3. Properties of reference plots from field measurements.

Plot Areas	Bioklim CZ	HTO	Bioklim	LAI
Park area	Šumava	Bavarian Forest	Bavarian Forest	Bavarian Forest
Number of plots	78	15	120	26
plot size (*ha*)	0.05	∅=0.21	0.05	0.05
Accuracy of plot centers	DGNSS	DGPS	DGNSS	DGNSS
Year of acquisition	2017	2007	2016	2017
Fir	6	10	157	1
Larch	1	0	5	0
Pine	237	0	0	0
Spruce	1284	677	1397	6
Alder	12	0	16	0
Ash	2	0	2	37
Beech	354	883	1414	570
Birch	195	0	21	0
Elm	1	0	1	0
Norway Maple	0	2	0	0
Rowan	11	5	31	53
Sycamore	23	19	55	14
Number of conifers	1528	687	1559	7
Number of broadleaf trees	598	909	1540	674
Conifers (%)	72	43	50	1
Broadleaf trees (%)	28	57	50	99
Total	2126	1596	3099	681
Forest type	coniferous	mixed	mixed	deciduous
Stem density (*stems/ha*)	542	550	516	524

Table 4. Properties of reference plots manually labelled.

Area	*A1, A2, A3*	*B1, B2*	*C1*
Year of acquisition	2017	2017	2017
Conifers	-	4772	1796
Broadleaf trees	-	813	141
Living trees	1345	-	-
Dead tree	706	-	-
Snag	131	-	-

2.5. Outline of the Method

In our approach, we classifed tree groups 'conifers' and 'broadleaf trees' as well as standing dead trees and snags using lidar and multispectral imagery. A fairly large feature set was generated comprising geometric features, intensity features, features for snags based on 3D shape contexts, and features generated from the multispectral images. Overfitting effects were reduced by automatically selecting prominent features. In our study, the following methodology was applied. We segmented the lidar point cloud into 3D segments representing single trees using the normalized cut algorithm [6,7] (see Section 2.6). The procedure recursively partitioned the lidar cloud into single tree segments until a predefined stopping criterion was achieved. The aim of the segmentation step was to provide tree segments that are used in the next steps to classify the segments with respect to the two tree groups, dead trees and snags. Figure 4 illustrates the entire procedure, which is explained in detail in the following subsections.

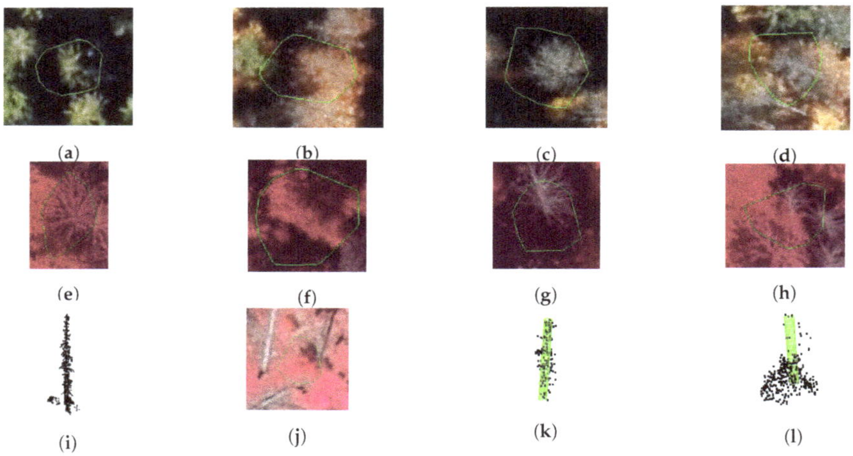

Figure 3. Subfigures (**a**) to (**d**): Examples for living spruce, broadleaf tree, dead spruce, and difficult case in National Park Šumava. (**a**) Living spruce. (**b**) Broadleaf tree. (**c**) Difficult example for a broadleaf tree in fall. Leaf-off. (**d**) Difficult example for a broadleaf tree. Partly leaf-off. Subfigures (**e**) to (**h**): Examples for living spruce, broadleaf tree, dead spruce, and difficult case in Bavarian Forest National Park. (**e**) Living spruce. (**f**) Broadleaf tree. (**g**) Good example for a dead spruce with a crown. (**h**) Difficult example for broadleaf tree. Overlapping with dead spruce. Subfigures (**i**) to (**l**): (**i**) Point cloud of snag. (**j**) Snag from (**i**) not visible in aerial image. (**k**) Point cloud of snag. Fitted cylinder is green. (see also Section 2.7) (**l**) Point cloud of non-snag. Fitted cylinder is green (see also Section 2.7).

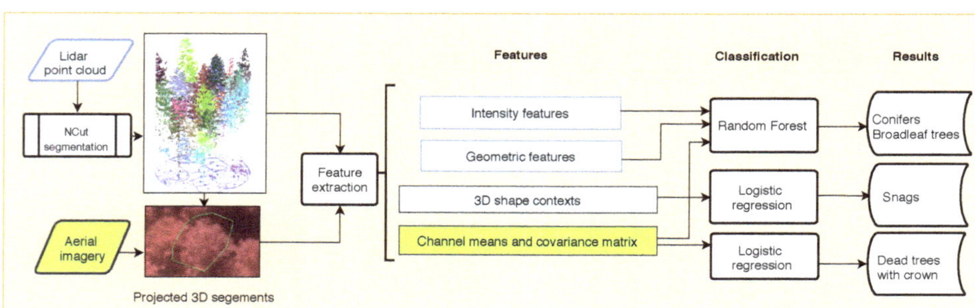

Figure 4. Overview of the approach for classification of conifers, broadleaf trees, standing dead trees, and snags.

2.6. Single Tree Detection

For single tree detection, we used the normalized cut segmentation which is a top-down method for segmenting objects over a discrete graph structure $G = (V, E)$, with V as the node set, E as the edge set, and $w_{ij}(o_i, o_j)$ as the symmetric, non-negative pairwise object similarity function. $W = \{w_{ij}(o_i, o_j)\}_{i=1...N, j=1...N}$ is the similarity matrix representing the pairwise weighted interrelationship between N primitives $O = \{o_i\}_{i=1...N}$ of a super-voxels, which is provided by a pre-segmentation with k-means. The normalized cut algorithm [7] recursively bisects the graph into disjoint K subgraphs representing clusters A and B by minimizing the objective function (=normalized nut)

$$NCut(A,B) = \frac{cut(A,B)}{Assoc(A)} + \frac{cut(A,B)}{Assoc(B)}, \quad (2)$$

where $Assoc(A) = \sum_{u \in A, v \in V} w_{ij}(A,V)$ is defined as the volume of subgraph A and is equivalent to the sum of weights of all edges ending up in cluster A. Equation (2) maximizes the similarity within clusters and minimizes the similarity between the disjoint clusters A and B.

The weighting matrix W is created from a similarity function

$$w_{ij}\,(o_i, o_j) = \prod_{k=1}^{M} e^{-\frac{d_k(o_i,o_j)^2}{\sigma_k^2}}, \quad (3)$$

where $d_k(o_i, o_j)^2$ denote the M features used as similarity measures. In this study, we used the feature weights σ_i for the values recommended in [6] (see also Section 2.10).

The recursive splitting of graph G into disjoint K subgraphs must normally be terminated by threshold value $NCut_{max}$. This parameter has no physical meaning and can be defined based on practical experience or using a trial-and-error strategy. In this study, a sensitivity analysis using training data was performed to find an optimum value for the objects to be segmented. However, the optimal value for $NCut_{max}$ turned out to be different for broadleaf and coniferous trees. Thus, we modified the segmentation and used two different stopping values for $NCut_{max}$, thereby making the segmentation dependent on the tree groups. In detail, we initially segmented the point cloud using the larger value of both $NCut_{max}$ parameters. In our case, this value was representative for conifers. Next, we applied the tree classification (see Section 2.8) to all tree segments. Third, the tree segments of the broadleaf trees were clustered with k-means clustering. Finally, the new tree segments potentially formed by broadleaf trees were newly segmented using the other $NCut_{max}$ parameter, which was the best for broadleaf trees.

2.7. Feature Extraction

In this study, for each tree segment, we calculated the following feature set from the lidar data and multispectral images. The feature set can be subdivided into four main groups.

(i) *Geometric features:*
Feature set S_g set includes (i) height-dependent variables at 10% intervals, (ii) density dependent variables at 10 intervals, and (iii) the two axis lengths of a paraboloid fitted to a tree crown (see Reitberger et al. [6]).

(ii) *Intensity features:*
The lidar point cloud contained, for each laser point, an amplitude value that was calibrated in advance (see Section 2.1). Based on this laser point attribute, the mean amplitudes per tree segment and in 10 tree layers were calculated to form feature subset S_I.

(iii) *3D shape contexts:*
For the snag classification, we followed the idea of Polewski et al. [24] and applied 3D shape contexts, which are constructed as cylinders with radius r_{seg} and length l_{seg} along the tree segments' axis (see Figure 3k,l). Features S_{3DC} are essentially histograms representing the spatial distribution of the points within the cylindrical volume.

(iv) *Covariance matrix:*
For this feature subset, we projected the tree segments into the next aerial color infrared image using the exterior orientation. Within the resulting enclosing polygon, we calculated the 3 × 3 covariance matrix and the mean of the three channels. The six independent elements of the covariance matrix and the mean channels form the nine elements of feature set S_{cov}.

2.8. Classification of Conifers and Broadleaf Trees

For classification of conifers and broadleaf trees, we used the feature sets S_g (= geometric features), S_I (= intensity feature), and S_{cov} (= features from aerial images). We chose Random Forest (RF) [38] as a binary classifier because of its robust ability to estimate the class probability. Note that RF provides a probability for each classified object.

2.9. Classification of Standing Trees and Snags

To classify both types of dead trees based on features S_{cov} and S_3DC we applied a multinomial logistic regression (MLR). We used a regularized-generalized model, which was implemented in the *glmnet* function as an *R* source package. Similar to RF, MLR classifies each dead tree with a probability p_{dead} for dead trees and p_{snag} for snags. Note that we also tested RF and Support Vector Machines (SVM) as classifiers. However, MLR turned out to be the best option for this classification task.

2.10. Choice of Parameters

The 3D tree segmentation based on the normalized cut was controlled by parameters listed in Table 5. According to Yao et al. [39], we introduced the quadratic distances $D^2_{XY_{ij}} = (X_i - X_j)^2 + (Y_i - Y_j)^2$ and $Z^2_{XY_{ij}} = (Z_i - Z_j)^2$ between super-voxels i and j as features. These two features were completed by parameter $F^{max}_{XY_{ij}}$, which defined the maximum distance of a super-voxel pair i and j to a tree position provided by the watershed segmentation. The amplitude of the laser points processed in Section 2.2 was neglected because this feature has practically no impact on the quality of the segmentation. Therefore, the weighting matrix W can be specified as

$$w_{ij} = \begin{cases} e^{-X(i,j)} \times e^{-Z(i,j)} \times e^{-F(i,j)} & \text{if } D_{XY_{ij}} < R_{XY} \\ 0 & \text{otherwise} \end{cases} \tag{4}$$

with

$$X(i,j) = \left(\frac{D_{XY_{ij}}}{\sigma_{xy}}\right)^2,\ Z(i,j) = \left(\frac{D_{Z_{ij}}}{\sigma_z}\right)^2 \tag{5}$$

$$F(i,j) = \left(\frac{F^{max}_{XY_{ij}}}{\sigma_F}\right)^2, \tag{6}$$

where R_{XY} defines the vertical cylinder within which the weights w_{ij} are calculated. The control parameters σ_{xy}, σ_z and σ_F weight the corresponding features in Equations (3) and (4). The values used, which are listed in Table 5, follow the recommendations of Yao et al. [39].

To calculate the features S_{3DSC} of the snag detection (see Section 2.7), we used the control parameters listed in Table 6.

Table 5. Control parameters of single tree segmentation.

Parameter	Symbols	Values
Min. number of super-voxels	N_{minV}	2
Weight for horizontal distance	σ_{xy}	1.35 m
Weight for vertical distance	σ_z	11.0 m
Weight for stem position	σ_F	3.5 m

Table 6. Control parameters of feature extraction with 3D shape contexts.

Parameter	Symbols	Values
Cylinder radius	r_{seg}	0.5 m
Cylinder length	l_{seg}	2.0 m

2.11. Evaluation and Classifier Training

We conducted four groups of experiments. First, we segmented single trees using the modified normalized cut segmentation. In this experiment, we showed that two different values $NCut_{max}$ can be successfully applied to segment trees in dependence on the tree classes "conifer" and "broadleaf tree". Second, we classified conifers and broadleaf trees using different feature sets. The third and fourth experiments focused on the classification of dead trees and snags, respectively. All four experiments utilized the reference data from the field measurements and the visual inspection (see Section 2.4).

In experiment #1, recall and precision quantities were used to evaluate the quality of the detected trees. The ratio of the reference tree numbers, which have at least one associated detected tree, to the total number of reference trees is called 'recall'. The number of detected trees that were successfully connected to a reference tree as a fraction of the total number of detected trees is defined as the 'precision'. Both recall and precision are calculated depending on stopping criterion $NCut_{max}$. Following the strategy proposed by Reitberger et al. [6], we considered two trees to be matched trees, if (i) the distance between the segmented tree to the reference single tree is smaller than 60% of the mean tree distance within the sample plot and (ii) the height difference between the segmented tree and the reference tree is less than 20% of the top height of the plot. If a reference tree was connected with more than one segmented trees, the tree position with the shortest distance to the reference tree was selected. In cases where a segmented tree could not be combined with a reference tree, we considered this to be a 'false positive' segment.

For classification of conifers and broadleaf trees (=experiment #2), we combined reference trees from field measurements and visual inspection to train the classifier. As a classifier, we used the RF classifier from Section 2.8. The reference data were split into two sets for training and testing. The RF classifier was also validated with reference data provided only by field measurements. For dead tree classification (=experiment #3 and #4), we only used reference trees determined from visual inspection. In both cases, the classifier was logistic regression (see Section 2.9). For these two latter experiments, the reference samples were also subdivided into two data sets for training and testing. We used a five-fold cross-validation to obtain the overall classification accuracy, precision, and recall as a compromise between the computational efficiency and reducing the effects of randomness. All classifiers were evaluated with test data not being used for training. All results are presented with confusion matrices calculated from reference data and test data.

3. Results

3.1. Single Tree Detection

A sensitivity analysis was conducted in the plots *HTO*, *Bioklim*, *LAI*, and *Bioklim CZ* to find the best value for parameter $NCut_{max}$. First, we applied the same value for parameter $NCut_{max}$ to conifers and broadleaf trees. An analysis of Table 7 with Tables 8 and 9 let us conclude that data set *HTO* was the best. In general, the recall of data set combination *Bioklim*, *LAI*, and *Bioklim CZ* was around 20% worse when compared with *HTO* data set (Table 8). When we compared the data set *Bioklim CZ* (Table 9) with *HTO* data set (Table 7), we also found a significant discrepancy in the recall, especially in the upper layer and, in general, for the total mean. The precision of this data set was 15% better than *HTO* in total mean. In summary, data set *HTO* was the most reliable, also because of the high absolute accuracy of the tree positions (see also Section 2.4.1). Thus, parameter $NCut_{max} = 0.16$ seemed to be the best compromise for controlling the segmentation for both conifers and broadleaf trees.

We further analyzed the segmentation with respect to conifers and broadleaf trees. Because we found that many broadleaf tree areas had a clear oversegmentation, we applied the new segmentation scheme discussed in Section 2.6. After varying parameter $NCut_{max}$ in a reasonable range, we found optimized values for this control parameter. Tables 10 and 11 demonstrate clearly that $NCut_{max} = 0.10$ increases the precision by 13% at a small cost of deterioration in recall. Figure 5c–f show nicely the reduction of oversegmentation with parameter $NCut_{max} = 0.10$ for the preclassified broadleaf trees.

Table 7. Sensitivity analysis for tree detection in reference areas. *HTO*. Total number of trees is 1596.

Parameter	Recall				Precision			
NCut	Lower	Middle	Upper	Total	Lower	Middle	Upper	Total
0.13	0.16	0.38	0.84	0.64	0.23	0.53	0.67	0.58
0.16	0.18	0.41	0.90	0.69	0.20	0.51	0.62	0.54
0.18	0.19	0.44	0.92	0.71	0.23	0.52	0.60	0.52
0.20	0.24	0.45	0.91	0.72	0.21	0.50	0.58	0.49

Table 8. Sensitivity analysis for tree detection in reference areas. HTO, Bioklim and LAI. Total number of trees is 5736.

Parameter	Recall				Precision			
NCut	Lower	Middle	Upper	Total	Lower	Middle	Upper	Total
0.16	0.16	0.38	0.67	0.48	0.30	0.48	0.57	0.58
0.18	0.17	0.39	0.68	0.49	0.30	0.45	0.56	0.56

Table 9. Sensitivity analysis for tree detection in reference areas. Bioklim CZ. Total number of trees is 2126.

Parameter	Recall				Precision			
NCut	Lower	Middle	Upper	Total	Lower	Middle	Upper	Total
0.16	0.13	0.37	0.67	0.51	0.29	0.56	0.70	0.73
0.18	0.13	0.36	0.67	0.52	0.28	0.53	0.69	0.72

Table 10. Sensitivity analysis of tree detection regarding conifers and broadleaf trees. The reference area is *HTO*. Control parameter $NCut_{max}$ = 0.16. Total number of trees is 1596.

Parameter	Recall				Precision			
Layer	Lower	Middle	Upper	Total	Lower	Middle	Upper	Total
Conifers	0.10	0.53	0.89	0.69	0.27	0.66	0.85	0.82
Broadleaf trees	0.30	0.43	0.91	0.72	0.16	0.52	0.60	0.54

Table 11. Sensitivity analysis of tree detection regarding conifers and broadleaf trees. The reference area is *HTO*. Control parameter NCut = 0.10. Total number of trees is 1596.

Parameter	Recall				Precision			
Layer	Lower	Middle	Upper	Total	Lower	Middle	Upper	Total
Broadleaf trees	0.18	0.33	0.84	0.63	0.15	0.50	0.73	0.66

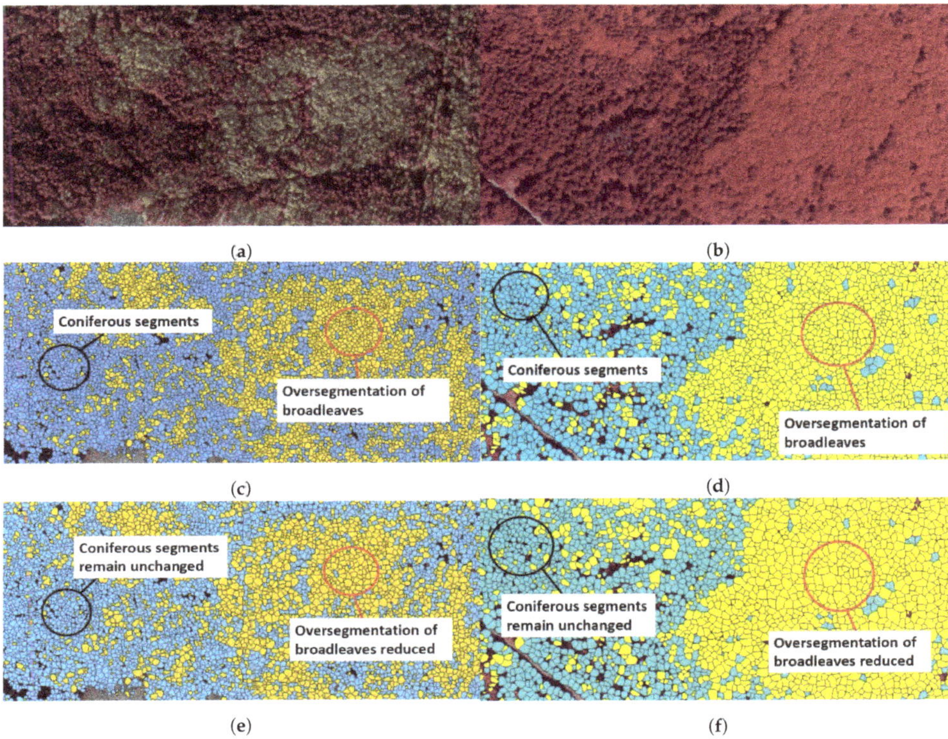

Figure 5. Impact of parameter $NCut_{max}$ calibrated for conifers and broadleaf trees. Coniferous tree segments are in blue; broadleaf tree segments are in yellow. (**a**) Cut-out of aerial image in forest area of Šumava National Park. Approximate location is 48°47′57″N, 13°51′27″E. Approximate size is 655 m × 310 m. (**b**) Cut-out of aerial image in forest area of Bavarian Forest National Park. Approximate location is 48°55′45″N, 13°20′15″E. Approximate size is 456 m × 216 m. (**c**) Tree segments in area (**a**) $NCut_{max}$ = 0.16. (**d**) Tree segments in area (**b**) $NCut_{max}$ = 0.16. (**e**) Tree segments in area (**a**) $NCut_{max}$ = 0.10. (**f**) Tree segments in area (**b**) $NCut_{max}$ = 0.10.

3.2. *Classification of Conifers and Broadleaf Trees*

For the Czech and German parts of the forest areas, we trained two separate classifiers for classification of conifers and broadleaf trees because the aerial images were captured at various times and with different cameras. We used 3D geometry features S_g describing crown shape and point distribution per tree and feature set S_{cov} generated from the aerial images (see Section 2.7). Interestingly, the intensity features S_I finally had to be ignored because, in some areas, the intensity distribution was not consistent and showed severe tiling effects (see Figure 6a,b).

Tables 12 and 13 show the validation of the classifiers on test data taken from four reference areas. In general, the classification results were excellent. For Šumava National Park, the overall accuracy (OA) of 97.5% was better by 5–7%. Moreover, the confusion between conifers and broadleaf trees was significantly lower, and broadleaf trees were better classified in terms of recall and precision. Very likely, the fall foliage helped to discriminate among conifers and broadleaf trees in Šumava National Park for this accuracy level.

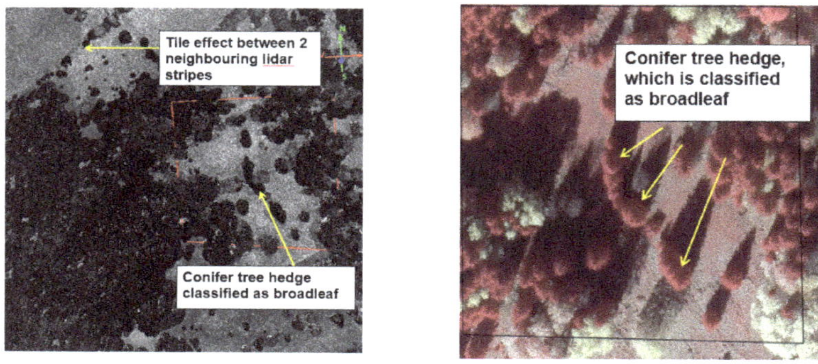

(a) (b)

Figure 6. Intensity issues in the mixed forest area. (**a**) Tile effect between neighboring laser strips (**b**) Mis-classified broadleaf trees.

Table 12. Results for the classification of conifers and broadleaf trees. Bavarian Forest National Park: Test data are from reference areas *HTO* and *C1*: 1637 conifers and 636 broadleaf trees for training. 722 conifers, 253 broadleaf trees for testing (=Reference #1). 293 conifers, 236 broadleaf trees for testing (=Reference #2 only from *HTO*).

		Reference #1			*Reference #2*		
		Conifer	Broadleaf	Precision	Conifer	Broadleaf	Precision
Predicted	Conifer	687	15	0.98	269	44	0.85
	Broadleaf	35	238	0.87	24	192	0.89
	Recall	0.95	0.94		0.92	0.81	
	OA: 94.9%	Kappa: 0.87			OA: 87.1%	Kappa: 0.89	

Table 13. Results for the classification of conifers and broadleaf trees. Šumava National Park: Test data are from reference areas *Bioklim CZ*, *B1* and *B2*: 3480 conifers and 712 broadleaf trees for training. 1493 conifers, 314 broadleaf trees for testing (=Reference #3). 138 conifers, 187 broadleaf trees for testing (=Reference #4 only from *Bioklim CZ*).

		Reference #3			*Reference #4*		
		Conifer	Broadleaf	Precision	Conifer	Broadleaf	Precision
Predicted	Conifer	1482	19	0.98	131	1	0.99
	Broadleaf	11	295	0.96	7	186	0.96
	Recall	0.99	0.94		0.95	0.99	
	OA: 97.5%	Kappa: 0.94			OA: 97.5%	Kappa: 0.95	

3.3. Classification of Standing Dead Trees

For both parks, we trained two separate classifiers to classify dead trees using the feature data set S_{cov} because of the different acquisition times for the aerial imagery. Note that the training data set was automatically balanced. Therefore, the number of dead trees and living trees in the confusion matrices differed from the relevant numbers indicated for training and testing in Tables 14 and 15. Moreover, a class probability p_{dead} = 90% was used to classify a dead tree.

For Bavarian Forest National Park, the OA of 91.7% calculated from the training data with five-fold cross validation was confirmed by the OA of 92.6% performed in the test area. Conversely, the OA of 82.8% for Šumava National Park was slightly reduced for the training data. Very likely, the classifier was influenced by the spectral similarity of the dead trees and the beeches. The OA of 100% for the test data was certainly too optimistic. Presumably, manually labelled trees could be classified easier.

3.4. Classification of Snags

In experiment #4, only one classifier was trained because the used feature set S_{3DSC} solely depends on lidar data, and the 3D structure of snags was assumed to be the same for both parks. Balancing of the training data and usage of class probability p_{snag} were the same as for the classification of dead trees. The excellent OA of 90.9% in the training area is in the same order of magnitude as the OA calculated with the test data (see Table 16). Figure 7 shows examples for labelled snags that were classified as snags with $p_{snag} > 90\%$. However, the analysis of the test reference revealed a significant confusion between snags and living trees.

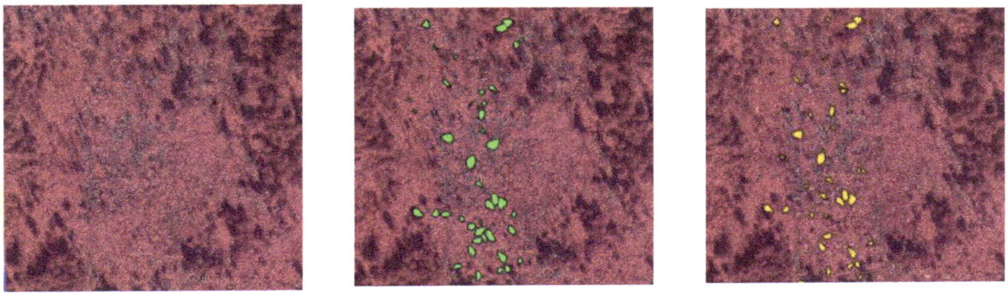

(a) (b) (c)

Figure 7. Verification of the snag classifier in Bavarian Forest National Park. (**a**) Small forest scene with living trees, dead trees and snags. (**b**) Labelled snags for verification (green). (**c**) Classified snags (orange) with threshold $p_{snag} > 90\%$.

Table 14. Classification results of dead trees. Labelled data are from areas $A1$, $A2$, and $A3$. Bavarian Forest National Park: 124 dead trees, 186 living trees for training. 310 dead trees, 761 living trees for testing. Confusion matrices for training (left): Number of living trees are reduced to the number of dead trees. Numbers refer to the average of the five test sets of the five-fold cross-validation.

		Training Reference			*Test Reference*		
		Dead	Living	Precision	Dead	Living	Precision
Predicted	Dead	22	2	0.92	255	24	0.91
	Living	2	22	0.92	55	737	0.93
	Recall	0.92	0.92		0.93	0.86	
	OA: 91.7%	Kappa: 0.93			OA: 92.6%	Kappa: 0.81	

Table 15. Classification results of dead trees. Labelled data are from areas $A1$, $A2$, and $A3$. Šumava National Park: 163 dead trees with crowns, 356 living trees for training. 109 dead trees, 142 living trees for testing. Tree segments are classified as dead trees if the threshold for class probability $p_{dead} > 90\%$.

		Training Reference			*Test Reference*		
		Dead	Living	Precision	Dead	Living	Precision
Predicted	Dead	27	6	0.82	109	0	1.0
	Living	5	26	0.84	0	142	1.0
	Recall	0.84	0.81		1.0	1.0	
	OA: 82.8%	Kappa: 0.65			OA: 100.0%	Kappa: 1.0	

Table 16. Results of the classification of snags for both parks. Labelled data are from areas $A1$, $A2$, and $A3$. 55 snags and 524 living trees for training. 76 snags and 1513 living trees for testing. Confusion matrices for training (left): Number of living are reduced to the number of snags. Numbers refer to the average of the five test sets of the five-fold cross-validation. Tree segments are classified as snags if the threshold for class probability $p_{snag} > 90\%$.

		Training Reference			*Test Reference*		
		Snag	Living	Precision	Snag	Living	Precision
Predicted	Snag	10	1	0.91	50	39	0.56
	Living	1	10	0.91	26	1474	0.98
	Recall	0.91	0.91		0.66	0.97	
	OA: 90.9%	Kappa: 0.81			OA: 95.9%	Kappa: 0.58	

4. Discussion

4.1. Main Goals of the Study

The main goals of the current study were to classify two tree classes (coniferous and deciduous), standing dead trees, and snags in a large forest area using lidar and multispectral imagery. We also showed a novel calibration of a control parameter that stops the iterative normalized cut segmentation.

4.2. Single Tree Detection

The experiments clearly showed that the segmentation accuracy depends on conifers and broadleaf trees. The best values for parameter $NCut_{max}$, which controls the stopping of the iterative normalized cut method, were 0.10 for conifers and 0.18 for broadleaf trees. Reitberger et al. [6] recommended an average value of 0.16 for both tree groups. Instead, we used two different parameters for $NCut_{max}$ and could improve the segmentation quality in terms of precision by 13%. However, we had to rerun the segmentation for broadleaf trees again. Thus, the improvement in accuracy was gained with an increased computation time of roughly one-third. Recently, Amiri et al. [9] replaced control parameter $NCut_{max}$ with an adaptive stopping criterion that analyses the remaining point cloud of the segmented clusters and allows a classifier to determine the termination of bisecting of the clusters. However, this method only works for conifers and is not applicable for broadleaf tree.

4.3. Classification of Conifers and Broadleaf Trees

In general, the results of tree classification were excellent, with values for OA beyond 95% for both parks. The laser intensity, calculated as a mean value for the upper part of the tree crown, turned out as useless. Very likely, the calibration parameter for intensity normalization was not representative for the entire project area. Interestingly, the spectral information of CIR aerial imagery successfully replaced the laser intensity as the salient feature in the tree classification process.

The OA in Tables 12 and 13 calculated from different test data sets (Reference #1 and #2 for Bavarian Forest National Park; Reference #3 and #4 for Šumava National Park) also demonstrated the transferability of the classifiers. For Bavarian Forest National Park, we noticed a difference of 7.8%, whereas the OA for Šumava National Park was essentially the same for both test areas.

4.4. Classification of Dead Trees and Snags

In both parks, the classification of dead trees performed with excellent accuracy, better than 90% in terms of recall and precision. In comparison, the accuracy in Šumava National Park was slightly reduced because the fall foliage of broadleaf trees was spectrally similar to the crowns of the dead trees.

Finally, the training of the snag classifier showed a fairly good OA of 91%. Although the OA for the test data was even better with 96%, the recall and precision for snags was only 66% and 56%, respectively. Because of the smooth transition from dead spruces with crowns and without crowns (=snags), we could not find an optimal classifier for snags. A detailed analysis revealed that some false positives were dead trees with only a few branches. Thus, they could also be considered snags in the labelling process.

4.5. Comparison with Related Work

First, we contrasted our results for tree segmentation with the findings of Reitberger et al. [6], which also investigated a forest site in Bavarian Forest National Park. In general, our recall for the upper layer (89% for conifers and 84% for broadleaf trees) was in the same order of magnitude. However, for the middle layer, we obtained a better recall of 43% (mean for conifers and broadleaf trees) compared to 32% reported in [6]. For the lower layer, the mean recall of 14% corresponded well with 17% shown in [6]. Our total precision of 74% was worse than the precision of 90% in [6].

Second, we compared the averaged results of the tree segmentation for the *HTO* data set with four studies: see Strîmbu and Strîmbu [4], Li et al. [40], Holmgren and Lindberg [41] and Silva et al. [42] (see Table 17). Note that two study areas are dominated by conifer pine trees, and that the stem densities of three studies are by a factor of 1.4 to 3.2 lower. In Table 17, the results of our study are presented with regard to (i) coniferous plots and (ii) all plots (=mixed) for the upper layer (see also Table 10 and Table 11). Li et al. [40] demonstrated in a Sierra mixed conifer pine forest a detection rate of 86% and precision of 94%. This method segmented trees using a top-to-bottom region growing approach. Strîmbu and Strîmbu [4], using a three-dimensional approach in which a weighted graph was built from hierarchical topologies, reported for a pine dominated forest a recall of 84% and precision of 90%. Silva et al. [42], applying the method to an open canopy longleaf pine forest area, segmented trees with a detection rate of 82% and precision of 85%. Recently, Holmgren and Lindberg [41] demonstrated a new watershed-based tree detection method that used a tree crown density model trained from the laser returns and some arbitrarily selected reference trees. The evaluation of a managed hemi-boreal forest showed a recall of 85% and precision of 82%. Although the stem density of our study was significantly higher in three cases, the tree segmentation detected coniferous trees slightly better. However, the precision turned out to be somewhat worse. For all the reference plots of our study, representing a mixed temperate forest, the recall was also still better or similar. Clearly, the corresponding precision was lower by 7-16% because some of the deciduous trees, which are generally more difficult to segment, could not be found in the point cloud.

Table 17. Parameters and accuracy performance of tree segmentations methods compared with our study. Note that recall and precision values were evaluated from field measurements of *HTO* plots.

Study	Recall	Precision	Points/m^2	Trees/ha	Coniferous (%)	Deciduous (%)
Li et al. [40]	0.86	0.94	6	400	100	0
Strîmbu and Strîmbu [4]	0.84	0.90	10	206	63	37
Silva et al. [42]	0.82	0.85	5	170	100	0
Holmgren and Lindberg [41]	0.85	0.82	83	626	>50	<50
Our study (coniferous)	0.89	0.85	55	550	100	0
Our study (mixed)	0.86	0.78	55	550	43	57

Third, we focused on the quality of the classification of conifers and broadleaf trees and how it compared to the results published in Yao et al. [21]. In this other study, which was also conducted in Bavarian Forest National Park, the lidar point density was 25 points/m^2. Several feature sets were investigated, including height- and density-dependent features, as well as intensity features. Very interestingly, our OA of 94.9% (see Table 12) obtained with feature sets S_g (=geometric tree features; density- and height-dependent features) and S_{cov} (=features from aerial images) was equivalent to OA = 94% reported by Yao et al. [21]. Very likely, the radiometric content of aerial imagery could successfully substitute for the lidar intensity in our case.

Forth, we compared our classification of dead tree and snags with other approaches obtained from the same forest site. Yao et al. [21] reported a classification accuracy between 71% to 73%, which was significantly worse than ours. The study in Polewski [23] reported an accuracy of around 88%, which is slightly worse than our findings. Although both of these other studies measured the reference data manually using orthophotos superimposed on the tree crown polygons, the combination of lidar data and multispectral optical imagery in a two-class classification made the classification of dead trees possible with high accuracy. Lastly, our snag detection was better by more than 10% when compared with the results presented in [24]. Again, we concluded that the increased point density of 55 points/m^2 better reflects the 3D structure of a snag.

Finally, we concluded that all the main findings of our large-scale experiment confirmed the results of several studies that were conducted in a temperate forest.

4.6. Practical Implementation

We present some interesting statistical numbers for this large-scale mapping of forest parameters. All the methods of Section 2.4.2 and related algorithms are coded in C++ and run under Windows (© Microsoft) and Ubuntu. The tree segmentation and the tree classification are parallelized, thereby taking advantage of any multi core PC systems. In total, we used eight PC workstations in parallel, each of them equipped with up to 22 cores and 128 GByte RAM. Therefore, we split up the forest areas into 6 km × 6 km macro blocks. Each workstation processed one or two super macro blocks comprising 2 × 2 macro blocks. In total, we processed 27.321.962 single trees on a forest area 924 km^2 in size. Table 18 gives an overview of the number of tree types classified with respect to conifers, broadleaf trees, dead trees, and snags. The time critical process was the tree segmentation due to the time-consuming solving of the inherent general eigenvalue problem. Our most powerful workstation, with 22 cores, 2.2 GHz performance, and 128 GByte RAM, processed the tree segmentation and tree classification at 25 sec/ha and 0.38 sec/ha, respectively (see Table 19). By contrast, the processing times of the other two processes are significantly higher because the code is not parallelized.

Table 18. Number of classified trees.

Tree Type	Conifer	Broadleaf	Dead Trees	Snag
Šumava	16.693.391 (=84%)	2.396.850 (=12%)	641.098 (=3%)	224.533 (=1%)
Bavarian Forest	4.446.373 (=60%)	2.453.498 (=33%)	312.058 (=4%)	154.161 (=2%)

Table 19. Time performance of implemented algorithms.

Process	Speed (sec/ha)
Tree segmentation	25
Classif. of conifers/broadleaf trees	0.38
Dead tree classification	79
Snag classification	75

We also have to address the time and effort needed for manual interactions to label additional tree samples, to train the classifiers, and to check the results of the individual processes. Interestingly, we had to add the reference areas $A1$, $A2$, $A3$, $B1$, $B2$, and $C1$ because the other reference areas from the original field measurements did not fully represent the forest characteristics and were somewhat imprecise in parts. This entire project could be finalized by one single person within 11 months. This included one month for data preprocessing (quality check of the lidar data, intensity calibration, and structuring the data into a block structure (see Section 2.2) and around three months for the manual editing part. The remaining seven months were dedicated to the processing of the remote sensing data. Note that we had to process the segmentation twice on roughly half of the entire forest because of another $NCut_{max}$ value useful for broadleaf trees. Moreover, the retraining of relevant classifiers instigated a second run for the classification of conifers, broadleaf trees, and dead trees as well. In addition, half the workstations were only equipped with processors using eight cores.

The tree segmentation and the classification of conifers and broadleaf trees were processed with the TreeFinder software package [43]. The detection of dead trees and snags was performed with an in-house software.

Finally, we show in Figure 8 the mapping of conifers, broadleaf trees, standing dead trees, and snags for a small part of the Bavarian Forest National Park.

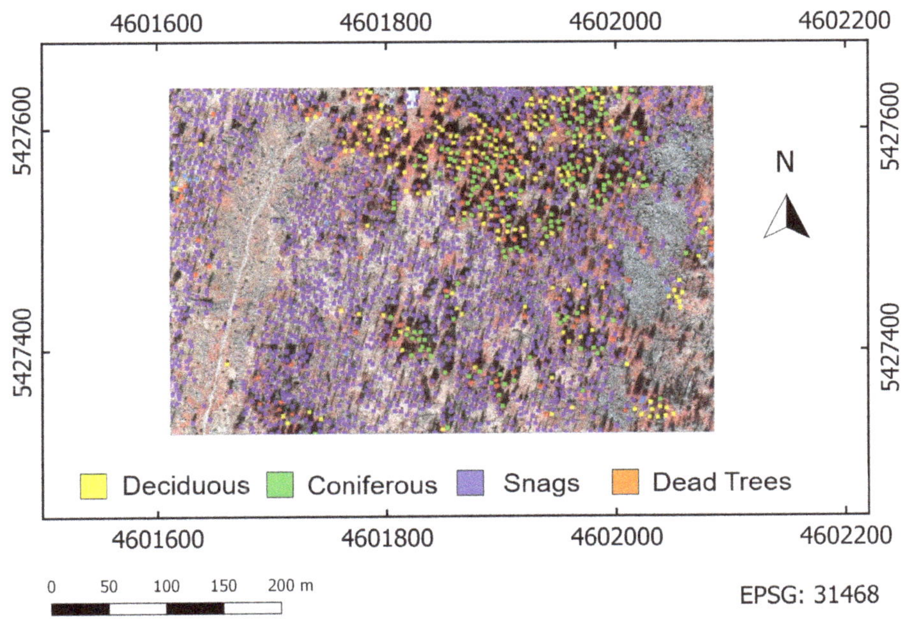

Figure 8. Mapping of conifers, broadleaf trees, dead trees, and snags superimposed with orthophoto. Image source is from Bavarian Forest National Park.

5. Conclusions

In summary, we were able to demonstrate that tree segmentation based on normalized cut can be successfully performed with a stopping criterion specifically calibrated for tree groups 'conifers' and 'broadleaf trees', thereby significantly reducing the effect of under- and oversegmentation. In our case, the precision for tree segments of broadleaf trees could be improved by 13%. However, this concept to calibrate parameter $NCut_{max}$ regarding the tree groups bears the drawback that the segmentation needs to be performed manifold in the case of multiple tree species. We further showed the successful mapping of conifers, broadleaf trees, and standing dead trees (with and without crowns) on a large forest 924 km^2 in size. Conifers and broadleaf trees could be classified in both parks with an accuracy better than 90%. Furthermore, the classification of standing dead trees was also possible with a high accuracy of more than 90%. The large consistent data set can be used by both parks for future preparation of cross-border management plans and studies on biodiversity [44].

Author Contributions: P.K., C.S. and M.H. conceived and designed the experiments; A.S. and P.K. performed the experiments; all the authors analyzed the data and results; P.K. wrote the paper; all the authors reviewed and edited the paper. All authors have read and agreed to the published version of the manuscript.

Funding: This research was funded by the cross-border cooperation programme Czech Republic–Bavaria Free State ETC goal 2014–2020, the Interreg V project No. 99: 'Přeshranični mapovani lesnich ekosystemů – cesta ke společnemu managementu NP Šumava a NP Bavorsky les /Grenzuberschreitende Kartierung der Waldokosysteme – Weg zum gemeinsamen Management in NP Sumava und NP Bayerischen Wald'.

Acknowledgments: We would like to thank MILAN Flug Gmbh, Olaf Schubert, Torben Hilmers, Joe Premier, and Horst Burghart for the field measurements in Bavarian Forest National Park. We appreciate the support of Jiří Roubínek, Jitka Zenáhlíková, Filip Chrapan, and Pavel Němčák for the field measurements in Šumava National Park. Sebastian Dersch (Munich University of Applied Sciences) and Miguel Köhling (UNIQUE) helped us a lot check and process the data.

Conflicts of Interest: The authors declare no conflicts of interest.

References

1. Latifi, H.; Heurich, M. Multi-Scale Remote Sensing-Assisted Forest Inventory: A Glimpse of the State-of-the-Art and Future Prospects. *Remote. Sens.* **2019**, *11*, 1260. doi:10.3390/rs11111260. [CrossRef]
2. Latifi, H.; Fassnacht, F.E.; Müller, J.; Tharani, A.; Dech, S.; Heurich, M. Forest inventories by LiDAR data: A comparison of single tree segmentation and metric-based methods for inventories of a heterogeneous temperate forest. *Int. J. Appl. Earth Obs. Geoinf.* **2015**, *42*, 162–174. doi:10.1016/j.jag.2015.06.008. [CrossRef]
3. Eysn, L.; Hollaus, M.; Lindberg, E.; Berger, F.; Monnet, J.M.; Dalponte, M.; Kobal, M.; Pellegrini, M.; Lingua, E.; Mongus, D.; et al. A Benchmark of Lidar-Based Single Tree Detection Methods Using Heterogeneous Forest Data from the Alpine Space. *Forests* **2015**, *6*, 1721–1747. doi:10.3390/f6051721. [CrossRef]
4. Strîmbu, V.F.; Strîmbu, B.M. A graph-based segmentation algorithm for tree crown extraction using airborne LiDAR data. *ISPRS J. Photogramm. Remote. Sens.* **2015**, *104*, 30–43. doi:10.1016/j.isprsjprs.2015.01.018. [CrossRef]
5. Dai, W.; Yang, B.; Dong, Z.; Shaker, A. A new method for 3D individual tree extraction using multispectral airborne LiDAR point clouds. *ISPRS J. Photogramm. Remote. Sens.* **2018**, *144*, 400–411. doi:10.1016/j.isprsjprs.2018.08.010. [CrossRef]
6. Reitberger, J.; Schnörr, C.; Krzystek, P.; Stilla, U. 3D segmentation of single trees exploiting full waveform LIDAR data. *ISPRS J. Photogramm. Remote. Sens.* **2009**, *64*, 561–574. doi:10.1016/j.isprsjprs.2009.04.002. [CrossRef]
7. Shi, J.; Malik, J. Normalized cuts and image segmentation. *IEEE Trans. Pattern Anal. Mach. Intell.* **2000**, *22*, 888–905. doi:10.1109/34.868688. [CrossRef]
8. Yao, W.; Krull, J.; Krzystek, P.; Heurich, M. Sensitivity Analysis of 3D Individual Tree Detection from LiDAR Point Clouds of Temperate Forests. *Forests* **2014**, *5*, 1122–1142. doi:10.3390/f5061122. [CrossRef]
9. Amiri, N.; Polewski, P.; Heurich, M.; Krzystek, P.; Skidmore, A.K. Adaptive stopping criterion for top-down segmentation of ALS point clouds in temperate coniferous forests. *ISPRS J. Photogramm. Remote. Sens.* **2018**, *141*, 265–274. doi:10.1016/j.isprsjprs.2018.05.006. [CrossRef]
10. Erikson, M. Species classification of individually segmented tree crowns in high-resolution aerial images using radiometric and morphologic image measures. *Remote. Sens. Environ.* **2004**, *91*, 469–477. doi:10.1016/j.rse.2004.04.006. [CrossRef]
11. Fassnacht, F.E.; Latifi, H.; Stereńczak, K.; Modzelewska, A.; Lefsky, M.; Waser, L.T.; Straub, C.; Ghosh, A. Review of studies on tree species classification from remotely sensed data. *Remote. Sens. Environ.* **2016**, *186*, 64–87. doi:10.1016/j.rse.2016.08.013. [CrossRef]
12. Hirschmüller, H. Stereo processing by semiglobal matching and mutual information. *IEEE Trans. Pattern Anal. Mach. Intell.* **2008**, *30*, 328–341. doi:10.1109/TPAMI.2007.1166. [CrossRef]
13. White, J.C.; Coops, N.C.; Wulder, M.A.; Vastaranta, M.; Hilker, T.; Tompalski, P. Remote Sensing Technologies for Enhancing Forest Inventories: A Review. *Can. J. Remote. Sens.* **2016**, *42*, 619–641. doi:10.1080/07038992.2016.1207484. [CrossRef]
14. Ullah, S.; Dees, M.; Datta, P.; Adler, P.; Koch, B. Comparing airborne laser scanning, and image-based point clouds by semi-global matching and enhanced automatic terrain extraction to estimate forest timber volume. *Forests* **2017**, *8*, 215–230. doi:10.3390/f8060215. [CrossRef]
15. Brandtberg, T. Classifying individual tree species under leaf-off and leaf-on conditions using airborne lidar. *Isprs J. Photogramm. Remote. Sens.* **2007**, *61*, 325–340. doi:10.1016/j.isprsjprs.2006.10.006. [CrossRef]

16. Holmgren, J.; Persson, Å.; Söderman, U. Species identification of individual trees by combining high resolution LiDAR data with multi-spectral images. *Int. J. Remote. Sens.* **2008**, *29*, 1537–1552. doi:10.1080/01431160701736471. [CrossRef]
17. Ørka, H.O.; Næsset, E.; Bollandsås, O.M. Classifying species of individual trees by intensity and structure features derived from airborne laser scanner data. *Remote. Sens. Environ.* **2009**, *113*, 1163–1174. doi:10.1016/j.rse.2009.02.002. [CrossRef]
18. Lindberg, E.; Briese, C.; Doneus, M.; Hollaus, M.; Schroiff, A.; Pfeifer, N. Multi-wavelength airborne laser scanning for characterization of tree species. In proceedings of the SilviLaser, La Grande Motte, France, 28–30 September 2015; pp. 271–273.
19. Shang, X.; Chazette, P. Interest of a full-waveform flown UV lidar to derive forest vertical structures and aboveground carbon. *Forests* **2014**, *5*, 1454–1480. doi:10.3390/f5061454. [CrossRef]
20. Hovi, A.; Korhonen, L.; Vauhkonen, J.; Korpela, I. LiDAR waveform features for tree species classification and their sensitivity to tree-and acquisition related parameters. *Remote. Sens. Environ.* **2016**, *173*, 224–237. doi:10.1016/j.rse.2015.08.019. [CrossRef]
21. Yao, W.; Krzystek, P.; Heurich, M. Tree species classification and estimation of stem volume and DBH based on single tree extraction by exploiting airborne full-waveform LiDAR data. *Remote. Sens. Environ.* **2012**, *123*, 368–380. doi:10.1016/j.rse.2012.03.027. [CrossRef]
22. Shi, Y.; Wang, T.; Skidmore, A.K.; Heurich, M. Important LiDAR metrics for discriminating forest tree species in Central Europe. *ISPRS J. Photogramm. Remote. Sens.* **2018a**, *137*, 163–174. doi:10.1016/j.isprsjprs.2018.02.002. [CrossRef]
23. Polewski, P. Reconstruction of Standing and Fallen Single Dead Trees in Forested Areas from LiDAR Data and Aerial Imagery. Ph.D. Thesis, Technische Universität München, München, Germany, 2017. Available online: https://www.pf.bgu.tum.de/pub/2017/polewski_phd17_dis.pdf (accessed on 16 January 2020).
24. Polewski, P.; Yao, W.; Heurich, M.; Krzystek, P.; Stilla, U. Free Shape Context descriptors optimized with genetic algorithm for the detection of dead tree trunks in ALS point clouds. *ISPRS Ann. Photogramm. Remote. Sens. Spat. Inf. Sci.* **2015**, *II-3/W5*. doi:10.5194/isprsannals-II-3-W5-41-2015. [CrossRef]
25. Bače, R.; Svoboda, M.; Janda, P.; Morrissey, R.; Wild, J.; Clear, J.; Čada, V.; Donato, D. Legacy of Pre-Disturbance Spatial Pattern Determines Early Structural Diversity following Severe Disturbance in Montane Spruce Forests. *PLoS ONE* **2015**, *10*, e0139214. doi:10.1371/journal.pone.0139214. [CrossRef] [PubMed]
26. Müller, J.; Bußler, H.; Gossner, M.; Rettelbach, T.; Duelli, P. The European spruce bark beetle Ips typographus in a national park: From pest to keystone species. *Biodivers. Conserv.* **2008**, *17*, 2979–3001. doi:10.1007/s10531-008-9409-1. [CrossRef]
27. Zenáhlíková, J.; Červenka, J.; Čížková, P.; Bečka, P.; Starý, M.; Marek, P.; Krenova, Z.; Svoboda, M. The Biomonitoring project-monitoring of forest ecosystems in non-intervention areas of the Šumava National Park. *Silva Gabreta* **2015**, *21*, 95–104. doi:10.1111/1365- 2664.12504. [CrossRef]
28. Zielewska-Büttner, K.; Heurich, M.; Müller, J.; Braunisch, V. Remotely Sensed Single Tree Data Enable the Determination of Habitat Thresholds for the Three-Toed Woodpecker (Picoides tridactylus). *Remote. Sens.* **2018**, *10*, 1972. doi:10.3390/rs10121972. [CrossRef]
29. Cailleret, M.; Heurich, M.; Bugmann, H. Reduction in browsing intensity may not compensate climate change effects on tree species composition in the Bavarian Forest National Park. *For. Ecol. Manag.* **2014**, *328*, 179–192. doi:10.1016/j.foreco.2014.05.030. [CrossRef]
30. Hexagon. ISAT—ImageStation Automatic Triangulation. Available online: https://support.hexagonsafetyinfrastructure.com/infocenter/index?page=product&facRef=1_3PDRJ2 (accessed on 12 December 2019).
31. Trimble Inpho. DTMaster Inpho Software. Available online: http://trl.trimble.com/docushare/dsweb/Get/Document-696442/022516-017C_Inpho_MATCH-AT_TS_USL_0516_LR.pdf (accessed on 12 December 2019).
32. Trimble Inpho. Release Notes for Inpho 10. Available online: https://trl.trimble.com/docushare/dsweb/Get/Document-932009/RelNotes-Inpho-v.10-English.pdf (accessed on 12 December 2019).

33. Bässler, C.; Förster, B.; Moning, C.; Müller, J. The BIOKLIM-Project: biodiversity research between climate change and wilding in a temperate montane forest–the conceptual framework. *Waldökologie Landschaftsforschung und Naturschutz* **2008**, *7*, 21–33. Available online: http://publikationen.ub.uni-frankfurt.de/files/13682/waldoekologie_online_heft_7_2.pdf (accessed on 16 January 2020).
34. Heurich, M.; Fischer, F.; Knörzer, O.; Krzystek, P. Assessment of Digital Terrain Models (DTM) from data gathered with airborne laser scanning in temperate European beech (Fagus sylvatica) and Norway spruce (Picea abies). *Photogramm. Fernerkund. Geoinf.* **2008**, *6/2008*, 473–488.
35. Meyer, L.H.; Heurich, M.; Beudert, B.; Premier, J.; Pflugmacher, D. Comparison of Landsat-8 and Sentinel-2 Data for Estimation of Leaf Area Index in Temperate Forests. *Remote. Sens.* **2019**, *11*, 1160. doi:10.3390/rs11101160. [CrossRef]
36. Kramer, H.; Akça, A. *Leitfaden zur Waldmeßlehre*, 3rd ed.; Sauerländer: Frankfurt am Main, Germany, 1995; p. 266.
37. Heurich, M. Evaluierung und Entwicklung von Methoden zur Automatisierten Erfassung von Waldstrukturen aus Daten Flugzeuggetragener Fernerkundungssensoren. Ph.D. Thesis, Technische Universität München, München, Germany, 2006. Available online: https://mediatum.ub.tum.de/doc/603731/603731.pdf (accessed on 16 January 2020).
38. Breiman, L. Random Forests. *Mach. Learn.* **2001**, *45*, 5–32. doi:10.1023/A:1010933404324. [CrossRef]
39. Yao, W.; Krzystek, P.; Heurich, M. Enhanced detection of 3D individual trees in forested areas using airborne full-waveform LiDAR data by combining normalized cuts with spatial density clustering. *ISPRS Ann. Photogramm. Remote. Sens. Spat. Inf. Sci.* **2013**, *1*, 349–354. doi:10.5194/isprsannals-II-5-W2-349-2013. [CrossRef]
40. Li, W.; Guo, Q.; Jakubowski, M.; Kelly, M. A New Method for Segmenting Individual Trees from the Lidar Point Cloud. *Photogramm. Eng. Remote. Sens.* **2012**, *78*, 75–84. doi:10.14358/PERS.78.1.75. [CrossRef]
41. Holmgren, J.; Lindberg, E. Tree crown segmentation based on a tree crown density model derived from Airborne Laser Scanning. *Remote. Sens. Lett.* **2019**, *10*, 1143–1152. doi:10.1080/2150704X.2019.1658237. [CrossRef]
42. Silva, C.A.; Hudak, A.T.; Vierling, L.A.; Loudermilk, E.L.; O'Brien, J.J.; Hiers, J.K.; Jack, S.B.; Gonzalez-Benecke, C.; Lee, H.; Falkowski, M.J.; Khosravipour, A. Imputation of Individual Longleaf Pine (Pinus palustris Mill.) Tree Attributes from Field and LiDAR Data. *Can. J. Remote. Sens.* **2016**, *42*, 554–573. doi:10.1080/07038992.2016.1196582. [CrossRef]
43. PRIMAVISION Technologies GbR. 3D Tree Segmentation from Point Clouds (LiDAR, DSM) for Forest Inventory. Available online: http://primavision-tec.de/products/prod_tree-finder (accessed on 12 December 2019).
44. Corona, P.; Chirici, G.; McRoberts, R.E.; Winter, S.; Barbati, A. Contribution of large-scale forest inventories to biodiversity assessment and monitoring. *For. Ecol. Manag.* **2011**, *262*, 2061–2069. doi:10.1016/j.foreco.2011.08.044. [CrossRef]

remote sensing

MDPI

Article

The Effect of Topographic Correction on Forest Tree Species Classification Accuracy

Chao Dong [1,2], Gengxing Zhao [2,*], Yan Meng [3], Baihong Li [4] and Bo Peng [1]

[1] College of Information Science and Engineering, Shandong Agricultural University, Tai'an 271018, China; dongchao@sdau.edu.cn (C.D.); 2018110265@sdau.edu.cn (B.P.)

[2] College of Resources and Environment, Shandong Agricultural University, Tai'an 271018, China

[3] Forestry College, Shandong Agricultural University, Tai'an 271018, China; 2017010113@sdau.edu.cn

[4] Natural Resources and Planning Bureau, Tai'an 271018, China; bhlkey@163.com

* Correspondence: zhaogx@sdau.edu.cn; Tel.: +86-8243939

Received: 26 January 2020; Accepted: 25 February 2020; Published: 1 March 2020

Abstract: Topographic correction can reduce the influences of topographic factors and improve the accuracy of forest tree species classification when using remote-sensing data to investigate forest resources. In this study, the Mount Taishan forest farm is the research area. Based on Landsat 8 OLI data and field survey subcompartment data, four topographic correction models (cosine model, C model, solar-canopy-sensor (SCS)+C model and empirical rotation model) were used on the Google Earth Engine (GEE) platform to carry out algorithmic data correction. Then, the tree species in the study area were classified by the random forest method. Combined with the tree species classification process, the topographic correction effects were analyzed, and the effects, advantages and disadvantages of each correction model were evaluated. The results showed that the SCS+C model and empirical rotation model were the best models in terms of visual effect, reducing the band standard deviation and adjusting the reflectance distribution. When we used the SCS+C model to correct the remote-sensing image, the total accuracy increased by 4% when using the full-coverage training areas to classify tree species and by nearly 13% when using the shadowless training area. In the illumination condition interval of 0.4–0.6, the inconsistency rate decreased significantly; however, the inconsistency rate increased with increasing illumination condition values. Topographic correction can enhance reflectance information in shaded areas and can significantly improve the image quality. Topographic correction can be used as a pretreatment method for forest species classification when the study area's dominant tree species are in a low light intensity area.

Keywords: illumination correction; GEE; forest species; Mount Taishan

1. Introduction

Forests are among the important components of the global terrestrial ecosystem. Whether from the perspective of forest ecology or from a service perspective in which the function of a forest is to provide wood and other products, it is necessary to obtain information on forest resources over time. Many countries around the world, such as Austria, Sweden and the United States, have carried out regular forest resource surveys to provide data for sustainable forest management [1]. The content of forest resource investigations includes various forestland area investigations, forest tree accumulation and many other aspects, among which the identification and distribution of forest types are the basis for and an important direction of forest resource monitoring.

Remote sensing is particularly useful for forest species classification, as it provides information on large areas at a high level of detail [2]. There are many sensors that provide the image data required for tree species classification, and these sensors typically feature either high spatial resolutions or multiple spectral bands. High-resolution satellite images and synthetic aperture radar (SAR) data have recently been used to classify forest tree species [3,4]. The results show that the classification works well but is only suitable for small areas due to its high cost. Although data from satellites such as Landsat have lower spatial resolutions, they capture more spectral information. Spectral information from satellite imagery can also be used to effectively identify forest tree species [5–8]. Such information is widely used in the investigation of forest species resources because of its effectiveness in species classification and low-cost advantages. Lu et al. [9] improved the accuracy of forest classification results based on Thematic Mapper data and adjusted entropy. Schuck et al. [10] focuses on the approach of combining the information from both remote sensing and forest inventory statistics in order to produce a forest proportion map. Immitzer et al. [11] mapped seven different deciduous and coniferous tree species in Germany based on Sentinel-2 data. Grabska et al. [12] evaluated the utility of the Sentinel-2 time series for mapping tree species in the complex, mixed forests. Chiang et al. [13] suggested that integrating topographic information and optical satellite image classification can improve mapping accuracy for tree species. Remote-sensing technology reduces manpower and material resource consumption, improves the quantitative description of resource information and improves survey timeliness better than traditional field investigations. The key to such forest tree resource investigation work is the analysis of spectral information from remote-sensing images and the determination of the tree species category according to the different spectral characteristics of tree resources.

The influences of the atmosphere, electromagnetic waves and topography are inevitable when remote-sensing satellites acquire information. Among these influences, the change in spectral reflectance caused by topography is one of the main problems affecting the quality of satellite data. The method of reducing these influences is referred to as topographic correction or terrain illumination correction. Therefore, various models have been developed for topographic correction of remote-sensing satellite data. Early models included cosine correction, Minnaert correction and C correction. After correction with each model, the accuracy of land cover classification can be improved [14]. C correction and Minnaert correction were excellent for land cover mapping, and the overall accuracy increased by 10% [15]. C correction introduced the semiempirical coefficient C to correct the overcorrection of the cosine model. C correction established a connection between the original image and the corrected image. This correction method also described the quantitative relationship between the spectral value and incidence angle. Therefore, the correction effectiveness was improved [16,17]. Previous correction models were based on a sun-surface-sensor scheme and did not consider the relationship between vegetation and the ground after topographic correction. The improved solar-canopy-sensor (SCS) correction makes the change in illumination direction more realistic during the process of reflectance change from sloping to horizontal surfaces. SCS correction can also introduce C correction and determine C parameters for each band. This correction method can markedly improve the effectiveness of topographic correction [18,19]. Compared with cosine correction, C correction and Minnaert correction, another method that is based on empirical statistics, have higher accuracies in forest classification [20], and combined with the preclassification/layering method, these methods have good robustness for topographic correction [21]. Many models are effective for the topographic correction of remote-sensing images, and all of these models can improve the accuracy of land cover classification. However, the topographic correction process is complex, and all of these models can only explain this process to a certain extent. In forest tree species classification, it is difficult to classify different tree species because of their high spectral similarity. Furthermore, machine learning methods, such as the support vector machine (SVM) and random forest (RF) techniques, have been used to improve the classification accuracy [22]. However, these methods have some problems, such as overfitting, and the models are not easy to interpret. In terms of classifying tree species, the relationship between the impact of topography and the classification results of remote-sensing images has not been studied. Consequently, the impact of topographic correction on forest resource surveys in complex terrain, especially in forest tree species resource surveys, is still unclear. Therefore, it is necessary to further research the relationship between existing typical topographic correction models and forest tree species classification.

Given the above discussion, four typical topographic correction models were used to correct Landsat satellite images, which were then used for tree species classification with the random forest method. The goal in this study is to determine the quantitative effects of topographic correction on remote-sensing bands and associated indexes by analyzing the relationship between correction effects and tree species. This study explores effective methods for accurate tree species classification in mountainous areas.

2. Materials and Methods

2.1. Study Area

The study area was Mount Taishan, which is located in Tai'an City, Shandong Province, China. The study area was delimited by the boundaries of the Mount Taishan forest farm (Figure 1). Mount Taishan is in the central part of Shandong Province. This area represents one of China's five great mountainous regions and has an important ecological and cultural status [23]. This region features a warm-temperate semihumid monsoon climate. The climate changes with elevation. The terrain tends to high in the north and west and low in the south and east, and the topography is undulatory, with a height difference between the plain and piedmont of over 1300 m. The forest coverage rate of Mount Taishan is more than 80%, which is mainly composed of plantation and secondary forest, and the proportion of pure forest is 55%. This area features coniferous forest and coniferous-broad-leaved mixed forest, which is typical in warm-temperate mountain areas. The Taishan forest farm, which covers 11,730 ha, was selected as the research area. The main tree species in the study area are pine, Chinese arborvitae (*Platycladus orientalis*), oak (*Quercus*) and black locust (*Robinia pseudoacacia*). Pine can be divided into *Pinus tabulaeformis*, *Pinus thunbergii* and *Pinus densiflora*, and oak can be divided into *Quercus acutissima* and *Quercus variabilis*. *Pinus tabulaeformis* accounts for the largest proportion and is the dominant vegetation species on Mount Taishan. Secondary forests of *Pinus tabulaeformis* are mainly distributed in the upper part of Mount Taishan. Secondary forests of oak and Chinese arborvitae are mainly distributed in the middle and lower parts of Mount Taishan [24,25].

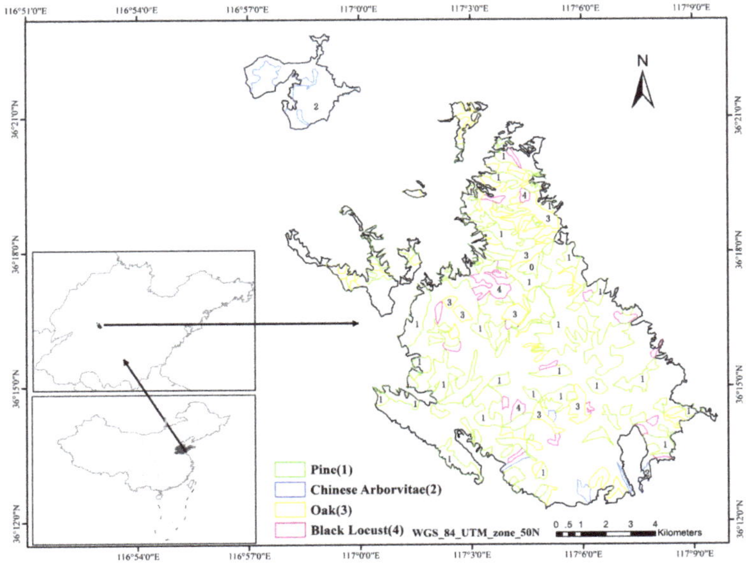

Figure 1. Location map of the study area and tree species distribution of pure forest area (produced from subcompartment data).

2.2. Data

2.2.1. Satellite Data

Landsat data is suitable for long-term and large-scale resource exploration. This study used surface reflectance (SR) data from Landsat 8 OLI (Landsat 8 OLI/TIRS surface reflectance) [26] Considering the effects of weather conditions and forest phenology, the data were consistent with the ground survey data. The dates of the images were March 26, 2016, September 2, 2016, October 4, 2016 and November 5, 2016. We chose Landsat 8 bands 2 through 7 for our analysis but focused our efforts on band 4 (red) and band 5 (near-infrared, NIR), as these bands are shown to respond most strongly to surface vegetation conditions (Table 1). The research was based on the implementation of the Google Earth Engine (GEE) platform [27], so all data were accessed on GEE.

Table 1. Spectral bands of Landsat 8 used for forest tree species classification.

Name	Units Scale	Wavelength (μm)	Description
Band2	0.0001	0.452–0.512	blue
Band3	0.0001	0.533–0.590	green
Band4	0.0001	0.636–0.673	red
Band5	0.0001	0.851–0.879	near infrared (NIR)
Band6	0.0001	1.566–1.651	shortwave infrared 1 (SWIR1)
Band7	0.0001	2.107–2.294	shortwave infrared 2 (SWIR2)

2.2.2. Digital Elevation Model

Digital elevation model (DEM) data are important for topographic correction. The Shuttle Radar Topography Mission (SRTM) [28] digital elevation data are an international research effort designed to obtain DEM data on a near-global scale. The SRTM V3 product (SRTM Plus) is provided by NASA JPL at a resolution of 1 arc-second (approximately 30 m). DEM data are the basic data for calculating illumination conditions (ICs), and their accuracy directly affects the topographic correction effectiveness. The slope data and aspect data in the correction model were accessed by calculating the DEM data.

2.2.3. Subcompartment Data

In China, the subcompartment is the basic unit of forest resource protection planning, investigation, statistics and management [29]. The division of subcompartments should be based on obvious topography and object boundaries as much as possible and should take into account the needs for resource investigation and management. Based on the subcompartment map, a field investigation was carried out in this study. The investigation included tree species and other information. In the geographic information system (GIS), the study area is divided into many polygons representing individual, homogeneous forest stands. The investigation results based on the polygon were imported in the form of fields. The subcompartment map used in this study was the result of the 2016 survey, which contains 1713 features and 68 fields. It was provided by the Mountain Tai Management Committee. Field information included tree species, dominant tree species, composition of tree species and other detailed survey data. The tree species field describes the tree species in the subcompartment feature with a tree species code. We can also see whether the subcompartment feature is a single tree species according to the number of tree species codes. There are many coniferous and broad-leaved species in the Mount Taishan region, but pine, oak, Chinese arborvitae and black locust account for the majority of the area. In this study, tree species were divided into these four categories. We created training samples and verification samples based on subcompartment data.

2.3. Methods

2.3.1. Topographic Correction Model

Four commonly used topographic correction models (cosine model [30], C model [31], SCS+C model [32] and empirical rotation model [33]) were used in this study.

Illumination conditions (IC) are the basis of all reflectivity compensation correction models. The IC has a proportional relationship that is determined by the cosine of the angle between the solar zenith and the normal line of the slope, and the model is defined as follows:

$$\cos\gamma_i = \cos\theta_z \cos\theta_s + \sin\theta_z \sin\theta_s \cos(\varphi_z - \varphi_s), \tag{1}$$

where θ_z is the solar zenith angle, θ_s is the topographic slope angle, φ_z is the solar azimuth angle and φ_s is the slope direction of the topographic surface.

(1) Cosine model:

$$\rho_H = \rho_I(\cos\theta_z / \cos\gamma_i), \tag{2}$$

where ρ_H is the reflectance of the horizontal surface or corrected reflectance, and ρ_I is the reflectance of the slope surface, or observed reflectance. The model considers that the correction has nothing to do with the wavelength, underestimating the reflectance of the topographic illumination surface and overestimating the reflectance of the topographic backlight surface, resulting in abnormal values [31].

(2) C model:

$$\rho_{H(\lambda)} = \rho_{I(\lambda)} \frac{\cos\theta_z + C_\lambda}{\cos\gamma_i + C_\lambda} \tag{3}$$

where C_λ is the correction coefficient for the λ band, C = b/a, and a and b are the slopes and intercepts of linear regression calculated between the IC data and this particular band, respectively.

$$\rho_{I(\lambda)} = \mathrm{a}(\lambda)\cos\gamma_i + b(\lambda), \tag{4}$$

The C model avoids wavelength independence of the cosine model to a certain extent and avoids overcorrection in lower $\cos\gamma_i$ value regions.

(3) The SCS+C model:

This is based on the relationship among the sun, canopy and sensor; because tree growth is geotropic, the topography cannot affect the geometric relationship between the sun and the tree. The topography affects the positional relationship between the tree and the surface. The SCS+C model is a model based on the canopy, which enables a change in illumination direction that is more consistent with practice during the process of light canopy correction from sloping to horizontal surfaces. This model is defined as follows:

$$\rho_{H(\lambda)} = \rho_{I(\lambda)} \frac{\cos\theta_z \cos\theta_S + C_\lambda}{\cos\gamma_i + C_\lambda}, \tag{5}$$

(4) Empirical rotation model:

The model is defined as follows:

$$\rho_{H(\lambda)} = \rho_{I(\lambda)} - \mathrm{a}(\lambda) * (\cos\gamma_i - \cos\theta_z), \tag{6}$$

where a is the interception of the band linear regression calculated in Formula (5). This model eliminates the linear dependence of reflectivity on $\cos\gamma_i$. The results show that this model can accurately obtain top-of-atmosphere and top-of-canopy reflectivities from Landsat data.

These four models can be used on the GEE platform. Topographic information was obtained by SRTM, and satellite metadata were used to obtain information such as the angle of the satellite on the solar zenith. The C_λ parameters used in the model were calculated for each band in Landsat data, which made the models more applicable [34].

2.3.2. Evaluation of the Topographic Correction

The evaluation involved the use of standard deviation (SD) and histograms to evaluate the effectiveness of the topographic correction [35]. First, the SD of all bands before and after correction were counted, and the contrast histogram of the NIR band before and after correction was created. Based on all data from the whole image, the effectiveness of the correction model was evaluated. Then, we chose a reservoir in the study area, which was very obvious in the remote-sensing image, and drew the region of interest to evaluate the stability of the correction model by calculating the reflectance changes before and after correction. Finally, the SDs of the NIR and red bands before and after correction and the correlation coefficients of the NIR band, red band and IC before and after correction were calculated according to each tree species by using the tree species survey data.

2.3.3. Method of Tree Species Classification

Tree species classification includes three parts: topographic correction of multitemporal remote-sensing images, training data production and random forest algorithm classification (Figure 2). First, according to the forest phenology and cloud amount factors, we chose four representative Landsat images. Multiseasonal remote-sensing data help to improve classification accuracy [36]. We use the four methods in the dotted box of Figure 2 to correct the remote-sensing image shadow and create the correction dataset. Then, the training data were created as random points in the pure forest data, which were extracted from the subcompartment data by forest structure fields. The field describes the composition of forest species in the subcompartment data. These data are full-coverage training data. To study the impact of training data distribution on classification, we created shadowless training data that used the shadow region extracted by the IC data to mask the full-coverage training data. We used these training data and a random forest algorithm to train classification models. The random forest algorithm [37] is a machine-learning algorithm that contains multiple decision trees [3,38]. Model training is the process of establishing the relationship between remote-sensing images and tree species labels. Since machine-learning algorithms have overfitting problems, shadowless training data avoid the algorithm, covering up the effectiveness of topographic correction, because the correct classification is based on spectral consistency, not by providing tree species information in shaded areas to the classifier. Finally, the training dataset was used to train the random forest classifier. We use a training model to classify the remote-sensing data.

2.3.4. Evaluation of the Effectiveness of Different Topographic Correction Models on Tree Species Classification

The original data and tree species classification results after correction were visually evaluated, and the accuracy was evaluated based on the validation data. Based on the area size of the subcompartment data, 240 sampling points are generated by using the function of randomly generating points in GIS. A total of 240 sampling points was randomly generated in the study area. Based on subcompartment data and high-resolution Google Earth images, tree species labels were added to the sampling points. Some sampling points were moved to the shaded area to evaluate the impact of the shaded area on tree species classification. To study the influence of topography in topographic correction on tree species classification, we used 1 to subtract the area ratio of correctly classified tree species in the pure forest area to calculate the inconsistency rate, zonal statistics of the inconsistency rate and the IC data, and we drew the distribution histogram of the inconsistency rate and evaluated the relationship between topographic correction and tree species.

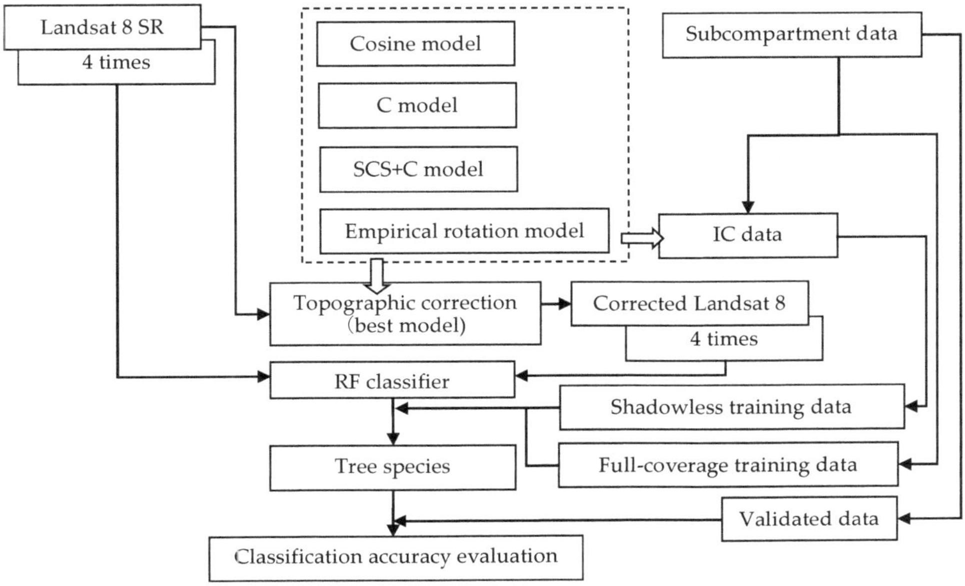

Figure 2. Methodological framework for the classification accuracy of topographic correction on tree species. IC: illumination conditions and RF: random forest.

3. Results

3.1. *Effectiveness of the Topographic Correction Models*

3.1.1. Visual Evaluation of the Topographic Correction

A qualitative evaluation was conducted by visual comparison of the correction effectiveness of the topographic correction model in the study area. The visual comparison is mainly based on three aspects: (1) the effectiveness of removing shadows, (2) whether the color is consistent before and after and whether there is an overcorrection problem and (3) texture features of the images. Through color synthesis, we can see the effect before and after correction (Figure 3). The original reflectance data have three-dimensional characteristics because of topographic fluctuations. IC data can reflect the change in brightness caused by the topographic influence. The comparison of the two pictures demonstrated that the three-dimensional characteristics of the original data were consistent with the change in brightness and shade of the ICs.

All four models can remove most of the topographic shadows and restore the reflectivity information of the shaded areas, which was consistent with the visual impression of adjacent nonshaded areas. The cosine model had many obvious bright spots in the ridge area, which was quite different from the original data and had the problem of overcorrection. There was no obvious overcorrection in the other models. However, after being stretched with SD enhancement, the color depth of the cosine model was found to have decreased, and the color depth of the C model was found to have increased, while the results of the SCS+C model and empirical rotation model were more consistent with and closer to the original data. All images can retain the texture features of objects well, but the corrected images were flatter, and the stereoscopic effect disappeared. Ridges, valleys and other topographic features were difficult to recognize. In summary, from the visual effect, the SCS+C model and empirical rotation model were better than the other models.

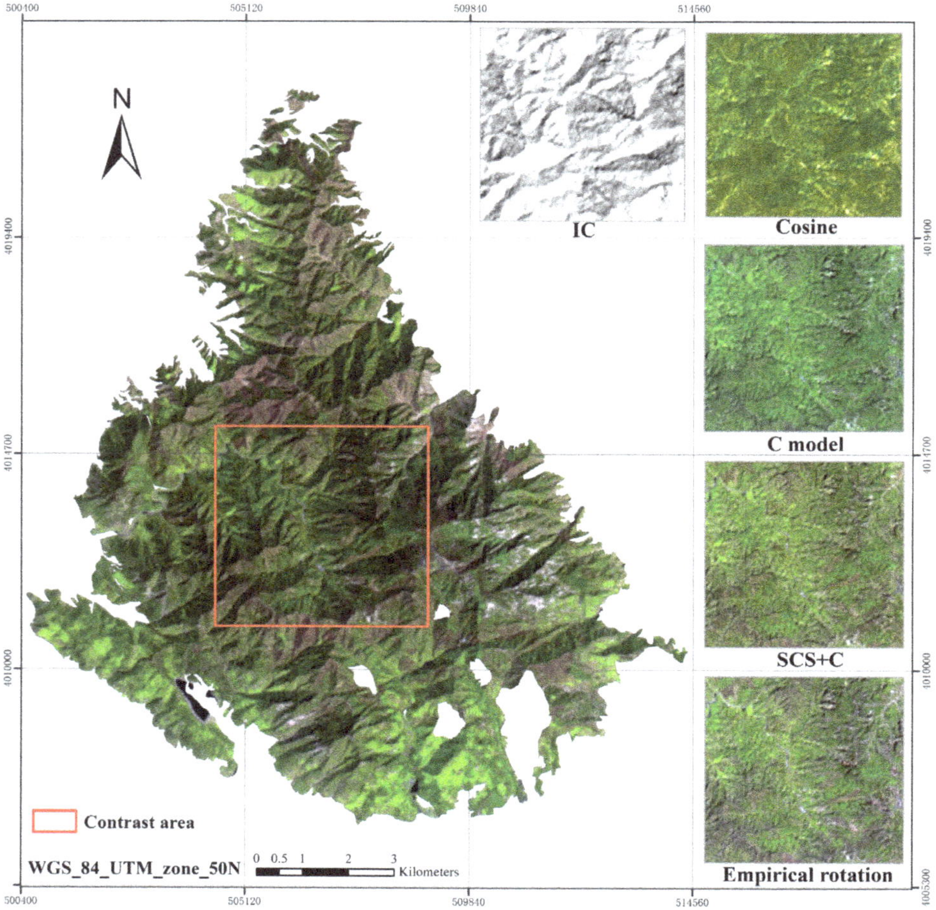

Figure 3. Comparison of Landsat 8 surface reflectances before and after topographic correction. The contrast area represents the region of Illumination conditions' data and topographic correction data (R: shortwave infrared 1 (SWIR1), G: near infrared (NIR) and B: red; data date: 26 March 2016).

3.1.2. Analysis of Band Information before and after Topographic Correction

The SD reflects the degree of data dispersion. Generally, the spectral value of the same ground feature should be the same, but due to the influence of topography, there will be great variations. We calculated the SD of different bands of the image before and after correction (Table 2). SWIR1 band reflectance data had the largest SD. After correction, the cosine model and empirical rotation model had the best correction effectiveness, and the SDs of the other two models were also reduced. The NIR and red bands are important bands for vegetation monitoring. The smallest SD after the NIR band correction was observed in the empirical rotation model, followed by the SCS+C model and C model, and the worst SD was observed in the cosine model. The empirical rotation model had the smallest SD after the red band correction, and the effectiveness of the four models was very close. All correction models can effectively reduce the SD of each band, and the empirical rotation model (followed by the SCS+C model) was the best in all bands.

Table 2. Standard deviation statistics of spectral bands before and after topographic correction. SR refers to "surface reflectance" and represents precorrected conditions. The model represents the conditions after topographic corrected by the specify model.

	SR	Cosine Model	C Model	SCS+C Model	Empirical Rotation Model
Blue	0.013	0.008	0.009	0.009	0.008
Green	0.016	0.012	0.011	0.011	0.009
NIR	0.042	0.037	0.030	0.029	0.027
Red	0.022	0.014	0.015	0.015	0.013
SWIR1	0.055	0.035	0.040	0.039	0.034
SWIR2	0.044	0.030	0.033	0.032	0.028

The NIR band responds well to the changes in vegetation. We obtained histogram statistics before and after the NIR band correction (Figure 4). The graph shows that the NIR band reflectance histogram of the remote-sensing image did not show a normal distribution due to topographic factors. It formed two peaks: a low value (0.12) and a high value (0.20). This showed that due to the influence of topographic slope and slope direction, some pixels received insufficient illumination, while the other parts showed a saturation trend. After the treatment of the four correction models, the NIR band histogram showed an approximate normal distribution, which was consistent with the random features of natural phenomena, showing the reflection characteristics of objects in the real state. In the four models, the cosine model greatly compressed the original data, gathered many values in the 0.15–0.2 reflectivity region and the left and right sides were not completely symmetrical. There were also data outside the original range, which was considered to be the result of overcompressing. Compared with the C model, the SCS+C model had the same area with values higher than 0.2. In the range of 0.12–0.17, the SCS+C model was smoother, which was related to the canopy correction rather than the topographic correction of the SCS+C model. The empirical rotation model was smooth on both sides of the peak. Compared with the former two models, the reflectance distribution was concentrated at the peak. All models can correct the NIR band reflectivity to approximate a normal distribution. In terms of comprehensive distribution morphology, continuity and deformation, the SCS+C model and empirical rotation model had better correction effectiveness.

3.1.3. Effectiveness of Topographic Correction on Tree Species Reflectivity

Based on forest species composition information in pure forest data, we counted the SD changes for pine, Chinese arborvitae, oak and black locust in the NIR and red bands before and after correction, respectively (Figure 5). When we examined a single tree species, the SD of the cosine model increased, and the pine species stretched abnormally in the NIR band. The empirical rotation model reduced the SDs of all tree species. When we examined pines, the SD of the blue, green and red bands of the C model was higher than the original SD of reflectance. The SD of the SCS+C model-corrected image was higher than the original SD in the blue band for pines and the red band for oak. Statistical results showed that different correction models had different effects on different tree species, which was related to the distribution of tree species and the inconsistencies in the reflections of different bands. The empirical rotation model was very stable, and the SD of all bands of all tree species was less than the original SD of reflectance. The second-best results were obtained by the SCS+C model.

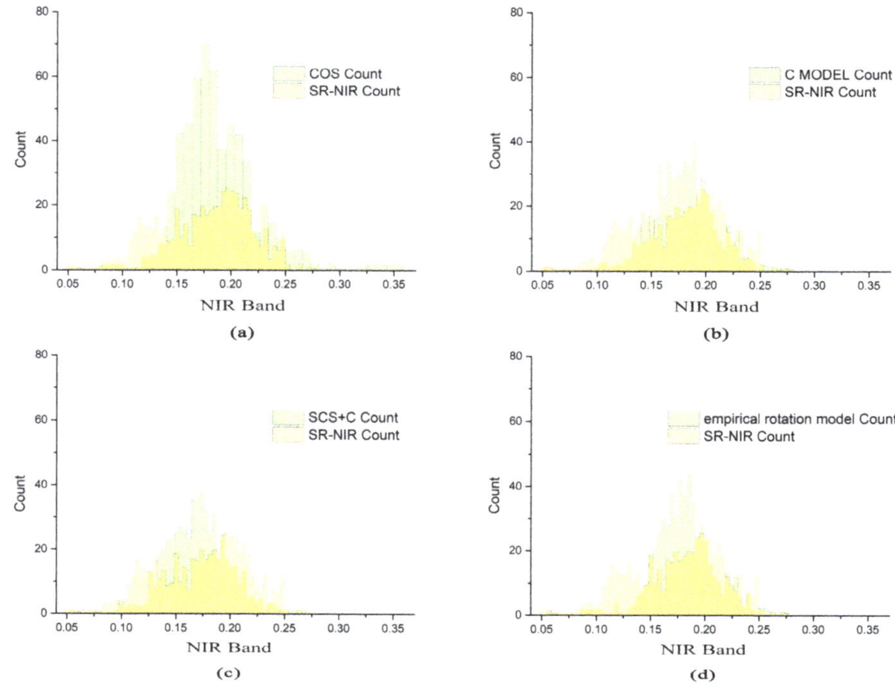

Figure 4. Frequency histograms before and after topographic correction in the NIR band by four models. Surface reflectance (SR)-NIR represents the precorrected conditions of the NIR band. The topographic correction model by using (**a**) the cosine model, (**b**) C model, (**c**) SCS+C model and (**d**) empirical rotation model.

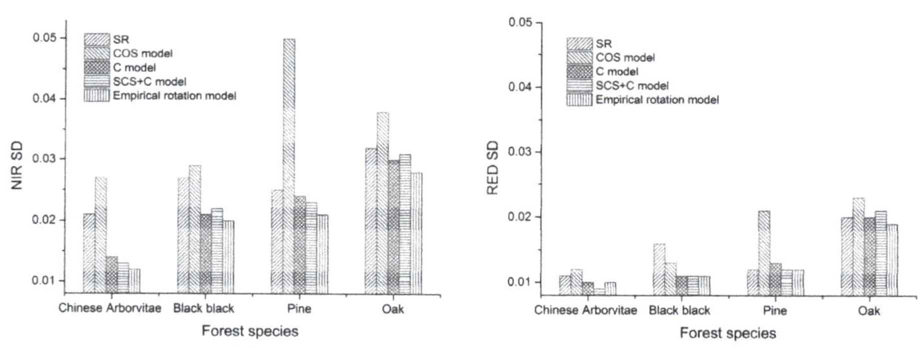

Figure 5. Comparison of tree species NIR band (**a**) and red band (**b**) standard deviation before and after topographic correction by the four models. SR represents the precorrected conditions of the spectral band. The model represents the conditions after topographic corrected by the specific model.

3.1.4. Stability Analysis of Topographic Correction

Changes in the water body reflectance can be used to evaluate the validity of a topographic correction model by statistical comparisons of this reflectance before and after correction has been applied to the study area (Figure 6). Water bodies are horizontal, so the topographic correction model should not have a large impact on them. The change in reflectance of all models was less than 10%, and the change in the SCS+C model was the smallest, at less than 2%. The empirical rotation model is an experience-based rotation model. Compared with other models, the reflectivity of the water surface changed greatly, especially in the NIR and short-wavelength infrared (SWIR) bands. The changes induced by the other two correction models were between 3% and 5%.

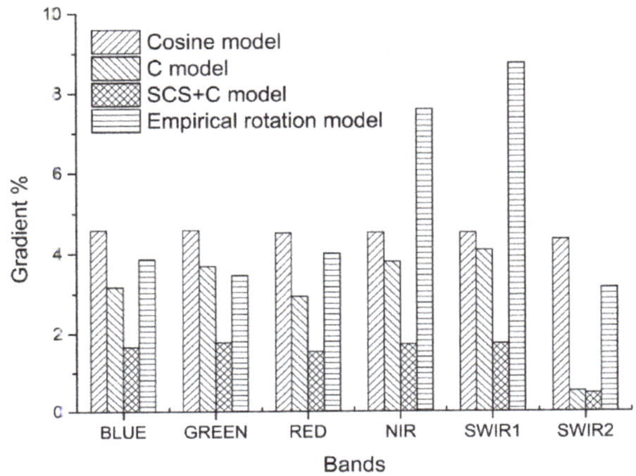

Figure 6. Percent change in band reflectance over water bodies after topographic correction by each of the four models.

In summary, compared with the other two models, the cosine model and C model did not perform as well in terms of the correction effectiveness and stability. Therefore, the SCS+C model and empirical rotation model were the main models used in the further study of tree species classification.

3.1.5. Analysis of Correlations among the IC, NIR Band and Red Band of Different Tree Species before and after Correction

The tree species distributed in the shaded area will affect the remote-sensing image, resulting in a correlation between the remote-sensing spectral value and topography. We used the tree species information of subcompartment data to calculate the correlation between the remote-sensing spectral value and the IC value in the region (Table 3). Before correction, both the NIR and red bands of the image had a high correlation with the IC. The highest correlations were observed in the NIR bands of pines and Chinese arborvitae, which were 0.58 and 0.69, respectively. The lowest correlations were found for the red bands of Chinese arborvitae and oak, which were 0.29 and 0.30, respectively. The correlation of the NIR band was higher than that of the red band. The correlation coefficients of pines and Chinese arborvitae after SCS+C correction were lower than those of the empirical rotation method, whereas those of oak exhibited the opposite pattern. For black locust, the SCS+C model was better than the empirical rotation model in the NIR band, and vice versa in the red band. The correlation reflects the degree of influence of the topography on the spectral reflectance of trees. Except for the red band of Chinese arborvitae, the correlation of the NIR band and red band of all tree species was significantly reduced by the models.

Table 3. Correlation coefficient of spectral band and illumination conditions (IC) before and after correction in the NIR band and red band. SR represents the coefficient before correction, SCS+C represents the coefficient after correction by SCS+C model, and empirical rotation represents the coefficient after correction by the empirical rotation model.

Tree Species	**NIR**			**Red**		
	SR	SCS+C	Empirical rotation	SR	SCS+C	Empirical rotation
Pine	0.58	0.79×10^{-3}	0.15×10^{-1}	0.43	0.28×10^{-3}	0.77×10^{-2}
Chinese arborvitae	0.69	0.13×10^{-1}	0.24×10^{-2}	0.29	0.22	0.22
Oak	0.41	0.94×10^{-2}	0.61×10^{-2}	0.30	0.33×10^{-2}	0.58×10^{-3}
Black locust	0.47	0.80×10^{-5}	0.85×10^{-2}	0.44	0.13×10^{-2}	0.60×10^{-3}

By constructing density plots between the IC dataset and the individual bands for pine, we can better understand the process of spectral value processing of the topographic correction method (Figure 7). As shown in the Figure 7, IC values of 0.3–1.0 were distributed over a large area. The original reflectance data were highly correlated with the IC values. After correction, the correlation coefficient between reflectance and IC was significantly reduced. As shown in the Figure 7, both the SCS+C model and empirical rotation model rotated the reflectance distribution of the original data from tilted to nearly horizontal, while the empirical rotation model rotated the distribution only on the basis of the original data, and its interior distribution was very close to that of the original data. The distribution of points in the scatter plot for the SCS+C model changed obviously to different degrees. In the NIR band, the density of red points with high densities was reduced after correction with the SCS+C model. The ed band exhibited clear changes: the empirical rotation model rotated the data as a whole, while the SCS+C model transformed one density center into two distinct density centers. Due to the relatively small distribution area of other tree species, the rotation changes were not obvious.

3.2. *Topographic Correction Effectiveness on Tree Species Classification*

3.2.1. Classification Results of Tree Species and Accuracy Evaluation

Whether the training data were covered with shaded areas had a great impact on the accuracy of the classification results (Table 4). The classification results of the full-coverage training data showed little difference in terms of accuracy before and after correction; the SCS+C model performed better, and the empirical rotation model showed slightly improved accuracy. The classification results of shadowless training data had great differences before and after topographic correction. The SCS+C model performed the best, and its accuracy was 13% higher than that before correction. The performance of the empirical rotation model also improved significantly. Due to the inconsistency in the reflectance of the original data in the shaded area, the accuracy was low. Therefore, when the survey data in the study area are comprehensive, the influence of topography can be avoided to some extent by including training sample points in shaded areas; when the amount of data in the study area is not adequate, the classification must rely on visual or partial data, and the topographic correction of the data should be included in the pretreatment step to improve the classification accuracy.

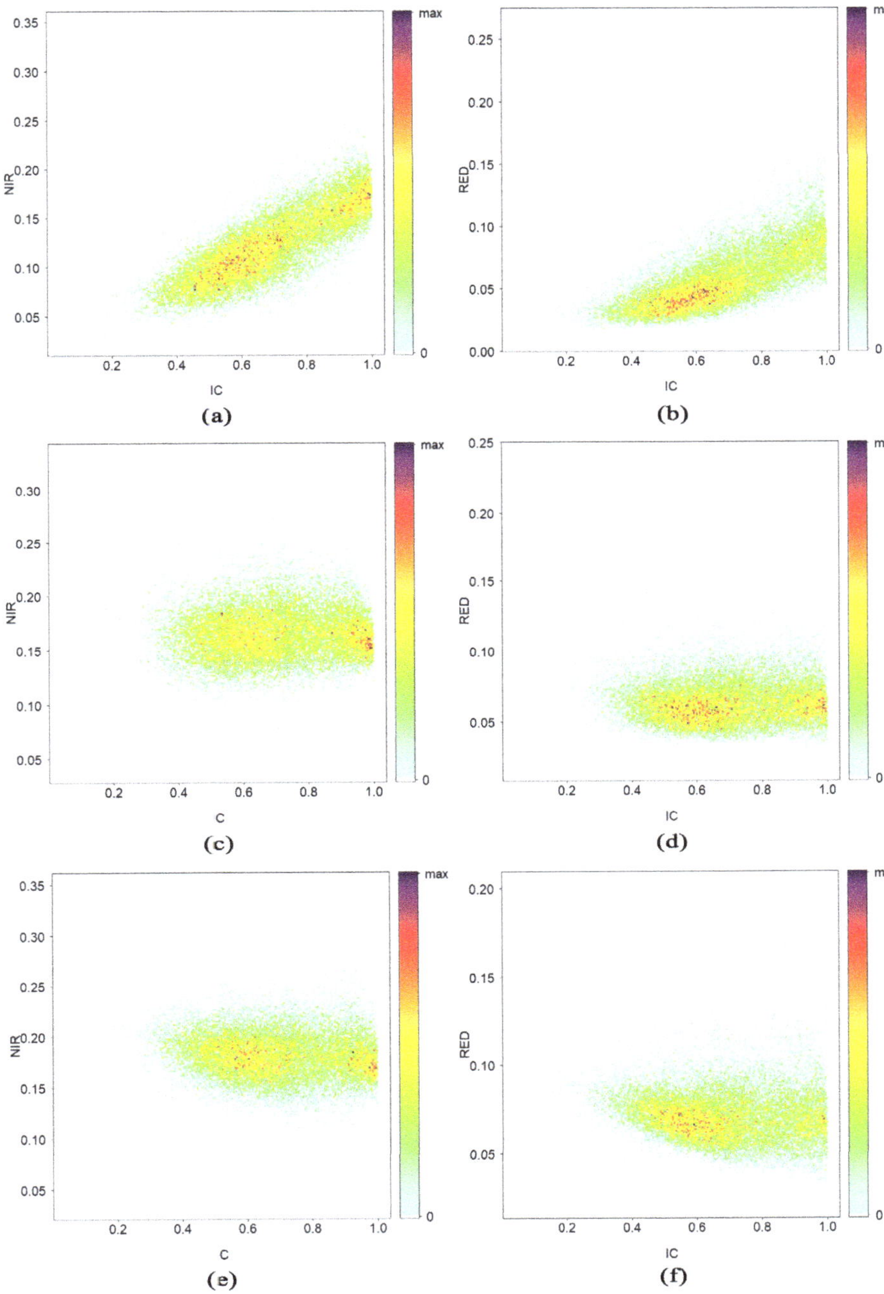

Figure 7. Scatter density map of the spectral band and IC of pine. (**a**) NIR band before correction (**b**) red band before correction, (**c**) NIR band corrected by SCS+C model, (**d**) red band corrected by SCS+C model, (**e**) NIR band corrected by empirical rotation model and (**f**) red band corrected by empirical rotation model.

Table 4. Classification accuracy with different training data. SR represents the data before correction, and SCS+C model and empirical rotation model represent the data after correction.

Shadowless Training Data	Overall Accuracy	Kappa	Full-Coverage Training Data	Overall Accuracy	Kappa
SR	0.65	0.52	SR	0.76	0.68
SCS+C model	0.74	0.67	SCS+C model	0.79	0.73
Empirical rotation model	0.72	0.64	Empirical rotation model	0.77	0.70

The random forest algorithm was used to classify the data before and after correction, and maps of the resulting forest species distribution were created. As shown in Figure 8, Chinese arborvitae is mainly distributed in the northwestern and southern parts of Mount Taishan. Oak species are mainly distributed in the northern part of Mount Taishan, while a small amount is distributed in the central and southern parts. Black locust covers the least area, and the distribution is fragmentary. Pines are widely distributed throughout the whole area. The Landsat 8 data corrected by the four models can effectively distinguish among these four tree species, but the classification results of the four models differed in terms of specific distributions due to the different corrections of each model (Table 5).

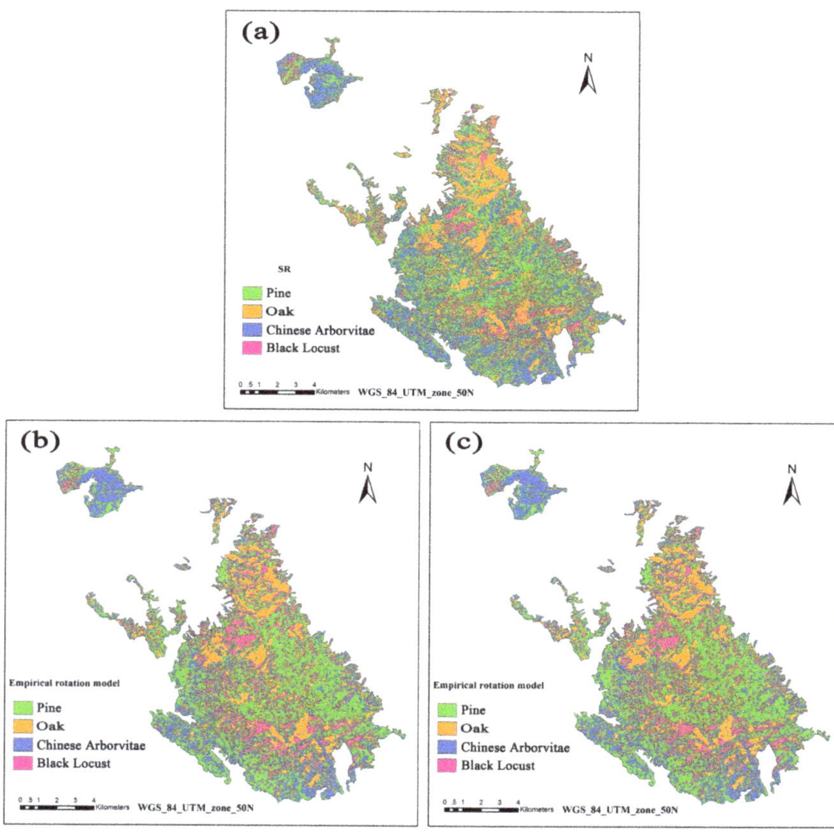

Figure 8. Classification map of the tree species with full-coverage training data. (**a**) Classification result with the precorrected data, (**b**) classification result with topographic correction data of the SCS+C model and (**c**) classification result with topographic correction data of the empirical rotation model.

Table 5. Area statistics derived from classification of the tree species. SR represents the data of the precorrected data classification results. The SCS+C model and empirical rotation model represent the data after topographic correction. The gradient represents the change rate between the before and after correction data.

Tree Species	SR	SCS+C		Empirical Rotation	
	Area (m^2)	Area (m^2)	Gradient (%)	Area (m^2)	Gradient (%)
Pines	57,028	58,647	2.84	58,981	3.424
Chinese arborvitae	27,515	26,101	−5.14	26,779	−2.67
Oak	21,438	22,631	5.56	21,801	1.69
Black locust	18,987	18,376	−3.22	20,535	8.15

3 2.2. Topographic Correction Effectiveness on Tree Species Classification in Shaded Areas

The tree species in the area with low ICs were entirely composed of Chinese arborvitae (Figure 9). In the original SR data classification results, most of these trees were identified as pine species and only the edges were identified as Chinese arborvitae. However, after correction, both the SCS+C model and empirical rotation model correctly recognized most of the species as Chinese arborvitae by recovering their reflectivities. Even so, there were still some errors in the two correction models. On the one hand, the spectral characteristics of these two tree species are similar, which can easily lead to errors. On the other hand, although the correction maintains the spectral information in shaded areas, the information recovery is not complete due to DEM error and the influence of the distribution of the vegetation itself. Complete restoration is only an ideal situation.

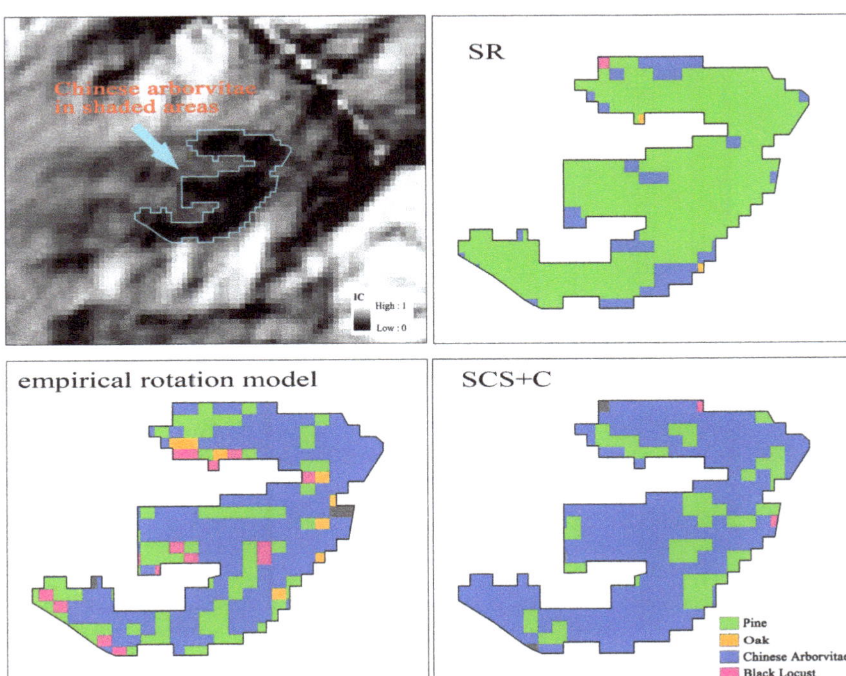

Figure 9. Contrast map of Chinese arborvitae classification in shaded areas by using shadowless training data. SR represents the pre-corrected conditions of the classification result. The model represents the classification result after topographic corrected by the specific model.

The classification results for oak were slightly different from those for Chinese arborvitae. The original data can be better distinguished by using training sample points in no-shadow areas. Even in areas with low illumination values, the classification accuracy of oak was high (Figure 10). The reason is that the spectral characteristics of oak are quite different from those of pine and Chinese arborvitae and are therefore easy to distinguish. After the correction, the results of the SCS+C model were better than those of the original image. After correction by the empirical rotation model, oak was easily distinguished, and a few areas were recognized as other tree species.

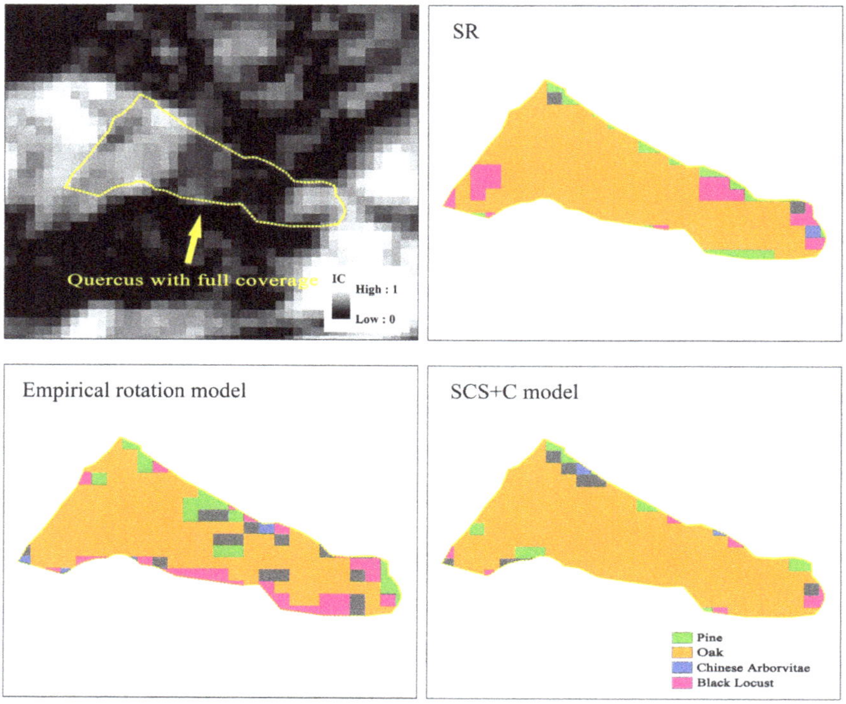

Figure 10. Contrast map of oak classification in full-coverage areas by using shadowless training data. SR represents the precorrected conditions of the classification result. The model represents the classification result after topographic corrected by the specific model.

3.2.3. Effect of Topographic on Classification Inconsistency Rate

The black symbols in Figure 11 are the normalized values of the area of tree species in pure forests in different intervals of the IC. The four types of tree species had different topographic distributions based on observation of the standard IC distribution. Pine had two peaks at IC values of 0.6 and 1.0. Black locust had similar peaks at IC values of 0.6 and 0.9. Chinese arborvitae and oak were mostly distributed in the IC range of 0.8–1.0. The red symbols in Figure 11 show the inconsistency rate of tree species classification results by remote sensing. After correction, the inconsistency rate of interval 0.4–0.6 increased and that of interval 0.7–0.9 decreased. After correction, the inconsistency rate of Chinese arborvitae increased in the 0.5–0.7 interval and decreased in the 0.9–1.0 interval. After correction, the inconsistency rate of oak increased in the 0.4–0.7 interval and decreased in the 0.7–0.9 interval. The inconsistency rate of black locust increased from 0.7 to 0.8 after correction. The inconsistency rate of pine changed greatly after correction and was greatly affected by correction. Black locust showed the opposite result. The relationship between the inconsistency rate and IC was determined by the combination of tree species and their topographic distributions. Topographic correction mainly reduced the inconsistency rate in the 0.4–0.6 interval. Obviously, the correction model maintained the reflectance information of the low IC area throughout the correction, but the model also introduced uncertainty in the classification of the high illumination area.

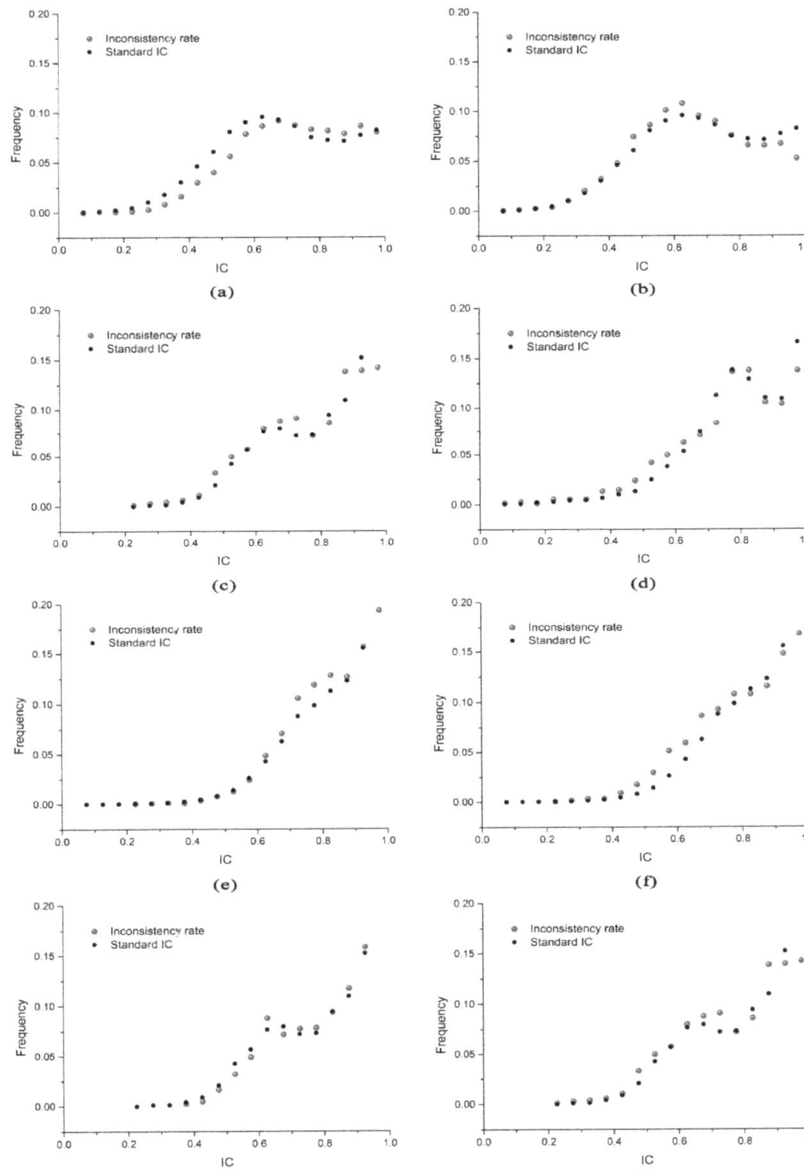

Figure 11. Inconsistency rate distribution histogram with different tree species before and after topographic correction (**a**) Pine before correction, (**b**) Pine corrected by the SCS+C model, (**c**) Chinese arborvitae before correction, (**d**) Chinese arborvitae corrected by the SCS+C model, (**e**) Oak before correction, (**f**) Oak corrected by the SCS+C model, (**g**) Black locust in SR data and (**h**) Black locust corrected by the SCS+C model.

4. Discussion

Topographic correction can reduce the effects of varying topography/terrain surfaces and associated shadowing on spectral reflectance. Although there is no effective method to completely remove this effect, researchers have attempted to improve the effectiveness of topographic correction using various methods. The influence on the topographic correction effectiveness is controlled mainly by the selection of the correction model and the accuracy of the DEM. At present, none of the models can fully explain the relationship between the illumination coefficient and the band reflectivity. Of the four models discussed in this study, the cosine model is a simple optical function, the C model is used to calculate the empirical relationship between the band reflectance and illumination coefficient, the SCS+C model considers the relationship between the sun canopy and sensor and uses the C model to avoid overcorrection and the empirical rotation model eliminates the linear relationship between the band reflectance and illumination coefficient. Although these models all achieve the effectiveness of topographic correction, they are not able to remove all illumination differences resulting from topographic variability. This requires more in-depth research to find a model that fully explains the relationship between reflectivity and the care factor.

The accuracy of the DEM data is another aspect that has a great influence on the correction effectiveness. In the topographic correction model, the illumination coefficient is an important parameter. The IC is calculated based on the slope and aspect, and the slope and aspect are calculated based on the DEM. Thus, the accuracy of the DEM has a great influence on the parameters of the topographic correction model. In this study, we used the elevation data provided by the SRTM, which has a spatial resolution of 30 m. We also used the data of TanDEM-X 90 m for comparison [39]. For TanDEM-X, the overall accuracy and kappa coefficients of the classification results increased by 0.03–0.08 and 0.03–0.09, respectively, but the accuracy of the empirical rotation model results decreased. Since TanDEM-X only obtains 90-m resolution data, the IC calculation involves both slope and aspect. In addition to elevation, horizontal distance is also needed in the calculation. Hence, the spatial resolution affects the accuracy. Although there is no in-depth study on how the DEM accuracy affects the classification accuracy, the higher the DEM accuracy is, the better the correction effectiveness. If the area is small, we can use photogrammetry technology to generate a more accurate surface. We believe that if we can obtain accurate ground data, the topographic correction effectiveness will be greatly improved. Topographic correction adjusts the value and distribution of the band reflectance according to the relationship between the band reflectance and IC. After the four models are corrected, the SD of the band reflectivity can be reduced. However, according to the histogram, the cosine model with different methods has greatly changed the distribution of the band histogram, while the changes associated with the other three models are more moderate. The difference in the histogram after correction is the difference in the interpretation of the band reflectance and the illumination coefficient by the correction method. Different topographic correction models produced different patterns (Section 3.1.5). In the density maps rotated by the topographic correction models, the overall structure of the graph has not changed, which shows that the models are based on the IC and adjust the spectral values in a predictable fashion. Small changes in the form of different shapes and centers are evident after adjustment. This is because both the SCS+C model and the empirical rotation model use coefficients derived from IC data and band regression. In this way, the model adaptability is good, but there are also problems. This regression value is only an empirical value, and there is great uncertainty. The topographic corrections affect different bands differently, and the empirical coefficients of different bands calculated by the C model can eliminate this difference. The empirical rotation model uses the slope, and the SCS+C model uses the slope and intercept.

The distribution of training data and topographic distribution of tree species also affects the accuracy of tree species classification. The spectra of objects in the shaded area definitely vary; thus, training our model with this knowledge will certainly improve the classification accuracy. However, this will also increase the uncertainty, as it may make the spectral values of a feature close to those of other features. Therefore, the selection of training samples is always a process of regional applicability. This is also demonstrated by the influence of the distribution of tree species on the accuracy of tree species classification. Different species have different IC intervals with good correction effects. This may be related to the spectral reflection, canopy structure or canopy direction of the tree species. However, the influence on the spectral reflectance of different tree species is more complicated. It is not always possible to retain all the information. The best way is to eliminate the impact of the topography. In fact, ratio indexes such as the normalized difference vegetation index (NDVI) can eliminate the influence of topography [40,41]. However, the influence of topography on the spectral band is not completely consistent, so we think that, although the index can reduce the influence, it cannot completely eliminate the influence in the same way that system error can be eliminated. It is necessary to correct the band reflectance. When we perform quantitative research on the band reflectance, we want to know the absolute value of the band reflectance rather than the relative value. This value may be more useful in revealing the influence of topography on the features.

5. Conclusions

Based on Landsat 8 data, we developed four topographic correction models on the GEE platform, and their effectiveness on forest tree species classification were compared in detail. The SCS+C model and empirical rotation model were the best models in terms of visual effects, reducing band SD and adjusting the reflectance distribution. The empirical rotation model had the best effect on reducing the SD of the tree species in the NIR and red bands, while the SCS+C model had a high level of consistency with the original scatter density in adjusting the correlation between the NIR and red bands and the IC. All of the corrected images were associated with improved tree species classification accuracy. When using the full-coverage training areas, the accuracy increased by 4%, and when using the shadowless training area, the total accuracy of the SCS+C model increased by nearly 13%. The inconsistency rate distribution histogram showed that the relationship between the inconsistency rate and the IC was determined by the comprehensive effect of tree species and topographic distribution. When forest species were concentrated in the IC interval of 0.4–0.6, the inconsistency rate decreased significantly after correction, whereas with increasing IC values, the inconsistency rate increased. In other words, the corrected image significantly improved the quality and maintained the reflectance information of the shaded area. This technique can be used as a pretreatment method for forest species classification in mountain areas.

The change in the reflectance of forest areas shaded by topography is the result of complex and multifactor interactions, so it is difficult to maintain an exact value with a single mathematical model. The DEM data used in this study have a resolution of 30 m. In this study, it was found that there were some dislocations between the DEM data and the remote-sensing data. Although the effect can be reduced by resampling, DEM data with higher accuracy should provide better topographic correction results. In addition, the complexity of the ground objects is another factor affecting the topographic correction. The uncertainty in the mixed pixels caused by the 30-m resolution of the Landsat data also led to a reduction in the classification accuracy.

In this study, the effectiveness of four commonly used topographic correction models on tree species classification were analyzed. This work provides a basis for the use of satellite data topographic correction in tree species classification. In future research, we will seek a more accurate model and a more accurate DEM to eliminate the influence of the topographic to the greatest extent. The application of topographic correction can produce more consistent spectral characteristics of ground objects, reduce misclassifications and improve the accuracy of forest tree species investigations, which are very important for the promotion of forest resource investigation methods using remote sensing.

Author Contributions: Data curation, C.D.; formal analysis, C.D.; funding acquisition, G.Z.; methodology, C.D.; project administration, G.Z.; resources, Y.M.; validation, B.P.; visualization, B.P.; writing—original draft, C.D. and writing—review and editing, G.Z. and Y.M. All authors have read and agreed to the published version of the manuscript.

Funding: This research was funded by the National Natural Science Foundation of China (41877003), the Chinese Science and Technology Projects (2015BAD23B0202), and the Funds of Shandong "Double Tops" Program (SYL2017XTTD02).

Conflicts of Interest: The authors declare no conflict of interest.

References

1. Scott, C.T.; Köhl, M.; Schnellbächer, H.J. A comparison of periodic and annual forest surveys. *For. Sci.* **1999**, *45*, 433–451.
2. Immitzer, M.; Atzberger, C.; Koukal, T. Tree species classification with random forest using very high spatial resolution 8-band worldview-2 satellite data. *Remote Sens.* **2012**, *4*, 2661–2693. [CrossRef]
3. Yu, Y.; Li, M.Z.; Fu, Y. Forest type identification by random forest classification combined with spot and multitemporal Sar data. *J. For. Res.* **2018**, *29*, 1407–1414. [CrossRef]
4. Ke, Y.; Quackenbush, L.J.; Im, J. Synergistic use of quickbird multispectral imagery and lidar data for object-based forest species classification. *Remote Sens. Environ.* **2010**, *114*, 1141–1154. [CrossRef]
5. Iizuka, K.; Tateishi, R. Estimation of CO_2 sequestration by the forests in Japan by discriminating precise tree age category using remote sensing techniques. *Remote Sens.* **2015**, *7*, 15082–15113. [CrossRef]
6. Kozoderov, V.V.; Dmitriev, E.V.; Melnik, P.G.; Donskoi, S.A. Evaluation of the species composition and the biological productivity of forests based on remote sensing data with high spatial and spectral resolution. *Izv. Atmos. Ocean. Phys.* **2018**, *54*, 1374–1380. [CrossRef]
7. Liu, A. Study on the Theory and Method of Annual Monitoring of Forest Resources. Ph.D. Thesis, Nanjing Forestry University, Nanjing, China, 2006.
8. Yu, X.Y.; Zhao, G.X.; Chang, C.Y.; Yuan, X.J.; Heng, F. Bgvi: A new index to estimate street-side greenery using baidu street view image. *Forests* **2019**, *10*, 3. [CrossRef]
9. Lu, D. Integration of vegetation inventory data and Landsat Tm image for vegetation classification in the Western Brazilian Amazon. *For. Ecol. Manag.* **2005**, *213*, 369–383. [CrossRef]
10. Schuck, A.; Päivinen, R.; Häme, T.; Van Brusselen, J.; Kennedy, P.; Folving, S. Compilation of a European forest map from Portugal to the Ural Mountains based on earth observation data and forest statistics. *For. Policy Econ.* **2003**, *5*, 187–202. [CrossRef]
11. Immitzer, M.; Vuolo, F.; Atzberger, C. First experience with Sentinel-2 data for crop and tree species classifications in Central Europe. *Remote Sens.* **2016**, *8*, 166. [CrossRef]
12. Grabska, E.; Hostert, P.; Pflugmacher, D.; Ostapowicz, K. Forest stand species mapping using the Sentinel-2 time series. *Remote Sens.* **2019**, *11*, 1197. [CrossRef]
13. Chiang, S.-H.; Valdez, M. Tree species classification by integrating satellite imagery and topographic variables using maximum entropy method in a Mongolian forest. *Forests* **2019**, *10*, 961. [CrossRef]
14. Vanonckelen, S.; Lhermitte, S.; Van Rompaey, A. The effect of atmospheric and topographic correction methods on land cover classification accuracy. *Int. J. Appl. Earth Obs. Geoinf.* **2013**, *24*, 9–21. [CrossRef]
15. Moreira, E.P.; Valeriano, M. Application and evaluation of topographic correction methods to improve land cover mapping using object-based classification. *Int. J. Appl. Earth Obs. Geoinf.* **2014**, *32*, 208–217. [CrossRef]
16. Zhu, X.Y.; Sun, M.W.; Wang, S.G.; Yu, L.; Chen, Q.; Wang, Y. Correction of false topographic perception phenomenon based on topographic correction in satellite imagery. *IEEE Trans. Geosci. Remote Sens.* **2017**, *55*, 468–476. [CrossRef]
17. Galvao, L.S.; Breunig, F.M.; Teles, T.S.; Gaida, W.; Balbinot, R. Investigation of terrain illumination effects on vegetation indices and Vi-derived phenological metrics in subtropical Deciduous forests. *Gisci. Remote Sens.* **2016**, *53*, 360–381. [CrossRef]
18. Vazquez-Jimenez, R.; Romero-Calcerrada, R.; Ramos-Bernal, R.N.; Arrogante-Funes, P.; Novillo, C.J. Topographic correction to Landsat imagery through slope classification by applying the Scs plus C method in mountainous forest areas. *ISPRS Int. J. Geo Inf.* **2017**, *6*, 287. [CrossRef]
19. Ghasemi, N.; Mohammadzadeh, A.; Sahebi, M.R. Assessment of different topographic correction methods in alos Avnir-2 data over a forest area. *Int. J. Digit. Earth* **2013**, *6*, 504–520. [CrossRef]
20. Pimple, U.; Sitthi, A.; Simonetti, D.; Pungkul, S.; Leadprathom, K.; Chidthaisong, A. Topographic correction of Landsat Tm-5 and Landsat Oli-8 imagery to improve the performance of forest classification in the mountainous terrain of Northeast Thailand. *Sustainability* **2017**, *9*, 258. [CrossRef]

21. Szantoi, Z.; Simonetti, D. Fast and robust topographic correction method for medium resolution satellite imagery using a stratified approach. *IEEE J. Sel. Top. Appl. Earth Obs. Remote Sens.* **2013**, *6*, 1921–1933. [CrossRef]
22. Ba, A.; Laslier, M.; Dufour, S.; Hubert-Moy, L. Riparian trees genera identification based on leaf-on/leaf-off airborne laser scanner data and machine learning classifiers in Northern France. *Int. J. Remote Sens.* **2020**, *41*, 1645–1667. [CrossRef]
23. Liu, S.; Peng, G.; Panwei, L.; Xiang, N.; Bing, W. An ecosystem services assessment of Tai mountain. *Acta Ecol. Sin.* **2017**, *37*, 3302–3310.
24. Lian, X. Study on Natural Regeneration and Sapling Distribution of Platycladus Orientalis Plantation in Mount Tai. Master's Thesis, Shandong Agricultural University, Shandong, China, 2014.
25. Meng, Y.; Cao, B.H.; Dong, C.; Dong, X.F. Mount Taishan forest ecosystem health assessment based on forest inventory data. *Forests* **2019**, *10*, 657. [CrossRef]
26. Schmidt, G.; Jenkerson, C.B.; Masek, J.; Vermote, E.; Gao, F. Landsat ecosystem disturbance adaptive processing system (Ledaps) algorithm description. In *Open-File Report*; U.S. Geological Survey: Reston, VA, USA, 2013.
27. Gorelick, N.; Hancher, M.; Dixon, M.; Ilyushchenko, S.; Thau, D.; Moore, R. Google earth engine: Planetary-Scale geospatial analysis for everyone. *Remote Sens. Environ.* **2017**, *202*, 18–27. [CrossRef]
28. Farr, T.; Paul, G.; Rosen, A.; Caro, E.; Crippen, R.; Duren, R.; Hensley, S.; Kobrick, M.; Paller, M.; Rodriguez, E.; et al. The shuttle radar topography mission. *Rev. Geophys.* **2007**, *45*, 361. [CrossRef]
29. Meng, Y.; Cao, B.; Mao, P.; Dong, C.; Cao, X.; Qi, L.; Wang, M.; Wu, Y. Tree species distribution change study in Mount Tai based on Landsat remote sensing image data. *Forests* **2020**, *11*, 130. [CrossRef]
30. Teillet, P.M.; Guindon, B.; Goodenough, D.G. On the slope-aspect correction of multispectral scanner data. *Can. J. Remote Sens.* **1981**, *8*, 84–106. [CrossRef]
31. Duguay, C.; Ledrew, E.F. Estimating surface reflectance and albedo over rugged terrain from Landsat-5 thematic mapper. *Photogramm. Eng. Remote Sens.* **1992**, *58*, 551–558.
32. Gu, D.; Gillespie, A. Topographic normalization of Landsat Tm images of forest based on subpixel sun-canopy-sensor geometry. *Remote Sens. Environ.* **1998**, *64*, 166–175. [CrossRef]
33. Tan, B.; Jeffrey, G.; Masek, R.W.; Gao, F.; Huang, C.; Eric, F.; Vermote, J.; Sexton, O.; Ederer, G. Improved forest change detection with terrain illumination corrected Landsat images. *Remote Sens. Environ.* **2013**, *136*, 469–483. [CrossRef]
34. Zhang, Z.M.; He, G.J.; Zhang, X.M.; Long, T.F.; Wang, G.Z.; Wang, M.M. A coupled atmospheric and topographic correction algorithm for remotely sensed satellite imagery over mountainous terrain. *Gisci. Remote Sens.* **2018**, *55*, 400–416. [CrossRef]
35. Zhong, Y.; Liu, L.; Wang, J.; Yan, G. Application and analysis of Scs + C topographic radiometric correction model. *Remote Sens. Land Resour.* **2006**, *14*, 18–79.
36. Macintyre, P.; van Niekerk, A.; Mucina, L. Efficacy of multi-season Sentinel-2 imagery for compositional vegetation classification. *Int. J. Appl. Earth Obs. Geoinf.* **2020**, *85*, 10. [CrossRef]
37. Breiman, L. Random forests. *Mach. Learn.* **2001**, *45*, 5–32. [CrossRef]
38. Li, M.; Im, J.; Beier, C. Machine learning approaches for forest classification and change analysis using multi-temporal Landsat Tm images over Huntington wildlife forest. *Gisci. Remote Sens.* **2013**, *50*, 361–384. [CrossRef]
39. Becek, K.; Akgül, V.; Inyurt, S.; Mekik, Ç.; Pochwatka, P. How well can spaceborne digital elevation models represent a man-made structure: A runway case study. *Geosciences* **2019**, *9*, 387. [CrossRef]
40. Conese, C.; Gilabert, M.A.; Maselli, F.; Bottai, L. Topographic normalization of Tm scenes through the use of an atmospheric correction method and digital terrain models. *Photogramm. Eng. Remote Sens.* **1993**, *59*, 1745–1753.
41. Becek, K.; Borkowski, A.; Mekik, Ç. A study of the impact of insolation on remote sensing-based landcover and landuse data extraction. *ISPRS Int. Arch. Photogramm. Remote Sens. Spat. Inf. Sci.* **2016**, 65–69. [CrossRef]

remote sensing

MDPI

Article

CNN-Based Tree Species Classification Using High Resolution RGB Image Data from Automated UAV Observations

Sebastian Egli [1,*] and Martin Höpke [2]

[1] Laboratory for Climatology and Remote Sensing, Philipps-University of Marburg, Deutschhausstr. 12, 35032 Marburg, Germany

[2] Map-Site, Rossweg 15b, 35094 Lahntal/Gossfelden, Germany; info@map-site.de

* Correspondence: eglis@staff.uni-marburg.de

Received: 26 October 2020; Accepted: 26 November 2020; Published: 27 November 2020

Abstract: Data on the distribution of tree species are often requested by forest managers, inventory agencies, foresters as well as private and municipal forest owners. However, the automated detection of tree species based on passive remote sensing data from aerial surveys is still not sufficiently developed to achieve reliable results independent of the phenological stage, time of day, season, tree vitality and prevailing atmospheric conditions. Here, we introduce a novel tree species classification approach based on high resolution RGB image data gathered during automated UAV flights that overcomes these insufficiencies. For the classification task, a computationally lightweight convolutional neural network (CNN) was designed. We show that with the chosen CNN model architecture, average classification accuracies of 92% can be reached independently of the illumination conditions and the phenological stages of four different tree species. We also show that a minimal ground sampling density of 1.6 cm/px is needed for the classification model to be able to make use of the spatial-structural information in the data. Finally, to demonstrate the applicability of the presented approach to derive spatially explicit tree species information, a gridded product is generated that yields an average classification accuracy of 88%.

Keywords: tree species classification; CNN; UAV; RGB

1. Introduction

Forests provide crucial ecosystem services such as biomass production, air purification, and carbon storage. However, their efficiency and resilience is closely linked to tree species richness [1]. Especially in the context of climate change and its potential impact on the distribution of forest compositions [2–4], accurate mapping of tree species will remain a fundamental challenge for large scale forest management as well as small scale local forest inventories and biodiversity assessments [5].

To meet the growing demand for spatially explicit data on the distribution of tree species, several classification approaches based on a variety of satellite data were proposed [6–10]. Since these data have a relatively coarse spatial resolution, they are, however, not well suited for classifying tree species at single-tree-level in complexly structured species-rich forests.

Tree species classification approaches based on passive multispectral or hyperspectral imaging data and active LiDAR or RADAR scans recorded with airplanes or unmanned aerial vehicles (UAVs) reach higher resolution levels and thus perform better on single-tree-level [11–15]. For a comprehensive overview of the latest advances in tree species classification methodologies based on airborne observations, see Fassnacht et al. [5] and Parkan [16]. Despite the general success of these approaches, they usually share the following insufficiencies: Most studies are based on the use of more than one data collection system which complicates data acquisition. As they are also often dependent on expensive sensor technology, the economic viability of the approaches is not always guaranteed [5].

Although the collected data have a high spatial resolution, in most cases pixel-based-only classification approaches are chosen. Hence, information on spatial patterns in the data are ignored. Varying weather conditions and seasonal as well as diurnal variations in the solar inclination cause changing illumination characteristics. This again, results in variations in the signal patterns reaching the sensor during different data acquisition times. Unfortunately, pixel-based classification models trained on data of passive collection approaches (e.g., multi- and hyperspectral data) cannot handle these variations adequately as they can only make their classification decision on the spectral signal of one single pixel at a time. This results in pixel-based models not being transferable to unstudied regions or data collection times that differ from the acquisition time of the training data.

Another limitation results from the fact that the remote sensing data collection was often only carried out at one or very few specific dates (often even temporally separated from the collection of ground-based training and validation data). As the season-dependent phenological stage significantly influences the spectral characteristics of forests, models trained on these data are therefore not directly applicable to data collected at a different time of the year. A further critical aspect of recent classification approaches concerns the acquisition of field data obtained during ground level site inspections for training and validation of the classification models. Geolocating field data in forests is not trivial due to poor GPS reception. Also, linking single-tree data obtained at ground level to canopy level (as seen from a UAV/airplane/satellite) is complicated and time-consuming and thus makes co-locating ground-based and remote sensing data even more difficult.

To carry out a tree species classification independent of the prevailing illumination conditions, it is advantageous not to perform a pixel-based classification, but to use spatial patterns within the canopy data, the recognition of which is essentially feasible independent of the illumination situation. We also know that, to a large extent, species information is encoded within the leaf shape and branch structure of a tree, which is why forest managers mainly use these features to identify tree species. Since these characteristics are contained within the structural patterns of the canopy (provided the resolution in the remote sensing data is sufficiently high), they can be used to improve the accuracy of tree species classifications. Provided that the training data were collected appropriately, this approach also largely avoids the problems of existing pixel-based methods and allows the training and application of classification models independently of the phenological stage, time of day and year, tree vitality and prevailing atmospheric conditions.

Spatially zero-dimensional machine learning models that were used in recent studies (mainly random forests (RF) or support vector machines (SVM)) are not able to recognize the spatial-structural patterns of leaves and branches in two-dimensional image data. On the other hand, modern convolutional neural networks (CNN), which were developed within the research field of computer vision, are capable of classifying image data on the basis of their spatial-structural content with accuracies that exceed human capabilities [17]. These models can also be used profitably in the classification process of plants in general. CNN-models were already trained on single-leaf data sets such as Flavia [18] or ICL [19] and were shown to achieve good performance [20–23]. However, these models are not transferable to RGB image data automatically gathered over forests as these are much more heterogeneous than the images from Flavia/ICL which were taken under laboratory conditions (white background, just one leave per image, no complex illumination and shadow conditions). Community based plant classification approaches on the basis of photos taken with smartphones such as Pl@ntNet [24,25] or Flora Incognita [26] also showed to be very successful. However, these approaches can only be used for selectively identifying plants but they cannot be used directly to produce spatially explicit data on plant occurrence.

A few recent studies use the high spatial resolution of remote sensing data acquired by satellite [27], airplane [28,29] or UAV [30,31] and apply CNNs in their classification process. Although all these studies show convincing results, they share the insufficiencies of previous studies: As their classification approach is dependent on multi- or hyperspectral image data, expensive hardware has to be employed during data acquisition. Also, the data dimensionality has to be decreased due to the Hughes phenomenon preliminary to the classification process [32]. Furthermore, data was acquired only on a single or very few dates and thus the models were not able to learn the characteristics of different phenological stages and weather conditions. Due to the chosen flight altitudes, the resolution of the data was also not sufficient to make leaf and branch structures recognizable. The only recent UAV-based approach that does not depend on hyperspectral data and that takes into account different phenological stages and changing illumination conditions is that of Natesan et al. [33]. Unfortunately, only two coniferous tree species are distinguished here and the validation results of 51% accuracy under strict temporal separation of training and test data set are in need of improvement.

Additionally, only very few studies have dealt with the examination of an optimal resolution in the input data for the classification task so far [5]. This is problematic as a high image resolution is at the expense of the flight altitude and thus the area that can be covered within a single flight mission.

In summary, we identified three main problems associated with tree species classification in remote sensing: (1) Most studies are based on expensive sensor ensembles. (2) Very few studies have dealt with changes in the spectral signal due to variations in the diurnal and seasonal illumination conditions as well as due to phenological modifications. (3) The influence of the spatial resolution of the input data on the classification result has not yet been investigated thoroughly.

To overcome these shortcomings, the aim of this study was to investigate the feasibility of a tree species classification based on high-resolution photos (simple RGB data) obtained with a low-price ready-to-fly UAV. In order to take into account different illumination conditions and phenological stages, the photos were taken at different times of day and during different seasons throughout the entire vegetation period. In order to be able to exploit the spatial-structural properties of the image material, a CNN model was used for the classification procedure.

Two broadleaf and two coniferous tree species were investigated, namely *Quercus robur* and *Fagus sylvatica* as well as *Larix decidua* and *Picea abies*. In the following, only the common English terms *beech*, *oak*, *larch* and *spruce* will be used.

In the context of this work, the following two hypotheses were tested:

Hypotheses 1 (H1). *It is possible to classify tree species on the basis of high-resolution RGB image data independent of phenological stage, time of day, season and prevailing atmospheric conditions.*

Hypotheses 2 (H2). *With decreasing spatial resolution below the detectability of leaf and branch structures, the quality of the classification result drops significantly.*

Finally, the feasibility of our approach to create a gridded product for the distribution of tree species was tested on an exemplary test site within the study area.

The paper is structured as follows: Section 2 describes the data acquisition and the applied methodology in detail. In Section 3 the results are presented and discussed. A final conclusion is given in Section 4.

2. Data and Methods

2.1. Study Area

This study was conducted in the Marburg Open Forest, a teaching and research forest of the Philipps-University of Marburg, located in Central Germany (see Figure 1). The forest is a mixed forest typical for this region. Besides occasional birches, ash trees and a few other rarely occurring tree species, it mainly consists of the leading tree species beech, oak, larch, spruce, fir, and Douglas fir. The forest is well mixed in terms of age, with average tree ages varying between 40 and 160 years.

Although it is partly managed for the purpose of timber production, logging activities are compensated for by appropriate sustainability measures.

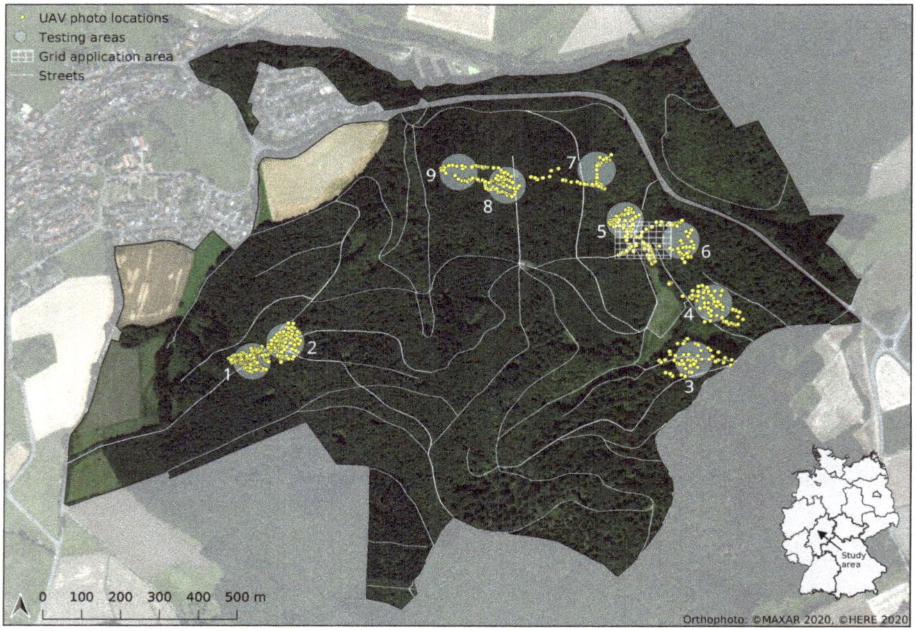

Figure 1. Study area: The Marburg Open Forest - Teaching and Research Forest of the Philipps-University of Marburg, State of Hesse, Germany. UAV photo locations of all conducted flight missions are denoted as yellow points. Testing areas 1 to 9 are delineated by cyan-colored circles. The area used for generating a gridded product is depicted as white raster box.

2.2. Data Collection

Image data were collected using a DJI Phantom 4 Pro UAV. A high-resolution RGB-camera with a 1" CMOS sensor and effectively 20 mega pixels with an edge length of 5472 × 3648 pixels is integrated into the vehicle. Although DJI has since expanded its consumer UAV series with additional models, the camera sensor with its 20 mega pixel resolution is still at the top of the technical range in the consumer drone market. The camera's field of view is specified as 84°, which corresponds approximately to the 35 mm format. With an initial aperture of 2.8, the camera is also relatively light-sensitive [34]. Images were taken vertically (top-down perspective) in auto-mode to automatically adjust to different illumination conditions and stored in JPEG format. No further image calibration was conducted.

To be able to perform automated flights close to the canopy, in a preliminary step grid flights were conducted at an altitude of 100 m. Using the open source software OpenDroneMap [35], a digital surface model (DSM) was derived from the image data obtained in these flights via the structure-from-motion approach. On the basis of the DSM, single tree locations were then manually identified and marked with waypoints. These waypoints were subsequently used for the preparation of automated flights close to the canopy level (10 m distance). In order to account for different phenological stages as well as changing weather situations and thus altering illumination conditions, image data were gathered during multiple flights in different forest divisions during most part of the vegetation period of 2020.

In total, 1556 photos were taken from 477 individual trees during six flight missions. Exemplary photos of each tree species assessed are shown in Figure 2. For a detailed overview of all conducted flight missions see Table 1 and Figure 1.

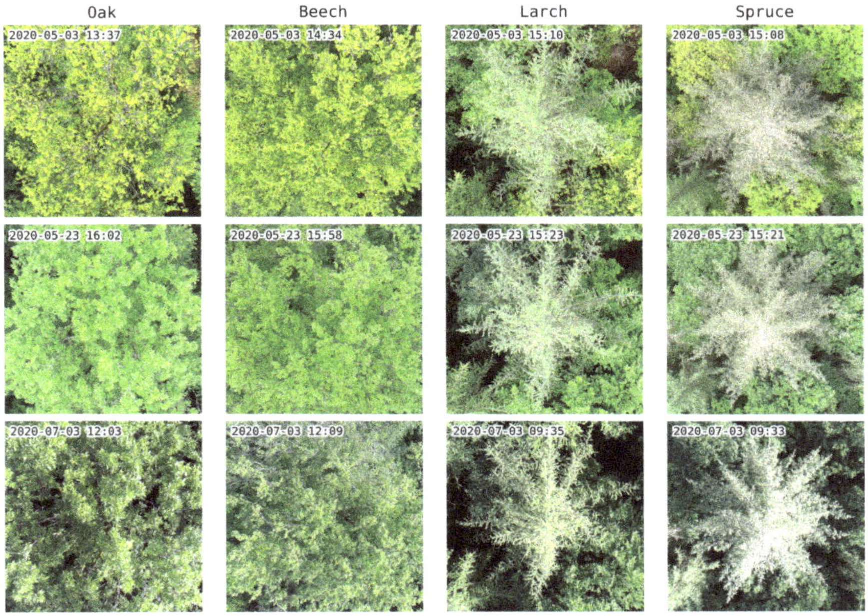

Figure 2. Exemplary UAV photos of the four tree species considered in this study. For each category the same individual tree is shown at different flight dates, weather conditions and phenological stages.

Table 1. List of UAV flights conducted within the scope of this study. Photo counts indicate the amount of all photos taken per flight date. Tile counts refer to the number of actually tagged tiles per category.

Date	Local Time	Weather Conditions	Photo Count	Tile Count			
				Oak	Beech	Larch	Spruce
3 May 2020	13:37–15:24	Sunny/cloudy (cumulus)	179	3745	2022	1663	1362
23 May 2020	14:31–16:09	Cloudy (low stratus)	250	2051	2383	1602	2921
23 June 2020	09:18–10:24	Sunny	217	1981	4254	2105	2036
3 July 2020	08:48–10:13	Sunny	292	2851	2186	2403	2743
9 July 2020	10:18–12:15	Cloudy (low stratus)	369	3253	2440	2403	2566
27 August 2020	09:44–13:18	Sunny/cloudy (cumulus)	249	2463	3408	2401	2745

2.3. Data Preparation

The performance of machine learning models at predicting space-time related properties is often strongly overestimated due to spatial and temporal auto-correlation in the data when using conventional cross-validation methods with simple random picks [36]. Thus, in order to evaluate the true performance of the classification model and its generalizability, care was taken to ensure that the training/validation data set was strictly separated from the testing data set both in time and space, following the leave-location-and-time-out (LLTO) cross validation procedure as described in Meyer et al. [36]. This means that image data used for testing the model performance was not recorded in the same region nor at the same time as the training/validation data set. For this purpose, nine different testing regions were defined (see Figure 1). During model training, successively data from all combinations of each of these areas and each flight date were completely excluded from the training process and used exclusively for testing.

A complete pixel-accurate segmentation of tree species within the UAV photos is not necessary from a remote sensing point of view, since a resolution in the cm range is not useful for the creation of a spatially explicit product on tree species occurrence. Therefore, and in order to simplify the manual tagging process, to keep the computational effort within limits and to be able to generate a larger training data set, each photo was divided into 216 tiles with an edge length of 304 pixels prior to further processing. An edge length of 304 pixels was chosen as a reasonable compromise for the trade-off between recognizability of leaf and branch structure (smaller tile sizes would have resulted in too few species-specific leaf and branch structures within a tile, especially in less leafy trees) and the assignability of a tile to one specific tree (larger tile sizes would have resulted in many tiles containing several different tree species).

It should also be mentioned here that, in contrast to existing classification approaches, we did not first georeference the original data and then perform the classification task on the resulting orthophoto. Instead, we classified the image data in its original resolution in order to exploit the maximum information content of the original data and to avoid distortion effects during orthophoto generation.

The previously generated tiles were then manually grouped into the four tree species categories oak, beech, larch and spruce. A complex and error-prone co-localization of training and remote sensing data was virtually unnecessary during this process, since the tiles could be tagged almost exclusively on the basis of the leaf and branch structures recognizable in the tiles themselves. Since individual tree locations were also visited during the flight missions on site, the type of the centrally located tree of each photo could explicitly be linked to the respective photo. Where a clear tree species assignment on the basis of a single tile was not possible, the information from the photo belonging to the respective tile could thus be used to tag the tile. Only at the tree crown edges a clear assignment of a tile to one specific tree species was not always feasible. These tiles were excluded from the study. In total, 59,987 tiles were tagged with this approach. Detailed numbers per category and flight date are listed in Table 1.

To assess the influence of the spatial resolution of the image data on the model performance, all tiles were downsampled with bilinear interpolation to ten different resolution levels corresponding to flight altitudes from 10 m to 2000 m above canopy level in logarithmic increments. These flight altitudes correspond to ground sampling densities between 0.27 cm/px and 54.78 cm/px. With each increment, the pixel count per sample decreases at a rate of 69%, starting from 92,416 px to 1 px. At higher flight altitudes than 2000 m, the original tile contents at 10 m flight altitude would be smaller than a single pixel in the image data which makes further downsampling pointless.

Differently sized dimensions in the input data require different CNN model architectures. To ensure that differences in the model performance are not due to changes in the model architecture but exclusively due to changes in the input data, the artificially downsampled tiles were thus scaled to the original tile resolution of 304 × 304 on all levels. An overview of all data preparation steps described here is shown in Figure 3.

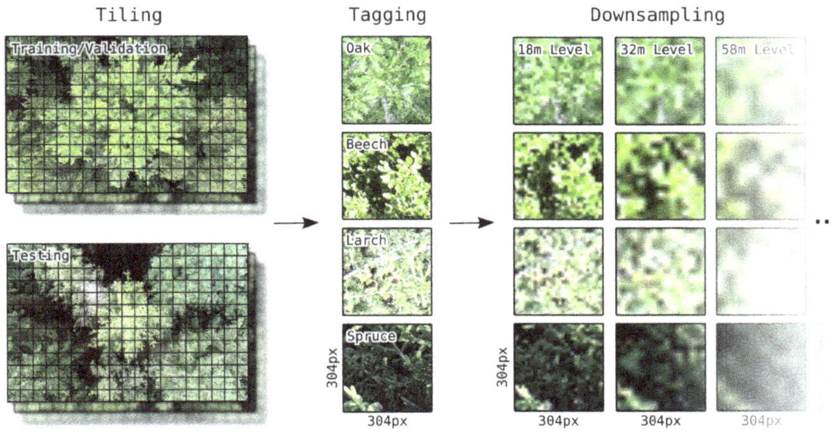

Figure 3. Image data preparation steps: Schematic depiction of the tiling, tagging and downsampling procedure of the image tiles.

2.4. Classification

Current state-of-the-art CNN-models such as VGG16/19, Inception or ResNet that are widely used for general image classification tasks (>1000 classes) are overly complex and thus unnecessarily computationally expensive when it comes to classification problems with very few classes (4 in our case). Since it has also been shown that these large CNN models do not perform better in classification tasks used in forestry applications than smaller models [37], a more compact model architecture was designed for the classification of the UAV image tiles. After testing several different architectures, a sequential model design with four consecutive convolution/pooling layers has proven to be the one with the best performance based on the available data set. All convolutional layers as well as the fully connected layer use ReLu activation functions whereas the last classification layer uses a softmax activation function to classify the tiles into the four tree species. With 2,982,484 trainable parameters, the model is much smaller and computationally less heavy than its large counterparts, the smallest of which comprises more than 25 million trainable parameters. The model implementation and its training were realized with the large-scale machine learning library TensorFlow [38] and the deep learning API Keras [39]. The complete model architecture is shown in Figure 4.

For all model training/validation runs in this study, Adam optimization with a learning rate of 0.001 was used as gradient descent method. Kernel regularizers were used in each convolutional layer to prevent the model from overfitting to the training data. Image tiles were augmented with horizontal and vertical flips. Models were trained in batches of 128 tiles for 50 epochs. Further training did not result in any enhancement of the validation accuracy.

The models were trained on a platform with twelve Intel(R) Xeon(R) CPU E5-1650 v4 units each with 3.60 GHz and four GeForce GTX TITAN X graphics cards. With this setup, about 120 s of computing time were needed for one epoch which resulted in training times of about 100 min per model.

To be able to assess the generalizability of the model performance and to test hypothesis H1, the model training and testing procedure was iteratively repeated for each combination of the nine testing areas and all flight dates as described in Section 2.3. As photos were not taken for each of all possible 54 area/date combinations, in total only 34 different training/testing runs were conducted. In each run, a new model was trained using only data that was not collected within the respective testing area nor at the respective testing date. Sample counts for the train/validation data set were equally distributed and limited to 10,000 per category but could be lower for certain area/date-combinations with a minimum value of 5875. Sample counts of the testing data set varied between 191 and 3039 tiles. Model weights were saved after every training epoch. Each model was then tested using the data from within the respective testing area that was collected on the respective testing date. For testing, the weights with the best performance in terms of validation accuracy were loaded.

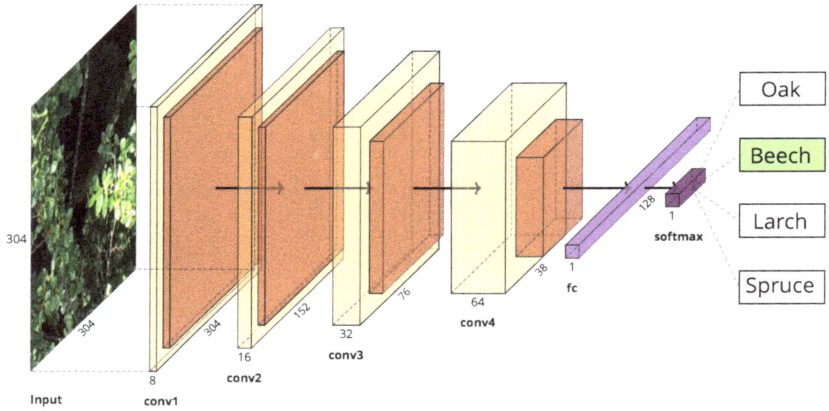

Figure 4. Model architecture of the CNN used in this study. The model consists of four consecutive convolution/pooling blocks (conv1 to conv4), one fully connected layer (fc) and a final output layer with four classes that can be mapped to the four tree species considered in this study. (Plot generated with the PlotNeuralNetwork software by Iqbal [40]).

To assess the influence of the spatial resolution of the image data on the accuracy of the classification results and to test hypothesis H2, in a second step the model was iteratively trained, validated and tested with artificially downsampled image data as described in Section 2.3. To this end, the image data set was randomly split into training, validation and testing subsets of size 8000, 2000 and 1000 tiles for each category.

To get a benchmark for the model performance, a subset of the testing data set was also manually classified by people who were not directly involved in the project. The subset consisted of 10 tiles per resolution level per person. With a total of 20 participants this resulted in 200 tiles per resolution level and 2000 tiles across all levels. During the manual tagging process, the participants did not have any information about the geo-location of the tiles, nor did they get any information about the original photos the tiles were clipped from. The only help was provided by showing 10 sample images per tree species from the training data set for each resolution level to the participants prior to the tagging process. To be able to compare the results of the human assessment directly with the model performance, the model was additionally applied to the tiles of the subset only.

2.5. Grid Application

To demonstrate the usability of the presented approach for the derivation of spatially explicit tree species information, a gridded product was generated for an exemplary region within the study area (see Figure 1). The region was selected in such a way that all tree species considered in this study occurred in it in sufficient abundance. Image data were recorded during an automated flight in a 5 m × 5 m grid 10 m above canopy level on 27 August 2020. An orthophoto of the region that was recorded during a preliminary UAV survey on 2 July 2020 is shown in Figure 5 together with the 5 m × 5 m grid flight pattern.

Figure 5. Orthophoto of the grid application region recorded on 2 July 2020. White dots mark points where photos were taken during the automated 5 m × 5 m grid flight. Coordinates are given in UTM zone 32N.

To minimize the offset between real-world coordinates and image pixels, the images were clipped to the central 1216 pixel × 1216 pixel area which roughly corresponds to a 5 m × 5 m area at a flight altitude of 10 m above canopy level. This central image area was then cut into 16 square tiles with the previously chosen 304 pixel edge length. A CNN was trained using only image data that were not taken on 27 August 2020 and outside the chosen region. This model was then applied to classify all 16 central tiles of each image taken within the region. In this way, a final product with 120 pixel × 80 pixel at 1.25 m spatial resolution was generated. In order to be able to evaluate the quality of the product, all tiles were also manually tagged and compared with the model predictions. Tiles that could not be assigned to any of the four tree species under consideration were not included in this analysis.

3. Results and Discussion

3.1. Generalizability

The results of the model training and testing procedure for all testing area and flight date combinations are shown in Figure 6. Validation accuracies of all 34 runs are depicted in the left plot. After approximately 20 epochs, only slight accuracy enhancements could be observed. After 50 epochs, average validation accuracy values peaked at 93% with the best model reaching 96% and the worst model reaching 92%. It should be noted that the best model performance did not necessarily occur after epoch 50, but often earlier with the earliest peak performance occurring after epoch 38. These results show that the chosen model architecture is essentially very well suited to differentiate the tree species treated in this study on the basis of simple RGB images. It can also be deduced from the plot that the variation of the validation accuracy becomes increasingly smaller as training progresses. This means that regardless of the date/area combination, the model was able to derive the essential information for the differentiation of tree species from the data.

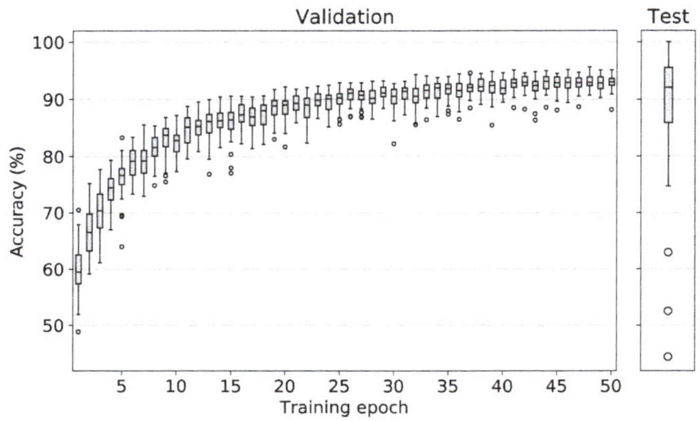

Figure 6. Left: Validation accuracy after each epoch of the 34 different models that were trained during the LLTO procedure. Right: Accuracy values reached during testing. Boxes range from the lower to the upper quartile values of the data with a line at the median. Whiskers extend to 1.5 interquartile range. Outliers are marked as circles.

Similar studies in the literature report highly varying accuracy results with values between 59% and 95% [28]. With an average validation accuracy of 93% after 50 epochs our model thus lies in the upper range. However, most of these study results cannot be directly compared with the results of this study, as too little focus was put on spatial and temporal independence in the validation procedure. The only work we are aware of, which is also based on RGB data collected with a UAV and which was validated spatially and temporally independently, is the work of Natesan et al. [33]. Here an average accuracy of 80% is reported.

To be able to assess the true spatiotemporal generalizability of the model, the testing accuracies must be analyzed. The right plot shows the testing accuracy results which were generated by loading the best weights for each model and then applying it to its respective testing data set. In comparison to the validation accuracies, here a similar median value of 92% was observed. However, the value range of the testing accuracies is much larger than the value range of the validation accuracies of the best models with a minimum of 44% and a maximum of 100% .

The median value of 92% in the test accuracies is very promising and shows that trained models can indeed be successfully applied in temporally and spatially independent areas where they are able to achieve prediction accuracies of 92% on average. However, the large value range of 56 pp indicates that in a few exceptional cases significantly higher/lower accuracy values are reached. The maximum accuracy of 100% was achieved for testing area 8 on 27 August 2020. However, only 191 test tiles were available here, all of which fell into the oak category, thus making this test sample unrepresentative. The same applies to the three lower outliers with accuracy values between 44% and 63% . With average tile counts of only 526 and a maximum of three tree species categories being represented, these test samples are not representative either.

In summary, it can be stated that the presented method made it possible to differentiate between the four tree species oak, beech, larch and spruce. Apart from a few non-representative test cases, the CNN models were able to achieve promising classification accuracies temporally and spatially independently. Hypothesis H1 is thus accepted.

3.2. Resolution Dependence

The model performance data collected during training of the model at different resolution levels of the image data are shown in Figures 7 and 8. Figure 7 shows the model accuracies after each training epoch. Starting from accuracy values between 50% and 60% after the first training epoch, all models were able to improve till the end of the training procedure. However, with increasing flight altitude and thus with lower information content in the image data, it became increasingly difficult for the models to classify the image data into the correct categories. This manifests itself in a successive flattening of the learning curves with maxima being achieved after less training epochs but at considerably lower levels. While for the original resolution at 10 m flight altitude a maximum validation accuracy of 95% was achieved after 50 epochs, at a theoretical flight altitude of 2000 m a maximum validation accuracy value of only 56% was reached already after 36 epochs.

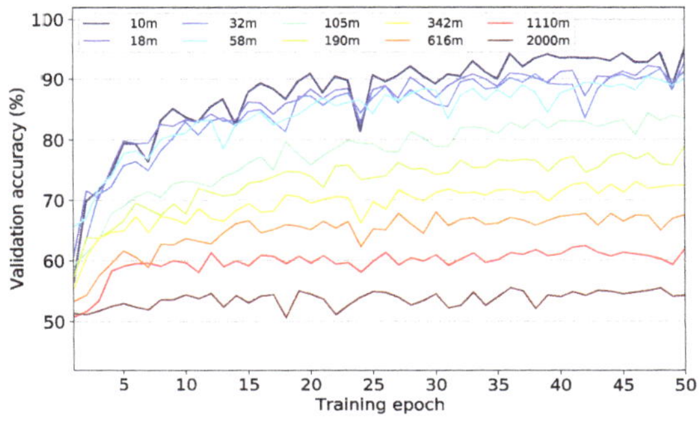

Figure 7. Validation accuracy after each epoch of the ten different models that were trained with successively decreased data resolutions corresponding to the flight altitudes listed.

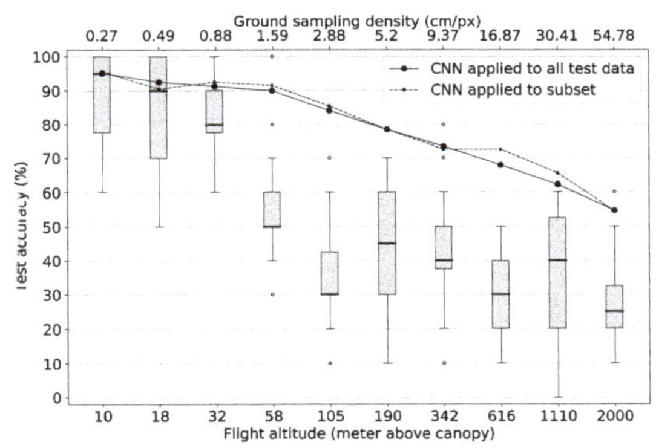

Figure 8. Test accuracies of model and human classification results at different resolution levels corresponding to the flight altitudes and ground sampling densities depicted on the horizontal axes (logarithmically scaled). Lines denote model results. Boxplots show human results.

The reason for the early reaching of maximum performance on a low level for models with low-resolution data is presumably the reduced information content in the data. This allows the models to learn the complete information content that is still contained in the data more quickly. However, this is also inevitably accompanied by a decrease in maximum accuracy values. On the other hand, models based on high-resolution data need longer to reach maximum performance due to the larger information content within the data. Nevertheless, the higher information content also allows the models to achieve higher maximum accuracies.

It should be pointed out that even at the maximum theoretical flight altitude of 2000 m accuracy values of 56% are still achieved. Since only one color value per tile remains, i.e., 3 features per sample in contrast to 277,248 features in the original samples, it can be concluded that the mean spectral values of the tiles already provide an explanatory content of 31%, assuming a base rate accuracy of 25%

of a random model with 4 equally distributed classes. Beyond this level, another 39% of additional explanatory content can be obtained from the spatial structure in the image data, resulting in maximum accuracies of 95% at the highest resolution level.

In Figure 8 accuracy values are shown that were obtained by applying each run's best model to the test data together with the human performance results. The solid line depicts the performance results of the models when they were tested against the complete testing data set of 4000 tiles per resolution level. The dashed line marks model results when they were tested against the subset that was used in the "human" tagging process (200 tiles per resolution level). With the only exceptions of flight altitude levels 616 m and 1110 m which correspond to ground sampling densities of 16.87 cm/px and 30.41 cm/px, the deviation between the two curves is minimal. These slight deviations can be explained by the size of the subset with only 200 samples per altitude level compared to 4000 in the complete testing data set. In general, both curves follow the same scheme:

The first four models trained with theoretical data acquisition altitudes of up to 58 m all reach accuracy values above 90%, only showing a very slow accuracy decrease of about 1.6 pp per flight altitude level increase. This can probably be explained by the fact that many images in the data set at these high resolution levels contain redundant information regarding the determinability of the four tree species investigated. For instance, the leaf shape is still "over-resolved" at these levels. Thus, despite the absolute loss of information of 97.1% from the 10 m level to the 58 m level (277,248 vs. 8112 features) the models are still able to distinguish between the four categories in most cases.

Starting from 105 m theoretical flight altitude and above, which is equivalent to a ground sampling density of 2.88 cm/px, accuracy values decrease more rapidly at a rate of about 5 pp per logarithmically increased altitude level. A possible explanation for this is that from this resolution on, important information in the image data required for the determination of tree species, such as the leaf and branch structure of the trees, can no longer be recognized by the models. Thus, the models can only rely on coarser structures and average spectral values, which do not allow for a clear identification anymore. Nevertheless, the decrease of the model performance is much slower, than a decrease proportional to the number of features per sample on all resolution levels. This again emphasizes the suitability of the CNN application to the tree species classification problem investigated in this work which becomes particularly obvious when comparing the model results with the results of the human classification:

Results of the manual tagging process are depicted as boxplots in Figure 8. Despite large variations at each level, the same basic pattern as for the models can be observed: The human classification results are at a similarly high level as the model results at the beginning and decrease with increasing theoretical flight altitudes. However, there are some differences to the model performance: The human classification accuracy significantly decreases already at the flight altitude level of 32 m (0.88 cm/px ground sampling density) where it drops relatively abruptly from median accuracy values of 80% over the 58 m level with 50% to 30% at level 105 m (2.88 cm/px ground sampling density). From this flight altitude level up to the maximum flight altitude level of 2000 m the classification accuracy then varies between median values of 45% and 25% with a slight downward trend towards coarser resolution levels.

This shows that humans can keep up with the models in the beginning where they reach a median accuracy value of 95% at the 10 m level. However, as soon as the visual characteristics essential for humans to differentiate between tree species can no longer be recognized, the performance drops abruptly and settles at values just above a random hit probability of 25%. Thus, humans are not able to draw meaningful information from the mean spectral characteristics and the coarse spatial structures, which are still recognizable at the coarser resolution levels, in order to make a meaningful differentiation. Besides this, the trend of the human performance supports the assumption that at a flight altitude level between 32 m and 105 m the recognition of important spatial-structural properties for the differentiation of tree species is no longer possible, given the UAV setup used in this study.

In summary, it can be stated that the predictive power of humans and the CNN models continuously decreases with decreasing resolution in the image data. The fact that the model performance decrease with logarithmically increasing flight altitudes is steeper for flight altitudes above 58 m (1.59 cm/px ground sampling density) and that the human performance drops significantly around this level too, indicates that essential spatial-structural information needed for the classification of tree species can no longer be recognized above this level. Hypothesis H2 can thus also be accepted. It can further be concluded that with the given UAV and camera sensor technology, acquisition altitudes of about 58 m above canopy level are sufficient to still achieve very good accuracy results with values of more than 90%. This is particularly noteworthy as at flight altitudes of around 58 m above canopy a much larger area can be covered per flight than at 10 m flight altitudes.

While Fassnacht et al. [5] hypothesize that classification accuracies of non-pixel-based approaches are likely to benefit from higher spatial resolutions, they emphasize that more systematic research is needed in this area. Most importantly, until now it remained unclear, where a possible upper limit of the resolution enhancement with respect to the classification accuracy improvement may be found. With the help of the results of this study, this upper limit can now be assumed to be somewhere around a ground sampling density of 1.6 cm/px which corresponds to flight altitudes of about 60 m with the UAV setup used in this study.

3.3. Grid Application

The result of the grid-based application of the CNN approach to an exemplary region is shown in Figure 9. Areas covered by oaks, beeches, larches and spruces are colored yellow, brown, coral and blue, which are the colors used for these tree species in map illustrations of the Hessian forestry management. White areas are regions that could not be assigned to one of the four categories either due to image data deficiencies such as motion blur or because the respective image tile did not contain any of the four tree species. This was the case when either no tree was visible in the tile at all (e.g., forest floor or shrubs) or because the tile showed another tree species than the four considered here. Apart from the white areas in the central part of the domain resulting from the presence of ash trees (*Fraxinus excelsior*), the optical comparison with the orthophoto shows that the approach has produced reasonable results. Beeches are relatively evenly distributed over the complete domain. Larch trees, on the other hand, are concentrated in the northwestern part, while oak trees are most common in the southeast. Spruces form a strip from the southern to the central part of the domain and show a few scattered occurrences in the southwest.

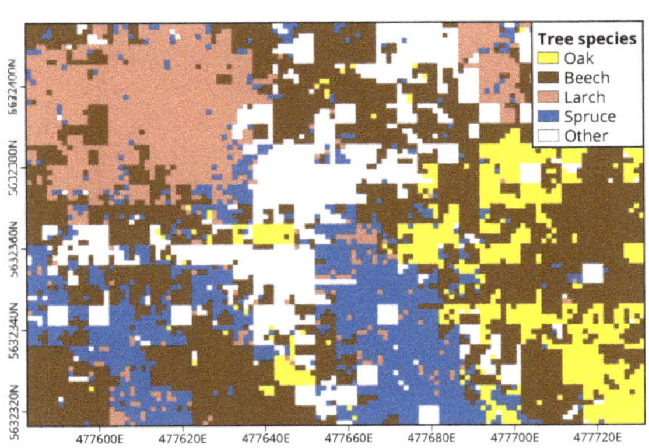

Figure 9. Result of the grid application. Tree colors correspond to the colors used in map illustrations of the Hessian forestry management. White areas mark regions where the model was not applicable. Coordinates are given in UTM zone 32N.

The validation against manually tagged tiles which were visually assignable to one of the four tree species resulted in an accuracy of 88%. This is within the range of the third quartile of the testing accuracies from the generalizability analysis in Section 3.1 and shows that the grid product achieves a good overall spatially explicit classification accuracy. It must be mentioned, however, that the geographical location of the individual pixels of the product is not very accurate. Although only the central section of each UAV photo was used for the classification procedure, an exact assignment of an image tile to a product pixel was not always unambiguously possible. This was due to optical distortions towards the edge of the image, irregularities in the height of the canopy and because of deviations between the measured and the actual position and orientation of the UAV.

Furthermore, due to the low flight altitude of the UAV, this approach cannot yet be applied on a large spatial scale. Nevertheless, the procedure may be used to generate spatially explicit data on tree species as seen from the aerial perspective for selected example areas or it can help to get an overview of tree species distribution during local forest inventories. Independently validated spatially explicit reference data on tree species occurrence are hardly available due to the difficulties in locating exact tree positions and crown extensions [5]. Locally gathered and gridded products using the method proposed in this study, could help to close this gap. These data sets could then also be used as "ground truth" to support spatial upscaling procedures based on coarser satellite data in order to produce large scale spatially explicit data on tree species occurrence.

4. Conclusions

In this study, we presented a novel CNN-employing tree species classification approach based on low-cost UAV-RGB image data. It was shown that validation results of 92% could be achieved with this technique on spatially and temporally independent data. This demonstrates that, in contrast to existing pixel-based classification methods, this approach enables the determination of tree species largely independent of the prevailing illumination conditions, i.e., independent of daytime and season, phenological stage and prevailing weather conditions. However, since only four tree species were classified in this work, a reasonable next step is to extend the data set by further tree species.

An analysis of the effect of the spatial resolution on the classification accuracy has shown that human and model performance continuously decrease with decreasing spatial resolution. However, it could be shown that up to a theoretical flight altitude of about 60 m, equivalent to a ground sampling density of 1.6 cm/px, the model's classification accuracy remained above 90% and that only at higher flight altitudes or lower spatial resolutions the classification accuracy decreased significantly. Although human performance shows large variances in the individual resolution levels, a clear trend is discernible which shows that human classification accuracy drops faster and earlier than the model's accuracy at decreasing resolution in the input image data. It should be added that the resolution analysis was only performed on the basis of artificially downsampled image data. Thus, to confirm our results, they should be compared to classification results generated on the basis of original data gathered in exemplary flights performed at the respective flight altitude levels.

Finally, it was shown that the presented approach can be used for the generation of gridded tree species products, which in turn could be made available as ground truth reference data for various further applications. A remaining deficiency of the grid generation is the inaccurate assignment of image tiles to grid pixels. This assignment could be improved by using RTK-UAVs, which can measure their location with high accuracy, or by including a simultaneously generated orthophoto in the post-processing of the collected image data for automatic georeferencing.

As the RGB image data required for the classification can be obtained at reasonable costs and since the UAV employed was an easy-to-use consumer model, the input data could potentially be directly collected by people who do not need to have remote sensing expertise, e.g., forest managers or rangers. This again increases the probability of a broad applicability of the approach in many different contexts.

Author Contributions: Conceptualization, S.E. and M.H.; methodology, S.E.; software, M.H.; validation, M.H.; formal analysis, S.E.; investigation, S.E.; resources, S.E.; data curation, S.E.; writing–original draft preparation, S.E.; writing–review and editing, M.H.; visualization, M.H.; supervision, S.E.; project administration, S.E. All authors have read and agreed to the published version of the manuscript.

Funding: This research received no external funding.

Acknowledgments: We would like to thank Ansgar Leonhardt for the productive cooperation, especially for helping us with his expertise in the field of forest inventory. We would also like to thank the diligent LCRS staff for manually classifying over 60,000 image tiles. This project was carried out as part of the LOEWE joint project "Nature 4.0" subproject UM2.

Conflicts of Interest: The authors declare no conflict of interest.

Abbreviations

The following abbreviations are used in this manuscript:

CNN	Convolutional Neural Network
DSM	Digital Surface Model
LiDAR	Light Detection and Ranging
LLTO	Leave Location and Time Out Cross Validation
RADAR	RAdio Detection And Ranging
RF	Random Forest
RGB	Red Green Blue
RTK	Real Time Kinematic
SVM	Support Vector Machine
UAV	Unmanned Aerial Vehicle

References

1. Gamfeldt, L.; Snäll, T.; Bagchi, R.; Jonsson, M.; Gustafsson, L.; Kjellander, P.; Ruiz-Jaen, M.C.; Fröberg, M.; Stendahl, J.; Philipson, C.D.; et al. Higher levels of multiple ecosystem services are found in forests with more tree species. *Nat. Commun.* **2013**, *4*, 1340. [CrossRef]
2. Chen, I.C.; Hill, J.K.; Ohlemüller, R.; Roy, D.B.; Thomas, C.D. Rapid range shifts of species associated with high levels of climate warming. *Science* **2011**, *333*, 1024–1026. [CrossRef]
3. Iverson, L.R.; McKenzie, D. Tree-species range shifts in a changing climate: Detecting, modeling, assisting. *Landsc. Ecol.* **2013**, *28*, 879–889. [CrossRef]
4. Thompson, I.; Mackey, B.; McNulty, S.; Mosseler, A. *Forest Resilience, Biodiversity, and Climate Change. A Synthesis of the Biodiversity/Resilience/Stability Relationship in Forest Ecosystems*; Secretariat of the Convention on Biological Diversity: Montreal, QC, Canada, 2009; p. 67.
5. Fassnacht, F.E.; Latifi, H.; Stereńczak, K.; Modzelewska, A.; Lefsky, M.; Waser, L.T.; Straub, C.; Ghosh, A. Review of studies on tree species classification from remotely sensed data. *Remote Sens. Environ.* **2016**, *186*, 64–87. [CrossRef]
6. Immitzer, M.; Atzberger, C.; Koukal, T. Tree Species Classification with Random Forest Using Very High Spatial Resolution 8-Band WorldView-2 Satellite Data. *Remote Sens.* **2012**, *4*, 2661–2693. [CrossRef]
7. Mora, B.; Wulder, M.A.; White, J.C. Identifying leading species using tree crown metrics derived from very high spatial resolution imagery in a boreal forest environment. *Can. J. Remote Sens.* **2010**, *36*, 332–344. [CrossRef]
8. Ortiz, S.M.; Breidenbach, J.; Knuth, R.; Kändler, G. The Influence of DEM Quality on Mapping Accuracy of Coniferous- and Deciduous-Dominated Forest Using TerraSAR-X Images. *Remote Sens.* **2012**, *4*, 661–681. [CrossRef]
9. Sheeren, D.; Fauvel, M.; Josipović, V.; Lopes, M.; Planque, C.; Willm, J.; Dejoux, J.F. Tree Species Classification in Temperate Forests Using Formosat-2 Satellite Image Time Series. *Remote Sens.* **2016**, *8*, 734. [CrossRef]
10. Wagner, F.H.; Sanchez, A.; Tarabalka, Y.; Lotte, R.G.; Ferreira, M.P.; Aidar, M.P.M.; Gloor, E.; Phillips, O.L.; Aragão, L.E.O.C. Using the U-net convolutional network to map forest types and disturbance in the Atlantic rainforest with very high resolution images. *Remote Sens. Ecol. Conserv.* **2019**, *5*, 360–375. [CrossRef]

11. Dalponte, M.; Bruzzone, L.; Gianelle, D. Tree species classification in the Southern Alps based on the fusion of very high geometrical resolution multispectral/hyperspectral images and LiDAR data. *Remote Sens. Environ.* **2012**, *123*, 258–270. [CrossRef]
12. Dalponte, M.; Orka, H.O.; Gobakken, T.; Gianelle, D.; Naesset, E. Tree Species Classification in Boreal Forests with Hyperspectral Data. *IEEE Trans. Geosci. Remote Sens.* **2013**, *51*, 2632–2645. [CrossRef]
13. Shen, X.; Cao, L. Tree-Species Classification in Subtropical Forests Using Airborne Hyperspectral and LiDAR Data. *Remote Sens.* **2017**, *9*, 1180. [CrossRef]
14. Shi, Y.; Skidmore, A.K.; Wang, T.; Holzwarth, S.; Heiden, U.; Pinnel, N.; Zhu, X.; Heurich, M. Tree species classification using plant functional traits from LiDAR and hyperspectral data. *Int. J. Appl. Earth Obs. Geoinf.* **2018**, *73*, 207–219. [CrossRef]
15. Yang, G.; Zhao, Y.; Li, B.; Ma, Y.; Li, R.; Jing, J.; Dian, Y. Tree Species Classification by Employing Multiple Features Acquired from Integrated Sensors. *J. Sens.* **2019**, *2019*, 1–12. [CrossRef]
16. Parkan, M.J. Combined use of Airborne Laser Scanning and Hyperspectral Imaging for Forest Inventories. Ph.D. Thesis, École Polytechnique Fédérale de Lausanne, Lausanne, Switzerland, 2019. [CrossRef]
17. He, K.; Zhang, X.; Ren, S.; Sun, J. Delving Deep into Rectifiers: Surpassing Human-Level Performance on ImageNet Classification. In Proceedings of the 2015 IEEE International Conference on Computer Vision (ICCV), Santiago, Chile, 7–13 December 2015; pp. 1026–1034. [CrossRef]
18. Wu, S.G.; Bao, F.S.; Xu, E.Y.; Wang, Y.X.; Chang, Y.F.; Xiang, Q.L. A Leaf Recognition Algorithm for Plant Classification Using Probabilistic Neural Network. In Proceedings of the 2007 IEEE International Symposium on Signal Processing and Information Technology, Giza, Egypt, 15–18 December 2007; pp. 11–16. [CrossRef]
19. Silva, P.F.B.; Marçal, A.R.S.; da Silva, R.M.A. Evaluation of Features for Leaf Discrimination. In *Lecture Notes in Computer Science (Including Subseries Lecture Notes in Artificial Intelligence and Lecture Notes in Bioinformatics)*; Springer: Berlin/Heidelberg, Germany, 2013; Volume 7950 LNCS, pp. 197–204. [CrossRef]
20. He, X.; Wang, G.; Zhang, X.p.; Shang, L.; Huang, Z.k. *Intelligent Computing Theories and Application*; Lecture Notes in Computer Science; Springer International Publishing: Cham, Switzerland, 2016; Volume 9772, pp. 332–340. [CrossRef]
21. Jeon, W.S.; Rhee, S.Y. Plant Leaf Recognition Using a Convolution Neural Network. *Int. J. Fuzzy Log. Intell. Syst.* **2017**, *17*, 26–34. [CrossRef]
22. Vilasini, M.; Ramamoorthy, P. CNN Approaches for Classification of Indian Leaf Species Using Smartphones. *Comput. Mater. Contin.* **2020**, *62*, 1445–1472. [CrossRef]
23. Wu, Y.H.; Shang, L.; Huang, Z.K.; Wang, G.; Zhang, X.P. Convolutional Neural Network Application on Leaf Classification. In *Lecture Notes in Computer Science (Including Subseries Lecture Notes in Artificial Intelligence and Lecture Notes in Bioinformatics)*; Springer: Berlin/Heidelberg, Germany, 2016; Volume 9771, pp. 12–17. [CrossRef]
24. Joly, A.; Goëau, H.; Bonnet, P.; Bakić, V.; Barbe, J.; Selmi, S.; Yahiaoui, I.; Carré, J.; Mouysset, E.; Molino, J.F.; et al. Interactive plant identification based on social image data. *Ecol. Inform.* **2014**, *23*, 22–34. [CrossRef]
25. Joly, A.; Bonnet, P.; Goëau, H.; Barbe, J.; Selmi, S.; Champ, J.; Dufour-Kowalski, S.; Affouard, A.; Carré, J.; Molino, J.F.; et al. A look inside the Pl@ntNet experience. *Multimed. Syst.* **2016**, *22*, 751–766. [CrossRef]
26. Wäldchen, J.; Rzanny, M.; Seeland, M.; Mäder, P. Automated plant species identification—Trends and future directions. *PLoS Comput. Biol.* **2018**, *14*, e1005993. [CrossRef]
27. Hartling, S.; Sagan, V.; Sidike, P.; Maimaitijiang, M.; Carron, J. Urban Tree Species Classification Using a WorldView-2/3 and LiDAR Data Fusion Approach and Deep Learning. *Sensors* **2019**, *19*, 1284. [CrossRef]
28. Fricker, G.A.; Ventura, J.D.; Wolf, J.A.; North, M.P.; Davis, F.W.; Franklin, J. A Convolutional Neural Network Classifier Identifies Tree Species in Mixed-Conifer Forest from Hyperspectral Imagery. *Remote Sens.* **2019**, *11*, 2326. [CrossRef]
29. Knauer, U.; von Rekowski, C.S.; Stecklina, M.; Krokotsch, T.; Pham Minh, T.; Hauffe, V.; Kilias, D.; Ehrhardt, I.; Sagischewski, H.; Chmara, S.; et al. Tree Species Classification Based on Hybrid Ensembles of a Convolutional Neural Network (CNN) and Random Forest Classifiers. *Remote Sens.* **2019**, *11*, 2788. [CrossRef]
30. Miyoshi, G.T.; Arruda, M.D.S.; Osco, L.P.; Marcato Junior, J.; Gonçalves, D.N.; Imai, N.N.; Tommaselli, A.M.G.; Honkavaara, E.; Gonçalves, W.N. A Novel Deep Learning Method to Identify Single Tree Species in UAV-Based Hyperspectral Images. *Remote Sens.* **2020**, *12*, 1294. [CrossRef]

31. Nezami, S.; Khoramshahi, E.; Nevalainen, O.; Pölönen, I.; Honkavaara, E. Tree species classification of drone hyperspectral and RGB imagery with deep learning convolutional neural networks. *Remote Sens.* **2020**, *12*. [CrossRef]
32. Hughes, G. On the mean accuracy of statistical pattern recognizers. *IEEE Trans. Inf. Theory* **1968**, *14*, 55–63. [CrossRef]
33. Natesan, S.; Armenakis, C.; Vepakomma, U. ResNet-based tree species classification using UAV images. *ISPRS Int. Arch. Photogramm. Remote. Sens. Spat. Inf. Sci.* **2019**, *XLII-2/W13*, 475–481. [CrossRef]
34. DJI. Phantom 4 Pro Specs. Available online: https://www.dji.com/de/phantom-4-pro/info (accessed on 26 November 2020).
35. Toffanin, P.; Benjamin, D.; Riba, E.; Gargallo, P.; Mather, S.; Fitzsimmons, S.; Gelder, F.; Bargen, D.; Cesar de Menezes, J.; Joseph, D.; et al. OpenDroneMap/ODM: 1.0.1 2020. Available online: https://github.com/OpenDroneMap/ODM (accessed on 26 November 2020).
36. Meyer, H.; Reudenbach, C.; Hengl, T.; Katurji, M.; Nauss, T. Improving performance of spatio-temporal machine learning models using forward feature selection and target-oriented validation. *Environ. Model. Softw.* **2018**, *101*, 1–9. [CrossRef]
37. Safonova, A.; Tabik, S.; Alcaraz-Segura, D.; Rubtsov, A.; Maglinets, Y.; Herrera, F. Detection of Fir Trees (Abies sibirica) Damaged by the Bark Beetle in Unmanned Aerial Vehicle Images with Deep Learning. *Remote Sens.* **2019**, *11*, 643. [CrossRef]
38. Abadi, M.; Agarwal, A.; Barham, P.; Brevdo, E.; Chen, Z.; Citro, C.; Corrado, G.S.; Davis, A.; Dean, J.; Devin, M.; et al. TensorFlow: Large-Scale Machine Learning on Heterogeneous Systems. *arXiv* **2015**, arXiv:1603.04467.
39. Chollet, F. Keras. 2015. Available online: https://github.com/fchollet/keras (accessed on 26 November 2020).
40. Iqbal, H. PlotNeuralNet v1.0.0. 2018. Available online: https://github.com/HarisIqbal88/PlotNeuralNet/tree/v1.0.0 (accessed on 26 November 2020).

remote sensing

Article

How Spatial Resolution Affects Forest Phenology and Tree-Species Classification Based on Satellite and Up-Scaled Time-Series Images

Kaijian Xu [1,2,3,*], Zhaoying Zhang [4], Wanwan Yu [1], Ping Zhao [1,2], Jibo Yue [4,5], Yaping Deng [1] and Jun Geng [1]

1 School of Resources and Environmental Engineering, Hefei University of Technology, Hefei 230009, China; 2019110660@mail.hfut.edu.cn (W.Y.); njuzhp@hfut.edu.cn (P.Z.); dengyaping@hfut.edu.cn (Y.D.); gengj@hfut.edu.cn (J.G.)
2 Institute for Spatial Information Intelligence Analysis and Application, Hefei University of Technology, Hefei 230009, China
3 Intelligent Interconnected Systems Laboratory of Anhui Province, Hefei 230009, China
4 International Institute for Earth System Science, Nanjing University, Nanjing 210023, China; dg1727038@smail.nju.edu.cn (Z.Z.); yuejibo@smail.nju.edu.cn (J.Y.)
5 College of Information and Management Science, Henan Agricultural University, Zhengzhou 450002, China
* Correspondence: kaijianxu@hfut.edu.cn

Citation: Xu, K.; Zhang, Z.; Yu, W.; Zhao, P.; Yue, J.; Deng, Y.; Geng, J. How Spatial Resolution Affects Forest Phenology and Tree-Species Classification Based on Satellite and Up-Scaled Time-Series Images. *Remote Sens.* **2021**, *13*, 2716. https://doi.org/10.3390/rs13142716

Academic Editor: Markus Immitzer

Received: 11 May 2021
Accepted: 7 July 2021
Published: 10 July 2021

Abstract: The distribution of forest tree species provides crucial data for regional forest management and ecological research. Although medium-high spatial resolution remote sensing images are widely used for dynamic monitoring of forest vegetation phenology and species identification, the use of multiresolution images for similar applications remains highly uncertain. Moreover, it is necessary to explore to what extent spectral variation is responsible for the discrepancies in the estimation of forest phenology and classification of various tree species when using up-scaled images. To clarify this situation, we studied the forest area in Harqin Banner in northeast China by using year-round multiple-resolution time-series images (at four spatial resolutions: 4, 10, 16, and 30 m) and eight phenological metrics of four deciduous forest tree species in 2018, to explore potential impacts of relevant results caused by various resolutions. We also investigated the effect of using up-scaled time-series images by comparing the corresponding results that use pixel-aggregation algorithms with the four spatial resolutions. The results indicate that both phenology and classification accuracy of the dominant forest tree species are markedly affected by the spatial resolution of time-series remote sensing data ($p < 0.05$): the spring phenology of four deciduous forest tree species first rises and then falls as the image resolution varies from 4 to 30 m; similarly, the accuracy of tree species classification increases as the image resolution varies from 4 to 10 m, and then decreases as the image resolution gradually falls to 30 m ($p < 0.05$). Therefore, there remains a profound discrepancy between the results obtained by up-scaled and actual remote sensing data at the given spatial resolutions ($p < 0.05$). The results also suggest that combining phenological metrics and time-series NDVI data can be applied to identify the regional dominant tree species across different spatial resolutions, which would help advance the use of multiscale time-series satellite data for forest resource management.

Keywords: time series; trees species identification; phenological metrics; scale effect; up-scaling

1. Introduction

Accurate spatial distribution information of forest tree species is a precondition for almost all questions dealing with regional forest ecology and is quite important for understanding the land-surface phenology (LSP) processes and the refinement of ecological and atmospheric models [1–3]. In addition, accurate mapping of the dominant tree species is critical for forest-environment monitoring, forest management, and associated decision-making and planning [4,5]. In recent years, remote sensing images with medium to high spatial resolution (≤30 m) have been widely used to map and classify dominant tree species in the different regional forest ecosystems, because the viewing field of the image is close to the size of tree species and of tree stands [6–9]. However, given the limitations to a few bands with wide central bandwidths, it remains challenging to map forests finely and accurately by solving the common problem of foreign bodies with spectra similar to that of tree species with multispectral images of differing resolution [10–13].

With today's continuous improvement in multispectral satellite revisit cycles, temporal characteristics have proven to have high application value for forest classification [14,15]. The time-series normalized differential vegetation index (NDVI) was generally considered to be applicable for identifying vegetation and extracting phenological information [16–18], which plays a key role in classifying the dominant tree species and overcoming the problem of foreign bodies with similar spectra [19,20]. Given this advantage, some studies have confirmed the use of time-series NDVI data and phenological characters would raise the mapping accuracy of forest types in both urban areas [13,21] and rural areas [18,22–24].

MODIS and NOAA satellites can provide images with very short revisit intervals (less than 2 days). Therefore, most previous researches have discussed the results of forest composition mapping and dynamic detection based on time-series NDVI data from a regional to global scale [5,25]. However, the spatial resolution of traditional time-series images such as NOAA AVHRR [26], SPOT Vegetation [27], and MODIS imagery data [28] generally range from hectometers to kilometers, resulting in poor spectral purity and limited identification of broad forest types (such as coniferous or broad-leafed forests). Hence, medium-to-low resolution remote sensing images were considered to be applicable to classify general forest-cover types [29,30] in which finer tree species composition information is not involved [9,31]. As a rule, medium-high resolution satellite images can be used to produce more accurate results of forest species composition by providing detailed spectral features of the canopy of dominant tree species [32,33].

An increasing number of medium- and high-resolution Earth-observation satellites have now entered use in recent years (e.g., Sentinel-2, SPOT-6 and -7, Gaofen-1 and -2, etc.). Some multitemporal medium-high-resolution NDVI images have proven to be applicable to classify the dominant tree species in forests [11,34–36], but less attention has been focused on spatial mapping [37]. In addition to the time-series spectral dynamics, plant phenology should be equally considered for tree species mapping [38–41]. Therefore, it is essential to use multiscale time-series data to explore the effect of image resolution and phenology information on the accuracy of forest tree species classification. The study of vegetation-phenology dynamics by remote sensing mainly involves extracting the corresponding key time nodes and characteristic values by analyzing the significant variations in vegetation index time series data, which are commonly called LSP metrics [11,17,42,43]. LSP metrics are important status parameters for land-vegetation ecosystems, especially at the start of spring (SOS) [44], also called the start of the growing season [45] or the green-up date [46]. These parameters are the most commonly extracted because of their importance in determining the growing season, and they are also powerful indicators of an ecosystem's response to climate change [44].

To date, LSP metrics have been used to identify different vegetation species. For example, Liu et al. (2018) used the combination of fused 30 m time-series NDVI data and phenological metrics to map rice paddies, and they found that using both of the selected NDVI and phenological features could allow the highest accuracy in extracting rice paddy areas [47]. Another recent study suggested that phenological metrics can help improve the accuracy of remote-sensing recognition of different forest stands [17]. In addition, Schwieder et al. (2018) demonstrated that adding phenological information from multitemporal imagery improves the estimation of aboveground biomass through modified discrimination of vegetation types [48]. However, few studies involving the mapping and phenology of forest tree species have integrated different spatial observation scales with multiscale satellite images, which is necessary to compensate for the lack of research on higher-resolution time-series data.

The mapping results of forest types or tree species through remote sensing would be influenced by the spatial resolution of remote sensing data obviously [6,7,49,50]. In addition, most previous studies have focused on single time-phase imagery, which led to no results available on temporal data [51]. In fact, both species classification and phenological metrics are affected by the spatial resolution of remote sensing data. For example, Zhang et al. (2017) explored how spatial resolution affects the start of vegetation season (SOS) by comparing observations obtained from images with 30 m and 500 m spatial resolution, and found the difference of SOS at the high and low resolution was larger in heterogeneous regions [52]. Tian et al. (2020) investigated how spatial resolution affects the difference in spring phenology of rural-urban vegetation as detected by resampling the satellite data from 30 m to 8 km spatial resolution. They found that coarser images overestimate the urbanization effects and that the SOS obtained by coarser time-series NDVI data would be earlier than that of the actual result [53]. It should be noted that the multiresolution imagery data used in the above researches consists of resampled multispectral images by using up-scaled algorithms (e.g., Nearest Neighbor (NN) and pixel aggregation (PA)), which generate a set of lower spatial resolution images using the high-resolution image through pixel resampling. However, the imaging mechanism of these up-scaled images differs from that of satellite imagery data obviously, which may lead to some uncertainty in the final results of relevant studies [51,54,55]. In consequence, it could be assumed that the phenological monitoring results obtained by up-scaled time-series NDVI data may differ from that detected from time-series satellite imagery, which accounts for the necessity to re-evaluate the results associated with time-series resampled images [53].

In this study, we systematically evaluate how temporal images with a medium-high spatial resolution (4, 10, 16, and 30 m) affect the accuracy of forest tree species mapping and the variability of spring phenological information. The multiscale time-series satellite remote sensing data were obtained from Gaofen-2 (4 m), Sentinel-2 (10 m), Gaofen-1 (16 m), and Landsat-8 (30 m), respectively, acquired over seasonal deciduous forests in 2018 from a state forest farm region in Harqin Banner, China. In addition, we evaluate the performances obtained by using up-scaled time-series images of different spatial resolutions. Eight phenological metrics are extracted to classify forest tree species based on the random forest (RF) algorithm combined with time-series NDVI data. Accordingly, we tried to find out (1) what is the regular pattern for monitoring spring phenology within the medium-high remote sensing spatial resolution range; (2) which scales of time-series images most effectively reflect spectral differences in each forest tree species and the relationship between these phenological metrics and tree species classification; and (3) whether the spatial patterns in up-scaled images are competitive to map the distribution of forest tree species from multiscale satellite datasets. Answers to these points should improve the credibility of using time-series remote sensing images to monitor and map forest tree species for use in forest resource management or regional eco-climate models.

2. Materials and Methods

2.1. Study Area

The study area is in northeast China near the city of Chifeng and has a size of approximately 55,100 ha of which over 45,500 ha are forested landscape. An overview of the area is given in Figure 1. The study area belongs to the border area of the Greater Hinggan Mountains and Yanshan Mountains within a range of 878 to 1890 m, and the climate here is temperate monsoon. The dominant tree species occurring in the local forest are deciduous forest dominated by *Quercus mongolica* (Qm), *Populus davidiana* (Pd), *Betula platyphylla* (Bp), and *Larix gmelinii* (Lg), accounting for about 84% of the total forest area in 2018; there is also an evergreen forest, which is dominated by *Pinus tabulaeformis* (Pt). The five dominant tree species mentioned above account for over 95% of the total forest area in 2018 [19].

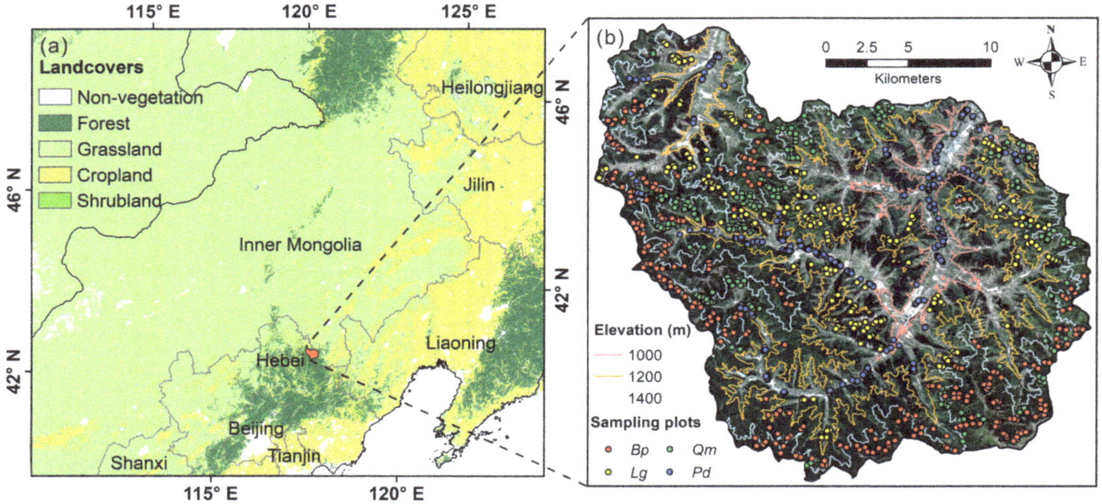

Figure 1. The study area located in northeast China with major vegetation cover types in the background (**a**), and sampling plots of the four deciduous tree species, including *Betula platyphylla* (Bp; n = 330), *Larix gmelinii* (Lg; n = 330), *Quercus mongolica* (Qm; n = 331), and *Populus davidiana* (Pd; n = 329), overlaid on a Landsat-8 imagery (21 June 2018) in the background (**b**).

2.2. Methods

The methodological workflow is illustrated by the flowchart presented in Figure 2. The research process includes four main steps. First, the collection and preprocessing of both remote sensing data and ground observation data were completed. Next, we calculated forest phenological metrics based on multiscale time-series data and analyzed the scale effect of spring phenology. Then, forest tree-species mapping was performed using satellites and up-scaled time-series NDVI data and phenological metrics with different spatial resolutions. Finally, comparisons of multiple considerations on the accuracy of forest mapping were conducted to illustrate how spatial resolution affects tree-species classification.

2.2.1. Data Resources and Preprocessing

We selected four sets of multiresolution remote sensing images acquired once per month to make a series of spatial scale images respectively (Table A1 in Appendix A) and acquired with cloudiness less than 15%. The main parameters of the images used in this study can be seen in Table 1. This study used a total of 48 satellite remote sensing images covering the entire study area; there were 46 of them acquired in 2018 while the other two with poor imaging quality were replaced by the data of the same period in 2017, which was proven to be feasible [19]. All Sentinel-2 images were obtained from the European Space Agency (ESA; https://scihub.copernicus.eu; accessed on 18 December 2019); Landsat-8 images were obtained from the United States Geological Survey (USGS; http://glovis.usgs.gov; accessed on 15 January 2020); Gaofen-1 and Gaofen-2 images were obtained from the China Center for Resources Satellite Data and Application (CCRSDA; http://www.cresda.com/CN; accessed on 12 September 2020).

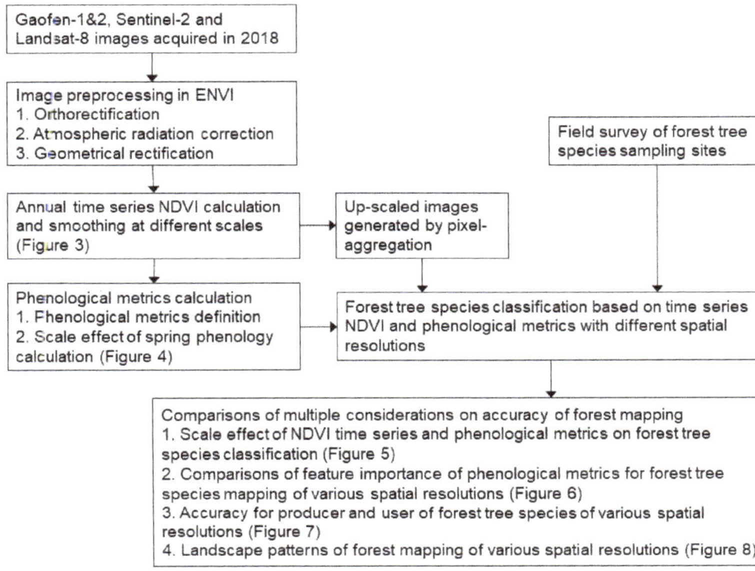

Figure 2. The methodological framework developed in this study.

Table 1. The main parameters of satellite images used in this study.

Satellite Sensor	Revisit Interval (day)	Spatial Resolution (m)	Swath Width (km)	Radiometric Resolution (bit)	Blue (μm)	Green (μm)	Red (μm)	Near-Infrared (μm)
Gaofen-2 PMS	4	4	45	10	0.45–0.52	0.52–0.59	0.63–0.69	0.77–0.89
Sentinel-2 MSI	5	10	290	12	0.46–0.52	0.54–0.58	0.65–0.68	0.79–0.90
Gaofen-1 WFV	4	16	800	10	0.45–0.52	0.52–0.59	0.63–0.69	0.77–0.89
Landsat-8 OLI	16	30	185	12	0.45–0.52	0.53–0.60	0.63–0.68	0.85–0.89

Before analysis, the four sets of time-series datasets with different spatial resolutions were carefully harmonized. A Savitzky–Golay (SG) algorithm served for polynomial filtering of the time-series NDVI data because it better maintains the temporal vegetation dynamics and minimizes atmospheric effects [56,57]. This also implied a smoothing and filtering of the four time-series could remove undesired artifacts due to poor atmospheric conditions and undetected clouds [58]. The four sets of multiscale time-series data preprocessing by using a series of related models and algorithms in ENVI (Environment for Visualizing Images; version 5.3) and IDL (Interactive Data Language; version 8.5) software platform (Research Systems Inc., Boulder, CO, USA). The orthorectification, atmospheric radiation correction, and geometrical rectification were carried out in sequence to preprocess all images used in this study, to transform the digital value into the actual surface spectral reflectance at different image resolutions (Figure 2). The radiometric correction methods include the incident causational matrix (ICAM), and pseudo-invariant feature (PIF) models [59], and the corresponding parameters of radiation correction of different satellite remote sensing data were obtained from their official websites, that is CCRSDA, ESA, and USGS, respectively. In addition, the geometric root means squared error (RMSE) of all images was controlled within 0.5 pixels.

The PA algorithm was selected for up-scaled images (Figure 2). PA assigns different weights to each pixel based on mapping distance which is integrated at the ENVI software, and it is widely used in researches of multiscale remote sensing [46,53]. One related study suggests that PA better maintains pixel values than do other up-scaled algorithms (e.g., nearest neighbor, bilinear, and cubic convolution) [51]. The NDVI data then were calculated as Equation (1). To ensure that different time series can be compared to each other, the up-scaled datasets were smoothed in the same way as the satellite data. To better analyze the timing of (phenological) events, the (smoothed) monthly NDVI series were saved as 30 daily values on the 15th of each month were assigned as DOY (day of the year; Julian date) for the compositing period of the month.

$$\mathrm{NDVI} = \frac{\rho_{\mathrm{NIR}} - \rho_{\mathrm{red}}}{\rho_{\mathrm{NIR}} + \rho_{\mathrm{red}}} \quad (1)$$

where ρ_{NIR} and ρ_{red} are the surface reflectance of near-infrared (NIR) and red bands.

2.2.2. Collection of Forest Inventory Data

The field surveys in the study area were undertaken in July 2018 and May 2019. A total of 1320 sample plots (30 m× 30 m) were selected randomly. Approximately 330 sample sites for each deciduous dominant tree species (Table A2 in Appendix A) are shown in Figure 1 (more than 85% of the tree population is single species). Then, 70% of these data were randomly set as training samples and the rest as verification samples. Therefore, the spectral reflectance of different dominant tree species was extracted from the center of each stand plaque at each sample point in the imagery to guarantee the spectral purity of the different image pixels.

2.2.3. Calculation of Forest Phenological Metrics

From the time series NDVI images with different spatial resolutions for related functions in TIMESAT [60], eight typical phenological metrics were obtained: the start of growth season (SOS), end of growth season (EOS), length of growth season (LOS), peaking time of growth season (POS), length of the peak-time (LOP), middle of SOS (SOS_m), middle of EOS (EOS_m), and amplitude (AMP). All the mentioned phenology-related metrics are noted as the Julian date. Here, both the SOS and EOS were calculated from the fitting function when the trees grow to a certain fixed time phase, e.g., the time position ranges from the position where 10–30% of the left (right) minimum of the NDVI value and the maximum NDVI value [61,62], which was indicated as a suitable time point. Nevertheless, the threshold of the determined phenological time points ranged around 20% according to the geographical locations and tree species selected [19,63,64]. The LOS is further calculated as the difference between the EOS and SOS, the POS is identified as the date when the trees grew to the maximum NDVI value at the fitted temporal dynamic curve, then the LOP is the length between 80% of the left and right maximum NDVI of the temporal dynamic curve. The amplitude (AMP) presents the maximum ranges of the tree photosynthetic dynamics across the whole growing season, and SOS_m and EOS_m stand for the average days of the growing season when the trees grow above their 80% level on the right and left side of the peaking time. A more detailed description of the calculated metrics can be found in Schwieder et al. (2018) [48] and Lebrini et al. (2019) [65].

2.2.4. Classification and Accuracy Assessment

Supervised classifiers are believed to be more clearly preferable while the prior knowledge for ground objects is enough [66]. In this study, we used the Random Forest (RF) classifier to identify deciduous tree species, and set its *ntree* to 500 and the *mtry* to the square root of the number of input features, as suggested [67–69]. RF is derived from statistical learning theory to process high-dimension datasets and reduce the overfitting issues [69–71], which was considered to be one of the most robust classifiers compared with other algorithms, such as maximum likelihood (ML), linear discriminant analysis (LDA), support vector machine (SVM), and decision tree classifiers (DTC) [72–74]. Therefore, it integrates hundreds of decision-making trees and encapsulates an important ranking predictor (the mean decreases of accuracy; MDA), which was designed to be used to evaluate the importance of variables in the accuracy efficiently [75]. Then, we extracted the forest boundary and used winter images to separate the dominant evergreen forest landscape.

Based on this, we focused on the remote-sensing classification results for the four deciduous forest tree species: *Betula platyphylla* (Bp), *Populus davidiana* (Pd), *Quercus mongolica* (Qm), and *Larix gmelinii* (Lg) based on time-series NDVI data and the combination of NDVI and phenological metrics.

In addition, we calculate the overall accuracy (OA), kappa, and the accuracies of producer and user to examine and compare each result from different image datasets by constructing corresponding error matrices [76]. By randomly selecting training samples and validation samples, each set of classification results was cross-validated 10 times and averaged. The one-way analysis of variance (ANOVA) was used to decide significant differences in the classification results among the different spatial resolutions and methods. This was followed by multiple comparisons using the least-significant difference (LSD) to identify where the differences lay ($p < 0.05$).

3. Results

3.1. Multiscale Sequence NDVI Curve of Different Deciduous Forest Stands

The annual NDVI curves of the four deciduous forest species with spatial resolution range from 4 to 30 m show obvious unimodal characteristics, with clear forest stand growth stage characteristics (Figure 3). However, the mean NDVI values of four dominant deciduous tree species in the study area were significantly different, in especial, the spectral reflectance captured in 4 m high-resolution images is significantly less than that in lower-spatial-resolution images ($p < 0.05$). In addition, the NDVI curves of different dominant tree species differ, where the NDVI curve of Qm is higher than that of the other deciduous tree species, and that of Pd is the lowest.

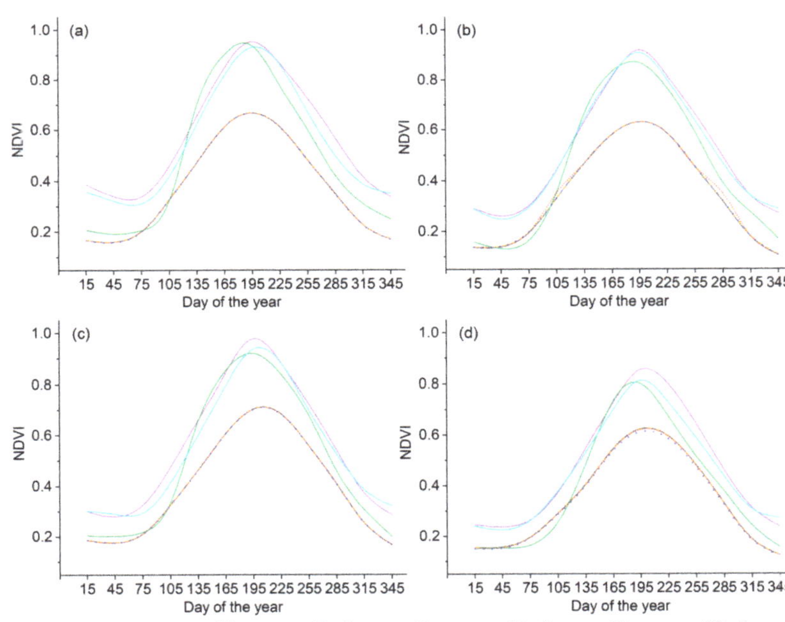

Figure 3. Time-series image training NDVI curves for different deciduous tree species of (**a**) *Betula platyphylla*, (**b**) *Larix gmelinii*, (**c**) *Quercus mongolica*, (**d**) *Populus davidiana*. The symbol # refers to the up-scaled images, the same below.

In terms of the up-scaled NDVI curves, there are no significant changes in almost all NDVI data observed ($p > 0.05$), which is significantly below the NDVI value from multiscale satellite images at each given spatial resolution ($p < 0.05$), meaning the up-scaled NDVI curves were underestimated compared to that obtained from actual satellite data at the same resolution.

3.2. Multiscale SOS Results by Satellite Images

The SOS results of multiscale images of different dominant tree species differ significantly ($p < 0.05$; Figure 4). For different tree species, the mean SOS for Lg (109–121) is the earliest, Bp (117–129) is relatively later than all species (including four deciduous tree species; 116–128), and Pd (118–134) is slightly earlier than Qm (122–132), showing the phenological difference among the four deciduous tree species in this study region. For different spatial resolutions, almost all dominant deciduous tree species reveal a significant difference in the mean SOS based on native multiscale images at the same spatial resolution from 4 to 30 m ($p < 0.05$). Overall, the mean SOS increases from 4 to 16 m and then decreases slightly to 30 m ($p < 0.05$). Similarly, the SOS results of all species first increase and reach a peak at 16 m resolution and then begin to decrease as the resolution comes to 30 m ($p < 0.05$), where the mean SOS delay time ranges from 8.02 to 11.95 days.

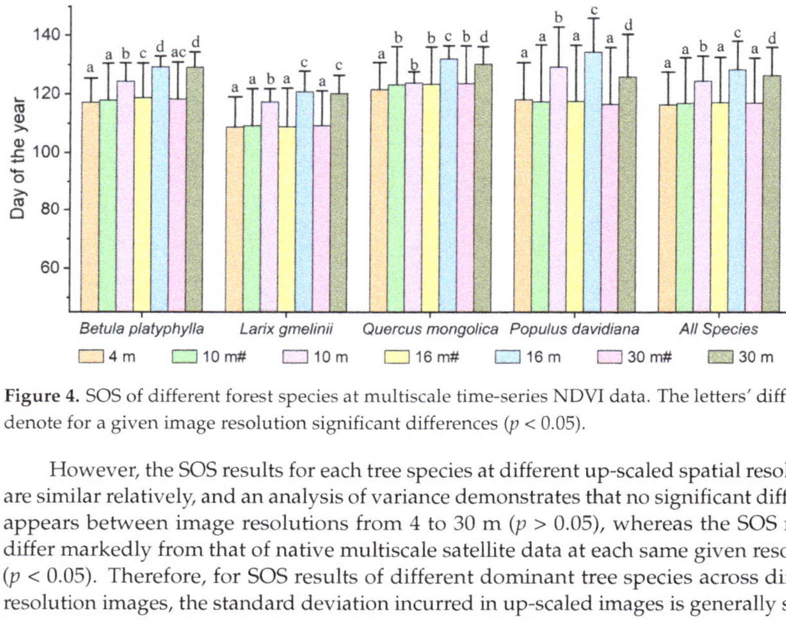

Figure 4. SOS of different forest species at multiscale time-series NDVI data. The letters' differences denote for a given image resolution significant differences ($p < 0.05$).

However, the SOS results for each tree species at different up-scaled spatial resolutions are similar relatively, and an analysis of variance demonstrates that no significant difference appears between image resolutions from 4 to 30 m ($p > 0.05$), whereas the SOS results differ markedly from that of native multiscale satellite data at each same given resolution ($p < 0.05$). Therefore, for SOS results of different dominant tree species across different resolution images, the standard deviation incurred in up-scaled images is generally several times greater than that of native satellite images with the same resolution, ranging from 1.33 to 3.39 ($p < 0.05$).

3.3. Multiscale Classification by Satellite Images and Up-Scaled Images

There were significant differences among the OA values of the dominant forest species at different spatial resolutions ($p < 0.05$; Figure 5a), while the Kappa values show consistent results ($p < 0.05$; Figure 5b). For multiscale satellite time-series NDVI images, the mean OA (Kappa) first increases while image resolution goes from 4 to 10 m ($p < 0.05$) at the highest accuracy of 83.63% (0.7825), and then decreases to the minimum accuracy of 78.60% (0.7154) at 30 m resolution ($p < 0.05$). The classification accuracies of forest tree species are closely related to the image resolutions, and an analysis of variance reveals significant differences between classification accuracies based on time-series satellite data of differing resolution ($p < 0.05$). Combining phenological metrics and time-series NDVI data can improve the mapping of the regional dominant tree species at each given resolution ($p < 0.05$; Figure 5).

When the contrast is integrated with time-series NDVI and LSP metrics, the best overall accuracy (Kappa) of forested landscape identification with 10 m resolution increases to 86.05% (0.8147; $p < 0.05$).

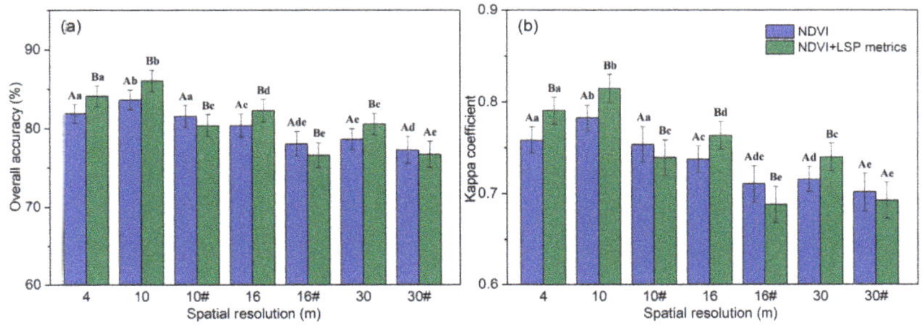

Figure 5. The OA (**a**) and kappa (**b**) of classification results of deciduous species at different image resolutions. The uppercase letters differences above the error bar refer to a significant difference between the two-given datasets, while lowercase letters differences denote for a given image resolution significant differences ($p < 0.05$).

However, for up-scaled time-series images, the classification accuracy of the dominant tree species decreases from 81.88% (0.7583) to 77.23% (0.7014) with the spatial resolution ranging from 4 to 30 m# ($p < 0.05$). Therefore, there is no significant difference appears between the classification accuracies of dominant tree species of up-scaled time-series data at the resolution of 16 and 30 m ($p > 0.05$). Finally, combining phenological metrics and time-series NDVI data does not increase the recognition accuracy but leads to the decline from 10 to 30 m# resolution, which differs obviously from the results of native multiscale satellite data at each same given resolution ($p < 0.05$).

3.4. Contributions of Different LSP Metrics to Multiscale Tree Species Classification

The feature importance for identifying forest tree species using the eight LSP metrics of four image resolutions are different (Figure 6). Before analysis, each final feature importance for variables was averaged after ten cross-verifications. The results show that the LSP metrics are valuable for differentiating among forest tree species in the study region ($p < 0.05$). The various LSP metrics produce different positive effects on the classification of dominant tree species, although the magnitude of the influence varies ($p < 0.05$; Figure 6). When considering the four different spatial resolutions, the importance of each LSP metric turns out to be inhomogeneous (Figure 6). The phenological information such as SOS_m, EOS, and POS are more important at higher image resolution, whereas only the LOS achieves the highest importance at lower spatial resolution. In addition, AMP and SOS show roughly the same degree of importance at different resolutions.

Testing the cumulative importance of each LSP metric for individual spatial resolutions reveals the importance of each variable from medium to high resolutions (Figure 6). At first, the LSP metrics in AMP, SOS_m, and SOS are by far the three most important features ($p < 0.05$), showing the middle and early growth period account for the most important contribution in tree-species phenological observations. They are followed by LOP and LOS, which control the length of the growth period and the peak period. Subsequently, the remaining EOS_m and EOS metrics represent the period of declining growth which has less importance. Finally, POS is the least important metric which indicates the peak time, and the differences of POS among the different dominant tree species are small.

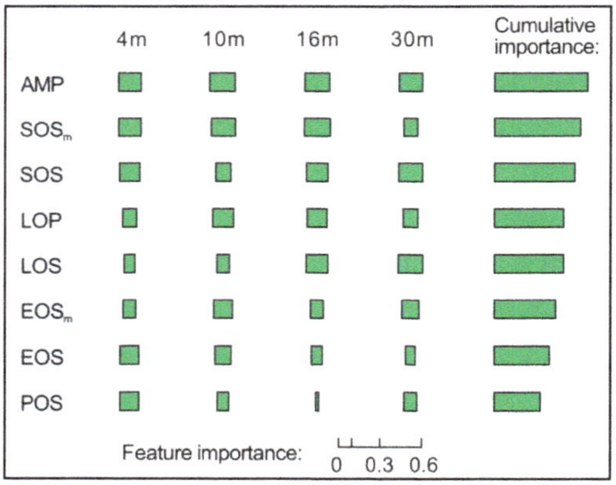

Figure 6. Feature importance for the forest tree species identification using the eight LSP metrics across different image resolutions. Cumulative bars on the side illustrate the sum of the importance for each phenological metric.

3.5. Comparisons of Accuracy for Producer and User of Dominant Tree Species

In native satellite images, the PA of Bp, Lg, Qm, Pd, and the UA of Bp, Lg, Qm show a downward trend on the whole with the decrease in spatial resolution goes from 4 to 30 m; except for the UA of Pd which increases at first and then decreases (Figure 7). None of the dominant classes of tree species have both a low UA and PA. The confusion matrix shows that many Bp, Lg, and Qm are misclassified into Pd in different spatial resolutions, which also leads to an over-mapping for Pd, so the UA of Pd is less than the PA from 4 to 30 m resolution, especially at 4 m. For the mapping results of forest tree species landscape (Figure 8), the comparison shows that the worst classification result occurs in imagery with 30 m resolution, where a portion of Qm and Lg is misclassified as Pd. In addition, some Lg are misclassified as Qm at 16 m resolution and the overall forest mapping at 10 m resolution is better than at 4 m resolution, which is mainly due to the higher mapping accuracy of Qm and Bp.

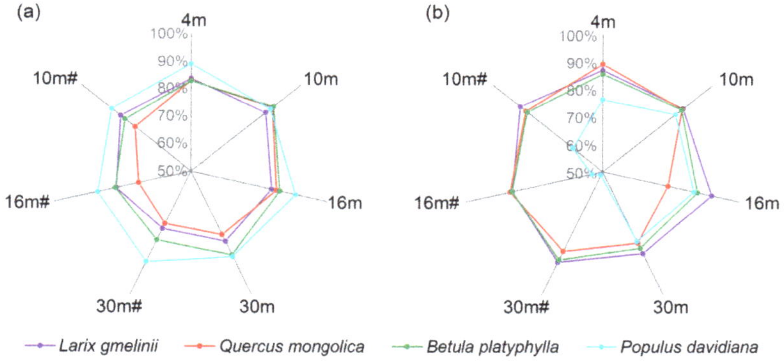

Figure 7. Comparison of accuracy for the producer (**a**) and user (**b**) of four deciduous species combining LSP metrics and NDVI from multiscale time-series images.

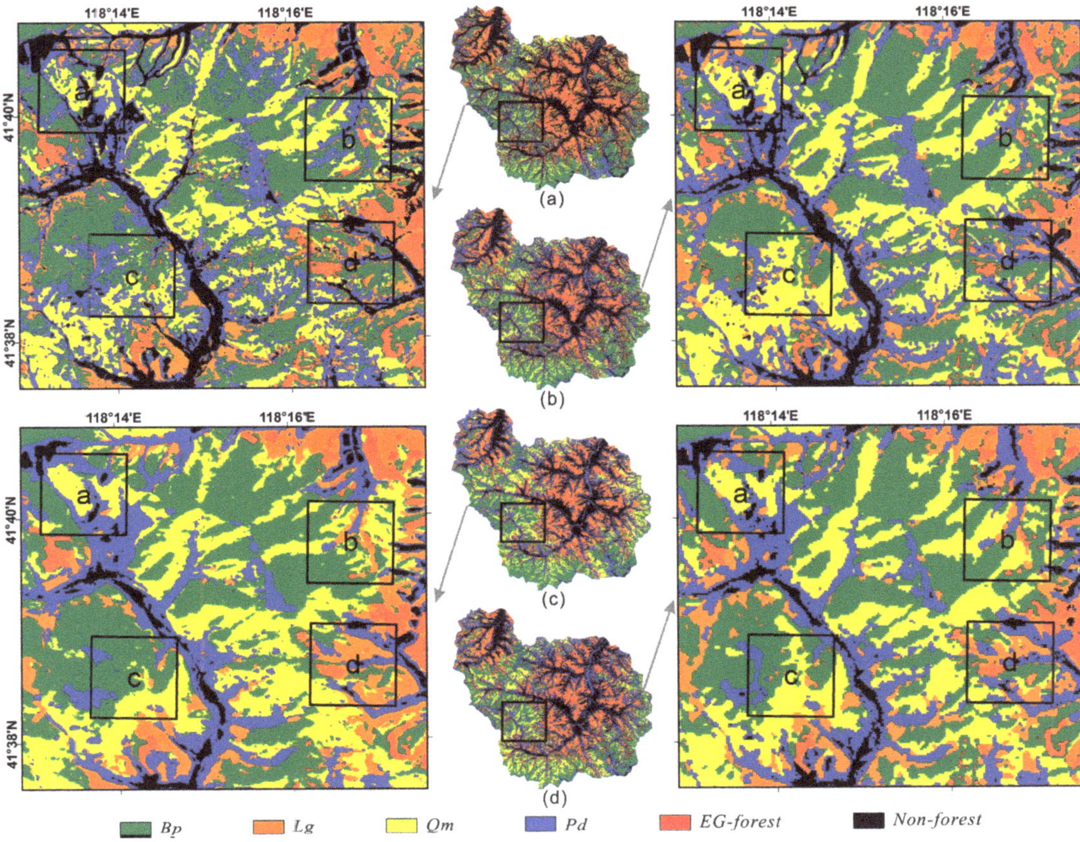

Figure 8. The distributions of forest tree species landscape in the study region based on the combination of time-series NDVI and LSP metrics in different resolutions, including (**a**) 4 m, (**b**) 10 m, (**c**) 16 m, and (**d**) 30 m. EG-forest means evergreen forest.

However, for up-scaled images, both PA and UA values of different deciduous species change only slightly when the spatial resolution goes down (Figure 7). The UA of Bp, Lg, and Qm and the PA of Pd change slightly from 4 to 30 m# resolution. At the same time, the UA of Bp, Lg, and Qm are significantly higher than their PA, except for Pd, implying over-mapping occurred. On the contrary, the forest tree species with high UA but low PA refer to under-mapping [77]. In this study, low PA of Bp, Lg, and Qm show that large areas of Bp, Lg, and Qm are not properly classified into their respective class, and their PA decrease as the image resolution goes down (Figure 7). In general, PA and UA of the most tree species for up-scaled data are often diminished and more unstable than that for real satellite data, furthermore, almost all tree species reaching their maximum value at the image resolution of 4 m.

4. Discussion

Spatial resolution significantly affects the identification of forest species and mapping based on single-date images [7,12,50,78]. Some studies have indicated that the NDVI temporal variation of stand canopy shows the detailed dynamics of timeline-based spectra of different forest types [14,15]. However, less attention has focused on the spatial scale of the uncertainty in the results. The present study confirms that the overall accuracy gradually extends from 78.59% to 83.63%, while the kappa synchronously increases from

0.7154 to 0.7825, which indicates that the classification accuracy of the dominant tree species is strongly linked to the spatial resolution of the time-series satellite data ($p < 0.05$). In this study, for a spatial resolution of 4 and 10 m, the classification accuracy is high (>81.88%), where the forest landscape identification is apparent, and each dominant species are classified correctly ($p < 0.05$). Overall, the time-series images with higher resolution (4–10 m) providing higher and more stable OA and kappa perform better in identifying the regional forest tree species than those with lower resolution (16–30 m). However, it does not mean that images with finer resolution can always provide more accurate results because it also depends on the selected feature datasets and classification methods.

The mean NDVI values of 4 m images are less than that of 10 m ($p < 0.05$; Figure 3), which we attribute to the NDVI being prematurely saturated at the higher resolution because a higher resolution correlates with less energy incident on the sensor, which decreases the spectral resolution [51,79]. Hence, the annual peak NDVI is mostly below 0.7 for the various dominant deciduous species at 4 m resolution, which is significantly less than the NDVI at the resolution of 10, 16, and 30 m, respectively ($p < 0.05$; Figure 3). Furthermore, the result is also assisted by the narrower near-infrared band and higher radiation resolution of Sentinel-2 image (12 bit, whereas 10 bit of the Gaofen-2 image), which could provide a better capability of vegetation information monitoring. It actually may overcome the negative effects of its coarser resolution to some extent. In addition, the multiscale annual mean NDVI of the sample data for different forest land remains essentially stable for up-scaled data (Figure 3). The spectral information in up-scaled images has a strong dependence on the input high-resolution images, which leads to discrepancies in comparison to actual imaging results at the given spatial resolution (Figure 3). As the image resolution decreases, the spectral information related to the forest tree species was smoothed which would increase omission and commission errors [80]. These results imply that the resampled time-series data have different spectral reflectance in comparison to native images acquired at the same resolution evidently ($p < 0.05$), which reduces the susceptibility to spatial resolution and barely represents the satellite image features of the same scale due to the spectral distortion [51]. Therefore, the classification results of tree species based on up-scaled images differ from those of native images for a given resolution ($p < 0.05$), indicating that native satellite data cannot be accurately replaced by up-scaled images.

Satellite remote sensing imagery is widely applied to monitor forest phenology on global and regional scales [42–44]. Plant phenology provides valuable information for classifying vegetation types [43], and some studies use spectral and phenological parameters to map crops [52,65,81]. However, research on mapping forest tree species has not considered the phenological metrics with images at a medium-high spatial resolution, which might introduce uncertainties in identifying and mapping different tree species [19]. Our classification results combined with NDVI and LSP metrics increases the OA (Kappa) by 2.24% (0.0322), 2.42% (0.0322), 1.89% (0.0258), and 1.94% (0.0243) at 4, 10, 16, and 30 m, respectively. For the 10 m time-series data, all deciduous tree species are correctly classified into their respective categories (Figure 8), and their UA and PA can reach more than 83.12%. These results show that using spatial resolution NDVI data combined with temporal phenological characteristics improves the accuracy of the results at each scale from 4 to 30 m ($p < 0.05$), which indicates that combining the spectral curve and the LSP metrics is a good way to identify and monitor the main forest ecosystem and may improve the precision of spatial mapping in temperate regions from medium to high spatial resolution.

Although differences appear in the phenological metrics at different resolutions, to some extent they have positive effects on the classification of forest tree species in this study region (Figure 6). The three highest phenological variables are AMP, SOS_m, and SOS, based on the contribution of different phenological metrics and the classification of tree species. This indicates that the growth stage from germination to maximum growth and the amplitude of the growth spectrum contribute the most to the identification of the tree species. In addition, after combining the phenology metrics, the accuracy of the classification results based on the up-scaled data decreases by different degrees compared with the previous results (Figure 5). This is mainly contributed to the decrease in effective spectral information after the smoothing of up-scaled NDVI data and the increase in invalid band information, which would increase feature redundancy, leading to the further degradation of classification results [51,82].

Considering the special significance of phenology in spring [38,44,53], many previous studies were conducted to investigate the effects of forest spring phenology with different resolutions. However, most of them used up-scaled images due to a lack of multiresolution satellite data covering the area with a shorter revisit period [83]. Therefore, we investigate herein how scaling affects spring phenology (SOS) via a comparison between results from low- and high-resolution data. The results show that retrievals of SOS date reflect the difference in spatial scale, resulting in a delay as the resolution worsens. In this study, the SOS results of all species of Landsat-8 data (30 m) are earlier than that of Genfen-1 data (16 m). This result is mainly attributed to the better radiometric precision of Landsat-8 OLI (12 bit, whereas 10 bit of the Genfen-1) and the spectral range refinement of the near-infrared band (Table 1), which improves the overall signal to noise ratio for vegetation monitoring [51] and somewhat avoid overestimation [44] and further reasons need to be verified in follow-up studies. Relative to data with a medium-low resolution that is commonly used to monitor vegetation phenology, medium-high resolution satellites can achieve higher classification accuracy. This is because the observation of deciduous vegetation phenological characteristics is vulnerable to the interference of the evergreen forests or other background factors such as the stand age and topography [84,85]. As a consequence, medium-high resolution satellites can observe more spatial detail information and therefore capture the spectral variation of vegetation in phenology which is not normally detectable in data from sensors with lower spatial resolution.

However, the results show that estimations of spring phenology from up-scaled imagery have little noticeable change, corresponding less to the results derived from multiscale satellite remote sensing data. These findings are consistent with related studies based on up-scaled data. For example, Tian et al. (2021) resampled time-series images with 10 m to a series of lower spatial resolution images from 30 m to 8 km and obtained relatively stable vegetation SOS data in rural areas with a resolution ranging from 10 m to 8 km [53]. Similarly, another study found that the overall SOS averaged from images with a resolution ranging from 1 × 250 m to 35 × 250 m is generally similar but with a difference of fewer than five days [44]. Therefore, Zhang et al. (2017) suggested SOS datasets should be calculated from actual time-series images of different spatial scales but not the time-series data up-scaled from finer resolution imagery data using simple resampling methods [52].

In this study, both phenological monitoring and the classification of forest tree species differ significantly when using multiscale satellite images and up-scaled images because of the differences in the imaging mechanism. Related research revealed the difference between the native satellite data and up-scaled data is determined by the mechanism of remote sensing imaging, in which the most up-scaled results are spatial insensitive because of not considering the spatial adjacent information of landscapes [51,86], nor adjusting as the variation of spectral reflection properties at different spatial resolutions [87]. Therefore, the up-scaled images cannot be directly used to replace remote sensing images in time-series monitoring or quantitative analysis. Therefore, methods are gradually being developed to scale-transformation model involving physical mechanism [86–89] and temporal–spatial fusion algorithms [23,53,90], to ensure image features from different spatial resolutions can be integrated into applications.

This study expands on previous work in two important ways: First, we continue to compare tree species classification for multiscale time-series images and explore the applicability of remote sensing images to the study of LSP metrics. Next, we discuss how spatial-scale uncertainty affects phenological observations and the classification of dominant tree species. In addition, we try to construct herein monthly time-series multiscale satellite images with resolutions ranging from 4 to 30 m. In fact, the monthly time-series data has proven to suffice to describe the one-year spectral curve dynamics of forest tree species when cloud-free imagery is available at specific phenological stages [17,24,70]. However, some studies have shown that more intensive time-series images would be better at capturing fine dynamic differences to separate information on similar forest tree species [19,91]. With recent breakthroughs in multispectral remote-sensing, the temporal-spatial resolution of satellite data is improved obviously. Using imagery from more advanced sensors will enhance the precision of forest phenology monitoring and tree species classification. Therefore, more intensive time-series multiscale images are required to extend the results in this paper.

5. Conclusions

The spatial resolution of satellite remote sensing data has a major impact on forest information extraction and dynamic monitoring, and it is critical to the success of regional dominant tree species mapping. This study develops a process for reconstructing time-series satellite images with medium-high spatial resolution and applies it to monitor forest phenology and map tree species. The study also examines how time-series spectral data responds to the scaling effect obtained from various spatial resolution satellites. Moreover, we discuss how spatial resolution and LSP metrics affect the classification of dominant tree species. The results indicate that remote sensing images with 10 m resolution are more appropriate for the time-series-based forest tree-species classification with superior performance in the study region. In addition, the use of additional LSP metrics further improves the classification results, so they are highly recommended.

Nowadays, though the temporal resolution of remote sensing data is improving, few pieces of research have been designed to investigate how up-scaled time-series images affect forest mapping on landscapes. The present results elaborate the general pattern of time-series remote-sensing tree species identification and spring phenology extraction with medium-high spatial resolution. Therefore, a significant difference was found by contrasting the forest landscape classification results obtained from real multiscale time-series satellite data with that of up-scaled time-series images. This work thus provides a primary understanding of how to use and compare time-series remote sensing images with different spatial resolutions for forest monitoring.

Author Contributions: Conceptualization and Methodology, K.X.; Software and Data Curation, Z.Z. and W.Y.; Formal analysis, J.Y. and W.Y.; Validation, P.Z. and Y.D.; Writing—Original Draft, K.X.; Writing—Review and Editing, Z.Z. and J.Y.; Data collection and processing K.X. and J.G. All authors have read and agreed to the published version of the manuscript.

Funding: This research was funded by the Anhui Provincial Natural Science Foundation of China, grant number 2008085QD193, and the Fundamental Research Funds for the Central Universities of China, grant number JZ2020HGQA0189 and JZ2021HGTA0166.

Data Availability Statement: The data that we used in this study can be requested by contacting the corresponding author.

Acknowledgments: The authors thank Shengnan Jiang for her assistance with the revision of the manuscript.

Conflicts of Interest: The authors declare no conflict of interest.

Appendix A

Table A1. The information of satellite images used in this study.

Month	Imaging Date (Gaofen-2)	Imaging Date (Sentinel-2)	Imaging Date (Gaofen-1)	Imaging Date (Landsat-8)
January	20180121	20180120	20180117	20180111
February	20180215	20180214	20180214	20180212
March	20180316	20180311	20180311	20180316
April	20180430	20180415	20180418	20180417
May	20180510	20180515	20180515	20180519
June	20180608	20180614	20180621	20180620
July	20180722	20180729	20180722	20180706
August	20180816	20180823	20180820	20180823
September	20170920	20170922	20180916	20180924
October	20181019	20181017	20181019	20181025
November	20181118	20181121	20181120	20181126
December	20181213	20181216	20181215	20181213

Table A2. Characteristics of the dominant tree species in this study.

Species Name	Picture	Main Characteristics
Larix gmelinii (Lg)		A deciduous tree (30 m), D.B.H. 90 cm or so. The barks are taupe, and the branches consist of obvious long branches and short branches. Light-loving, cold-resistant, drought-resistant. Mainly distributed in Northeast China as well as the mountainous areas from eastern Inner Mongolia to eastern Siberia of Russia.
Populus davidiana (Pd)		A deciduous tree (25 m), D.B.H. 60 cm or so. The barks are gray or gray-green, and the leaves are ovate or nearly round with incised margin. Light-loving, cold-resistant, and soil are adaptable. Mainly distributed in valleys of high mountains in Northeast China, Northwest China, North China and Southwest China.
Betula platyphylla (Bp)		A deciduous tree (27 m), D.B.H. 80 cm or so. The barks are white and smooth or local cracking, and the leaves are ovate with a jagged margin. Light-loving, cold-loving, and soil adaptable. Mainly distributed in Northeast China, North China and the mountainous areas on the outer edge of the Qinghai-Tibet Plateau.
Quercus mongolica (Qm)		A deciduous tree (30 m), D.B.H. 60 cm or so. The barks are dark taupe, the branches are purplish-brown, the leaves are obovate with around margin. Light-loving, soil adaptable. Mainly distributed in Northeast China and part of northern North China.

References

1. Sannier, C.; McRoberts, R.E.; Fichet, L. Suitability of global forest change data to report forest cover estimates at national level in Gabon. *Remote Sens. Environ.* **2016**, *173*, 326–338. [CrossRef]
2. Feng, Y.; Lu, D.S.; Chen, Q.; Keller, M.; Moran, E.; Dos-Santo, M.N.; Bolf, E.L.; Batistella, M. Examining effective use of data sources and modeling algorithms for improving biomass estimation in a moist tropical forest of the Brazilian Amazon. *Int. J. Digit. Earth* **2017**, *10*, 996–1016. [CrossRef]
3. Frenne, P.D.; Zellweger, F.; Rodríguez-Sánchez, F.; Scheffers, B.R.; Hylander, K.; Luoto, M.; Vellend, M.; Verheyen, K.; Lenoir, J. Global buffering of temperatures under forest canopies. *Nat. Ecol. Evol.* **2019**, *3*, 744–749. [CrossRef] [PubMed]
4. Heinzel, J.; Koch, B. Investigating multiple data sources for tree species classification in temperate forest and use for single tree delineation. *Int. J. Appl. Earth Obs.* **2012**, *18*, 101–110. [CrossRef]
5. Fassnacht, F.E.; Latifi, H.; Stereńczak, K.; Modzelewska, A.; Lefsky, M.; Waser, L.T.; Straub, C.; Ghosh, A. Review of studies on tree species classification from remotely sensed data. *Remote Sens. Environ.* **2016**, *186*, 64–87. [CrossRef]
6. Meddens, A.H.; Hicke, J.A.; Vierling, L.A. Evaluating the potential of multispectral imagery to map multiple stages of tree mortality. *Remote Sens.* **2011**, *115*, 1632–1642. [CrossRef]
7. Ghosh, A.; Fassnacht, F.E.; Joshi, P.K.; Koch, B. A framework for mapping tree species combining hyperspectral and LiDAR data: Role of selected classifiers and sensor across three spatial scales. *Int. J. Appl. Earth Obs. Geoinf.* **2014**, *26*, 49–63. [CrossRef]
8. Tang, L.N.; Shao, G.F. Drone remote sensing for forestry research and practices. *J. For. Res.* **2015**, *26*, 791–797. [CrossRef]
9. Immitzer, M.; Böck, S.; Einzmann, K.; Vuolo, F.; Pinnel, N.; Wallner, A.; Atzberger, C. Fractional cover mapping of spruce and pine at 1 ha resolution combining very high and medium spatial resolution satellite imagery. *Remote Sens. Environ.* **2017**, *204*, 690–703. [CrossRef]
10. Wu, H.; Li, Z.L. Scale issues in remote sensing: A review on analysis, processing and modeling. *Sensors* **2009**, *9*, 1768–1793. [CrossRef] [PubMed]
11. Zhu, X.L.; Liu, D.S. Accurate mapping of forest types using dense seasonal Landsat time-series. *ISPRS J. Photogramm. Remote Sens.* **2014**, *96*, 1–11. [CrossRef]
12. Roth, K.l; Roberts, D.A.; Dennison, P.E.; Peterson, S.H.; Alonzo, M. The impact of spatial resolution on the classification of plant species and functional types within imaging spectrometer data. *Remote Sens. Environ.* **2015**, *171*, 45–57. [CrossRef]
13. Pu, R.; Landry, S.; Yu, Q. Assessing the potential of multi-seasonal high resolution Pléiades satellite imagery for mapping urban tree species. *Int. J. Appl. Earth Obs. Geoinf.* **2018**, *71*, 144–158. [CrossRef]
14. Dudley, K.L.; Dennison, P.E.; Roth, K.L.; Roberts, D.A.; Coates, A.R. A multi-temporal spectral library approach for mapping vegetation species across spatial and temporal phenological gradients. *Remote Sens. Environ.* **2015**, *167*, 121–134. [CrossRef]
15. Grabska, E.; Hostert, P.; Pflugmacher, D.; Ostapowicz, K. Forest stand species mapping using the Sentinel-2 time series. *Remote Sens.* **2019**, *11*, 1197. [CrossRef]
16. Kempeneers, P.; Sedano, F.; Seebach, L.M.; Strobl, P.; San-Miguel-Ayanz, J. Data fusion of different spatial resolution remote sensing images applied to forest-type mapping. *IEEE T. Geocsi. Remote.* **2011**, *49*, 4977–4986. [CrossRef]
17. Atkinson, P.M.; Jeganathan, C.; Dash, J.; Atzberger, C. Inter-comparison of four models for smoothing satellite sensor time-series data to estimate vegetation phenology. *Remote Sens. Environ.* **2012**, *123*, 400–417. [CrossRef]
18. Granero-Belinchon, C.; Adeline, K.; Briottet, X. Impact of the number of dates and their sampling on a NDVI time series reconstruction methodology to monitor urban trees with Venµs satellite. *Int. J. Appl. Earth Obs. Geoinf.* **2021**, *95*, 102257. [CrossRef]
19. Xu, K.J.; Tian, Q.J.; Zhang, Z.Y.; Yue, J.B.; Chang, C.T. Tree species (genera) identification with GF-1 time-series in a forested landscape, Northeast China. *Remote Sens.* **2020**, *12*, 1554. [CrossRef]
20. Kollert, A.; Bremer, M.; Löw, M.; Rutzinger, M. Exploring the potential of land surface phenology and seasonal cloud free composites of one year of Sentinel-2 imagery for tree species mapping in a mountainous region. *Int. J. Appl. Earth Obs.* **2021**, *94*, 102208. [CrossRef]
21. Li, X.; Chen, W.Y.; Sanesi, G.; Lafortezza, R. Remote sensing in urban forestry: Recent applications and future directions. *Remote Sens.* **2019**, *11*, 1144. [CrossRef]
22. Hill, R.A.; Wilson, A.K.; George, M.; Hinsley, S.A. Mapping tree species in temperate deciduous woodland using time-series multi-spectral data. *Appl. Veg. Sci.* **2010**, *13*, 86–99. [CrossRef]
23. Jia, K.; Liang, S.; Wei, X.; Yao, Y.; Su, Y.; Jiang, B.; Wang, X. Land cover classification of Landsat data with phenological features extracted from time series MODIS NDVI data. *Remote Sens.* **2014**, *6*, 11518–11532. [CrossRef]
24. Masemola, C.; Cho, M.A.; Ramoelo, A. Sentinel-2 time series based optimal features and time window for mapping invasive Australian native Acacia species in KwaZulu Natal, South Africa. *Int. J. Appl. Earth Obs. Geoinf.* **2020**, *93*, 102207. [CrossRef]
25. Gessner, U.; Machwitz, M.; Conrad, C.; Dechab, S. Estimating the fractional cover of growth forms and bare surface in savannas. A multi-resolution approach based on regression tree ensembles. *Remote Sens. Environ.* **2013**, *129*, 90–102. [CrossRef]
26. Achard, F.; Estreguil, C. Forest classification of Southeast Asia using NOAA AVHRR data. *Remote Sens. Environ.* **1995**, *54*, 198–208. [CrossRef]
27. Xiao, X.M.; Boles, S.; Liu, J.Y.; Zhuang, D.F.; Liu, M.L. Characterization of forest types in Northeastern China, using multi-temporal SPOT-4 VEGETATION sensor data. *Remote Sens. Environ.* **2002**, *82*, 335–348. [CrossRef]

28. Yu, X.F.; Zhuang, D.F.; Chen, H.; Hou, X.Y. Forest classification based on MODIS time series and vegetation phenology. In Proceedings of the IEEE International Geoscience and Remote Sensing Symposium, Anchorage, AK, USA, 20–24 September 2004; Volume 4, pp. 2369–2372. [CrossRef]
29. Pimple, U.; Sitthi, A.; Simonetti, D.; Pungkul, S.; Leadprathom, K.; Chidthaisong, A. Topographic correction of Landsat TM-5 and Landsat OLI-8 imagery to improve the performance of forest classification in the mountainous terrain of Northeast Thailand. *Sustainability* **2017**, *9*, 258. [CrossRef]
30. Yin, H.; Khamzina, A.; Pflugmacher, D.; Martius, C. Forest cover mapping in post-Soviet Central Asia using multi-resolution remote sensing imagery. *Sci. Rep.* **2017**, *7*, 1–11. [CrossRef]
31. Townshend, J.R.; Masek, J.G.; Huang, C.; Vermote, E.F.; Gao, F.; Channan, S.; Sexton, J.O.; Feng, M.; Narasimhan, R.; Kim, D.; et al. Global characterization and monitoring of forest cover using Landsat data: Opportunities and challenges. *Int. J. Digit. Earth* **2012**, *5*, 373–397. [CrossRef]
32. Ota, T.; Mizoue, N.; Yoshida, S. Influence of using texture information in remote sensed data on the accuracy of forest type classification at different levels of spatial resolution. *J. For. Res.* **2011**, *16*, 432–437. [CrossRef]
33. Pu, R.L.; Bell, S. Mapping seagrass coverage and spatial patterns with high spatial resolution IKONOS imagery. *Int. J. Appl. Earth Obs. Geoinf.* **2017**, *54*, 145–158. [CrossRef]
34. Puzzolo, V.; Denatale, F.; Gianne, F. Forest species discrimination in an Alpine mountain area using a fuzzy classification of multi-temporal SPOT (HRV) data. *IEEE Int. Geosci. Remote Sens. Symp.* **2003**, *4*, 2538–2540. [CrossRef]
35. Persson, M.; Lindberg, E.; Reese, H. Tree species classification with multi-temporal Sentinel-2 data. *Remote Sens.* **2018**, *10*, 1794. [CrossRef]
36. Wessel, M.; Brandmeier, M.; Tiede, D. Evaluation of different machine learning algorithms for scalable classification of tree types and tree species based on Sentinel-2 data. *Remote Sens.* **2018**, *10*, 1419. [CrossRef]
37. Gomez, C.; White, J.C.; Wulder, M.A. Optical remotely sensed time series data for land cover classification: A review. *ISPRS J. Photogramm. Remote Sens.* **2016**, *116*, 55–72. [CrossRef]
38. Liu, J.H.; Pan, Y.Z.; Zhu, X.F.; Zhu, W.Q. Using phenological metrics and the multiple classifier fusion method to map land cover types. *J. Appl. Remote Sens.* **2014**, *8*, 083691. [CrossRef]
39. Michez, A.; Piegay, H.; Jonathan, L.; Claessens, H.; Lejeune, P. Mapping of riparian invasive species with supervised classification of Unmanned Aerial System (UAS) imagery. *Int. J. Appl. Earth Obs.* **2015**, *44*, 88–94. [CrossRef]
40. Xu, K.J.; Tian, Q.J.; Xu, N.X.; Yue, J.B.; Tang, S.F. Classifying forest dominant trees species based on high dimensional time-series NDVI data and differential transform methods. *Spectrosc. Spectr. Anal.* **2019**, *39*, 3794–3800. [CrossRef]
41. Kong, J.X.; Zhang, Z.C.; Zhang, J. Classification and identification of plant species based on multi-source remote sensing data: Research progress and prospect. *Biodivers. Sci.* **2019**, *27*, 796–812. [CrossRef]
42. Peng, D.; Wu, C.; Zhang, X.; Yu, L.; Huete, A.R.; Wang, F.; Luo, S.; Liu, X.; Zhang, H. Scaling up spring phenology derived from remote sensing images. *Agric. For. Meteorol.* **2018**, *256–257*, 207–219. [CrossRef]
43. Zeng, L.L.; Wardlow, B.D.; Xiang, D.X.; Hu, S.; Li, D. A review of vegetation phenological metrics extraction using time-series, multispectral satellite data. *Remote Sens. Environ.* **2020**, *237*, 111511. [CrossRef]
44. Peng, D.; Zhang, X.; Zhang, B.; Liu, L.; Liu, X.; Huete, A.R.; Huang, W.; Wang, S.; Luo, S.; Zhang, X.; et al. Scaling effects on spring phenology detections from MODIS data at multiple spatial resolutions over the contiguous United States. *ISPRS J. Photogramm. Remote Sens.* **2017**, *132*, 185–198. [CrossRef]
45. Ge, Q.; Dai, J.; Cui, H.; Wang, H. Spatiotemporal variability in start and end of growing season in China related to climate variability. *Remote Sens.* **2016**, *8*, 433. [CrossRef]
46. Liu, L.; Cao, R.; Shen, M.; Chen, J.; Zhang, X. How does scale effect influence spring vegetation phenology estimated from satellite-derived vegetation indexes? *Remote Sens.* **2019**, *11*, 2137. [CrossRef]
47. Liu, W.J.; Zeng, Y.N.; Zhang, M. Mapping rice paddy distribution by using time series HJ blend data and phenological parameters. *J. Remote Sens.* **2018**, *22*, 381–391. [CrossRef]
48. Schwieder, M.; Leitão, P.J.; Pinto, J.R.; Teixeira, A.C.; Pedroni, F.; Sanchez, M.; Bustamante, M.M.; Hostert, P. Landsat phenological metrics and their relation to aboveground carbon in the Brazilian Savanna. *Carbon Balanc. Manag.* **2018**, *13*, 1–15. [CrossRef]
49. Schaaf, A.; Dennison, P.; Fryer, G.; Roth, K.; Roberts, D. Mapping plant functional types at multiple spatial resolutions using imaging spectrometer data. *GISci. Remote Sens.* **2011**, *48*, 324–344. [CrossRef]
50. Peña, M.A.; Cruz, P.; Roig, M. The effect of spectral and spatial degradation of hyperspectral imagery for the sclerophyll tree species classification. *Int. J. Remote Sens.* **2013**, *34*, 7113–7130. [CrossRef]
51. Xu, K.J.; Tian, Q.J.; Yang, Y.J; Yue, J.B.; Tang, S.F. How up-scaling of remote-sensing images affects land-cover classification by comparison with multiscale satellite images. *Int. J. Remote Sens.* **2019**, *40*, 2784–2810. [CrossRef]
52. Zhang, X.; Wang, J.; Gao, F.; Liu, Y.; Schaaf, C.; Friedl, M.; Yu, Y.; Jayavelu, S.; Gray, J.; Liu, L.; et al. Exploration of scaling effects on coarse resolution land surface phenology. *Remote Sens. Environ.* **2017**, *190*, 318–330. [CrossRef]
53. Tian, J.Q.; Zhu, X.L.; Wu, J.; Shen, M.G.; Chen, J. Coarse-resolution satellite images overestimate urbanization effects on vegetation spring phenology. *Remote Sens.* **2020**, *12*, 117. [CrossRef]
54. Moody, A.; Woodcock, C.E. Scale-dependent errors in the estimation of land-cover proportions: Implications for global land-cover datasets. *Photogramm. Eng. Remote Sens.* **1994**, *60*, 585–594.

55. Luan, H.J.; Tian, Q.J.; Yu, T.; Hu, X.L.; Huang, Y.; Liu, L.; Du, L.T.; Wei, X. Review of up-scaling of quantitative remote sensing. *Adv. Earth Sci.* **2013**, *28*, 657–664. [CrossRef]
56. Chen, J.; Jönsson, P.; Tamura, M.; Gu, Z.; Matsushita, B.; Eklundh, L. A simple method for reconstructing a high-quality NDVI time-series data set based on the Savitzky-Golay filter. *Remote Sens. Environ.* **2004**, *91*, 332–344. [CrossRef]
57. Wang, Y.F.; Xue, Z.H.; Chen, J.; Chen, G.Z. Spatio-temporal analysis of phenology in Yangtze river delta based on MODIS NDVI time series from 2001 to 2015. *Front. Earth Sci.* **2019**, *13*, 92–110. [CrossRef]
58. Shao, Y.; Lunetta, R.S.; Wheeler, B.; Iiames, J.S.; Campbell, J.B. An evaluation of time-series smoothing algorithms for land-cover classifications using MODIS-NDVI multi-temporal data. *Remote Sens. Environ.* **2016**, *174*, 258–265. [CrossRef]
59. Xu, K.J.; Zeng, H.D.; Zhu, X.B.; Tian, Q.J. Evaluation of five commonly used atmospheric correction algorithms for multi-temporal aboveground forest carbon storage estimation. *Spectrosc. Spect. Anal.* **2017**, *37*, 3493–3498. [CrossRef]
60. Jönnson, P.; Eklundh, L. TIMESAT-a program for analyzing time-series of satellite sensor data. *Comput. Geosci.* **2004**, *30*, 833–845. [CrossRef]
61. Chang, C.T.; Wang, H.C.; Huang, C. Impact of vegetation onset time on the net primary productivity in a mountainous island in Pacific Asia. *Environ. Res. Lett.* **2013**, *8*, 05030. [CrossRef]
62. Yang, Y.T.; Guan, H.D.; Shen, M.G.; Liang, W.; Jiang, L. Changes in autumn vegetation dormancy onset date and the climate controls across temperate ecosystems in China from 1982 to 2010. *Glob. Chang. Biol.* **2014**, *21*, 1–14. [CrossRef]
63. Heumann, B.W.; Seaquist, J.W.; Eklundh, L.; Jönsson, P. AVHRR derived phenological change in the Sahel and Soudan, Africa, 1982-2005. *Remote Sens. Environ.* **2007**, *108*, 385–392. [CrossRef]
64. Jiao, F.S.; Liu, H.Y.; Xu, X.J.; Gong, H.B.; Lin, Z.S. Trend evolution of vegetation phenology in China during the period of 1981–2016. *Remote Sens.* **2020**, *12*, 572. [CrossRef]
65. Lebrini, Y.; Boudhar, A.; Hadria, R.; Lionboui, H.; Elmansouri, L.; Arrach, R.; Ceccato, P.; Benabdelouahab, T. Identifying agricultural systems using SVM classification approach based on phenological metrics in a semi-arid region of Morocco. *Earth Syst. Environ.* **2019**, *3*, 277–288. [CrossRef]
66. Sothe, C.; de Almeida, C.M.; Liesenberg, V.; Schimalski, M.B. Evaluating Sentinel-2 and Landsat-8 data to map sucessional forest stages in a subtropical forest in southern Brazil. *Remote Sens.* **2017**, *9*, 838. [CrossRef]
67. Breiman, L. Random forests. *Mach. Learn.* **2001**, *45*, 5–32. [CrossRef]
68. Immitzer, M.; Vuolo, F.; Atzberger, C. First experience with Sentinel-2 data for crop and tree species classifications in central Europe. *Remote Sens.* **2016**, *8*, 166. [CrossRef]
69. Belgiu, M.; Drăgut, L. Random forest in remote sensing: A review of applications and future directions. *ISPRS J. Photogramm. Remote Sens.* **2016**, *114*, 24–31. [CrossRef]
70. Immitzer, M.; Atzberger, C.; Koukal, T. Tree species classification with random forest using very high spatial resolution 8-band WorldView-2 satellite data. *Remote Sens.* **2012**, *4*, 2661–2693. [CrossRef]
71. Reyes-Palomeque, G.; Dupuy, J.M.; Portillo-Quintero, C.A.; Andrade, J.L.; Tun-Dzul, F.J.; Hernández-Stefanoni, J.L. Mapping forest age and characterizing vegetation structure and species composition in tropical dry forests. *Ecol. Indic.* **2021**, *120*, 106955. [CrossRef]
72. Chutia, D.; Bhattacharyya, D.K.; Sarma, K.K.; Kalita, R.; Sudhakar, S. Hyperspectral remote sensing classifications: A perspective survey. *Trans. GIS* **2016**, *20*, 463–490. [CrossRef]
73. Shang, X.; Chisholm, L. Classification of Australian native forest species using hyperspectral remote sensing and machine-learning classification algorithms. *IEEE J. Sel. Top. Appl. Earth Obs. Remote Sens.* **2014**, *7*, 2481–2489. [CrossRef]
74. Han, H.; Lee, S.; Im, J.; Kim, M.; Lee, M.I.; Ahn, M.; Chung, S.R. Detection of convective initiation using meteorological imager onboard communication, ocean, and meteorological satellite based on machine learning approaches. *Remote Sens.* **2015**, *7*, 9184–9204. [CrossRef]
75. Immitzer, M.; Neuwirth, M.; Bck, S.; Brenner, H.; Atzberger, C. Optimal input features for tree epecies classification in central Europe based on multi-temporal Sentinel-2 data. *Remote Sens.* **2019**, *11*, 2599. [CrossRef]
76. Janssen, L.L.F.; van der Wel, F.J.M. Accuracy assessment of satellite derived land-gover data: A review. *Photogramm. Eng. Remote Sens.* **1994**, *60*, 419–426. [CrossRef]
77. Yang, Y.; Xiao, P.; Feng, X.; Li, H. Accuracy assessment of seven global land cover datasets over China. *ISPRS J. Photogramm. Remote Sens.* **2017**, *125*, 156–173. [CrossRef]
78. Richter, R.; Reu, B.; Wirth, C.; Doktor, D.; Vohland, M. The use of airborne hyperspectral data for tree species classification in a species-rich Central European forest area. *Int. J. Appl. Earth Obs. Geoinf.* **2016**, *52*, 464–474. [CrossRef]
79. Wang, L.; Yang, R.; Tian, Q.; Yang, Y.; Zhou, Y.; Sun, Y.; Mi, X. Comparative analysis of GF-1 WFV, ZY-3 MUX, and HJ-1 CCD sensor data for grassland monitoring applications. *Remote Sens.* **2015**, *7*, 2089–2108. [CrossRef]
80. Li, N.; Xie, G.D.; Zhou, D.M.; Zhang, C.S.; Jiao, C.C. Remote sensing classification of marsh wetland with different resolution images. *J. Resour. Ecol.* **2016**, *7*, 107–114. [CrossRef]
81. Lessel, J.; Ceccato, P. Creating a basic customizable framework for crop detection using Landsat imagery. *Int J. Remote Sens.* **2016**, *37*, 6097–6107. [CrossRef]
82. Hu, Y.F.; Xu, Z.Y.; Liu, Y.; Yan, Y.; Wang, Q.Q. Accuracy analysis of up-scaling data: A case study with land use data in Xilin Gol of Inner Mongolia, China. *Geogr. Res.* **2012**, *31*, 1961–1972. [CrossRef]
83. Zhang, Z.; Tao, Y.U.; Meng, Q.; Xinli, H.U.; Chang, L.I. Image Quality Evaluation of Multi-Scale Resampling in Geometric Correction. *J. Huazhong Norm. Univ.* **2013**, *47*, 426–430. [CrossRef]

84. Melaas, E.K.; Friedl, M.A.; Zhu, Z. Detecting interannual variation in deciduous broadleaf forest phenology using Landsat TM/ETM+ data. *Remote Sens. Environ.* **2013**, *132*, 176–185. [CrossRef]
85. Tang, S.F.; Tian, Q.J.; Xu, K.J.; Xu, N.X.; Yue, J.B. Age information retrieval of *Larix gmelinii* forest using Sentinel-2 data. *J. Remote Sens.* **2020**, *24*, 1511–1524. [CrossRef]
86. Hay, G.J.; Niernann, K.O.; Goodenough, D.J. Spatial thresholds, image-objects, and upscaling: A multiscale evaluation. *Remote Sens. Environ.* **1997**, *62*, 1–19. [CrossRef]
87. Tian, Y.; Wang, Y.; Zhang, Y.; Knyazikhin, Y.; Bogaert, J.; Mynemi, R.B. Radiative transfer based scaling of LAI retrieval from reflectance data of different resolutions. *Remote Sens. Environ.* **2002**, *84*, 143–159. [CrossRef]
88. Wu, H.; Tang, B.H.; Li, Z.L. Impact of nonlinearity and discontinuity on the spatial scaling effects of the leaf area index retrieved from remotely sensed data. *Int. J. Remote Sens.* **2013**, *34*, 3503–3519. [CrossRef]
89. Jiang, J.; Xiao, Z.; Wang, J.; Song, J. Multiscale estimation of leaf area index from satellite observations based on an ensemble multiscale filter. *Remote Sens.* **2016**, *8*, 229. [CrossRef]
90. Wu, M.Q.; Huang, W.J.; Niu, Z.; Wang, C.Y.; Li, W.; Yu, B. Validation of synthetic daily Landsat NDVI time series data generated by the improved spatial and temporal data fusion approach. *Inf. Fusion* **2018**, *40*, 34–44. [CrossRef]
91. Vrieling, A.; Meroni, M.; Darvishzadeh, R.; Skidmore, A.K.; Wang, T.; Zurita-Milla, R.; Oosterbeek, K.; O'Connor, B.; Paganini, M. Vegetation phenology from Sentinel-2 and field cameras for a Dutch barrier island. *Remote Sens. Environ.* **2018**, *215*, 517–529. [CrossRef]

remote sensing

MDPI

Article

A New Individual Tree Species Recognition Method Based on a Convolutional Neural Network and High-Spatial Resolution Remote Sensing Imagery

Shijie Yan [1], Linhai Jing [2,*] and Huan Wang [3]

1 Aerospace Information Research Institute, Chinese Academy of Sciences, Beijing 100094, China; yansj@aircas.ac.cn
2 Key Laboratory of Digital Earth Science, Aerospace Information Research Institute, Chinese Academy of Sciences, Beijing 100094, China
3 College of Urban and Environmental Sciences, Peking University, Beijing 100871, China; 2001111811@stu.pku.edu.cn
* Correspondence: jinglh@aircas.ac.cn

Abstract: Tree species surveys are crucial to forest resource management and can provide references for forest protection policy making. The traditional tree species survey in the field is labor-intensive and time-consuming, supporting the practical significance of remote sensing. The availability of high-resolution satellite remote sensing data enable individual tree species (ITS) recognition at low cost. In this study, the potential of the combination of such images and a convolutional neural network (CNN) to recognize ITS was explored. Firstly, individual tree crowns were delineated from a high-spatial resolution WorldView-3 (WV3) image and manually labeled as different tree species. Next, a dataset of the image subsets of the labeled individual tree crowns was built, and several CNN models were trained based on the dataset for ITS recognition. The models were then applied to the WV3 image. The results show that the distribution maps of six ITS offered an overall accuracy of 82.7% and a kappa coefficient of 0.79 based on the modified GoogLeNet, which used the multi-scale convolution kernel to extract features of the tree crown samples and was modified for small-scale samples. The ITS recognition method proposed in this study, with multi-scale individual tree crown delineation, avoids artificial tree crown delineation. Compared with the random forest (RF) and support vector machine (SVM) approaches, this method can automatically extract features and outperform RF and SVM in the classification of six tree species.

Keywords: high-resolution remote sensing imagery; individual tree species recognition; individual tree crown delineation; convolutional neural network

Citation: Yan, S.; Jing, L.; Wang, H. A New Individual Tree Species Recognition Method Based on a Convolutional Neural Network and High-Spatial Resolution Remote Sensing Imagery. *Remote Sens.* **2021**, *13*, 479. https://doi.org/10.3390/rs13030479

Academic Editors: Javier Marcello and Henning Buddenbaum
Received: 14 December 2020
Accepted: 27 January 2021
Published: 29 January 2021

1. Introduction

Forests are among the most important terrestrial ecosystems and are essential for human development [1]. Well-managed forests provide renewable resources, protect biodiversity, maintain a stable energy cycle, and prevent soil degradation and erosion [2]. Precise tree species surveys are crucial to forest inventory and management because they provide managers with a better understanding of forest species composition, changes in forest species, quantity of forest resources, and references for the formulation and adjustment of forestry policies [3]. However, traditional survey methods are inefficient and their associated labor costs are high. Remote sensing-based methods are efficient when mapping forest types in areas with rough terrain or that are difficult to reach, and can significantly improve survey efficiency and reduce labor costs [4].

Many remote sensing-based forest classification studies have considered multi-scale remote sensing data sources. Early developments used medium-spatial resolution satellite remote sensing data, such as Landsat Thematic Mapper imagery, for regional-scale forest classification [5,6]. However, because the spatial resolution of Landsat data is relatively low, individual trees cannot be precisely mapped. Spaceborne hyperspectral data have rarely been used for individual tree species classification. Most applications of hyperspectral data to date have been airborne [7–9]. The high spatial resolution of airborne data meets the requirements for determining the locations of trees [10–13]; however, data acquisition costs are normally high and data processing is complex [14,15]. With the launch of IKONOS, QuickBird, GeoEye, and WorldView satellites, high-resolution optical images can be readily obtained and meet the requirements for locating individual trees. Using such high-resolution images to precisely classify tree species saves time and costs in tree distribution mapping.

Traditional classification methods, such as random forest (RF) and support vector machine (SVM), have been widely used to classify tree species [16–19]. These approaches generally require the artificial design and extraction of classification features, such as spectra, texture, and vegetation indices, in addition to linear transformations [20,21]. The classification accuracy depends largely on the rationality of the artificial feature design and selection, which is highly subjective; therefore, extensive professional knowledge is necessary.

The designed artificial features limit the information that can be used by the classifier [22]. New classification technologies are necessary to improve classification efficiency and accuracy. As a promising classification technology, convolutional neural networks (CNNs) perform well in image classification tasks [23,24]. A CNN method does not require feature engineering, and its multi-layer structure can fully use the information in the data to automatically extract abstract and higher-level features for classification. As a result, CNN methods tend to result in accurate classifications.

In recent years, CNNs have shown satisfactory results when applied to tree species classification [25]. CNNs have been applied to classify three-dimensional point clouds of trees [26,27], airborne hyperspectral data [28,29], and high-resolution data combined with LiDAR data [30]. These studies were almost all based on airborne imaging systems and multiple data sources, which are characterized by high data acquisition costs and complex data processing, preventing their wide application. To date, CNNs have rarely been applied to recognize individual tree species (ITS) from a single satellite data source. A method is needed for using satellite data to classify ITS for mapping forest tree species with a low data acquisition cost.

In this study, we explored the combination of high-resolution satellite remote sensing imagery and CNNs to recognize ITS. A CNN-based multi-scale ITS recognition (CMSIR) method was developed to improve the automation and accuracy of ITS mapping. In the CMSIR method, a tree crown delineation approach is used to quickly build an individual tree crown training dataset, and several popular CNN models are employed to automatically extract classification features and recognize ITS from a high-resolution satellite image. The multi-scale characteristic of different tree species was considered in tree crown delineation and ITS recognition.

2. Materials and Methodologies

2.1. Study Area

The study area (Figure 1) is located in the Olympic Forest Park (116°22′29″E–116°23′43″E, 40°0′30″–40°1′11″N), Chaoyang District, Beijing, China. The total area of the park is approximate 100 ha and the vegetation coverage rate is 95.6% [31,32]. The forests inside were manually planted and arranged in rows, clusters, and groups with a large number of single tree species. The study area was suitable for constructing a standard sample dataset due to the similarity of trees within specific zones, including tree height, size, and species. Shrubs were planted on the periphery of the arbor forest to act as a natural fence to separate roads from trees. Artificial buildings are sparsely distributed in the forest, and occupy a small portion of the area and have little impact on the forest.

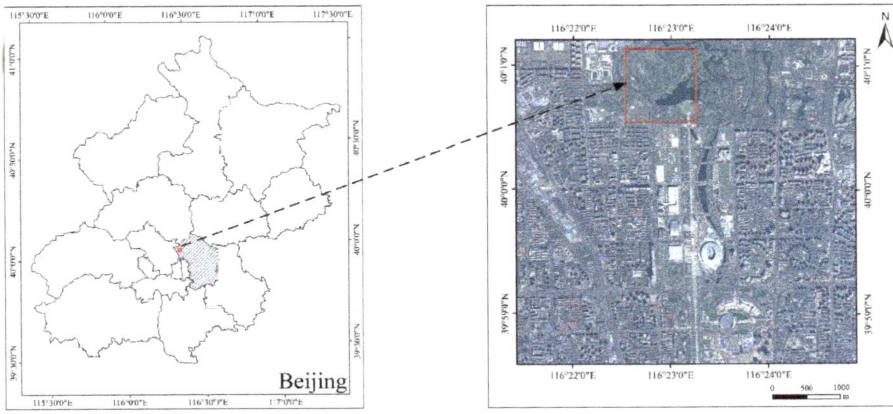

Figure 1. Location of the research area.

2.2. Data

2.2.1. Field Investigation Data

A field investigation of tree species in the study area was carried out in October 2018, for several days. Trees in the study area were manually planted in a monoculture scheme; thus, many zones in the study area had consistent tree species. The boundaries of zones with consistent tree species were marked on images. The orange boundaries in Figure 2 are an example of the marked zones. On the first day, we entered each marked zone and located 3–5 GPS points with a Trimble® Geo7X global positioning system (GPS) handheld device (Trimble Inc., Sunnyvale, CA, USA). The points in Figure 2 show the located GPS points in the marked zones. The position accuracy was improved by taking the mean value of the measured real-time position 10 times. Simultaneously, the tree species of each marked zone was identified by expert experience and recorded together with the located GPS points. After the first day's investigation, GPS points were overlaid on the image, and we found that the points were almost all located in the marked zones. Therefore, the GPS was accurate enough to locate the marked zones. The dominant tree species in the study area were determined based on the first day's investigation result, and the investigation was continued over the next several days to obtain more samples. In addition, understory trees in forests cannot be observed via remote sensing images and therefore were not sampled. All of the samples were distributed as shown in Figure 3; the dominant tree species included 175 ash (*Fraxinus chinensis* Roxb.), 132 poplar (*Populus tomentosa*), 128 cypress (*Sabina chinensis*), 127 pagoda (*Sophora japonica*), 127 willow (*Salix babylonica* L.), and 110 pine (*Pinus* L.). The field investigation data supported the manual labeling of tree crowns and the interpretation of tree species.

2.2.2. Remote Sensing Data and Pre-Processing

A cloud-free subset of a high-resolution WorldView-3 (WV3) Beijing scene acquired on 4 September 2014 was used in this study. The acquisition date of the image was within the fully developed season, which provides good conditions for the classification of tree species.

The dataset consisted of four 3145 × 3145 multispectral bands with 1.6 m spatial resolution, including blue (0.450–0.510 μm), green (0.510–0.580 μm), red (0.630–0.690 μm), and near-infrared (NIR; 0.770–0.895 μm) bands, and a corresponding 12,580 × 12,580 panchromatic band (0.450–0.800 μm) with 0.4 m spatial resolution. Although the WV3 sensor provides 16 multispectral bands, the image collected over the study area only had four standard bands. A digital elevation model (DEM) was used to orthorectify the panchromatic and multispectral images, which were fused using the Gram–Schmidt method in the ENVI 5.3 (L3Harris Geospatial, Broomfield, CO, USA) package to obtain a four-band fused multispectral image with a cell size of 0.4 m. The pre-processed image showed good overall geometric precision with less than 1 pixel offset compared with the control points in the field investigation.

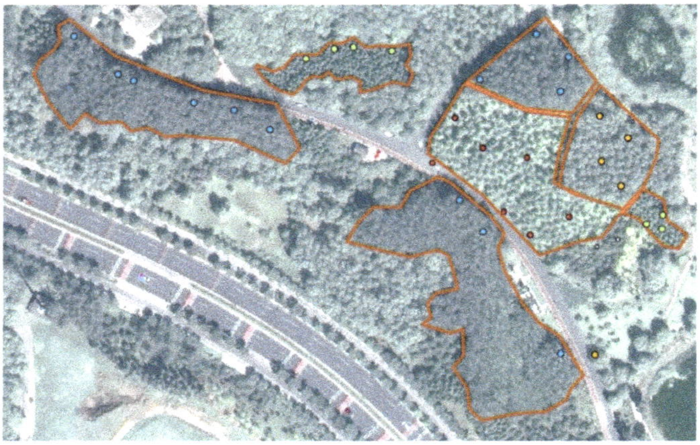

Figure 2. Example of marked investigation zones and located global positioning system (GPS) points.

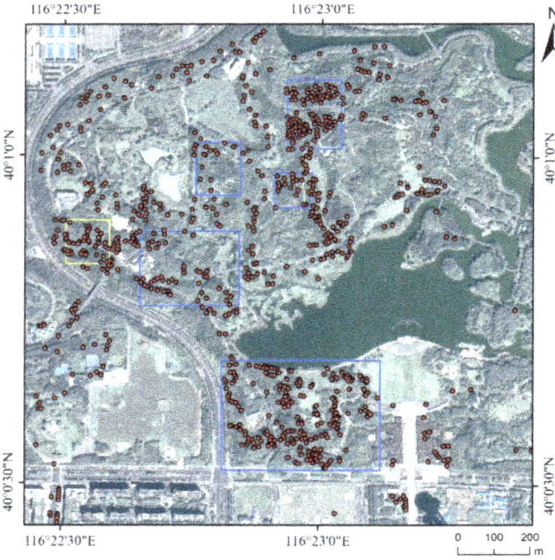

Figure 3. Locations of tree species samples (red dots) identified during the field investigation. Training sample collection regions (blue rectangles) and the test sample collection region (yellow rectangle).

2.3. Methodologies

As shown in Figure 4, the flowchart of the CMSIR method proposed in this study includes seven steps: (1) data pre-processing; (2) individual tree crown (ITC) delineation using the crown slices from imagery (CSI) tree crown delineation algorithm [33]; (3) tree species labeling based on field investigation; (4) ITS training dataset construction; (5) RF, SVM, and CNN model configuration and training; (6) ITS recognition using RF, SVM, and CNN; and (7) accuracy assessment with field investigation data.

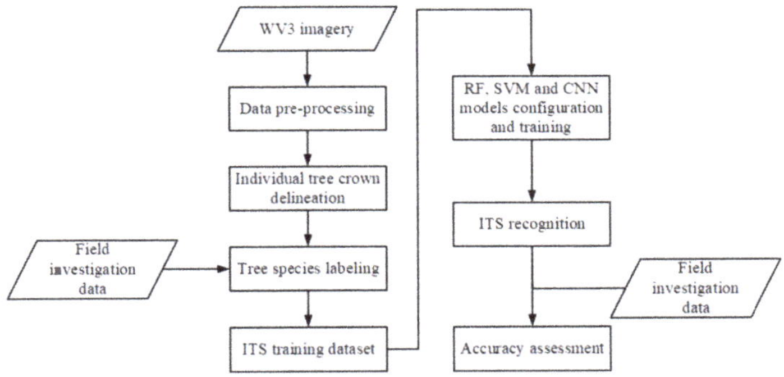

Figure 4. The technical flowchart of the convolutional neural network (CNN)-based multi-scale individual tree species (ITS) recognition (CMSIR) method. RF, random forest; SVM, support vector machine.

2.3.1. Training Dataset

The training samples were collected from five typical regions in the study area (blue rectangles in Figure 3). These five regions were selected because they included patches of consistent tree species, so it was easy to label tree species. The test sample collection region (yellow rectangle in Figure 3) was used to construct the test dataset to evaluate the performance of the trained models and generate the tree species classification maps. The test region included all tree species in the training dataset and few non-vegetation features and no grassland, so interference was reduced. In addition, we separated the training data from test data to ensure independence of the training and testing samples.

A training dataset with labeled trees was needed to train classification models, so a process was designed to quickly create an ITC training dataset. First, segmentation maps obtained from the CSI tree crown delineation algorithm were overlaid on the image. Second, tree species were labeled based on the field investigation data and manual interpretation. We judged the quality of the delineation during manual labeling: tree crowns without clearly visible appearances were discarded, and only the well-delineated tree crowns were labeled. Thus, the error caused by the CSI delineation was reduced. Third, the minimum outer cut rectangle of each labeled tree was taken to obtain the ITC slice images. Finally, the sliced images with the same labels were grouped into the same category. Figure 5 shows the training dataset construction process.

CSI Tree Crown Delineation Algorithm

To avoid manual delineation of tree crowns, the CSI tree crown delineation algorithm [33] was used to extract ITCs. This algorithm was developed for multi-scale ITC delineation from high-resolution optical imagery, and can effectively reduce image over-segmentation and provide fine tree crown maps. It consists of 5 steps: (1) The three-dimensional (3D) radiometric shape of a tree crown is considered as a half-ellipsoid. The half-ellipsoid can be horizontally sliced into a series of slices from top to bottom, and the slices represent multiple tree crown levels of different scales. The morphological opening operations are used to measure dominant sizes of tree crowns. (2) Multiple Gaussian filters are used to generate multi-scale representations of the forest image from the multiple tree crown levels obtained in step 1. (3) The watershed segmentation method [34] is used on the multi-scale representations of the forest image obtained in step 2 to generate multi-scale segmentation maps. The multi-scale segmentation maps represent the target forest's multi-scale tree crown sizes. (4) The boundaries of segments generated in step 3 are adjusted using the filtered image at a coarse scale based on the original image at the dominant size. (5) The multi-scale segmentation maps obtained in step 4 are integrated to generate the final tree crown map. In this study, the method of visual evaluation was adopted. Adjusting the parameters in the algorithm, the match between the delineation lines and the real tree crowns was visually assessed to obtain the most accurate delineation map [35].

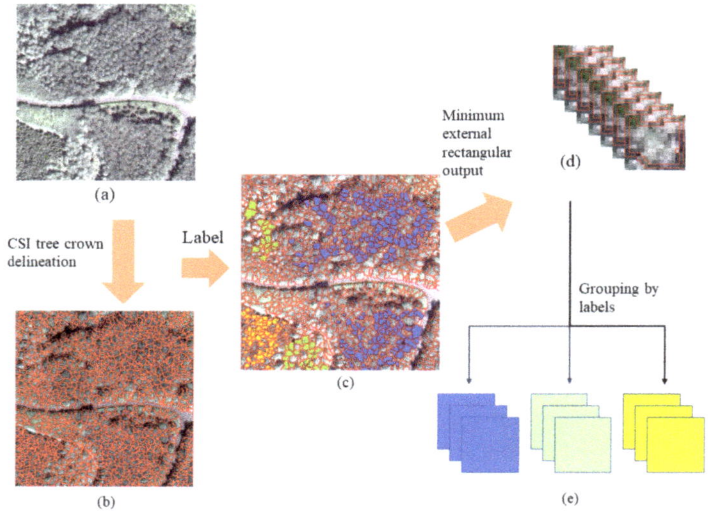

Figure 5. Schematic diagram of the training dataset construction. (**a**) Image of training sample collection region, (**b**) crown slices from imagery (CSI) algorithm-generated individual tree crown (ITC) map, (**c**) tree species labeled, (**d**) ITC slice images, (**e**) ITC training dataset.

Data Augmentation

Increasing the number of sample images can improve the generalization ability of the model. Due to the limited number of available field samples, data augmentation was used to increase the number of sample images. Data augmentation applies transformations on labeled images to increase the number of samples. Commonly used transformations include rotation, translation, random scaling, cropping, and flipping. Because the scale of each sample image was small, random scaling and clipping methods were not applied. Each labeled image was rotated by 90°, 180°, and 270°, and was flipped horizontally and vertically. After data augmentation, the number of samples increased by a factor of five. The number of training samples for each species is shown in Table 1, and the total number of training samples after augmentation was 14,976. Each image had the four R, G, B, and NIR channels.

Features Extraction

Feature extraction was performed with RF and SVM methods to classify tree crowns. Spectra, texture, and shape features were considered for extraction; they are described in Table 2. Texture and shape features were considered in this study because crowns of different tree species have various structures, shapes, and canopy densities [28,30].

Table 1. The statistics of the training samples.

Tree Species	Number of Training Samples	Number of Training Samples After Data Augmentation	Image
Pine	324	1944	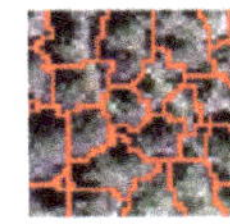
Cypress	227	1362	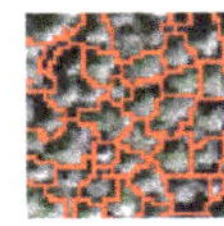
Poplar	430	2580	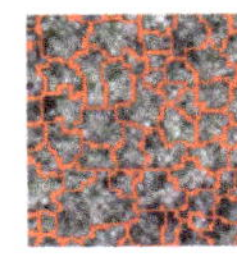
Ash	491	2946	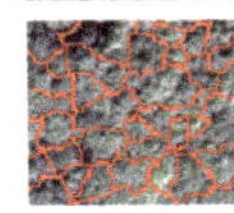
Pagoda	486	2916	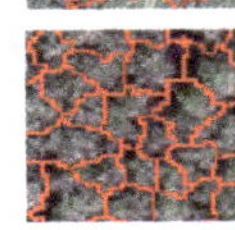
Willow	538	3228	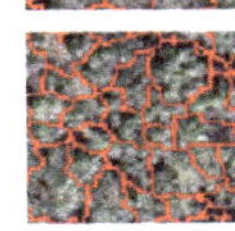
Total	2496	14,976	

Table 2. Features used in this study.

Group	Feature Name	Description
Spectra	Mean	The average value of all pixels contained in the object in band n
	Variance	The variance value of all pixels contained in the object in band n
	Maximum	The maximum value of all pixels contained in the object in band n
	Minimum	The minimum value of all pixels contained in the object in band n
Texture	Angular second moment (ASM)	Measures the number of repeated pairs
	Contrast	Measures the local contrast
	Entropy	Measures the randomness of a gray-level distribution
	Correlation	Measures the correlation between pairs of pixels
Shapes	Width	Width of an object
	Height	Height of an object
	Compactness	Indicates compactness of an object
	Boundary length	The sum of pixels of an object boundary.

2.3.2. CNN Model Configuration

CNN models were used to classify tree species in this study. CNN models were inspired by biological neural perception mechanisms, with the relationships between layers being similar to those used by the human vision system [36]. When processing multi-dimensional images, CNN models can automatically extract features that are beneficial to image interpretation and processing. CNN models have been widely used in image classification tasks [37], and have become a popular research topic in the field of recognition and classification [38,39]. A common CNN model is comprised of an input layer, stacks of convolution and pooling layers, a fully connected layer, and an output layer [40]. Inspired by the effects of CNN models on visual image classification, we considered several CNN models that have demonstrated excellent performance in image classification in recent years. ImageNet Large Scale Vision Recognition Challenge (ILSVRC) is a worldwide influential platform used to evaluate computer vision and artificial intelligence algorithms. Since 2010, several CNN-based deep learning models have won the image classification challenge, significantly contributing to the development of CNN models. We selected some models that won the ILSVRC to classify ITC images: AlexNet, GoogLeNet, and ResNet. Other winning network models are not listed because their structures are too complex to handle small-scale images.

AlexNet

AlexNet [23] was the first breakthrough CNN architecture and won the 2012 ILSVRC. AlexNet firstly applies rectified linear units (ReLU) as the activation function to successfully accelerate model convergence. The dropout function is used to prevent overfitting, and the addition of a local response normalization (LRN) layer increases its generalization ability. AlexNet has been widely used due to its relatively simple network structure and shallow depth. The AlexNet model was originally used to classify the 224 × 224 × 3 ImageNet dataset. For the dataset in this study, the convolution kernel and the output of the network are both large, so the scale of the network needs to be reduced.

We modified AlexNet as follows: First, the input layer was altered from 227 × 227 × 3 to 15 × 15 × 4. Secondly, the first convolution layer was reduced from 11 × 11 to 5 × 5 to prevent underfitting problems caused by an excessively large local receptive field. Next, all of the strides of the pooling layers were decreased from 2 × 2 to 1 × 1 to avoid a feature map that was too small. Then, the number of convolution filters was decreased by 6 to 7, and the output of the fully connected layer was reduced from 4096 to 140 or 70 to prevent overfitting. The modified AlexNet model structure is shown in Table 3.

Table 3. Modified AlexNet model structure. ReLU, rectified linear units.

Layer	Input Size	Output Size	Parameter
Conv1	15 × 15 × 4	8 × 8 × 16	kernel 5 × 5, filter 16, stride 2
Max pooling1	8 × 8 × 16	6 × 6 × 16	pool size 3 × 3, stride 1
Conv2	6 × 6 × 16	6 × 6 × 48	kernel 5 × 5, filter 48, stride 1
Max pooling2	6 × 6 × 48	4 × 4 × 48	pool size 3 × 3, stride 1
Conv4	4 × 4 × 48	4 × 4 × 54	kernel 3 × 3, filter 54, stride 1
Conv5	4 × 4 × 54	4 × 4 × 54	kernel 3 × 3, filter 54, stride 1
Conv6	4 × 4 × 54	4 × 4 × 36	kernel 3 × 3, filter 36, stride 1
Max pooling3	4 × 4 × 36	2 × 2 × 36	pool size 3 × 3, stride 1
Fully connected1	2 × 2 × 36	140	ReLU, dropout 0.5
Fully connected2	140	70	ReLU, dropout 0.5
Output	70	6	Softmax

GoogLeNet

GoogLeNet [41] was established by Szegedy's team, who won the title of ILSVRC classification task champion in 2014. The original GoogLeNet model was used to classify the ImageNet image set at 224 × 224 × 3. Thus, it was necessary to modify the network to adapt to the small-scale samples in this study.

The modified GoogLeNet consisted of one input layer, one convolutional layer, eight inception modules, two downsample modules, one pooling layer, and the classification layer (Table 4). The fully connected layer was replaced by the global average pooling layer to greatly reduce the number of parameters. Therefore, there were far fewer training parameters for the modified GoogLeNet, and the memory allocation was significantly reduced.

Table 4. Modified GoogLeNet model structure.

Layer Name	Input Size	Output Size	Parameter
Convolutional	15 × 15 × 4	15 × 15 × 96	kernel 3 × 3, filter 96, stride 1
Inception module 1a	15 × 15 × 96	15 × 15 × 64	I: filter 32, II: filter 32
Inception module 1b	15 × 15 × 64	15 × 15 × 80	I: filter 32, II: filter 48
Downsampling module1	15 × 15 × 80	7 × 7 × 160	III: filter 80, IV: maxpooling
Inception module 2a	7 × 7 × 160	7 × 7 × 160	I: filter 112, II: filter 48
Inception module 2b	7 × 7 × 160	7 × 7 × 160	I: filter 96, II: filter 64
Inception module 2c	7 × 7 × 160	7 × 7 × 160	I: filter 80, II: filter 80
Inception module 2d	7 × 7 × 160	7 × 7 × 144	I: filter 48, II: filter 96
Downsampling module2	7 × 7 × 144	3 × 3 × 240	III: filter 96, IV: maxpooling
Inception module 3a	3 × 3 × 240	3 × 3 × 336	I: filter 176, II: filter 160
Inception module 3b	3 × 3 × 336	3 × 3 × 336	I: filter 176, II: filter 160
Average Pooling	3 × 3 × 336	1 × 1 × 336	pool size 3 × 3
Fully connected	1 × 1 × 336	336	dropout 0.5
Output	336	6	Softmax

In the inception module, the output of the previous layer is passed to two parallel paths, called path I and path II. The two paths contain a convolution layer, a normalization layer, and an activation layer. The convolutional layer filter size in path I is 1 and 3 in path II. The inception module learns the 1 × 1 and 3 × 3 filters (which are computed in parallel) and concatenates the resulting feature maps along the channel dimensions. In general, this process allows the network to learn local features with small convolutions and abstract features with large convolutions. Table 5 lists the structure of the inception module 1a, which has 32 filters in the convolution layer and uses batch normalization in the normalization layer and ReLU in the activation layer. The other inception modules differ only in the number of filters. In the downsampling module, the output of the previous layer is passed into two parallel paths: path III and path IV. Path III is composed of a convolution layer, a normalization layer, and an activation layer, where the convolution layer filter size is 3, the stride is 2, and the downsampling module1 has 80 filters. Path III uses batch normalization in the normalization layer and ReLU in the activation layer. Path IV is a maxpooling layer with a pooling size of 3 × 3 and a stride of 2. Table 6 shows the structure of the downsampling module 1. The only difference in the downsampling module 2 is that it has 96 filters in the convolution layer. The output feature maps of paths III and IV are concatenated along the channel dimensions.

Table 5. Inception module 1a structure.

Path I	Path II
conv_1: kernel 1 × 1, filter 32, stride 1 Batch normalization ReLU	conv_2: kernel 3 × 3, filter 32, stride 1 Batch normalization ReLU
Concatenation	

Table 6. Downsampling module1 structure.

Path III	Path IV
conv: kernel 3 × 3, filter 80, stride 2 Batch normalization ReLU	Maxpooling size 3 × 3, stride 2
Concatenation	

ResNet

ResNet [42] was proposed to solve the problem of gradient disappearance after increasing the network depth and won first place at the 2015 ILSVRC. This model divides the output of the neurons into H(x) and F(x) in a certain layer and establishes a shortcut between convolution layers to form a residual unit (Figure 6). H(x) is the expected output after the input x passes through the convolution layers. The residual F(x) is the difference between the expected output H(x) and the input x, so F(x) = H(x) − x. The shortcut passes the input information to the subsequent layer to protect the integrity of the information and converts the learning object into residuals. The network only needs to learn the residuals F(x) rather than the complete output, and does not increase the network parameters or the computational complexity, but can greatly reduce the difficulty of network optimization and simplify the learning objectives. ResNet maintains training accuracy when the network depth is increased. Because our training image size was generally smaller than 20 × 20 pixels, the feature map would shrink after pooling. ResNet has only two pooling layers but has a deep network structure, making it suitable as a feature extractor for small images. The structures of the 18-, 34-, and 50-layer ResNet models are shown in Table 7. We changed the parameters of conv1 to make the network suitable for small sample training, and the other structures are consistent with those in He [42].

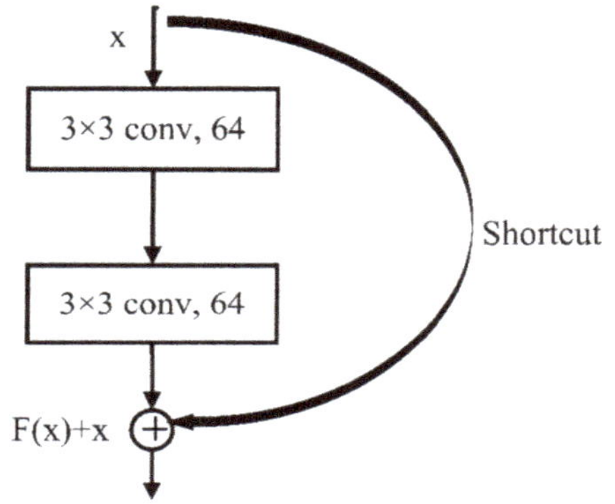

Figure 6. Building residual block. F(x), residual; x, input.

Table 7. ResNet model structures.

Layer Name	18-Layer	34-Layer	50-Layer
Conv1	kernel 5 × 5, filter 32, stride 2		
Max pooling	Pool size 3 × 3, stride 2		
Conv2_x	$\begin{bmatrix} 3\times3, 64 \\ 3\times3, 64 \end{bmatrix} \times 2$	$\begin{bmatrix} 3\times3, 64 \\ 3\times3, 64 \end{bmatrix} \times 3$	$\begin{bmatrix} 1\times1, 64 \\ 3\times3, 64 \\ 1\times1, 256 \end{bmatrix} \times 3$
Conv3_x	$\begin{bmatrix} 3\times3, 128 \\ 3\times3, 128 \end{bmatrix} \times 2$	$\begin{bmatrix} 3\times3, 128 \\ 3\times3, 128 \end{bmatrix} \times 4$	$\begin{bmatrix} 1\times1, 128 \\ 3\times3, 128 \\ 1\times1, 512 \end{bmatrix} \times 4$
Conv4_x	$\begin{bmatrix} 3\times3, 256 \\ 3\times3, 256 \end{bmatrix} \times 2$	$\begin{bmatrix} 3\times3, 256 \\ 3\times3, 256 \end{bmatrix} \times 6$	$\begin{bmatrix} 1\times1, 256 \\ 3\times3, 256 \\ 1\times1, 1024 \end{bmatrix} \times 6$
Conv5_x	$\begin{bmatrix} 3\times3, 512 \\ 3\times3, 512 \end{bmatrix} \times 2$	$\begin{bmatrix} 3\times3, 512 \\ 3\times3, 512 \end{bmatrix} \times 3$	$\begin{bmatrix} 1\times1, 512 \\ 3\times3, 512 \\ 1\times1, 2048 \end{bmatrix} \times 3$
Average pooling	Pool size 2 × 2, stride 1		
Output	Softmax		

2.3.3. Model Training

The training dataset was split randomly into training (75%) and validation (25%) samples. Each image was resized to 15 × 15 to ensure all of the sizes of images input into network were the same. The batch size was set to 60, which was the total number of training images used to train the model at each update. The deep learning platform was based on Tensorflow1.6 and Keras2.1.

A maximum of 500 epochs was set and an early stopping approach was used to avoid overfitting. The loss was calculated based on the validation samples after the operation of each epoch. If the loss did not decrease after 10 epochs, the training was stopped and the weights that provided the best validation accuracy were saved. Models in the actual training process usually converged within 150 epochs. The initial learning rate was set to 10^{-4}. The learning rate decay was set to ensure that the network could automatically reduce the learning rate based on the loss situation. The learning rate would decay when the loss was not reduced after the operation of five epochs. The learning rate factor for each reduction was 5×10^{-3}, with a lower limit of the learning rate of 5×10^{-6}.

Several models were trained by adjusting the parameters during the training process. The CNN model that had the highest validation accuracy was saved for later testing.

To verify the classification performance of CNN, two machine learning classifiers were tested for comparison: RF and SVM. These are widely used to classify tree species because they can accurately handle high-dimensional data [30]. RF is an ensemble learning algorithm that trains a classifier composed of multiple decision trees through a bagging strategy [43]. The training samples of each decision tree are obtained by random sampling, and the classification result is the majority classification result of all decision trees. During the test, we found that when the number of trees exceeded 500, the accuracy did not increase significantly. To save computing time, the number of trees was set to 500. SVM creates a model that is based on a user-defined kernel function and transforms the data into classes. Then, an optimal hyperplane that maximizes the margin distance between classes can be found [44]. The radial basis function was used as the kernel function, with a grid search method [45] to determine optimal classifier parameters in the study.

2.3.4. Accuracy Assessment

The test image in Figure 7 was used to assess the generalization error of the models. The test image was cropped by the yellow rectangle in Figure 3, and was completely independent of training samples. The separation of testing from training data is important to evaluate the classification results. Each tree crown in the test image was extracted using the CSI tree crown delineation algorithm, and the minimum outer cut rectangle of each crown was extracted as a test sample. The test samples were predicted with the CNN models as described in Section 2.3.2. Simultaneously, RF and SVM were used for comparison. The label and location of each test sample were combined to generate tree species classification maps by assigning the predicted labels to the delineated tree crowns. The test points were based on the field investigation and were used as true values to calculate the following indexes: (a) confusion matrices; (b) producer accuracy (PA); (c) user accuracy (UA); (d) average accuracy (AA); (e) overall accuracy (OA); (f) kappa coefficient.

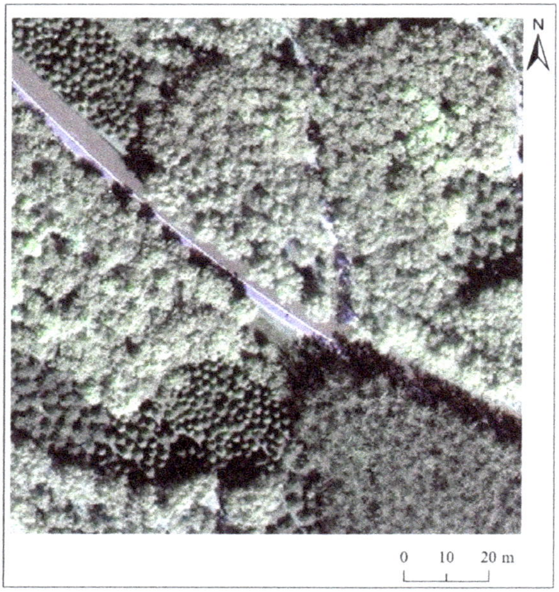

Figure 7. The test image used for the accuracy assessment.

3. Results

Table 8 shows confusion matrices and statistical measures of RF, SVM, AlexNet, GoogLeNet, and ResNet for ITS classification. Table 9 shows the classification accuracies and kappa coefficients for six species based on different models. For CNN models, GoogLeNet achieved the best OA (82.7%) with the highest kappa coefficient (0.79), and was the only model that achieved an OA over 80%. It was followed by ResNet-18 (74.8%), ResNet-50 (71.7%), ResNet-34 (70.9%), and AlexNet, which achieved the lowest OA (52.0%) and kappa coefficient (0.41). GoogLeNet achieved the highest OA for each tree species, with almost all exceeding 80%. AlexNet misclassified almost all pines as poplars, and its classification accuracy for cypress (58.3%), pagoda (45.1%), and willow (34.9%) was about 30% lower than those of GoogLeNet and ResNet. All ResNet misclassified many ashes into willows; its classification accuracy for ash was significantly lower (20%) than that of GoogLeNet. ResNet-18 outperformed ResNet-34 and ResNet-50 of OA by 3.9% and 3.1%, respectively (74.8% > 70.9% > 71.7%).

Compared with RF and SVM, the OA of GoogLeNet (82.7%) was significantly higher than that of RF (44.1%) and SVM (48.8%); even AlexNet had higher OA (52.0%) than RF (44.1%) and SVM (48.8%). Likewise, CNN models all achieved higher kappa coefficients than RF (0.32) and SVM (0.40). RF classified almost all pines as cypresses, making pines almost invisible in the classification map. The classification accuracy of RF for pine, cypress, poplar, and ash ranged from 20% to 30%, which is much lower than that of GoogLeNet. The classification accuracies of RF for pagoda and willow were only about 20%. SVM classified almost all willows as pagodas, making willows almost invisible in the classification map. The classification accuracy of SVM for pine, poplar, ash, and pagoda ranged from 20% to 30%, much lower than that achieved by GoogLeNet.

Table 8. Confusion matrix and statistical measures of RF, SVM, AlexNet, GoogLeNet, and ResNet for individual tree species classification. PA, producer accuracy; UA, user accuracy; AA, average accuracy; OA, overall accuracy.

			RF			
Classified as	**Pine**	**Cypress**	**Poplar**	**Ash**	**Pagoda**	**Willow**
Pine	2	0	0	0	0	0
Cypress	12	13	4	0	2	0
Poplar	1	1	14	0	4	1
Ash	0	0	0	18	0	13
Pagoda	0	1	4	4	4	10
Willow	0	0	0	6	8	5
PA (%)	13.3	86.7	63.6	64.3	22.2	17.2
UA (%)	100.0	41.9	66.7	58.1	17.4	26.3
AA (%)	56.7	64.3	65.2	61.2	19.8	21.8
OA (%)	44.1			Kappa	0.32	
			SVM			
Classified as	**Pine**	**Cypress**	**Poplar**	**Ash**	**Pagoda**	**Willow**
Pine	9	1	5	0	2	1
Cypress	1	14	0	0	0	0
Poplar	5	0	14	0	1	0
Ash	0	0	0	11	0	2
Pagoda	0	0	3	17	14	26
Willow	0	0	0	0	1	0
PA (%)	60.0	93.3	63.6	39.3	77.8	0.0
UA (%)	50.0	93.3	70.0	84.6	23.3	0.0
AA (%)	55.0	93.3	66.8	62.0	50.6	0.0
OA (%)	48.8			Kappa	0.40	
			AlexNet			
Classified as	**Pine**	**Cypress**	**Poplar**	**Ash**	**Pagoda**	**Willow**
Pine	0	0	0	0	0	0
Cypress	3	7	0	0	0	0
Poplar	10	7	20	0	4	0
Ash	0	0	0	26	2	21
Pagoda	1	0	2	0	5	0
Willow	1	1	0	2	7	8
PA (%)	0.0	46.7	90.9	92.9	27.8	27.6
UA (%)	0.0	70.0	48.8	53.1	62.5	42.1
AA (%)	0.0	58.3	69.9	73.0	45.1	34.9
OA (%)	52.0			Kappa	0.41	
			GoogLeNet			
Classified as	**Pine**	**Cypress**	**Poplar**	**Ash**	**Pagoda**	**Willow**
Pine	13	2	1	0	0	0
Cypress	0	13	0	0	0	0
Poplar	0	0	17	0	0	2
Ash	0	0	0	22	1	3
Pagoda	1	0	4	0	16	0
Willow	1	0	0	6	1	24
PA (%)	86.7	86.7	77.3	78.6	88.9	82.8
UA (%)	81.3	100.0	89.5	84.6	76.2	75.0
AA (%)	84.0	93.3	83.4	81.6	82.5	78.9
OA (%)	82.7			Kappa	0.79	

Table 8. *Cont.*

ResNet-18						
Classified as	**Pine**	**Cypress**	**Poplar**	**Ash**	**Pagoda Tree**	**Willow**
Pine	13	3	1	0	0	0
Cypress	0	12	0	0	0	0
Poplar	0	0	16	0	0	1
Ash	0	0	0	16	4	4
Pagoda	0	0	5	0	14	0
Willow	2	0	0	12	0	24
PA (%)	86.7	80.0	72.7	57.1	77.8	82.8
UA (%)	76.5	100.0	94.1	66.7	73.7	63.2
AA (%)	81.6	90.0	83.4	61.9	75.7	73.0
OA (%)	74.8			Kappa	0.69	
ResNet-34						
Classified as	**Pine**	**Cypress**	**Poplar**	**Ash**	**Pagoda**	**Willow**
Pine	13	3	2	0	0	0
Cypress	0	12	0	0	0	0
Poplar	0	0	15	0	1	2
Ash	0	0	0	13	2	4
Pagoda	1	0	5	0	14	0
Willow	1	0	0	15	1	23
PA (%)	86.7	80.0	68.2	46.4	77.8	79.3
UA (%)	72.2	100.0	83.3	68.4	70.0	57.5
AA (%)	79.5	90.0	75.8	57.4	73.9	68.4
OA (%)	70.9			Kappa	0.65	
ResNet-50						
Classified as	**Pine**	**Cypress**	**Poplar**	**Ash**	**Pagoda**	**Willow**
Pine	12	3	1	0	0	0
Cypress	0	12	0	0	0	0
Poplar	0	0	13	0	0	1
Ash	0	0	0	16	5	2
Pagoda	2	0	8	1	12	0
Willow	1	0	0	11	1	26
PA (%)	80.0	80.0	59.1	57.1	66.7	89.7
UA (%)	75.0	100.0	92.9	69.6	52.2	66.7
AA (%)	77.5	90.0	76.0	63.4	59.4	78.2
OA (%)	71.7			Kappa	0.65	

Table 9. Classification accuracies for six tree species using RF, SVM, AlexNet, GoogLeNet, and ResNet. All values except for kappa coefficient, are percentages.

Species	RF	SVM	AlexNet	GoogLeNet	ResNet-18	ResNet-34	ResNet-50
Pine	56.7	55.0	0.0	84.0	81.6	79.5	77.5
Cypress	64.3	93.3	58.3	93.3	90.0	90.0	90.0
Poplar	65.2	66.8	69.9	83.4	83.4	75.8	76.0
Ash	61.2	62.0	73.0	81.6	61.9	57.4	63.4
Pagoda	19.8	50.6	45.1	82.5	75.7	73.9	59.4
Willow	21.8	0.0	34.9	78.9	73.0	68.4	78.2
Kappa	0.32	0.40	0.41	0.79	0.69	0.65	0.65
OA	44.1	48.8	52.0	82.7	74.8	70.9	71.7

Figure 8 shows the ITS maps predicted by RF, SVM, and CNN. RF misclassified almost all pine as cypress. SVM did not separate willow from ash and pagoda. AlexNet classified many pines as poplars or cypresses.

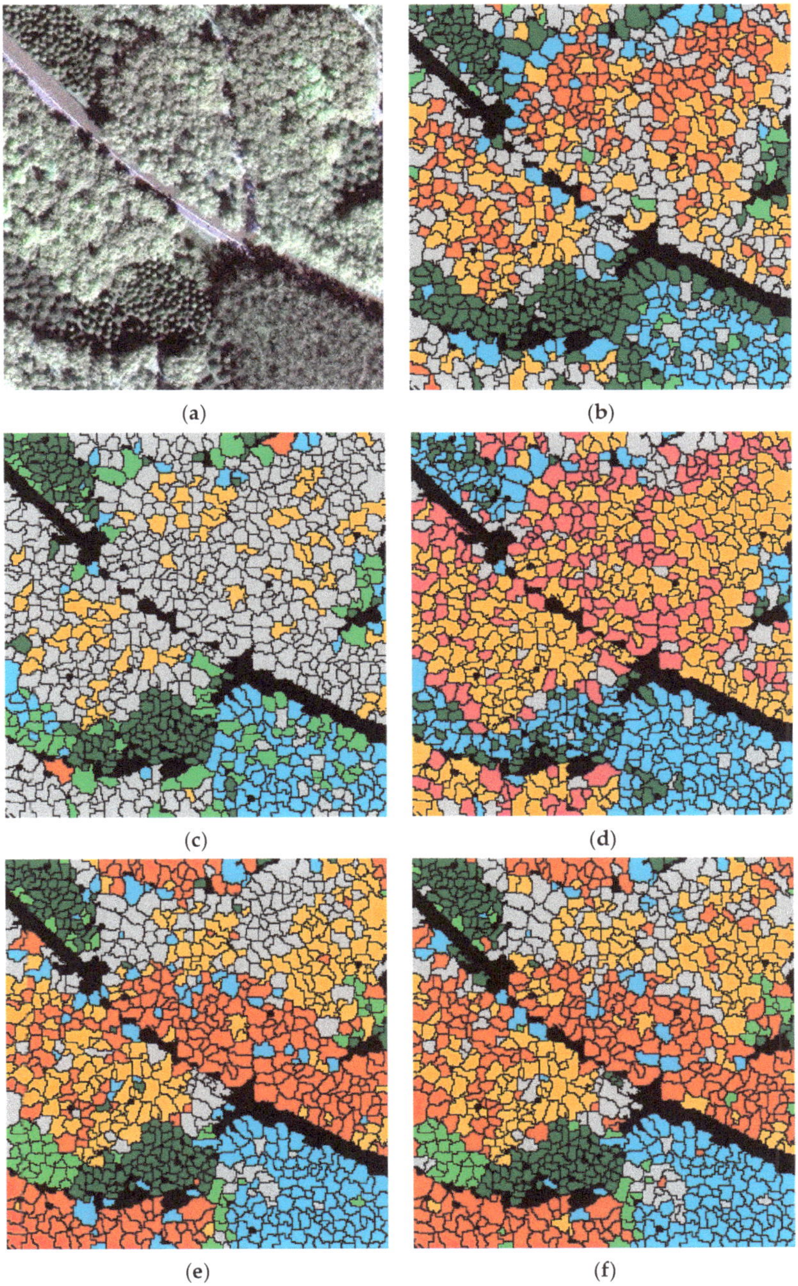

Figure 8. *Cont.*

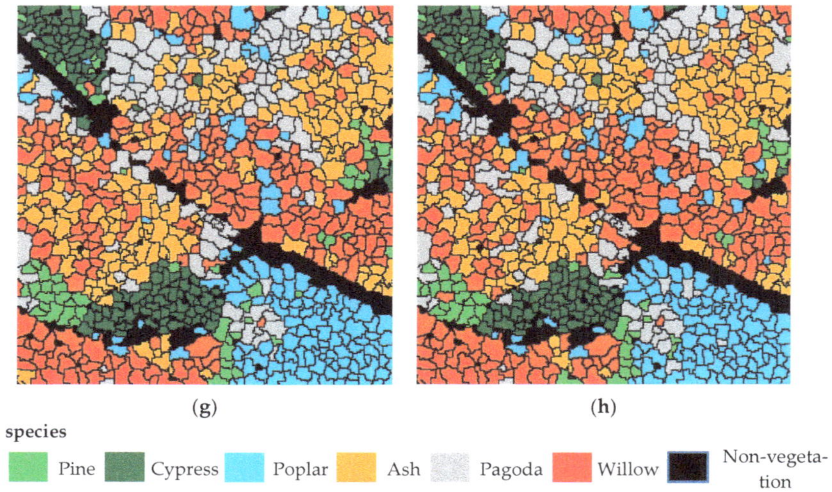

Figure 8. (**a**) The test image. Individual tree species map obtained by (**b**) RF, (**c**) SVM, (**d**) AlexNet, (**e**) GoogLeNet, (**f**) ResNet-18, (**g**) ResNet-34, and (**h**) ResNet-50.

The CSI tree crown delineation produced inevitable errors due to under- and over-segmentation, which are common problems in tree crown delineations [35], especially when only two-dimensional information is used. Some non-crown segments produced by the under- and over-segmentation affected the classification map and reduced classification accuracies. Some tree crowns were misclassified, especially on both sides of the road and the edge of the image, because some non-crown pixels were included in test samples when taking the minimum outer cut rectangles, which changed the classification features of the test samples.

4. Discussion

4.1. Classification Results of Different CNN Models

GoogLeNet achieved the best classification accuracy, followed by ResNet. In addition, AlexNet achieved the lowest accuracy. AlexNet is sequential, in which the output of the previous layer is directly input into the next layer. In contrast, GoogLeNet and ResNet have micro-architecture modules that enable networks to learn faster and more efficiently with increasing depth. The micro-architectural building blocks are stacked together with traditional layers, such as convolution and pool layers, to form the macro-architectures. The results showed that the micro-architecture can improve the classification accuracy. GoogLeNet achieved the highest classification accuracy (82.7%) and similar accuracies for each tree species, which were all higher than 80%, because GoogLeNet uses multi-scale convolution filters to extract features. Coniferous and deciduous tree crowns have different sizes; multi-scale features can be extracted by multi-scale convolution filters in GoogLeNet. The findings suggest that GoogLeNet, with a multi-scale feature extractor, is well-suited for tree species classification. As a comparison, Hartling et al. [30] used WV2/3 and LiDAR data and DenseNet to classify eight tree species; the maximum OA was 82.6%, which is similar to the accuracy in this study. However, the accuracy of different tree species varied greatly. Cypress and oak can be accurately classified with accuracy above 90%, but ash classification accuracy is only 60%. The GoogLeNet used in our study can provide similar accuracies, which were all higher than 80% for each tree species. Sothe et al. [28] used hyperspectral and photogrammetric data for classifying fourteen tree species in a rain forest; the forest environment was more complicated compared to our study area. The best result was achieved by the CNN with an OA of 84.4%, which outperformed SVM and RF.

Fricker et al. [29] used high-spatial resolution airborne hyperspectral imagery to identify tree species in a mixed-conifer forest and the OA reached 87% for the hyperspectral CNN model. However, the study area in Fricker et al. [29] was located in the Southern Sierra Nevada Mountains, CA, USA, so the distribution of trees across an elevation gradient could be analyzed and dead trees were recognized as an individual category. Our study area was located in a park with flat terrain, where trees were manually planted and taken care of by gardeners. In addition, compared with natural forests with abundant and disorganized species, the distribution of trees in our study had a certain pattern that was easy to classify. Thus, higher accuracies would be expected in the manually planted forest.

4.2. Classification Results of Different Tree Species

The ResNet classification results showed that the classification accuracy of different tree species varied widely. The classification accuracies of conifers (pine, cypress) and poplar were above 80%, whereas the classification accuracies of ash, pagoda, and willow were lower than 70%. The test image shows that the texture and shape of the pine, cypress, and poplar are unique and show obvious differences. However, the differences among ash, pagoda, and willow are not obvious. The unique crown shape of cypress creates clear boundaries among trees. Furthermore, accurate segmentation of the crown means that ITC images have large inter-class differences and small intra-class differences, which is conducive to feature extraction and classification. Therefore, the classification accuracy of cypress was highest, reaching 90%. Liu et al. [46] also concluded that the classification accuracies of conifers, which have obvious crown structures, are higher. Differences in the crowns among ash, pagoda, and willow are small; thus, the features extracted by the network are similar, which results in the misclassification of ash, pagoda, and willow. Liu et al. [46] concluded that if two species have similar crown structures, they are prone to misclassification, a finding that is supported by the results of our study. Deciduous tree crowns are dense and the gaps between crowns are small, which complicates delineating deciduous tree crowns. Inaccurate tree crown delineation would reduce differences between tree species and reduce classification accuracy.

4.3. Comparison with Other Machine Learning Models

Spectra, texture, and shape features were extracted for tree species classification based on machine learning methods such as RF and SVM; these features and classifiers have been widely used in other research related to tree species classification [28,30,47]. The methods based on RF or SVM require manual feature extraction, which is usually complicated and requires researchers to design a large number of features based on professional knowledge and experience. Conversely, the methods based on CNN use multi-layer neural networks to abstract from low- to high-level features, avoiding manual feature extraction, thus simplifying the process of feature extraction [22,48]. The results of this study also showed that the CNN obtained higher OA than RF or SVM. The OA of GoogLeNet for six species reached 82.7%, whereas those of RF and SVM were 44.09% and 48.82%, respectively. GoogLeNet significantly outperformed RF and SVM. Many studies compared the classification effects of CNN, RF, and SVM. For example, Hartling [30] showed that DenseNet, RF, and SVM for the classification of eight tree species had an OA of 82.6%, 51.8%, and 52%, respectively, with CNN showing higher accuracy. Sothe et al. [28] used CNN to achieve 22% and 26% higher OA than SVM and RF for hyperspectral data, respectively. The outstanding performance of CNN is due to its ability to enhance the texture, shape, and spatial information in images, and to use that information to detect generic structures in other images [49].

5. Conclusions

Based on practical application requirements, the CMSIR method focuses on two key issues in ITS recognition: fast and accurate construction of a training sample dataset and ITS classification methods. We proposed a method to construct ITC training samples suitable for CNN models. This method combines a multi-scale ITC delineation method, manual labeling of tree species, and sample enhancement techniques to build a training sample dataset. The CSI tree crown delineation algorithm was used to automatically describe tree crowns, which avoided manual sketching.

A method for ITS classification was explored using different CNN methods and high-resolution satellite remote sensing imagery. The data source was readily obtained and was low in cost and wide in scope. The CNN used in this study avoids manual feature extraction and its structures are relatively common, easy to build, and have a low training difficulty and fast training speed. GoogLeNet, which achieved the highest OA, can use multi-scale convolution filters to extract features in multi-scale tree crowns. The classification accuracies for tree crowns in different scales were very high (greater than 80%). Compared with other commonly used machine learning models, such as RF and SVM, the CNN models do not require manual feature extraction and achieve higher OA.

Our study area was located in a park, where trees were manually planted and pruned regularly by gardeners. The distribution of trees had a certain pattern. The crowns of the same trees were similar in size. There were intervals between the crowns, which made it easier to delineate the crowns. Thus, we expected higher accuracies in the manually planted forest. The CMSIR method could be improved in future research in the following ways: (1) Natural forests have a larger number of tree species and the distribution of tree species is more random. The tree crown delineation significantly influences the accuracy of the subsequent classification. Researchers need to effectively delineate tree crowns, especially in dense natural forests. (2) We used some classic CNN models, which were originally used for the ILSVRC dataset. Compared with ILSVRC images, tree crown images have smaller scales and multiple bands. Therefore, a network model suitable for small-scale and multiband images should be constructed for ITS classification.

Author Contributions: S.Y. designed and completed the experiments and drafted the manuscript. L.J. designed the methodology and provided feedback on the manuscript. H.W. assisted in investigating tree species and modified the manuscript. All authors assisted in writing and improving the paper. All authors have read and agreed to the published version of the manuscript.

Funding: This study was funded by the Aerospace Information Research Institute, Chinese Academy of Sciences (No. Y951150Z2F), the Science and Technology Major Project of Xinjiang Uygur Autonomous Region (No. 2018A03004), the Strategic Priority Research Program of the Chinese Academy of Sciences (No. XDA19030501, No. XDA19090300), and the National Natural Science Foundation of China (No. 41972308, No. 61731022).

Institutional Review Board Statement: Not applicable.

Informed Consent Statement: Not applicable.

Data Availability Statement: Data sharing is not applicable to this article.

Acknowledgments: The authors want to thank the China Remote Sensing Satellite Ground Station and Key Laboratory of Digital Earth Science for supporting this research with hardware devices. In addition, we are grateful to the anonymous reviewers who provided helpful comments and suggestions to improve the manuscript.

Conflicts of Interest: The authors declare no conflict of interest.

References

1. Zaki, N.A.M.; Abd Latif, Z. Carbon sinks and tropical forest biomass estimation: A review on role of remote sensing in aboveground-biomass modelling. *Geocarto Int.* **2017**, *32*, 701–716. [CrossRef]
2. Bravo, F.; LeMay, V.; Jandl, R. *Managing Forest Ecosystems: The Challenge of Climate Change*, 2nd ed.; Springer: New York, NY, USA, 2008; pp. 3–15.
3. Kangas, A.; Maltamo, M. *Forest Inventory: Methodology and Applications*; Springer: Dordrecht, The Netherlands, 2006; Volume 10, p. 368.
4. Liu, Y.A.; Gong, W.S.; Hu, X.Y.; Gong, J.Y. Forest type identification with random forest using Sentinel-1A, Sentinel-2A, multi-temporal Landsat-8 and DEM data. *Remote Sens.* **2018**, *10*, 946. [CrossRef]
5. Wulder, M.A.; Hall, R.J.; Coops, N.C.; Franklin, S.E. High spatial resolution remotely sensed data for ecosystem characterization. *Bioscience* **2004**, *54*, 511–521. [CrossRef]

6. Ke, Y.; Quackenbush, L.J.; Im, J. Synergistic use of QuickBird multispectral imagery and LIDAR data for object-based forest species classification. *Remote Sens. Environ.* **2010**, *114*, 1141–1154. [CrossRef]
7. Chen, E.; Li, Z.; Tan, B.; Liang, Y.; Zhang, Z. Validation of statistic based forest types classification methods using hyperspectral data. *Scientia Silvae Sinicae.* **2007**, *43*, 84–89.
8. Ghosh, A.; Fassnacht, F.E.; Joshi, P.K.; Koch, B. A framework for mapping tree species combining hyperspectral and LiDAR data: Role of selected classifiers and sensor across three spatial scales. *Int. J. Appl. Earth Obs. Geoinf.* **2014**, *26*, 49–63. [CrossRef]
9. Yu, L.; Yu, Y.; Liu, X.; Du, Y.; Zhang, H. Tree species classification with hyperspectral image. *J. Northeast For. Univ.* **2016**, *44*, 40–43, 57.
10. Reitberger, J.; Krzystek, P.; Stilla, U. Analysis of full waveform LIDAR data for the classification of deciduous and coniferous trees. *Int. J. Remote Sens.* **2008**, *29*, 1407–1431. [CrossRef]
11. Suratno, A.; Seielstad, C.; Queen, L. Tree species identification in mixed coniferous forest using airborne laser scanning. *ISPRS J. Photogramm. Remote Sens.* **2009**, *64*, 683–693. [CrossRef]
12. Yu, X.; Litkey, P.; Hyyppa, J.; Holcpainen, M.; Vastaranta, M. Assessment of low density full-waveform airborne laser scanning for individual tree detection and tree species classification. *Forests* **2014**, *5*, 1011–1031. [CrossRef]
13. Hovi, A.; Korhonen, L.; Vauhkonen, J.; Korpela, I. LiDAR waveform features for tree species classification and their sensitivity to tree- and acquisition related parameters. *Remote Sens. Environ.* **2016**, *173*, 224–237. [CrossRef]
14. Fassnacht, F.E.; Latifi, H.; Sterenczak, K.; Modzelewska, A.; Lefsky, M.; Waser, L.T.; Straub, C.; Ghosh, A. Review of studies on tree species classification from remotely sensed data. *Remote Sens. Environ.* **2016**, *186*, 64–87. [CrossRef]
15. Immitzer, M.; Atzberger, C.; Koukal, T. Tree species classification with random forest using very high spatial resolution 8-band WorldView-2 satellite data. *Remote Sens.* **2012**, *4*, 2661–2693. [CrossRef]
16. Somers, B.; Asner, G.P. Tree species mapping in tropical forests using multi-temporal imaging spectroscopy: Wavelength adaptive spectral mixture analysis. *Int. J. Appl. Earth Obs. Geoinf.* **2014**, *31*, 57–66. [CrossRef]
17. Lee, J.; Cai, X.; Lellmann, J.; Dalponte, M.; Malhi, Y.; Butt, N.; Morecroft, M.; Schonlieb, C.-B.; Coomes, D.A. Individual Tree Species Classification From Airborne Multisensor Imagery Using Robust PCA. *IEEE J. Sel. Top. Appl. Earth Obs. Remote Sens.* **2016**, *9*, 2554–2567. [CrossRef]
18. Le Louarn, M.; Clergeau, P.; Briche, E.; Deschamps-Cottin, M. "Kill Two Birds with One Stone": Urban Tree Species Classification Using Bi-Temporal Pleiades Images to Study Nesting Preferences of an Invasive Bird. *Remote Sens.* **2017**, *9*, 916. [CrossRef]
19. Mishra, N.B.; Mainali, K.P.; Shrestha, B.B.; Radenz, J.; Karki, D. Species-level vegetation mapping in a Himalayan treeline ecotone using unmanned aerial system (UAS) imagery. *ISPRS Int. Geo Inf.* **2018**, *7*, 445. [CrossRef]
20. Moisen, G.G.; Freeman, E.A.; Blackard, J.A.; Frescino, T.S.; Zimmermann, N.E.; Edwards, T.C., Jr. Predicting tree species presence and basal area in Utah: A comparison of stochastic gradient boosting, generalized additive models, and tree-based methods. *Ecol. Model.* **2006**, *199*, 176–187. [CrossRef]
21. Franklin, S.E.; Ahmed, O.S. Deciduous tree species classification using object-based analysis and machine learning with unmanned aerial vehicle multispectral data. *Int. J. Remote Sens.* **2018**, *39*, 5236–5245. [CrossRef]
22. Zhang, L.; Zhang, L.; Du, B. Deep Learning for Remote Sensing Data: A technical tutorial on the state of the art. *IEEE Geosci. Remote Sens. Mag.* **2016**, *4*, 22–40. [CrossRef]
23. Krizhevsky, A.; Sutskever, I.; Hinton, G.E. Imagenet classification with deep convolutional neural networks. In Proceedings of the Twenty-Sixth Annual Conference on Neural Information Processing Systems, Lake Tahoe, NV, USA, 3–8 December 2012; pp. 1097–1105.
24. Liu, B.; Yu, X.C.; Yu, A.Z.; Wan, G. Deep convolutional recurrent neural network with transfer learning for hyperspectral image classification. *J. Appl. Remote Sens.* **2018**, *12*, 17. [CrossRef]
25. Kattenborn, T.; Leitloff, J.; Schiefer, F.; Hinz, S. Review on Convolutional Neural Networks (CNN) in vegetation remote sensing. *ISPRS J. Photogramm. Remote Sens.* **2021**, *173*, 24–49. [CrossRef]
26. Guan, H.; Yu, Y.; Ji, Z.; Li, J.; Zhang, Q. Deep learning-based tree classification using mobile LiDAR data. *Remote Sens. Lett.* **2015**, *6*, 864–873. [CrossRef]
27. Zou, X.H.; Cheng, M.; Wang, C.; Xia, Y.; Li, J. Tree Classification in Complex Forest Point Clouds Based on Deep Learning. *IEEE Geosci. Remote Sens. Lett.* **2017**, *14*, 2360–2364. [CrossRef]
28. Sothe, C.; De Almeida, C.M.; Schimalski, M.B.; La Rosa, L.E.C.; Castro, J.D.B.; Feitosa, R.Q.; Dalponte, M.; Lima, C.L.; Liesenberg, V.; Miyoshi, G.T.; et al. Comparative performance of convolutional neural network, weighted and conventional support vector machine and random forest for classifying tree species using hyperspectral and photogrammetric data. *GISci. Remote Sens.* **2020**, *57*, 369–394. [CrossRef]
29. Fricker, G.A.; Ventura, J.D.; Wolf, J.A.; North, M.P.; Davis, F.W.; Franklin, J. A Convolutional Neural Network Classifier Identifies Tree Species in Mixed-Conifer Forest from Hyperspectral Imagery. *Remote Sens.* **2019**, *11*, 2326. [CrossRef]
30. Hartling, S.; Sagan, V.; Sidike, P.; Maimaitijiang, M.; Carron, J. Urban Tree Species Classification Using a WorldView-2/3 and LiDAR Data Fusion Approach and Deep Learning. *Sensors* **2019**, *19*, 1284. [CrossRef]
31. Hu, M.M. Preliminary Study on Plant Landscape and Eco-Efficiency of Beijing Olympic Forest Park. Master's Thesis, Beijing Forestry University, Beijing, China, 2009. (In Chinese)
32. Qin, C. Cooling and Humidifying Effects and Driving Mechanisms of Beijing Olympic Forest Park in Summer. Ph.D. Thesis, Beijing Forestry University, Beijing, China, 2010. (In Chinese)

33. Jing, L.; Hu, B.; Noland, T.; Li, J. An individual tree crown delineation method based on multi-scale segmentation of imagery. *ISPRS J. Photogramm. Remote Sens.* **2012**, *70*, 88–98. [CrossRef]
34. Chen, Q.; Baldocchi, D.; Gong, P.; Kelly, M. Isolating individual trees in savanna woodland using small footprint LIDAR data. *Photogramm. Eng. Remote Sensing* **2006**, *72*, 923–932. [CrossRef]
35. Qiu, L.; Jing, L.; Hu, B.; Li, H.; Tang, Y. A New Individual Tree Crown Delineation Method for High Resolution Multispectral Imagery. *Remote Sens.* **2020**, *12*, 585. [CrossRef]
36. Alom, M.Z.; Taha, T.M.; Yakopcic, C.; Westberg, S.; Sidike, P.; Nasrin, M.S.; Hasan, M.; Van Essen, B.C.; Awwal, A.A.S.; Asari, V.K. A State-of-the-Art Survey on Deep Learning Theory and Architectures. *Electronics* **2019**, *8*, 292. [CrossRef]
37. LeCun, Y.; Bengio, Y.; Hinton, G. Deep learning. *Nature* **2015**, *521*, 436–444. [CrossRef] [PubMed]
38. Dang, Y.; Zhang, J.; Deng, K.; Zhao, Y.; Yu, F. Study on the Evaluation of Land Cover Classification using Remote Sensing Images Based on AlexNet. *J. Geo-Inf. Sci.* **2017**, *19*, 1530–1537.
39. Liu, Y.; Huang, X.; Li, H.; Liu, Z.; Cheng, C.; Wang, X. Extraction of Irregular Solid Waste in Rural based on Convolutional Neural Network and Conditional Random Field Method. *J. Geo-Inf. Sci.* **2019**, *21*, 259–268.
40. Vedaldi, A.; Lenc, K. MatConvNet: Convolutional neural networks for MATLAB. In Proceedings of the 23rd ACM International Conference on Multimedia, Brisbane, Australia, 26–30 October 2015.
41. Szegedy, C.; Liu, W.; Jia, Y.; Sermanet, P.; Reed, S.; Anguelov, D.; Erhan, D.; Vanhoucke, V.; Rabinovich, A. Going Deeper with Convolutions. In Proceedings of the IEEE Conference on Computer Vision and Pattern Recognition, Columbus, OH, USA, 24–27 June 2014; pp. 1–9.
42. He, K.; Zhang, X.; Ren, S.; Sun, J. Deep Residual Learning for Image Recognition. In Proceedings of the IEEE Conference on Computer Vision and Pattern Recognition, Boston, MA, USA, 7–12 June 2015; pp. 770–778.
43. Gislason, P.; Benediktsson, J.A.; Sveinsson, J.M. Random forests for land cover classification. *Pattern. Recogn. Lett.* **2006**, *27*, 294–300. [CrossRef]
44. Cortes, C.; Vapnik, V. Support-vector networks. *Mach. Learn.* **1995**, *20*, 273–297. [CrossRef]
45. Chang, C.-C.; Lin, C.-J. LIBSVM: A Library for Support Vector Machines. *ACM Trans. Intell. Syst. Technol.* **2011**, *2*. [CrossRef]
46. Liu, L.; Coops, N.C.; Aven, N.W.; Pang, Y. Mapping urban tree species using integrated airborne hyperspectral and LiDAR remote sensing data. *Remote Sens. Environ.* **2017**, *200*, 170–182. [CrossRef]
47. Ferreira, M.P.; Zortea, M.; Zanotta, D.C.; Shimabukuro, Y.E.; de Souza Filho, C.R. Mapping tree species in tropical seasonal semi-deciduous forests with hyperspectral and multispectral data. *Remote Sens. Environ.* **2016**, *179*, 66–78. [CrossRef]
48. Gao, Q.; Lim, S.; Jia, X. Hyperspectral Image Classification Using Convolutional Neural Networks and Multiple Feature Learning. *Remote Sens.* **2018**, *10*, 299. [CrossRef]
49. Sylvain, J.D.; Drolet, G.; Brown, N. Mapping dead forest cover using a deep convolutional neural network and digital aerial photography. *ISPRS J. Photogramm. Remote Sens.* **2019**, *156*, 14–26. [CrossRef]

remote sensing

MDPI

Article

Spatio-Temporal Classification Framework for Mapping Woody Vegetation from Multi-Temporal Sentinel-2 Imagery

Jovan Kovačević [1,*], Željko Cvijetinović [1], Dmitar Lakušić [2], Nevena Kuzmanović [2], Jasmina Šinžar-Sekulić [2], Momir Mitrović [3], Nikola Stančić [1], Nenad Brodić [1] and Dragan Mihajlović [1]

1 Faculty of Civil Engineering, University of Belgrade, Bulevar kralja Aleksandra 73, 11000 Belgrade, Serbia; zeljkoc@grf.bg.ac.rs (Ž.C.); nstancic@grf.bg.ac.rs (N.S.); nbrodic@grf.bg.ac.rs (N.B.); draganm@grf.bg.ac.rs (D.M.)

2 Faculty of Biology, University of Belgrade, Institute of Botany and Botanical Garden "Jevremovac", Takovska 43, 11000 Belgrade, Serbia; dlakusic@bio.bg.ac.rs (D.L.); nkuzmanovic@bio.bg.ac.rs (N.K.); jsekulic@bio.bg.ac.rs (J.Š.-S.)

3 MapSoft d.o.o, Ustanička 64/7, 11000 Belgrade, Serbia; momir@mapsoft.rs

* Correspondence: jkovacevic@grf.bg.ac.rs

Received: 30 July 2020; Accepted: 28 August 2020; Published: 2 September 2020

Abstract: The inventory of woody vegetation is of great importance for good forest management. Advancements of remote sensing techniques have provided excellent tools for such purposes, reducing the required amount of time and labor, yet with high accuracy and the information richness. Sentinel-2 is one of the relatively new satellite missions, whose 13 spectral bands and short revisit time proved to be very useful when it comes to forest monitoring. In this study, the novel spatio-temporal classification framework for mapping woody vegetation from Sentinel-2 multitemporal data has been proposed. The used framework is based on the probability random forest classification, where temporal information is explicitly defined in the model. Because of this, several predictions are made for each pixel of the study area, which allow for specific spatio-temporal aggregation to be performed. The proposed methodology has been successfully applied for mapping eight potential forest and shrubby vegetation types over the study area of Serbia. Several spatio-temporal aggregation approaches have been tested, divided into two main groups: pixel-based and neighborhood-based. The validation metrics show that determining the most common vegetation type classes in the neighborhood of 5 × 5 pixels provides the best results. The overall accuracy and kappa coefficient obtained from five-fold cross validation of the results are 82.97% and 0.75, respectively. The corresponding producer's accuracies range from 36.74% to 97.99% and user's accuracies range from 46.31% to 98.43%. The proposed methodology proved to be applicable for mapping woody vegetation in Serbia and shows a potential to be implemented in other areas as well. Further testing is necessary to confirm such assumptions.

Keywords: classification; Sentinel-2; woody vegetation; probability random forest; forest inventory; Serbia

---

1. Introduction

Forests represent the most dominant land ecosystem, which covers about 31 percent of the Earth's total land area [1]. The forests can be sorted into number of different types, commonly defined using information about the composition of species, productivity or crown closure [2]. Mapping woody vegetation is of crucial importance to properly understand forest ecosystems. Providing accurate and cost-efficient information about the forest is critical for proper forest management, ecosystem preservation and climate change mitigation [3–5]. Traditional field survey methods provide highly accurate forest inventory, yet these methods are highly time-consuming and a lot of manual labor is required. The advancements of the remote sensing techniques try to fill such gaps by reducing the required amount of time and labor, yet with the preservation of the accuracy and the information richness [6–8].

Optical sensors are widely used to identify woody vegetation, although this type of forest inventory remains a challenge. In addition to common problems that affect all optical remote sensing, such as weather conditions, illumination and cloud cover, there are also some issues that particularly affect forest inventory. The spectral separability between the tree species can be influenced by forest maturity and tree density in combination with leaf biochemical properties and canopy structure. Thus, the spectral variability within species may be higher than the variability between species [7]. Furthermore, a continuous transition between vegetation types is characteristic for some tree species, making it difficult to spectrally differentiate such changes [9]. Sentinel-2 is a relatively new satellite mission equipped with MultiSpectral Instrument (MSI). Sentinel-2 collects data in 13 spectral bands, with spatial resolution from 10 m to 60 m, depending on the spectral channel. With constellation of two twin-satellites, Sentinel-2 has a very short revisit time, ideally providing land cover observations every 5 days under cloud-free conditions [10]. Such characteristics proved to be very useful when it comes to vegetation monitoring, with many studies specifically aiming at forest monitoring and inventory [11–13]. Multiple studies proved that the use of time-series improves the tree species identification compared to using only single date satellite images [12,14,15]. This way, the species phenology can be exploited, but it also induces some new problems, such as increasing the dimensionality of the data and the requirements for more complex algorithms, and longer processing times [16,17].

Machine learning (ML) techniques proved to be very useful when it comes to mapping woody vegetation from remotely sensed data [6,18]. Random forest (RF) is one of the vastly popular ML algorithms, which yields very high accuracy of results even with high-dimensional, noisy, and multi-source data [19]. Unfortunately, RF does not account for spatial and temporal characteristics per se, and therefore such information needs to be taken into account in some other way. A number of approaches have been proposed in order to adapt to the spatio-temporal aspects in the RF model. Temporal aspects are commonly accounted for implicitly by defining a separate set of variables for each timestamp [6,7], or explicitly by including such information as one or more variables (day of year, cumulative day of year, etc.) [20]. Spatial aspects can also be taken into account in the same way by including coordinates as variables in the model [20,21]. There are also more complicated solutions such as the one proposed by Sekulić et al., where the spatially and temporally nearest observations are included explicitly as variables [22]. In the case of land cover and land use classification problems, the neighborhood of the pixels proved to be very important. Aonpong et al. improved land cover prediction by involving the neighboring pixels [23]. Samardžić-Petrović et al. used the most common class in the Moore neighborhood for predicting the short term land-use changes, which proved to be one of the most informative variables in the created model [24]. To the best of the authors' knowledge, the applicability of neighborhood-based approaches has not yet been tested for woody vegetation mapping.

The more common approach to defining a woody vegetation prediction model is to omit temporal variables from the model and instead include temporal information implicitly by creating separate sets of variables of all spectral bands for each timestamp [6,7]. This means that the number of variables in the model grows as there are more timestamps available. For too many timestamps, this can lead to computational and performance issues. However, the bigger problem is that this approach, in most cases, requires that each part of the study area has observations available for all timestamps used. Therefore, especially in the case of using optical data, the timestamps with clouds must be avoided. As the study area increases, it becomes more difficult to obtain cloud-free data with similar timestamps. For some large areas, it is even impossible to have completely cloud-free data. There are several solutions for this problem, each with its pros and cons. One way is to decrease the temporal resolution and use monthly or yearly timestamps. Such generalization introduces additional errors in the model and can miss the fine phenological vegetation patterns. There have also been various attempts to build methods specifically to fill missing data gaps due to cloud cover using different kinds of techniques [25–27]. These approaches add an extra layer of complexity, reduce spatial resolution and/or are not good enough when the cloud gaps are too large.

In this study, the new spatio-temporal classification framework has been proposed for mapping woody vegetation from multi-temporal Sentinel-2 imagery. The method exploits the benefits of explicitly defining temporal variables in the model and the use of pixels' neighborhoods to determine the most probable class of woody vegetation. This was done by using each pixel's class probability prediction obtained from the RF model together with its neighboring pixels' prediction to improve output class estimation. The framework has been tested for mapping forests over Serbia, using a classification scheme with eight potential forest and shrubby vegetation types.

2. Materials and Methods

2.1. Study Area

The study area is the complete area of the Republic of Serbia, which covers 88,361 km^2 (Figure 1). The northern part of the country belongs to the Pannonian Plain, where elevation in most part is less than 200 m. The central parts of Serbia are dominated by hills, which turn into mountains in the south, with more than 30 mountain peaks higher than 2000 m. The humid continental climate is present throughout most of the country, whereas the alpine climate can be found in higher mountains. The average amount of precipitation is 896 mm.

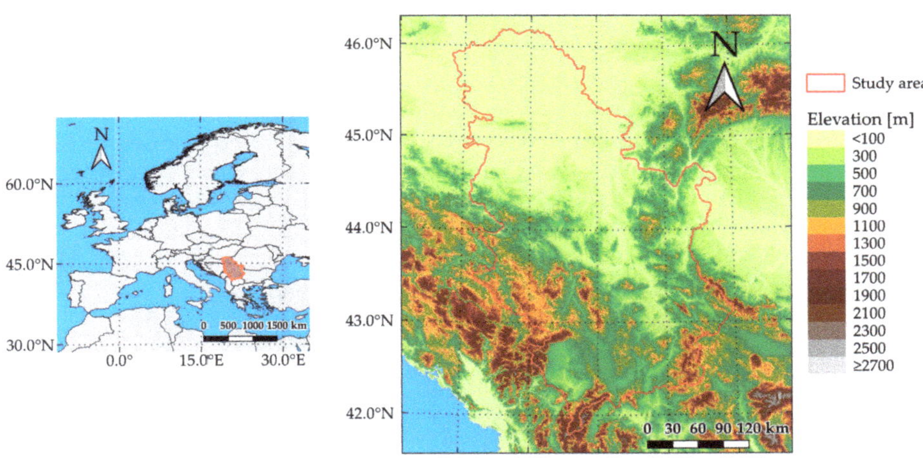

Figure 1. The location of study area—Republic of Serbia.

The land cover is in accordance with the relief of Serbia, where agricultural land covers most of the Pannonian Plain and the plains around larger rivers. The rest of the area is mainly covered by forests and grasslands. Wetlands and barren land area are also present in certain parts of the country. Forests cover more than a third of the area, with species of oak, beech, pines and firs being the most common [28]. According to Bohn et al. [29] the most common forest types in Serbia are thermophilous mixed deciduous broadleaved forests (G.1, G.2, G.3—forest dominated by *Quercus cerris, Q. frainetto, Q. petraea, Q. pubescens, Carpinus orentalis, Ostrya carpinifolia*) and mesophytic broadleaved deciduous (F.3, F5—forest dominated by *Fagus sylvatica, F. moesiaca, Carpinus betulus*). In addition, significant areas are covered with mesophytic coniferous forests (D.4, D.5, D.6—forest dominated by *Abies alba,*

Picea abies, P. omorika, Pinus sylvestris, Pinus peuce); xerophytic coniferous forests, coniferous woodland and scrub (K.1—forest dominated by *P. nigra* agg., *P. heldreichii*); forest steppes (L.1, L.2—forest dominated by *Quercus pubescens, Q. robur, Q. pedunculiflora, Acer tataricum*) and softwood alluvial forests (U.5—forest dominated by *Populus nigra, P. alba, Salix alba*).

2.2. Data

2.2.1. Sentinel-2 Imagery

The Sentinel-2 mission is a part of the Copernicus Programme developed by the European Space Agency (ESA) [10]. It is equipped with a MultiSpectral Scanner that has 13 spectral bands—4 bands with 10 m spatial resolution, 6 with 20 m spatial resolution and the remaining 3 bands with 60 m spatial resolution. In this research, Level-2A Sentinel-2 products have been used. They represent orthorectified Bottom-Of-Atmosphere (BOA) reflectance [30]. The product was disseminated as 100 × 100 km granules, with bands resampled to 10 m, 20 m and 60 m spatial resolution with a corresponding Scene Classification map (SCL). The SCL map differentiates several basic classes (cloud, cloud shadows, vegetation, soils/deserts, water, snow, etc.) [30]. All of the available 60 m resampled bands (B01, B02, B03, B04, B05, B06, B07, B08a, B09, B11 and B12) and 60 m SCL maps have been used in this research.

A total of 88 Sentinel-2 granules from the year of 2019 have been obtained from Copernicus Open Access Hub (https://scihub.copernicus.eu/). Only the year 2019 was chosen in order to reduce the ambiguities caused by forest changes, which occurred over multiple years. At the same time, this made the woody vegetation map very recent and applicable to other ongoing studies/research over the study area. Granules were selected to have less than 20% cloud cover and so that each part of the study area was covered by at least one granule in each of the time intervals: (1) June–July, (2) August, (3) September and (4) October. The months before June and after October were severely influenced by cloud contamination and snow cover at higher altitudes, which made them not particularly useful for the study area of interest. In addition, these four periods were selected in order to take advantage of phenological patterns for differentiating the types of woody vegetation. All types of woody vegetation were in full vegetation in June and their vegetation type specifically reduced as the colder period approached. The model aimed to capture these patterns to detect different types of woody vegetation. The timestamps and spatial coverage of the acquired granules can be seen in Figure 2.

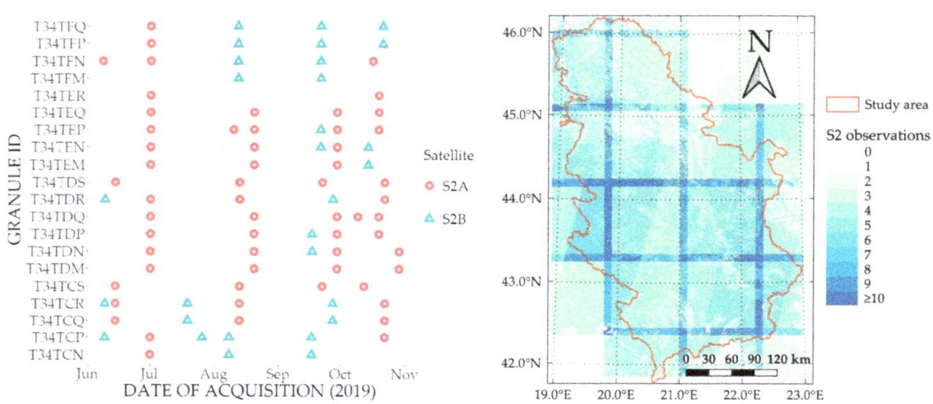

(**a**) (**b**)

Figure 2. The used Sentinel-2 granules over the study area: (**a**) Timestamps the granules; (**b**) Number of Sentinel-2 observations available over the study area.

2.2.2. Woody Vegetation Polygons

The "Database on the distribution of potentially endangered species and habitats of Serbia" [31], created through extensive field campaign and over multiple seasons, was used as a base source for the delineation of woody vegetation types. The locations of species were extracted from the database and used to create buffers with a radius of 500 m. These buffers were used as a "base map" to help in the manual vectorization of woody vegetation polygons using Google Earth software and high-resolution satellite imagery. The woody vegetation polygons were created using sufficiently large zoom to reduce vectorization errors and to be as homogenous as possible. The historical imagery (representing different seasons) was also used where necessary to solve ambiguities. The 8 potential forest and shrubby vegetation types classes have been chosen for this research, as they represent the main woody vegetation classes present in the study area. The resulting dataset consists of 549 polygons representing 7 forest and 1 shrubby vegetation classes. The definition of classes and their individual representation, expressed as number of polygons and covered area, are given in Table 1. For an easier interpretation of the results and discussion, informal phytosociological names, without a syntaxonomic meaning, were assigned to each class. The corresponding European Nature Information System (EUNIS) [32] and SrbHab code [33] was determined for each class (Appendix A).

Table 1. The definition of classification scheme and the number of vectorized training polygons for their representation.

Class ID	Vegetation Type	Informal Phytosociological Name	EUNIS Code	SrbHab Code	Number of Polygons
1	Thermophytic broadleaved deciduous (*Carpinus orientalis–Ostrya carpinifolia*) forests	Ostryo-Carpinion	G1.7C	A2.9	40
2	Thermophytic broadleaved deciduous (Quercus) forests	Quercion frainetto	G1.76	A2.1	63
3	Thermo–Mesophytic broadleaved deciduous (Quercus) forests	Quercion petraea-cerris	G1.76	A2.5	107
4	Mesophytic and hygromesophytic broadleaved deciduous (Quercus) forests	Quercion roboris	G1.2 and G1.A	A1.3 and A1.4	116
5	Mesophytic broadleaved deciduous (Fagus) forests	Fagion	G1.6	A3.2	93
6	Xerophytic relic coniferous (Pinus) forests	Pinion nigrae	G3.5	A5.1	55
7	Psychrophytic boreomontane coniferous (Picea) forests	Vaccinio-Piceion	G3.1	A6.1	50
8	Psychrophytic subalpine coniferous krummholz (Pinus) scrub	Pinion mugo	F2.4	B6.1	25

It was not possible to create the woody vegetation polygons with a specific spatial distribution, nor with a distribution that would respect the flat and mountainous areas in the study area. This is because the northern flat part of the study area has only a small portion of forest cover with only certain types of woody vegetation. Thus, some types of woody vegetation are found only in certain parts of the study area. In addition, the locations for woody vegetation polygons were limited by data available in the "Database on the distribution of potentially endangered species and habitats of Serbia". Although a number of clusters of polygons are present, there is no class that is represented by a single polygon cluster (Figure 3).

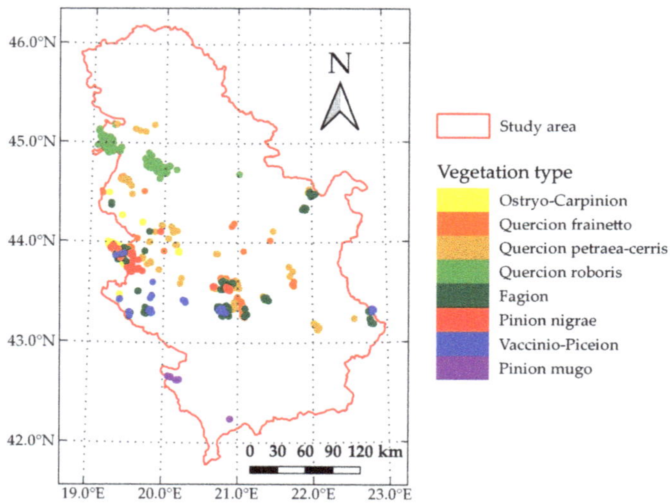

Figure 3. The overview of created woody vegetation polygons' locations over the study area.

2.2.3. Forest Mask

The Copernicus Land Monitoring Service—High Resolution Layer Forest product has been used in order to create a forest mask. This was necessary in order to remove non-forest areas from the study area. High Resolution Layer Forest consists of multiple products, which are created from the multitemporal Sentinel-2A and Landsat 8 data [34]. These products are: (1) Tree Cover Density, (2) Dominant Leaf Type, (3) Forest Type and (4) Forest Additional Support Layer. All of these products are available over the complete area of Europe in 20 m and 100 m spatial resolutions for the years 2012 and 2015.

From several available products, the Forest Type product and the Forest Additional Support Layer have been acquired, both in the spatial resolution of 20 m and for year 2015. All types of forest areas available in the Forest Type product have been reclassified into a binary mask. Trees in urban or agricultural contexts have been excluded from such a mask. This has been done by using information on such areas that is available in Forest Additional Support Layer products.

2.3. *Synthetic Minority Oversampling TEchnique (SMOTE)*

The unbalanced classes in the dataset, representing a situation where one class in the training set dominates the others, is a common classification problem that can severely influence the performance of the classification model. The Synthetic Minority Oversampling TEchnique (SMOTE) is a popular approach for dealing with unbalanced datasets [35]. This method works by artificially creating additional examples of the minority class using the nearest neighbors' examples. Some SMOTE implementations simultaneously undersample the majority classes, which produces a more balanced dataset in the end and reduces the amount of data. The SMOTE algorithm implemented in the DMwR R package has been used in this research for balancing the extracted dataset of woody vegetation classes [35,36].

2.4. *Probability Random Forest*

Random forest (RF) is a non-parametric supervised machine learning technique proposed by Breiman [37] that works for both classification and regression problems. The main idea behind RF is to build a set of decision trees in the training phase. Each tree is created by bootstrap sampling part of the input data while the rest of the input data is used to evaluate performance and adjust the building process of the trees. When the forest of decision trees is created, the final prediction is determined as mean prediction of all created decision trees in case of a regression, or as a most common predicted class in classification [37].

Probability random forest is an enhancement of the random forest method that provides consistent probability estimation [38]. In case of binary classification, this is done by using the proportion of "1s" in the output of the trained classification forest. Such an approach is generalized for multiclass classification problems as well. In this research, the probability random forest method implemented in the Ranger R package has been used [36,39].

The RF method was chosen over other ML techniques because of its simple implementation, robustness and interpretability of the results. Additionally, it has been proven to perform well for tree species and forest vegetation detection in similar studies [6,40,41]. Unlike the standard random forest classifier where the output is a plain class label, probability RF also provides the probabilities for every class to be voted. These probability estimates allow for more realistic representation of the ambiguity of phenomena and are suitable to perform the spatio-temporal aggregation.

3. Methodology

There are several processing steps of the proposed spatio-temporal framework for woody vegetation classification (Figure 4). First, the training polygons are overlaid with available Sentinel-2 granules in order to extract spectral information and to calculate the Normalized Difference Vegetation Index (NDVI) value for each observation. This was done for each pixel over the study area. The classes of the extracted dataset are then balanced using the SMOTE algorithm. Next, the probability random forest model is grown and used to predict the class probabilities for each pixel. The spatio-temporal aggregation was done in the final step in order to determine the most probable woody vegetation classes over the study area. The obtained results were assessed using five-fold and 10-fold leave-location out cross-validation (LLOCV), by calculating the parameters from the confusion matrix. These calculated parameters are user's accuracy, producer's accuracy, overall accuracy and kappa coefficient of agreement [42]. All these steps are further explained in the following sections.

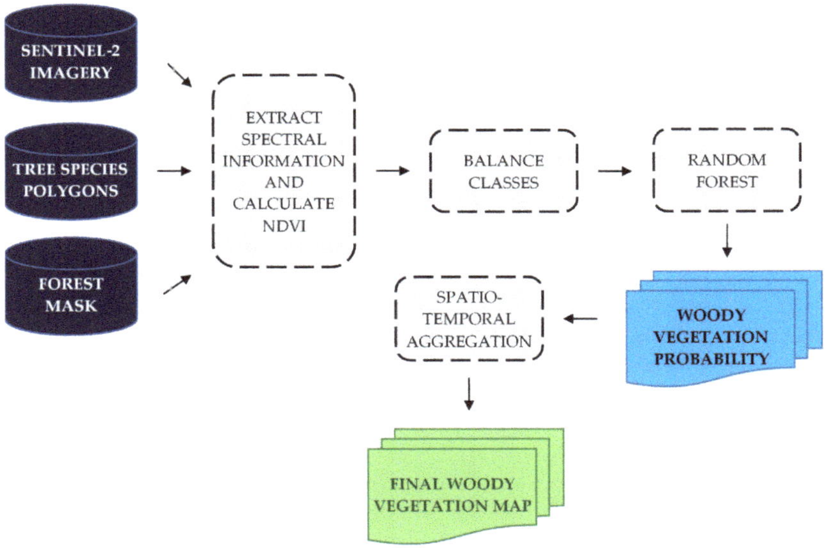

Figure 4. The flowchart diagram of the proposed spatio-temporal classification framework.

3.1. Extraction of Spectral Information, NDVI Calculation and Balancing of Classes

The spectral information was extracted, overlaying the woody vegetation polygons with the acquired Sentinel-2 granules. The extraction was done per granule, where all pixels within the training polygons were considered as independent observations. The spectral information of all resampled 60 m bands was extracted in this way for all pixels that were labeled as vegetation class on the Sentinel-2 scene classification map. Since the vegetation class did not represent only forest areas, the forest mask was additionally applied to filter out the non-forest pixels. Although there were many different vegetation indices (VI), the Normalized Difference Vegetation Index (NDVI) was chosen as it is one of the most widely used vegetation indices. Other spectral indices might have improved the model performance, but they would have required considerations regarding the selection criteria and their effect on the methodology performance. However, the conclusion was that this would extend the paper significantly, so it was decided to leave this topic for future studies. Therefore, only the corresponding NDVI value has been calculated for each observation using the following formula:

$$\mathrm{NDVI} = \frac{\mathrm{B08a} - \mathrm{B04}}{\mathrm{B08a} + \mathrm{B04}} \tag{1}$$

where B08a and B04 represent the narrow NIR and Red Sentinel-2 spectral bands, respectively. Band B08a was chosen instead of band B08 because it came already resampled in 60 m spatial resolution in the Level-2A Sentinel-2 product distribution. Additionally, the narrower spectral width of band B08a should help to better distinguish woody vegetation classes compared to the use of spectrally wider band B08.

In order to deal with the unbalanced classes, SMOTE has been applied. The SMOTE has been used in such a way that no under sampling was performed at all and that the balancing was only done by creating the examples of the minority classes. This was repeated until there were no more than 10% differences between the amounts of class examples.

3.2. Growing Probability Forest and Determination of the Probability for Each Class

The probability random forest for woody vegetation classification has been grown using the previously generated balanced dataset. In order to avoid data leakage in the time series data, the RF model was defined for each S2 observation (not per pixel) with the following formula:

$$\begin{aligned}\mathrm{WOODY}_{\mathrm{VEGETATION}} = {} & \mathrm{day} + \mathrm{month} + \mathrm{B01} + \mathrm{B02} + \mathrm{B03} + \mathrm{B04} + \mathrm{B05} + \mathrm{B06} + \mathrm{B07} + \\ & \mathrm{B08}a + \mathrm{B09} + \mathrm{B11} + \mathrm{B12} + \mathrm{NDVI}\end{aligned} \tag{2}$$

where day and month represent the timestamp of the granule that each observation belongs to, B_{xx} is the corresponding Sentinel-2 spectral band and NDVI is the previously calculated vegetation index. The same reasons for choosing band B08a over band B08 previously stated for NDVI calculation also apply to the RF model. Two variants of defining timestamps in the model were tested. The first one assumed the use of two variables (day and month) and the second one only one variable (Day of Year—DOY) in order to make model more robust. The test indicated that the first variant provides slightly better model performances, so it was selected as the final model in the study. Currently, the authors cannot provide a plausible explanation for this behavior of the model.

3.3. Spatio-Temporal Aggregation

Since each part of the study area was covered with multiple Sentinel-2 granules, there were several class probability predictions available for each pixel of the study area. Such multiple predictions per pixel had different timestamps and allowed for temporal aggregation. Spatial aspects were also taken into account, since the corresponding pixels and their neighborhoods had been used. All these aspects were taken into account in the last step and the aggregated woody vegetation class predictions were produced for every pixel of the study area.

Several types of spatio-temporal aggregations have been tested. They were separated into two main groups: pixel-based and neighborhood-based. The pixel-based method means that the aggregation function is applied using all available class probability predictions per pixel of the study area. The neighborhood-based aggregation uses the class probability predictions of the surrounding pixels together with predictions for the central pixel. In this research, the neighborhood-based method was implemented using focal raster operation with 3 × 3 and 5 × 5 kernels, for which the appropriate aggregation function was used. Several aggregation functions were tested, where the final woody vegetation class was determined using some of the following rules:

1. Most common class (MC);
2. Class with the highest probability—simple mean (SM);
3. Class with the highest probability—geometric mean (GM).

The most common class rule means that each prediction from the set of multiple predictions casts a vote to the class with highest probability. The final class is then set to the class that has the highest number of votes. The other two approaches are based on averaging expert predictions [43]. In these cases, each probability prediction was considered as a single expert prediction, which were then averaged by calculating the simple mean or geometric mean. No transformation on predictions was performed and no weights were used in the averaging. The final class was then set to the class that had the highest averaged probability.

3.4. Validation

The performance of the final woody vegetation class predictions was validated using K-fold leave-location out cross-validation (LLOCV). Leave-location-out was implemented in a way that the entire woody vegetation polygon (with all corresponding observations) was assigned to the same fold. The complete dataset was therefore randomly split into K groups, or folds, of similar size. The number of folds was chosen to be 5 and 10, according to the common practice, providing objective accuracy assessment and reasonable computational requirements [44]. The original proportion of the classes in the complete dataset was also preserved within the created folds. Once the folds were created, the first fold was treated as a validation set, and the remaining K-1 folds were used for training. The class balancing and growing of the probability forest was performed only on the training set. The grown probability forest was then applied to obtain class probability predictions and later to perform spatio-temporal aggregation only on the validation set. This procedure was repeated K times; each time, a different fold was treated as a validation set. In order to suppress errors introduced by creating folds, the K-fold cross-validation procedure was performed 10 times, each time randomly creating new folds. Random folds were preferred over spatially independent folds because, for some classes, woody vegetation polygons had to be clustered, due to the specific environment and climate conditions required for these types of vegetation. Having in mind the structure of the input dataset, it would be difficult to create spatially independent folds. Therefore, although such a random split was not implicitly spatially independent, the authors believe that the same objective accuracy assessment was obtained due to the randomization and leave-location-out procedure. The evaluation was performed on the final merged dataset by creating the confusion matrix and calculating the overall accuracy (OA), producer's accuracy (PA), user's accuracy (UA) and kappa coefficient of agreement (Kappa). These metrics are considered good practices commonly used for evaluation of land cover classification results [42].

4. Results

The available woody vegetation polygons have been used to extract spectral information for each of the woody vegetation class. This resulted in 115,059 pixels representing all eight woody vegetation classes (Table 2). Since the classes in the created dataset are very unbalanced, the SMOTE method has been applied. The resulting dataset has almost perfectly balanced classes.

Table 2. The definition of classification scheme and the number of vectorized training polygons for their representation.

Class ID	Informal Phytosociological Name	Extracted Number of Vegetation Pixels
1	Ostryo-Carpinion	2299
2	Quercion frainetto	6705
3	Quercion petraea-cerris	11,152
4	Quercion roboris	65,561
5	Fagion	14,971
6	Pinion nigrae	3235
7	Vaccinio-Piceion	10,871
8	Pinion mugo	265

The probability forest has been grown using the created balanced dataset. The output model was then used to the predict class probability of each woody vegetation class for all Sentinel-2 forest pixels. This was done for both pixel-based and neighborhood-based approaches and for all three tested aggregation functions. The validation metrics show that the performance of the proposed woody vegetation classification method depends both on the aggregation function, as well as on the number of pixels used for the prediction (Table 3). All tested pixel-based aggregation functions achieved almost the same performance, where simple mean and geometric mean functions slightly outperformed the approach based on the most common class rule. The neighborhood-based performances were always higher than the performances of matching pixel-based methods. Increasing the size of the neighborhood had a limited effect, with only marginal improvements. Among the tested aggregation functions, the most common class rule provided the best performance of all tested combinations, but only in combination with the neighborhood-based approach. This suggests that this approach most successfully recognizes the situations of wrong predictions where the central pixel prediction is different from the predictions for the surrounding pixels.

Table 3. Validation metrics obtained after 5-fold and 10-fold leave-location out cross-validation.

Aggregation Rule		5-Fold LLOCV		10-Fold LLOCV	
		OA [%]	Kappa	OA [%]	Kappa
pixel-based	MC	81.02	0.72	80.63	0.72
	SM	81.92	0.74	81.67	0.74
	GM	81.94	0.74	81.87	0.74
3 × 3	MC (3 × 3)	**83.09**	**0.76**	**82.89**	**0.75**
	SM (3 × 3)	82.27	0.74	81.68	0.74
	GM (3 × 3)	82.39	0.74	81.87	0.74
5 × 5	MC (5 × 5)	**82.97**	**0.75**	**83.10**	**0.76**
	SM (5 × 5)	82.30	0.74	81.98	0.74
	GM (5 × 5)	82.41	0.74	82.05	0.74

The supremacy of the most common class rule as an aggregation method is confirmed also by inspection of the Producer's and User's accuracy barplots (Figure 5). Tables with complete PA and UA values are available in the Appendix B.

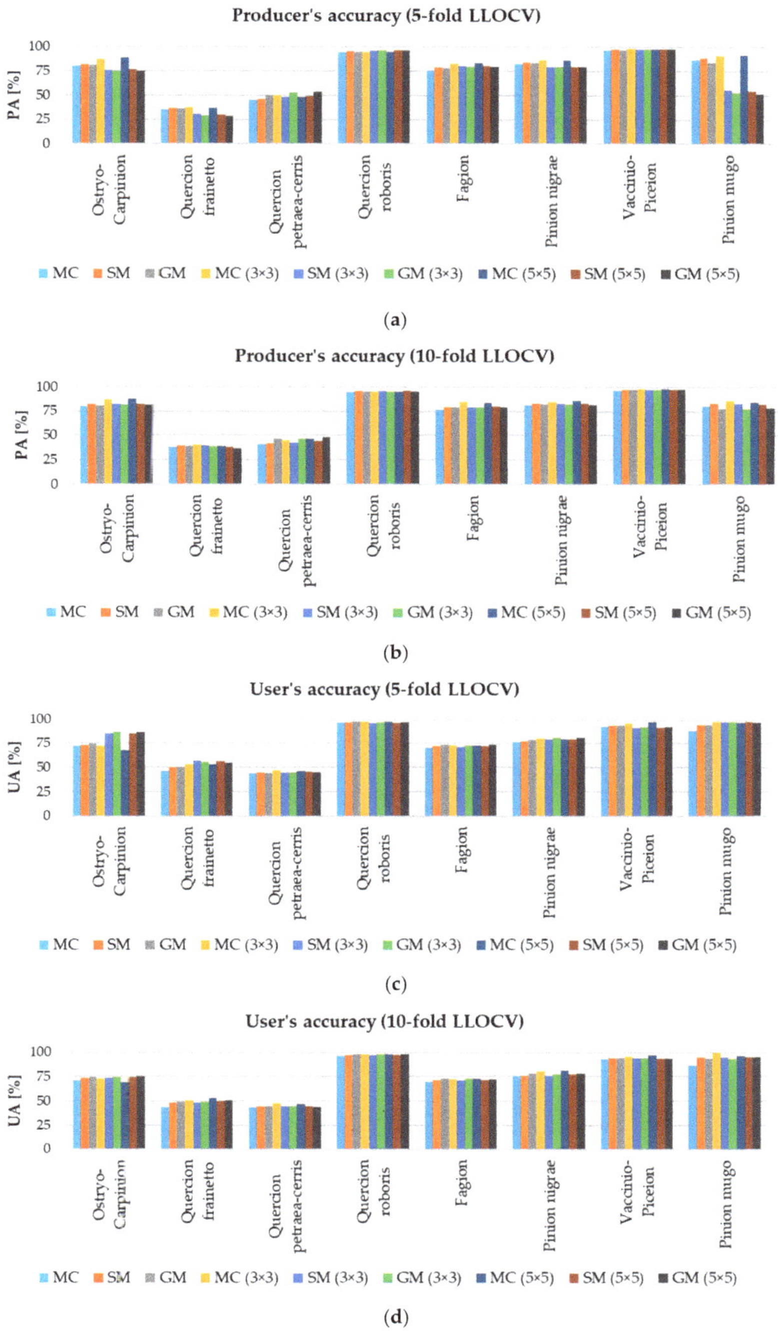

Figure 5. Producer's accuracy and user's accuracy barplots after K-fold leave-location out cross-validation (**a**) PA with K = 5 (**b**) PA with K = 10 (**c**) UA with K = 5 (**d**) UA with K = 10.

For most classes, the MC rule with either 3 × 3 or 5 × 5 neighborhood provides the best metrics or close to the best metrics. For half of the modeled types of woody vegetation, the difference in the performance of this approach is much greater than the rest of the methods. This is evident for coniferous woody vegetation (Pinion nigrae, Vaccinio-Piceion and Pinion mugo) with PA being higher from 2% up to 40% and with the preservation of high UA (greater than 80%). Improvement is also present for Fagion woody vegetation, although it is not as large as for coniferous woody vegetation. These four woody vegetation classes represent more than a third of the total forest area, which suggests that such improvements have a major impact on the final classification map. It is important to note that these improvements are obtained even with a considerable disproportion of classes in the original training dataset. This highlights the importance of class balancing using SMOTE, as some of these classes are minority classes (Pinion mugo) and some have a very large number of training observations (Fagion). This suggests that the improvements of this approach are mainly due to the ability to capture the specifics of these types of woody vegetation growing in large, homogenous and dense groups, without "small islands" of different types of woody vegetation. It should be noted that for some other woody vegetation classes, the neighborhood-based MC rule does not perform the best (like for Ostryo-Carpinion), where it provides the highest PA, but fails to provide the best UA.

The differences between five-fold and 10-fold LLOCV validation metrics are least noticeable for the MC rule in combination with the neighborhood approach. In contrast, performances of all other methods change with different numbers of folds. This is particularly evident for the Pinion mugo class. The MC rule with both neighborhoods provides the best metrics for its prediction, indicating significantly higher accuracy than other neighborhood-based aggregations in case of five-fold LLOCV. In case of 10-fold LLOCV, it is still the best approach, but without large differences in performances. This can be explained by the amount of data used for model training. In every step, 10-fold LLOCV uses ≈90% of the data of each class for model training, compared to ≈80% used by five-fold LLOCV. Even though the SMOTE is applied, this shows that such reduction can severely influence the model performances, particularly for minority classes (as Pinion mugo). This shows that the MC rule in combination with the neighborhood approach is the most robust regarding the amount of training data necessary for proper modeling of minority classes.

All validation metrics indicate that the best results are achieved by using the MC rule, with almost the same performance of both neighborhoods. Individually per woody vegetation class, the 5 × 5 neighborhood outperforms the 3 × 3 neighborhood for most of the classes for both PA and UA. Confusion matrix after five-fold and 10-fold LLOCV are given in Appendix C. Therefore, this approach can be considered to be the best one for woody vegetation classification using the proposed spatio-temporal framework. Consequently, the final classification map has been produced using the 5 × 5 neighborhood with the MC rule (Figure 6). The covered areas of each woody class are given in the Table 4.

Table 4. Area and share in the total forest area for each woody vegetation class after classification using the MC rule and 5 × 5 neighborhood for aggregation.

Class ID	Informal Phytosociological Name	Area [km^2]	Area [%]
1	Ostryo-Carpinion	4411	12.42
2	Quercion frainetto	1104	3.11
3	Quercion petraea-cerris	11,707	32.95
4	Quercion roboris	5528	15.56
5	Fagion	10,768	30.31
6	Pinion nigrae	949	2.67
7	Vaccinio-Piceion	907	2.55
8	Pinion mugo	153	0.43

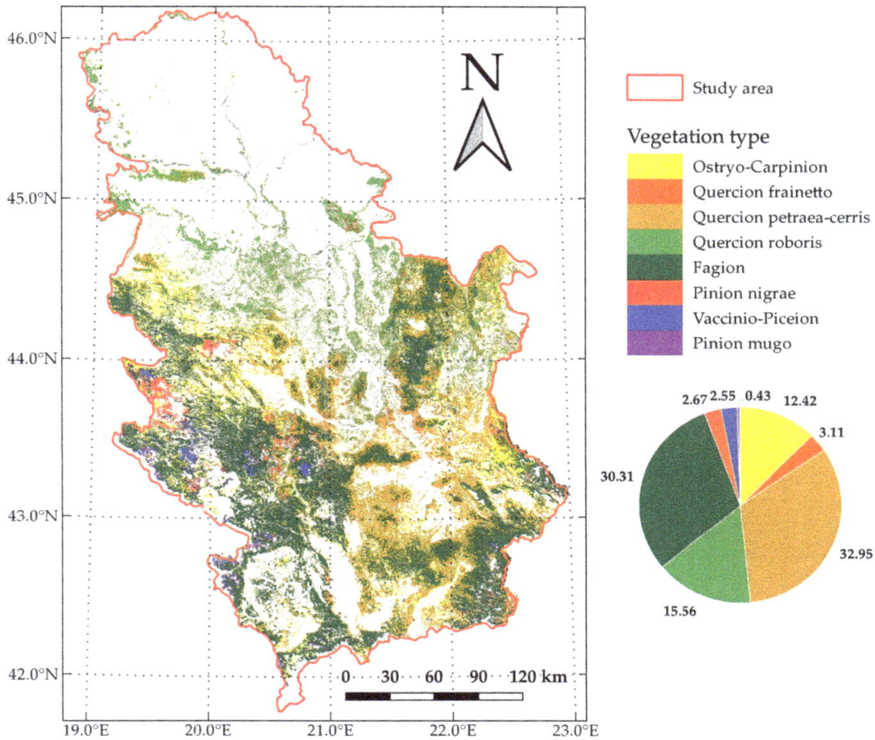

Figure 6. Final classification map of woody vegetation coverage over the study area.

5. Discussion

5.1. Final Classification Map of Woody Vegetation

The previously explained LLOCV procedure can give information about the model accuracy, but the accuracy metrics are generally biased for mapping the entire study area. Unfortunately, no comparable external data source is available for the purpose of assessing the accuracy of the final woody vegetation map. In addition, creating a validation set based on probability sampling requires extensive resources and field work, which are beyond the scope of this research. Therefore, validation of the final classification map of woody vegetation has been performed by using visual inspection and comparison with available statistical reports.

By visual inspection and comparison, the final classification map of woody vegetation produced in this paper corresponds to the existing general vegetation maps: Map of the Natural Vegetation of Europe (scale 1:2,500,000) [29] and the Map of Natural Potential Vegetation of SFR Yugoslavia (scale 1:1,000,000) [45]. Additionally, the proportional relationships of the analyzed vegetation types for most of the classes fit into the results obtained by the first National Forest Inventory of Serbia in the period 2004–2006, which was done using traditional field survey methods that provide a highly accurate forest inventory [46,47]. Significant deviations were obtained only in the proportions of thermophytic broadleaved deciduous forests and mesophytic and hygromesophytic broadleaved forests. The share of Ostryo-Carpinion forests in this research is 12.42%, compared to the share of *Carpinus orientalis* and *Ostrya carpinifolia* forests in [47], which is 3.9%. The results of this research show that Quercion roboris forests cover 16% of the area, while the corresponding forest stand category in [47] *Quercus robur* forests has the coverage of 1.4%.

In the first case (Ostryo-Carpinion), the discrepancy can be explained by the fact that the data from the National Forest Inventory of Serbia refer primarily to forests and do not include data on shrubs and lower degradation forms of forest vegetation. Since in the thermophilic forest zone, the degradation forms of forest vegetation are mainly represented by shrub vegetation belonging to the Ostryo-Carpinion type, the increased share of this type of vegetation in our research can be explained by the assumption that different degradation forms of thermophytic broadleaved were recognized in this class. In the second case (Quercion roboris), the deviation is explained by the fact that our class Quercion roboris is much wider than the class *Quercus robur* forests presented in the first National Forest Inventory of Serbia (Appendix A).

Since the spatial distribution of the training polygons cannot be considered ideal, it is possible that some discrepancies in the area assessments are also caused as a result. Therefore, additional woody vegetation polygons in the forest areas sparsely covered by polygons (like southern part of the study area) can significantly benefit in better modeling classes in RF model and more accurate area assessment.

5.2. Variable Usefulness and Importance

The usefulness and importance of variables used for the woody vegetation classification is determined first by inspection of the temporal phenological spectral patterns. Temporal phenological spectral patterns are created by plotting the average spectral responses of all woody vegetation types per each available date (Figure 7). This was done for all used Sentinel-2 spectral bands and the calculated NDVI value.

The created temporal phenological spectral plots show that patterns vary per woody vegetation type and per spectral band. The decreasing pattern can be observed across most of the bands. This is as expected, since all woody vegetation classes are in full vegetation at the start of the observation period, which later decreases as autumn and winter approach. This decreasing pattern is particularly noticeable for deciduous tree species, while it is less present for coniferous tree species. The separability between species also follows such a pattern, with the highest separability being in the June–August period. This separability highly depends on the type of woody vegetation. Independently per band, Bands B11 (Short-wave infrared 1) and B12 (Short-wave infrared 2) visually show the highest separability between types of woody vegetation. Deciduous and coniferous woody vegetation groups can be clearly observed across these spectral bands. All three red edge bands (B05, B06 and B07), as well as the NIR band, also show high spectral separability between vegetation types. Individual vegetation types also show specific distinguishing patterns on certain bands, such as Fagion woody vegetation on bands B02 (Blue) and B04 (Red) or Ostryo-Carpinion on band B03 (Green). These can be potentially used to subset spectral bands and further relieve the model complexity. Ottosen et al. also highlighted that the classification results are improved by using only subsets of Sentinel-2 spectral bands [48]. This was not taken into account in this research, even though it is important to understand how the classification results obtained by this methodology are influenced by the subset of used spectral bands. This needs to be addressed in the future.

The variable importance is additionally assessed through the Mean decrease Gini measure (Figure 8). The Mean decrease Gini is a measure of how much the Gini impurity metric is reduced by a variable for each class. The Mean decrease Gini metric was chosen as the importance measure because it was also used for growing the probability random forest. The scaled variable importance based on the Mean decrease Gini measure is produced and plotted. The results correspond to the observed temporal phenological spectral patterns. Bands B11 and B02 are the most important spectral variables, followed by B12. The day and month temporal variables turned out to be very important, being the second and third most important variables of all. This highlights the model's ability to capture woody vegetation phenological patterns. The other spectral variables are of approximately the same importance.

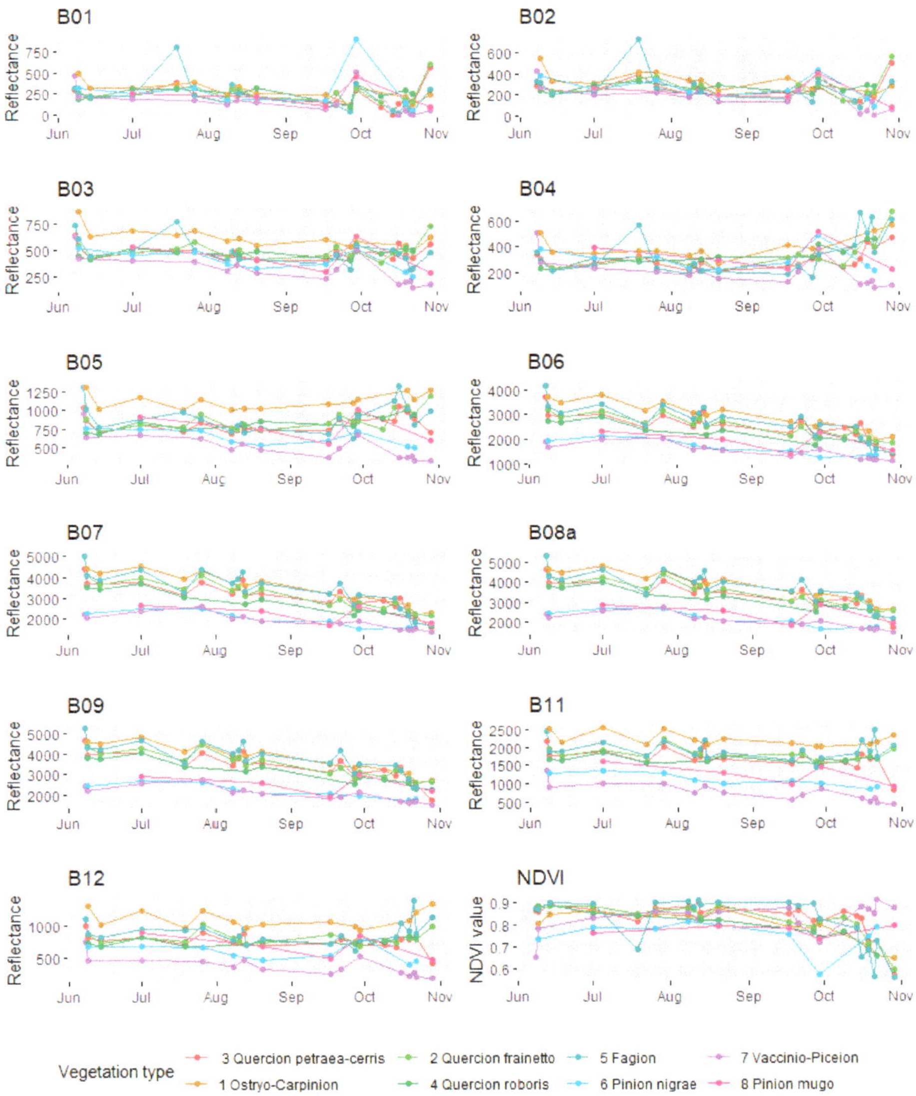

Figure 7. Temporal phenological patterns for the used Sentinel-2 spectral bands and NDVI values.

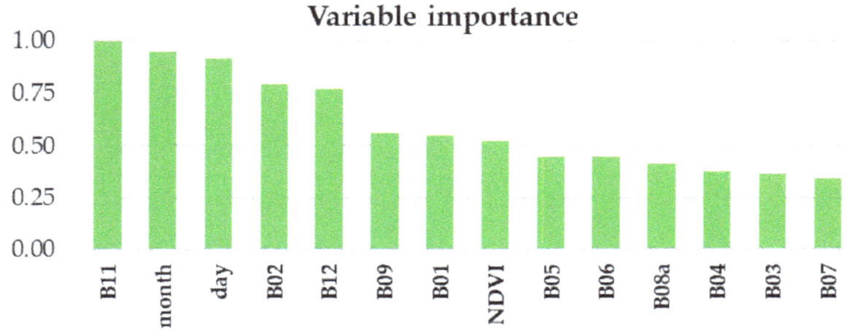

Figure 8. Variable importance barplot expressed through the Mean decrease Gini measure.

5.3. Spatio-Temporal Framework for Woody Vegetation Classification

The spatio-temporal framework proposed in this research defines the timestamp of the observation explicitly in the model through day and month variables. Such change in the model means that, in order to determine the woody vegetation class, the methodology requires only one cloud-free pixel available for each part of the study area. This change in the model allows timestamps to be freely combined and enables the entire methodology to be easily adjusted to any area of interest, regardless of its size. Additionally, the model is lighter, as every spectral band is represented only once through a single variable, as an alternative to multiple timestamp–spectral band variable combinations. There is also no data-wrangling and only the original observations are used for both training and prediction phases. As the number of cloud-free pixels over the study area increases, there will be multiple predictions for each pixel of the study area. This enables the full potential of spatio-temporal aggregation to be exploited. The spatio-temporal aggregation is not only pixel-based, but is expanded to take into account the neighboring pixel predictions as well. The main idea behind the neighborhood-based approach is to solve situations where surrounding pixels show strong probability of belonging to a certain class, which is different from the class probability of the central pixel. In these cases, there is a strong reason to believe that the central pixel should also belong to the class of surrounding pixels. It is more likely that the prediction based solely on spectral information for the central pixel is wrong, which is a consequence of some noise present in the spectral information for that pixel, than the probability that the central pixel's class differs from the class of its neighbors.

Applied spatio-temporal aggregation can be affected by temporal and/or spatial homogeneity. For areas with no temporal and spatial heterogeneity over the time period covered with satellite observations and the used neighborhoods, the prediction results should be very reliable. In other cases, the proposed methodology can deal with such heterogeneity up to a certain amount. This is because each of the used S2 observation influences the decision on the final woody vegetation class and any temporal or spatial change can potentially suppress other observations. Temporally, this means that changes that last longer within the used time interval are more likely to be recognized by the model. Similarly, only spatial changes that have larger spatial extents compared to the neighborhood size will be recognized. This means that the minimum mapping unit gets coarser as the size of neighborhood increases. At the same time, the neighborhood approach achieves smoothing effects, where isolated pixels are classified into the class of pixels in the neighborhood. Therefore, it is not advised to extend both the observation period and the neighborhood size more than necessary. A shorter observation period reduces the amount of changes that occur in the first place, and the appropriate neighborhood size limits the amount of smoothing which occurs.

The proposed methodology is applied using resampled 60 m Sentinel-2 bands. This spatial resolution may be too coarse, since the increase in neighborhood size has a limited effect on the performance. It is expected that using the original spatial resolution of 10 m and 20 m for available bands, should reduce the effects of mixed pixels and further increase the classification accuracy. In addition, this should help the previously discussed heterogeneity issues, where smaller changes could be modelled. Stronger neighborhood effects can also be expected to occur for higher spatial resolutions, so the neighborhood size could become a more important parameter in the proposed methodology. However, there are several motivations for choosing a coarser spatial resolution of 60 m for all bands. Firstly, this is done in order to match the chosen classification scheme, designed for country-wide woody vegetation mapping. The used classification scheme is defined according to the "Database on the distribution of potentially endangered species and habitats of Serbia" and the increase in spatial resolution would require the introduction of new woody vegetation classes for which the proper training data are lacking. Secondly and more importantly, the proposed methodology aims to be used on large areas and this way the computation and performance issues are reduced. At a spatial resolution of 60 m, the used dataset comprises more than 160 million S2 observations, which would increase about nine or 36 times in the case of 20 m and 10 m spatial resolutions, respectively. One solution to deal with such an increase in size of data is to select only more informative spectral bands. Since the proposed methodology has advantages and applicability, especially over larger areas, this can become a very important issue.

The results of this research match those obtained by other similar studies, which were based on high- or medium-resolution optical data. Grabska et al. have achieved an OA greater than 90% on the study area of 305 km^2 for nine tree species using 18 cloud-free sentinel-2 images [6]. Liu et al. identified forest types using Sentinel-1A, Sentinel-2A, Landsat-8 and DEM Data with OA of 82.78%. They extracted the four single-dominant forests and the four mixed forest types over the area of 2261 km^2 [40]. Lu et al. implemented Spatial–Spectral–Temporal Integrated Fusion to identify six tree species and two mixed forest types with an OA of 83.6% over the area of 1610 km^2 [49]. In addition to covering a significantly larger study area, this research and the proposed solution also completely relies only on optical satellite sources. Therefore, there is a potential for further improvement by using other data sources such as radar satellite data, LiDAR, multiple optical sources and/or other auxiliary data (topography, etc.).

This proposed spatio-temporal framework relies on Sentinel-2 scene classification maps and a forest mask to properly identify forest areas. This is potentially the weak point of the method, because the output quality is in a way limited by the quality of the used forest mask. It is important to use the most recent available forest mask as possible, especially in the areas of rapid forest change. The forest mask used in this research is from 2015, which creates a 4-year gap with the used Sentinel-2 data. Such a forest mask can be considered outdated since there is an active forest management over the study area, with the forest area being influenced by felling of trees, forest growing and silviculture. Additionally, some forest fires occur annually, usually in the warm summer period. However, these changes are not extensive and do not have large spatial extents. Visual inspection of forest mask showed that the misrepresentations happen only in a few local areas, so the authors believe that these errors do not significantly affect the overall results. The methodology can also be further expanded to address this issue, with the same spatio-temporal framework being used to firstly identify the forest and non-forest pixels. This issue was beyond the scope of this research due to the lack of proper training data for good representation of non-forest areas. This issue needs to be addressed in the future. Another setback of the proposed approach is that each pixel is always assigned to some class. In a situation where all woody vegetation types are well represented by training polygons, this does not pose a problem; however, when the study area is very complex or insufficiently researched in terms of woody vegetation, it might not be possible to represent all the woody vegetation types in the study area.

The study has revealed several problematic cases in which predictions do not achieve a satisfactory level of accuracy and reliability. Some of them are problems with: (1) identification of sufficiently representative learning polygons within the complex and mosaic area, such as alluvial forests in riparian and swamp galleries and woodland in alluvial plains; (2) insufficiently researched natural vegetation types, such as forests of birch (Betula pendula), common aspen (Populus tremula) or rowan (Sorbus aucuparia); (3) the natural absence of clear boundaries between ecologically and physiognomically close vegetation types that build phylogenetically close tree species (e.g., gradual transition between thermophytic (Quercion frainetto), thermo-mesophytic (Quercion petraea-cerris) and mesophytic to hygromesophytic (Quercion roboris) forests in which all three main oak species have such a wide ecological valence that their populations occur within all three types of oak forests); (4) a priori classification of mixed deciduous and coniferous woodlands in which a large number of tree species have different parts in the general coverage; (5) an absence of data on areas with highly artificial tree plantations or naturalized forest of alien species(e.g., black locust forests (Robinia pseudoaccacia), american ash (*Fraxinus* spp.) and maple (*Acer* spp.) forests, poplar plantations (*Populus* spp.), etc.

Given that eight analyzed woody vegetation classes cover a large portion of the study area, it can be expected that the omission of other classes should not significantly affect the spatial coverage of woody vegetation in the study area. Since there are multiple predictions for each pixel of the study area, as well as the probability of each prediction, there is a possibility to use this information to identify the weak pixels, which probably do not belong to any of the classes. This is also something to be considered in future research.

6. Conclusions

The novel spatio-temporal classification framework has been presented in this paper. The proposed methodology has been successfully applied for mapping eight potential types of woody vegetation over the study area of Serbia. This was done using 88 Sentinel-2 multitemporal granules dating from the June–October period of 2019. The method relies on forest mask to exclude non-forest areas and on training polygons to represent each woody vegetation class of interest. The used framework is novel because it includes temporal information explicitly in the probability random forest model that has been grown, which produces several probability predictions for each pixel of the study area. Multiple predictions enable specific spatio-temporal aggregation to be performed in order to determine most probable woody vegetation class for each pixel of the study area.

Several spatio-temporal aggregation approaches have been tested within this research. They can be divided into pixel-based and neighborhood-based methods. According to the validation metrics, the most common class rule together with the neighborhood of 5×5 pixels provided the best results. The most common class rule means that from the set of multiple predictions, each prediction casts a vote to the class with the highest probability. The final class is then set to the class that has the highest number of votes. The overall accuracy and kappa coefficient of agreement obtained from five-fold cross validation by this approach are 82.97% and 0.75, respectively. The corresponding producer's accuracies range from 36.74% to 97.99% and user's accuracies range from 46.31% to 98.43%. Such results match the performances of other similar studies, although most of them have been done for much smaller study areas.

Given the obtained accuracy, it can be said that the proposed methodology is applicable for woody vegetation mapping in Serbia. The classification scheme and the corresponding training polygons are prerequisites for its implementation. The methodology has been tested over the territory of Serbia, but it has the potential to be transferred to other areas if appropriate adjustments to the classification scheme are introduced and new training polygons are made available. Further testing is necessary to confirm such assumptions. It is implemented in such a way that it can be easily integrated into existing operational programs intended for forest inventory and mapping. The identified weak points of the framework are its reliability on the forest mask and the fact that each pixel of the study area is always assigned to some class. The second issue is more complex and maybe it can be overcome by using the probability of each prediction to identify the weak probability pixels, indicating that they probably do not belong to any of the classes. These two things are planned to be addressed in future research.

Author Contributions: All authors worked equally on the research conceptualization, investigation, discussion, and data modeling. J.K. was involved in data preparation and case study implementation. D.L., N.K. and J.Š.-S. covered the biological aspects of the research. All authors participated in the writing of the manuscript, however J.K. took the lead. All authors have read and agreed to the published version of the manuscript.

Funding: This research received no external funding.

Acknowledgments: This study was supported by the Serbian Ministry of Education, Science and Technological Development, projects TR 36020 and OI 173030. The authors also thank the anonymous reviewers for constructive and valuable comments that improved the quality of this study.

Conflicts of Interest: The authors declare no conflict of interest.

Appendix A

Table A1. Analyzed types of forests and shrubs according to the national classification of habitats in Serbia.

Class ID	SrbHab Code	Habitat Type Name
1	A2.9	Forests of Oriental hornbeam (*Carpinus orientalis*) and hop-hornbeam (*Ostrya carpinifolia*)
	A2.91	Forests of Oriental hornbeam (*Carpinus orientalis*)
	A2.92	Forests of hop-hornbeam (*Ostrya carpinifolia*)
2	A2.1	Forests of Hungarian (*Quercus frainetto*) and Austrian oak (*Quercus cerris*) forests
	A2.11	Forests of Hungarian (*Quercus frainetto*) and Austrian oak (*Quercus cerris*) forests
	A2.12	Forests of Hungarian oak (*Quercus frainetto*)
3	A2.5	Forests of sessile (*Quercus petraea*) and Austrian oak (*Quercus cerris*)
	A2.51	Forests of sessile oak (*Quercus petraea*)
	A2.52	Forests of Austrian oak (*Quercus cerris*)
4	A1.3	Forests of pedunculate oak (*Quercus robur*) and field ash (*Fraxinus angustifolia*)
	A1.31	Forests of pedunculate oak (*Quercus robur*)
	A1.33	Forests of pedunculate oak (*Quercus robur*) and field ash (*Fraxinus angustifolia*)
	A1.35	Forests of pedunculate oak (*Quercus robur*), hornbeam (*Carpinus betulus*) and field ash (*Fraxinus angustifolia*)
	A1.4	Forests of pedunculate oak (*Quercus robur*) and hornbeam (*Carpinus betulus*)
	A1.41	Forests of pedunculate oak (*Quercus robur*) and hornbeam (*Carpinus betulus*)
	A1.42	Forests of pedunculate oak (*Quercus robur*), hornbeam (*Carpinus betulus*) and Austrian oak (*Quercus cerris*)
5	A3.2	Beech forests (*Fagus moesiaca*)
	A3.22	Submontane beech forests (*Fagus moesiaca*)
	A3.23	Mountain beech forests (*Fagus moesiaca*)
	A3.27	Subalpine beech forests (*Fagus moesiaca*)
6	A5.1	Forests of black (*Pinus nigra*) and white pine (*Pinus sylvestris*)
	A5.11	Black pine forest (*Pinus nigra*)
	A5.12	Forest of black (*Pinus nigra*) and white pine (*Pinus silvestris*)
7	A6.1	Spruce (*Picea* spp.) and fir (*Abies* spp.) Forests
	A6.11	Spruce (*Picea abies*) and fir (*Abies alba*) forests
	A6.12	Spruce forests (*Picea abies*)
8	B6.1	Krummholz (*Pinus mugo*) scrub

Appendix B

Table A2. Producer's accuracy for all woody vegetation classes for 5-fold LLOCV.

Vegetation Type	MC	SM	GM	MC (3 × 3)	SM (3 × 3)	GM (3 × 3)	MC (5 × 5)	SM (5 × 5)	GM (5 × 5)
Ostryo-Carpinion	80.12	81.93	80.84	86.88	76.18	75.26	88.48	76.22	75.24
Quercion frainetto	35.59	37.17	36.31	37.73	30.41	29.29	36.74	29.87	28.73
Quercion petraea-cerris	45.25	46.34	50.04	49.70	48.65	53.00	48.54	49.13	53.44
Quercion roboris	95.17	95.32	94.94	94.92	96.55	96.30	94.88	96.59	96.33
Fagion	76.15	78.88	78.12	82.82	80.54	79.45	83.61	80.51	79.37
Pinion nigrae	82.46	84.06	83.29	86.58	79.82	79.40	86.12	79.85	79.43
Vaccinio-Piceion	97.18	97.57	97.48	98.53	98.21	98.02	97.99	98.20	98.00
Pinion mugo	87.16	88.69	84.25	91.44	55.96	52.75	91.98	54.17	51.54

Table A3. Producer's accuracy for all woody vegetation classes for 10-fold LLOCV.

Vegetation Type	MC	SM	GM	MC (3 × 3)	SM (3 × 3)	GM (3 × 3)	MC (5 × 5)	SM (5 × 5)	GM (5 × 5)
Ostryo-Carpinion	80.23	82.16	81.11	86.88	82.16	81.15	87.35	82.24	81.44
Quercion frainetto	37.81	39.29	38.57	40.60	39.30	38.58	38.42	37.80	36.83
Quercion petraea-cerris	40.95	42.22	46.32	44.52	42.45	46.49	46.61	44.23	48.01
Quercion roboris	95.59	95.97	95.60	95.43	95.95	95.58	95.25	95.75	95.35
Fagion	76.76	79.48	79.33	84.51	79.45	79.29	84.15	79.67	79.23
Pinion nigrae	81.17	82.93	82.10	84.67	82.93	82.10	85.81	82.79	81.82
Vaccinio-Piceion	97.13	97.42	97.52	98.63	97.42	97.51	98.48	97.68	97.58
Pinion mugo	80.65	83.04	77.61	85.87	83.04	77.61	84.83	82.07	78.39

Table A4. User's accuracy for all woody vegetation classes for 5-fold LLOCV.

Vegetation Type	MC	SM	GM	MC (3 × 3)	SM (3 × 3)	GM (3 × 3)	MC (5 × 5)	SM (5 × 5)	GM (5 × 5)
Ostryo-Carpinion	71.90	73.29	74.12	72.56	85.39	86.50	67.54	85.39	86.44
Quercion frainetto	46.21	50.35	50.06	53.11	56.84	55.60	53.17	56.09	54.84
Quercion petraea-cerris	43.79	44.55	43.83	47.18	44.94	44.45	46.31	45.17	44.58
Quercion roboris	97.02	97.27	97.95	98.29	96.30	97.23	98.43	96.27	97.20
Fagion	70.46	72.25	73.62	73.09	71.79	73.18	73.32	72.05	73.46
Pinion nigrae	76.96	77.87	78.95	80.90	79.84	81.01	80.07	80.02	81.13
Vaccinio-Piceion	93.57	94.25	94.33	96.27	92.25	92.48	97.86	92.30	92.53
Pinion mugo	89.20	95.39	95.00	98.36	98.12	98.01	97.39	98.04	97.66

Table A5. User's accuracy for all woody vegetation classes for 10-fold LLOCV.

Vegetation Type	MC	SM	GM	MC (3 × 3)	SM (3 × 3)	GM (3 × 3)	MC (5 × 5)	SM (5 × 5)	GM (5 × 5)
Ostryo-Carpinion	71.56	73.51	74.71	72.72	73.47	74.65	69.19	74.18	74.96
Quercion frainetto	43.75	48.88	49.45	50.63	48.85	49.40	52.89	50.20	50.43
Quercion petraea-cerris	43.67	44.69	44.33	47.61	44.82	44.43	46.80	44.98	44.26
Quercion roboris	96.97	97.13	97.90	98.22	97.18	97.94	98.30	97.30	97.98
Fagion	69.86	71.51	72.64	72.39	71.47	72.58	72.76	71.31	72.47
Pinion nigrae	75.93	76.14	77.92	80.43	76.14	77.89	81.67	77.59	78.59
Vaccinio-Piceion	93.36	94.13	94.20	95.70	94.12	94.20	97.07	93.57	93.84
Pinion mugo	86.89	94.79	93.46	99.50	94.79	93.46	96.60	95.45	95.25

Appendix C

Table A6. Confusion matrix after the 5-fold LLOCV using the MC rule and 5 × 5 neighborhood for aggregation (Legend for Class ID: 1 Ostryo-Carpinion, 2 Quercion frainetto, 3 Quercion petraea-cerris, 4 Quercion roboris, 5 Fagion, 6 Pinion nigrae, 7 Vaccinio-Piceion, 8 Pinion mugo).

		Reference								
	Class ID	1	2	3	4	5	6	7	8	Total
Prediction	1	4309	190	858	15	288	676	27	17	6380
	2	9	4495	2989	339	621	0	0	1	8454
	3	215	4589	11,442	4442	3765	206	48	3	24,710
	4	1	621	796	88,903	0	1	0	0	90,322
	5	336	2121	6247	0	23,979	0	22	0	32,705
	6	0	218	1024	0	5	5710	154	20	7131
	7	0	0	205	0	20	34	12,347	11	12,617
	8	0	0	11	0	0	3	2	596	612
	Total	4870	12,234	23,572	93,699	28,678	6630	12,600	648	182,931
	PA [%]	88.48	36.74	48.54	94.88	83.61	86.12	97.99	91.98	
	UA [%]	67.54	53.17	46.31	98.43	73.32	80.07	97.86	97.39	

Table A7. Confusion matrix after the 10-fold LLOCV using MC rule and 5 × 5 neighborhood for aggregation (Legend for Class ID: 1 Ostryo-Carpinion, 2 Quercion frainetto, 3 Quercion petraea-cerris, 4 Quercion roboris, 5 Fagion, 6 Pinion nigrae, 7 Vaccinio-Piceion, 8 Pinion mugo).

		Reference								
	Class ID	1	2	3	4	5	6	7	8	Total
Prediction	1	4254	157	702	8	343	637	27	20	6148
	2	16	4606	2980	353	753	0	0	1	8709
	3	260	4123	10,485	3921	3379	191	41	5	22,405
	4	0	681	801	85,803	2	2	0	0	87,289
	5	340	2207	6394	1	23,916	0	10	1	32,869
	6	0	216	826	0	7	5257	113	18	6437
	7	0	0	295	0	22	37	12,408	21	12,783
	8	0	0	10	0	0	2	1	369	382
	Total	4870	11,990	22,493	90,086	28,422	6126	12,600	435	177,022
	PA [%]	87.35	38.42	46.61	95.25	84.15	85.81	98.48	84.83	
	UA [%]	69.19	52.89	46.80	98.30	72.76	81.67	97.07	96.60	

References

1. *Global Forest Resources Assessment 2020*; FAO: Rome, Italy, 2020; ISBN 978-92-5-132581-0.
2. *Introduction to Forest Ecosystem Science and Management*, 3rd ed.; Wiley: New York, NY, USA, 2002. Available online: https://www.wiley.com/en-us/Introduction+to+Forest+Ecosystem+Science+and+Management%2C+3rd+Edition-p-9780471331452 (accessed on 13 June 2020).
3. Bouchard, M.; Aquilué, N.; Périé, C.; Lambert, M.-C. Tree species persistence under warming conditions: A key driver of forest response to climate change. *For. Ecol. Manag.* **2019**, *442*, 96–104. [CrossRef]
4. Modzelewska, A.; Fassnacht, F.E.; Stereńczak, K. Tree species identification within an extensive forest area with diverse management regimes using airborne hyperspectral data. *Int. J. Appl. Earth Obs. Geoinf.* **2020**, *84*, 101960. [CrossRef]
5. Jiang, X.; Huang, J.-G.; Stadt, K.J.; Comeau, P.G.; Chen, H.Y.H. Spatial climate-dependent growth response of boreal mixedwood forest in western Canada. *Glob. Planet. Chang.* **2016**, *139*, 141–150. [CrossRef]
6. Grabska, E.; Hostert, P.; Pflugmacher, D.; Ostapowicz, K. Forest Stand Species Mapping Using the Sentinel-2 Time Series. *Remote Sens.* **2019**, *11*, 1197. [CrossRef]
7. Sheeren, D.; Fauvel, M.; Josipović, V.; Lopes, M.; Planque, C.; Willm, J.; Dejoux, J.-F. Tree Species Classification in Temperate Forests Using Formosat-2 Satellite Image Time Series. *Remote Sens.* **2016**, *8*, 734. [CrossRef]
8. Madonsela, S.; Cho, M.A.; Ramoelo, A.; Mutanga, O.; Naidoo, L. Estimating tree species diversity in the savannah using NDVI and woody canopy cover. *Int. J. Appl. Earth Obs. Geoinf.* **2018**, *66*, 106–115. [CrossRef]
9. Vaglio Laurin, G.; Liesenberg, V.; Chen, Q.; Guerriero, L.; Del Frate, F.; Bartolini, A.; Coomes, D.; Wilebore, B.; Lindsell, J.; Valentini, R. Optical and SAR sensor synergies for forest and land cover mapping in a tropical site in West Africa. *Int. J. Appl. Earth Obs. Geoinf.* **2013**, *21*, 7–16. [CrossRef]
10. Drusch, M.; Del Bello, U.; Carlier, S.; Colin, O.; Fernandez, V.; Gascon, F.; Hoersch, B.; Isola, C.; Laberinti, P.; Martimort, P.; et al. Sentinel-2: ESA's Optical High-Resolution Mission for GMES Operational Services. *Remote Sens. Environ.* **2012**, *120*, 25–36. [CrossRef]
11. Heckel, K.; Urban, M.; Schratz, P.; Mahecha, M.D.; Schmullius, C. Predicting Forest Cover in Distinct Ecosystems: The Potential of Multi-Source Sentinel-1 and -2 Data Fusion. *Remote Sens.* **2020**, *12*, 302. [CrossRef]
12. Cheng, K.; Wang, J. Forest-Type Classification Using Time-Weighted Dynamic Time Warping Analysis in Mountain Areas: A Case Study in Southern China. *Forests* **2019**, *10*, 1040. [CrossRef]
13. Astola, H.; Häme, T.; Sirro, L.; Molinier, M.; Kilpi, J. Comparison of Sentinel-2 and Landsat 8 imagery for forest variable prediction in boreal region. *Remote Sens. Environ.* **2019**, *223*, 257–273. [CrossRef]
14. Mickelson, J.G. Delineating Forest Canopy Species in the Northeastern United States Using Multi-Temporal TM Imagery. *Photogramm. Eng. Remote Sens.* **1998**, *64*, 891–904.
15. Key, T.; Warner, T.A.; McGraw, J.B.; Fajvan, M.A. A Comparison of Multispectral and Multitemporal Information in High Spatial Resolution Imagery for Classification of Individual Tree Species in a Temperate Hardwood Forest. *Remote Sens. Environ.* **2001**, *75*, 100–112. [CrossRef]
16. Immitzer, M.; Atzberger, C.; Koukal, T. Tree Species Classification with Random Forest Using Very High Spatial Resolution 8-Band WorldView-2 Satellite Data. *Remote Sens.* **2012**, *4*, 2661–2693. [CrossRef]
17. Immitzer, M.; Vuolo, F.; Atzberger, C. First Experience with Sentinel-2 Data for Crop and Tree Species Classifications in Central Europe. *Remote Sens.* **2016**, *8*, 166. [CrossRef]
18. Marrs, J.; Ni-Meister, W. Machine Learning Techniques for Tree Species Classification Using Co-Registered LiDAR and Hyperspectral Data. *Remote Sens.* **2019**, *11*, 819. [CrossRef]
19. Xu, B.; Huang, J.Z.; Williams, G.; Wang, Q.; Ye, Y. Classifying Very High-Dimensional Data with Random Forests Built from Small Subspaces. *Int. J. Data Wareh. Min.* **2012**, *8*, 44–63. [CrossRef]
20. He, X.; Chaney, N.W.; Schleiss, M.; Sheffield, J. Spatial downscaling of precipitation using adaptable random forests. *Water Resour. Res.* **2016**, *52*, 8217–8237. [CrossRef]
21. Li, J.; Heap, A.D.; Potter, A.; Daniell, J.J. Application of machine learning methods to spatial interpolation of environmental variables. *Environ. Model. Softw.* **2011**, *26*, 1647–1659. [CrossRef]
22. Sekulić, A.; Kilibarda, M.; Heuvelink, G.B.M.; Nikolić, M.; Bajat, B. Random Forest Spatial Interpolation. *Remote Sens.* **2020**, *12*, 1687. [CrossRef]

23. Aonpong, P.; Kasetkasem, T.; Rakwatin, P.; Kumazawa, I.; Chanwimaluang, T. The Random forest approach for land cover mapping. In Proceedings of the 2016 7th International Conference of Information and Communication Technology for Embedded Systems (IC-ICTES), Bangkok, Thailand, 20–22 March 2016; IEEE: Bangkok, Thailand, 2016; pp. 1–6.
24. Samardžić-Petrović, M.; Kovačević, M.; Bajat, B.; Dragićević, S. Machine Learning Techniques for Modelling Short Term Land-Use Change. *ISPRS Int. J. Geo Inf.* **2017**, *6*, 387. [CrossRef]
25. Meraner, A.; Ebel, P.; Zhu, X.X.; Schmitt, M. Cloud removal in Sentinel-2 imagery using a deep residual neural network and SAR-optical data fusion. *ISPRS J. Photogramm. Remote Sens.* **2020**, *166*, 333–346. [CrossRef] [PubMed]
26. Pipia, L.; Muñoz-Marí, J.; Amin, E.; Belda, S.; Camps-Valls, G.; Verrelst, J. Fusing optical and SAR time series for LAI gap filling with multioutput Gaussian processes. *Remote Sens. Environ.* **2019**, *235*, 111452. [CrossRef]
27. Yan, L.; Roy, D.P. Large-Area Gap Filling of Landsat Reflectance Time Series by Spectral-Angle-Mapper Based Spatio-Temporal Similarity (SAMSTS). *Remote Sens.* **2018**, *10*, 609. [CrossRef]
28. Bankovic, S.; Medarevic, M.; Pantic, D.; Petrovic, N.; Sljukic, B.; Obradovic, S. The growing stock of the Republic of Serbia: State and problems. *Glas. Sumar. Fak.* **2009**, *2*, 7–29. [CrossRef]
29. Bohn, U.; Gollub, G.; Hettwer, C.; Neuhäuslová, Z.; Raus, T.; Schlüter, H.; Weber, H. *Karte der natürlichen Vegetation Europas/Map of the Natural Vegetation of Europe. Maßstab/Scale 1:2500,000*; Landwirtschaftsverlag: Münster, Germany, 2000.
30. Müller-Wilm, U.; Louis, J.; Richter, R.; Gascon, F.; Niezette, M. Sentinel-2 level 2a prototype processor: Architecture, algorithms and first results. In Proceedings of the ESA Living Planet Symposium, Edinburgh, UK, 9–13 September 2013; pp. 9–13.
31. Lakušić, D. *Database on the Distribution of Potentially Endangered Species and Habitats of Serbia—Results of the Project, Acquisition of Data and other Services for the Purpose of Establishing the Ecological Network of the European Union NATURA 2000 as Part of the Ecological Network of the Republic of Serbia 2016–2019, Centre for Biodiversity Informatics (CBI) of the University of Belgrade—Faculty of Biology (UBFB)*; Centre for Biodiversity Informatics of the University of Belgrade—Faculty of Biology: Belgrade, Serbia, 2019.
32. Davies, C.E.; Moss, D. *EUNIS Habitat Classification, February 2002*; European Environment Agency (EEA): Copenhagen, Denmark, 2002.
33. Pravilnik o Kriterijumima za Izdvajanje Tipova Staništa, o Tipovima Staništa, Osetljivim, Ugroženim, Retkim i za Zaštitu Prioritetnim Tipovima Staništa i o Merama Zaštite za Njihovo Očuvanje: 35/2010-29. Available online: http://www.pravno-informacioni-sistem.rs/SlGlasnikPortal/eli/rep/sgrs/ministarstva/pravilnik/2010/35/5/reg/ (accessed on 29 July 2020).
34. Forests—Copernicus Land Monitoring Service. Available online: https://land.copernicus.eu/pan-european/high-resolution-layers/forests (accessed on 13 July 2020).
35. Torgo, L. *Data Mining with R, Learning with Case Studies*; Chapman and Hall/CRC: Boca Raton, FL, USA, 2010.
36. R Core Team. *R: A Language and Environment for Statistical Computing*; R Foundation for Statistical Computing: Vienna, Austria, 2020.
37. Breiman, L. Random Forests. *Mach. Learn.* **2001**, *45*, 5–32. [CrossRef]
38. Malley, J.D.; Kruppa, J.; Dasgupta, A.; Malley, K.G.; Ziegler, A. Probability Machines: Consistent Probability Estimation Using Nonparametric Learning Machines. *Methods Inf. Med.* **2012**, *51*, 74–81. [CrossRef]
39. Wright, M.N.; Ziegler, A. ranger: A Fast Implementation of Random Forests for High Dimensional Data in C++ and R. *J. Stat. Softw.* **2017**, *77*, 1–17. [CrossRef]
40. Liu, Y.; Gong, W.; Hu, X.; Gong, J. Forest Type Identification with Random Forest Using Sentinel-1A, Sentinel-2A, Multi-Temporal Landsat-8 and DEM Data. *Remote Sens.* **2018**, *10*, 946. [CrossRef]
41. Yu, Y.; Li, M.; Fu, Y. Forest type identification by random forest classification combined with SPOT and multitemporal SAR data. *J. For. Res.* **2018**, *29*, 1407–1414. [CrossRef]
42. Olofsson, P.; Foody, G.M.; Herold, M.; Stehman, S.V.; Woodcock, C.E.; Wulder, M.A. Good practices for estimating area and assessing accuracy of land change. *Remote Sens. Environ.* **2014**, *148*, 42–57. [CrossRef]
43. Kivinen, J.; Warmuth, M.K. Averaging Expert Predictions. In Proceedings of the Computational Learning Theory, Nordkirchen, Germany, 29–31 March 1999; Fischer, P., Simon, H.U., Eds.; Springer: Berlin/Heidelberg, Germany, 1999; pp. 153–167.
44. James, G.; Witten, D.; Hastie, T.; Tibshirani, R. *An Introduction to Statistical Learning*; Springer Texts in Statistics; Springer: New York, NY, USA, 2013; Volume 103, ISBN 978-1-4614-7137-0.

45. Fukarek, P.; Jovanović, B. *Karta Prirodne Potencijalne Vegetacije SFR Jugoslavije 1:1000,000—Naučno Veće Vegetacijske Karte Jugoslavije*; Naučno Veće Vegetacijske Karte Jugoslavije, Šumarski Fakultet Univerziteta Kiril i Metodije u Skopju: Skopje, SFR Yugoslavia, 1983.
46. Banković, S.; Medarević, M.; Pantić, D.; Petrović, N. Nacionalna inventura šuma Republike Srbije. *Šumarstvo* **2008**, *3*, 1–16.
47. Banković, S.; Medarević, M.; Pantić, D.; Petrović, N. *Nacionalna Inventura Šuma Republike Srbije—Šumski Fond Republike Srbije*; Ministarstvo Poljoprivrede Šumarstva i Vodoprivrede Republike Srbije—Uprava za šume: Belgrade, Serbia, 2009.
48. Ottosen, T.-B.; Petch, G.; Hanson, M.; Skjøth, C.A. Tree cover mapping based on Sentinel-2 images demonstrate high thematic accuracy in Europe. *Int. J. Appl. Earth Obs. Geoinf.* **2020**, *84*, 101947. [CrossRef]
49. Lu, M.; Chen, B.; Liao, X.; Yue, T.; Yue, H.; Ren, S.; Li, X.; Nie, Z.; Xu, B. Forest Types Classification Based on Multi-Source Data Fusion. *Remote Sens.* **2017**, *9*, 1153. [CrossRef]

remote sensing

MDPI

Article

Evaluation of Hyperspectral Multitemporal Information to Improve Tree Species Identification in the Highly Diverse Atlantic Forest

Gabriela Takahashi Miyoshi [1,*], Nilton Nobuhiro Imai [1,2], Antonio Maria Garcia Tommaselli [1,2], Marcus Vinícius Antunes de Moraes [1] and Eija Honkavaara [3]

1. Graduate Program in Cartographic Sciences, São Paulo State University (UNESP), Roberto Simonsen 305, Presidente Prudente SP 19060-900, Brazil; nilton.imai@unesp.br (N.N.I.); a.tommaselli@unesp.br (A.M.G.T.); antunes.moraes@unesp.br (M.V.A.d.M.)
2. Department of Cartography, São Paulo State University (UNESP), Roberto Simonsen 305, Presidente Prudente SP 19060-900, Brazil
3. Finnish Geospatial Research Institute, National Land Survey of Finland, Geodeetinrinne 2, 02430 Masala, Finland; eija.honkavaara@nls.fi

* Correspondence: gabriela.t.miyoshi@unesp.br; Tel.: +55-18-32295515

Received: 13 November 2019; Accepted: 7 January 2020; Published: 10 January 2020

Abstract: The monitoring of forest resources is crucial for their sustainable management, and tree species identification is one of the fundamental tasks in this process. Unmanned aerial vehicles (UAVs) and miniaturized lightweight sensors can rapidly provide accurate monitoring information. The objective of this study was to investigate the use of multitemporal, UAV-based hyperspectral imagery for tree species identification in the highly diverse Brazilian Atlantic forest. Datasets were captured over three years to identify eight different tree species. The study area comprised initial to medium successional stages of the Brazilian Atlantic forest. Images were acquired with a spatial resolution of 10 cm, and radiometric adjustment processing was performed to reduce the variations caused by different factors, such as the geometry of acquisition. The random forest classification method was applied in a region-based classification approach with leave-one-out cross-validation, followed by computing the area under the receiver operating characteristic (AUCROC) curve. When using each dataset alone, the influence of different weather behaviors on tree species identification was evident. When combining all datasets and minimizing illumination differences over each tree crown, the identification of three tree species was improved. These results show that UAV-based, hyperspectral, multitemporal remote sensing imagery is a promising tool for tree species identification in tropical forests.

Keywords: tree species classification; semideciduous forest; hyperspectral multitemporal information; UAV

1. Introduction

Forests play an important role in biodiversity, carbon stocks, the water cycle, and feedstock, but they are rapidly being degraded. Knowledge about the tree species of a forest is fundamental information. Tree species mapping can be performed through fieldwork campaigns, but generally, this practice has limitations, since it is expensive and laborious. Remote sensing, together with automatic analysis techniques, has become a prominent tool for tree species mapping.

Satellite sensors [1] and airborne passive and/or active sensors [2,3], combined with the use of field spectroscopy [4], provide valuable information for the identification of tree species. Support vector machine (SVM) [5] and random forest (RF) [6] are examples of machine learning algorithms that have been successfully applied to identify tree species in urban environments [7], savannas [2], and different types of forests, including northern, boreal, temperate, and tropical forests [1,8–12]. Recently, these algorithms, with images acquired using unmanned aerial vehicles (UAVs), have become a powerful tool for forest monitoring [13–15].

UAVs enable fast information acquisition, and despite their constraints regarding the trade-off between resolution and coverage, they are low-cost alternatives for capturing information in areas that are endangered or need constant monitoring, such as mines or agricultural crops. UAVs can fly over many areas that are challenging for field data acquisition, such as water surfaces or dense forest areas. UAV missions can be quickly configured according to the user's needs. Furthermore, in the past few years, UAVs have been rapidly developed to fly for several hours; an example of such a platform is the fixed-wing Batmap II UAV, which is able to fly for more than 2 h [16]. UAVs can capture very high or ultrahigh spatial resolution data with ground sampling distances (GSD) ranging from centimeters to decimeters [17–20] using small-format multispectral or hyperspectral cameras, such as MicaSense RedEdge-MX [21], Rikola hyperspectral imager [22], and Cubert FireflEYE [21]. Beyond that, UAVs can repeatedly acquire information through surface targets, such as trees, which is a highly promising option in forest monitoring, since it can measure dynamic phenological behavior according to seasons and tree characteristics.

Although most of the previous studies conducted with multiseasonal information have not employed UAVs, they have shown spectral differences within tree species and reported whether the tree species classification was improved [7,23–30]. Moreover, most of them have been performed in northern forests [31]. There is a lack of studies in forests such as the Brazilian Atlantic forest, which encompasses different ecosystems, such as mixed ombrophilous, dense ombrophilous, open ombrophilous, semideciduous seasonal, and deciduous seasonal forests [32]. Sothe et al. [14] studied a mixed ombrophilous forest whose characteristics differ from those of other types of Brazilian Atlantic forest, especially the semideciduous and deciduous seasonal forests. Their floristic compositions and forest structure are different [32], which highlights the importance of studying them separately.

In addition to their application in northern forests in which coniferous tree species are predominant, most studies have investigated well-developed forests or forests in which trees with different heights are spatially distinguished. Plots containing tree species in different development stages can present similar heights, and thereby cause spectral mixing due to leaf mixture and the effect of neighborhood spectra, since the number of emergent trees is smaller. Notwithstanding the importance of monitoring mature forests, monitoring fragments that are in the initial regeneration process is considered a key element in the connection of forest patches, and contributes to the maintenance of biodiversity [33,34].

The objective of this study was to develop and assess UAV-based remote sensing techniques to improve tree species identification in Brazilian Atlantic forest areas with vegetation stages that range from initial to advanced restoration. A further objective was to study whether multitemporal spectral information acquired over multiple years in the same season but under different moisture conditions could improve tree species classification. UAV-based hyperspectral images were acquired over three years during the winter season. Additionally, the classification performance of two different spectral features was compared. The findings in this paper are expected to provide information about the use of spectral features with minimized effects from illumination differences to classify tree species in the highly diverse Brazilian Atlantic forest, and demonstrate the use of hyperspectral multitemporal information.

2. Materials and Methods

2.1. Study Site

The transect forest sample used in this study is centered at 22°23′55.21″S, 52°31′18.31″W, in the municipality of Euclides da Cunha Paulista, western part of São Paulo State, Brazil (Figure 1). The transect has an approximate length and width of 500 m × 130 m, and comprises a variety of tree species in different successional stages. The study area was established inside the Ponte Branca forest fragment, a protected area belonging to the Black Lion Tamarin Ecological Station. According to the Brazilian Institute for Geography and Statistics (IBGE), it is an area of the submontane, semideciduous seasonal forest [35]. The regional climate is considered a tropical zone with dry winters (Aw), according to the Köppen classification [36]. The mean temperature during the dry season is 21 °C, with less than 60 mm of total precipitation [37]. The weather patterns differed between the years 2017, 2018, and 2019 and the flight campaigns (Figure 2). The season was wetter in 2017, with precipitation of 69 mm before the flight campaign, whereas the precipitation was 18.6 mm before the flight campaign of 2018 and 51 mm before the flight campaign of 2019; however, rain did not occur for at least 8 days before image acquisition [37].

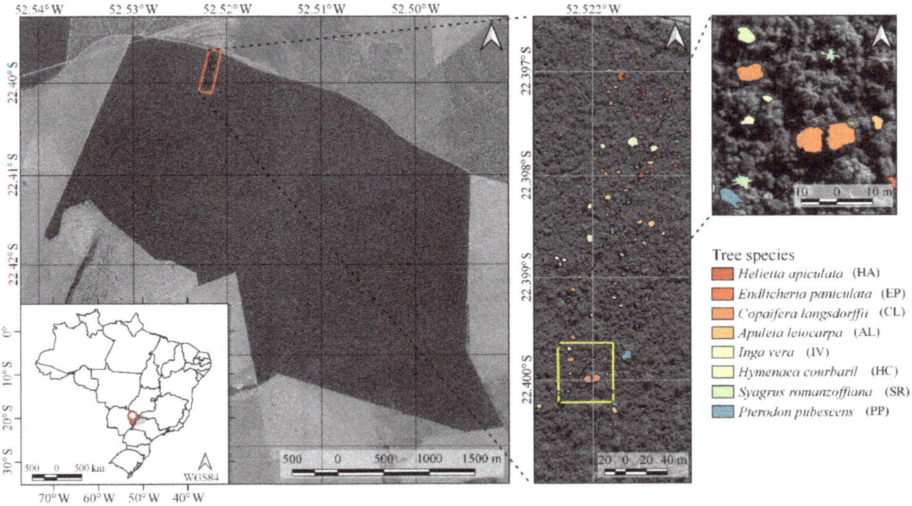

Figure 1. Study area and tree species recognized in the field in the 2017 dataset. The red rectangle represents the imaged area, and the yellow rectangle is a zoom of the study area.

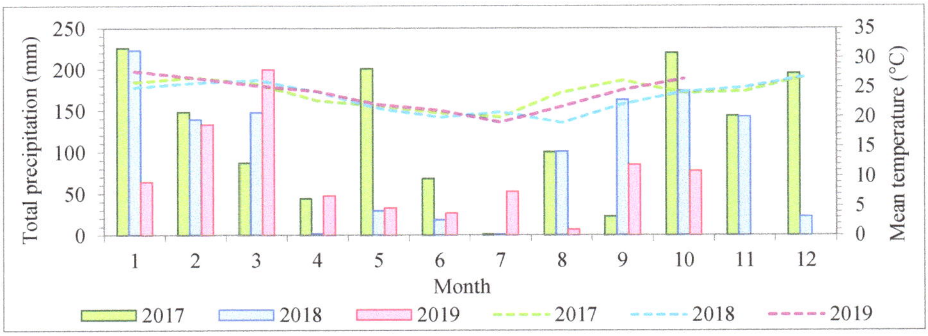

Figure 2. Climograph of Paranapoema station from the National Institute of Meteorology. Bars represent the accumulated rainfall per month (mm), and dashed lines represent the mean monthly temperature (°C).

2.2. Reference Data

More than 25 tree species with a diameter at breast height (DBH) greater than 3.8 cm were detected during fieldwork [38]. Tree species were in different successional stages, with the northernmost part of the study area containing trees in the initial stage of succession and the southernmost in a more advanced stage [39]. Moreover, we located 90 trees of eight species that emerged from the canopy (Tables 1 and 2). Their crowns were manually delineated through visual interpretation of RGB image composites of each dataset (R: 628.73 nm; G: 550.39 nm; B: 506.22 nm). Figure 1 shows examples of delineated individual tree crown (ITC) polygons in the 2017 dataset, and Figure 3 shows canopy examples of each tree species in the mosaic of images acquired during the 2017 flight campaign. These tree species were chosen because they are important for characterizing the successional stage of the forest, e.g., *Syagrus romanzoffiana* (SR), which can be associated with the faunal composition [40]. It is important to note that smaller tree species were excluded from analysis because the lianas that cover and mix among individuals negatively affect the classification accuracy. From now on, tree species will be called by their abbreviations from Tables 1 and 2.

Table 1. The number of samples recognized in the field and the average number of pixels per crown for each tree species and its sum.

Abbreviation	Specie	ITCs	Average Pixels/Crown	Sum of Pixels
AL	*Apuleia leiocarpa*	10	2328	23,278
CL	*Copaifera langsdorffii*	17	2148	36,520
EP	*Endlicheria paniculata*	7	1254	8776
HA	*Helietta apiculata*	10	1669	16,689
HC	*Hymenaea courbaril*	11	2800	30,799
IV	*Inga vera*	8	1288	10,302
PP	*Pterodon pubescens*	7	2715	19,007
SR	*Syagrus romanzoffiana*	20	1315	26,293

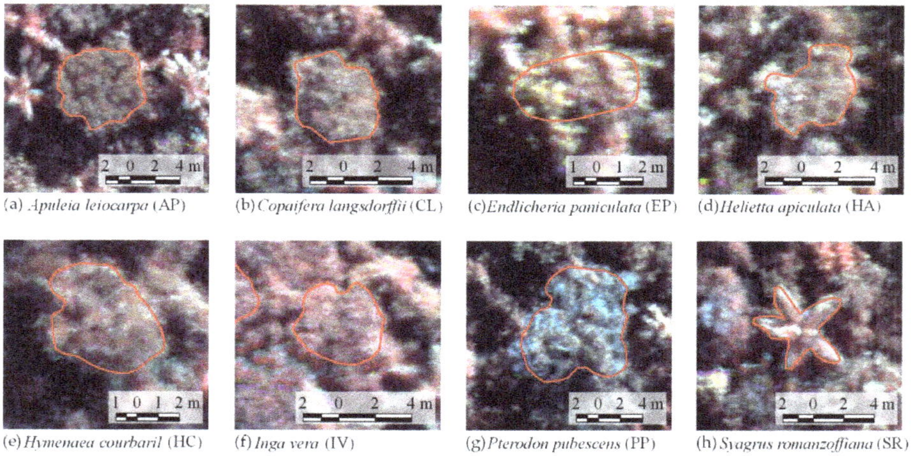

Figure 3. Canopy examples of each tree species identified in the field and delineated in the images acquired in 2017 (R: 780.49 nm; G: 565.10 nm; B: 506.22 nm; automatic contrast from QGIS software, version 3.0.0). (**a**) *Apuleia leiocarpa* (AP), (**b**) *Copaifera langsdorffii* (CL), (**c**) *Endlicheria paniculata* (EP), (**d**) *Helietta apiculata* (HA), (**e**) *Hymenaea courbaril* (HC), (**f**) *Inga vera* (IV), (**g**) *Pterodon pubescens* (PP), (**h**) *Syagrus romanzoffiana* (SR).

Table 2. Tree species identified in the field and their characteristics [1].

Abbreviation	Species	Family	Height (m)/Trunk Diameter (cm)	Characteristics
AL	*Apuleia leiocarpa*	Fabaceae: Caesalpinioideae	25–35/60–90	Deciduous and heliophyte. Blooms with the tree completely without leaves, usually in August–September
CL	*Copaifera langsdorffii*	Fabaceae: Caesalpinioideae	10–15/50–80	Semideciduous, heliophyte, xerophyte selective. Blooms between December to March, fruits ripen in August–September with the tree almost without leaves
EP	*Endlicheria paniculata*	Lauraceae	5–10/30–50	Evergreen, cyophyte, and hygrophyte selective. Blooms during the summer, January–March, and fruits ripen in May–July depending on the season
HA	*Helietta apiculata*	Rutaceae	10–18/30–50	Evergreen, heliophyte, and hygrophytic selective. Blooms between November–December and fruits ripen in March to May, outside the dry season
HC	*Hymenaea courbaril*	Fabaceae: Caesalpinioideae	15–20/up to 100	Semideciduous, heliophyte, xerophyte selective. Blooms in October–December and fruits ripen from July
IV	*Inga vera*	Fabaceae: Mimosoideae	5–10/20–30	Semideciduous, heliophyte, pioneer, and hygrophyte selective. Blooms in August–November and fruits ripen during the Summer, December to February
PP	*Pterodon pubescens*	Fabaceae: Faboideae	8–16/30–40	Deciduous, heliophyte, xerophyte selective. Blooms between September to October and the fruits ripen with the tree almost without leaves
SR	*Syagrus romanzoffiana*	Arecaceae	10–20/30–40	Evergreen, heliophyte, and hygrophyte selective. Blooms almost during the entire year and fruits ripen mainly in February to August

[1] Information extracted from [41,42].

The number of samples was low for some species because of challenges in acquiring reference data. First, our study area comprised different successional stages; thus, the species composition varied over the area. Second, we used tree samples that emerged from the canopy. In tropical forests, the tree heights can be modeled by the DBH, which can be related to the number of trees per hectare [43–45]. According to Lima et al. [44] and d'Oliveira et al. [45], the relationship between the DBH and the frequency of trees in tropical forests has an "inverted J shape", because the number of trees per hectare decreases substantially as the DBH increases, so the number of taller trees also decreases.

2.3. Remote Sensing Data

Hyperspectral images were acquired with a 2D-format camera based on the tunable Fabry–Pérot Interferometer (FPI) from Senop Ltd, model DT-0011 [46–48]. The camera has two sensors, both of which have 1017 × 648 pixels with a pixel size of 5.5 µm. The total weight of the camera is around 700 g with its accessories, which include an irradiance sensor and a Global Positioning System (GPS) receiver. Spectral bands can be selected from the visible (VIS) to near-infrared (NIR) region (500–900 nm), which are acquired sequentially, i.e., the air gap of the FPI moves to acquire the different spectral bands of the same image. The spectral range of the first and second sensors are 647–900 nm and 500–635 nm, respectively. A total of 25 spectral bands were chosen, with the Full-Width at Half Maximum (FWHM) varying from 12.44 to 20.45 nm (Table 3 and Figure 4). For this spectral setting, each image cube needs 0.779 s to be acquired. The exposure time was set to 5 ms, and the image blocks were divided into two flight strips, ensuring more than 70% and 50% forward and side overlaps, respectively.

Table 3. Spectral settings of an FPI camera, model DT-0011, with the respective FWHM, both in nm (λ represents the central wavelength of the band).

Band	λ (nm)	FWHM (nm)	Band	λ (nm)	FWHM (nm)	Band	λ (nm)	FWHM (nm)	Band	λ (nm)	FWHM (nm)
1	506.22	12.44	8	609.00	15.08	14	679.84	20.45	20	740.42	17.98
2	519.94	17.38	9	620.22	16.26	15	690.28	18.87	21	750.16	17.97
3	535.09	16.84	10	628.73	15.30	16	700.28	18.94	22	769.89	18.72
4	550.39	16.53	11	650.96	14.44	17	710.06	19.70	23	780.49	17.36
5	565.10	17.26	12	659.72	16.83	18	720.17	19.31	24	790.30	17.39
6	580.16	15.95	13	669.75	19.80	19	729.57	19.01	25	819.66	17.84
7	591.90	16.61									

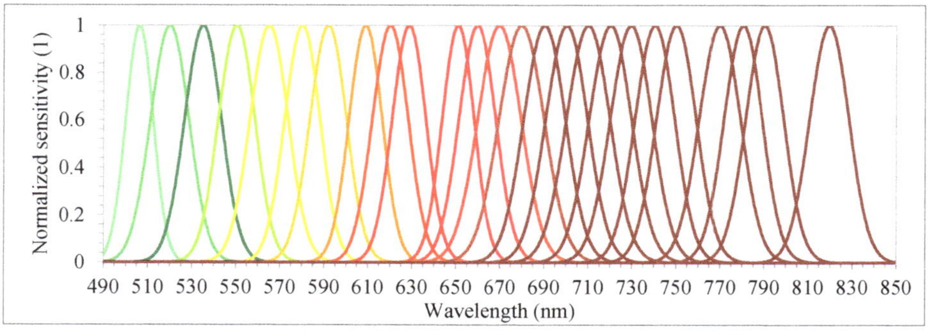

Figure 4. Normalized sensitivities of each spectral band set in the FPI camera. Responses were calculated from the central wavelength and FWHM.

The FPI camera was mounted onboard the UX4 UAV, which is a rotary-wing quadcopter developed by the company Nuvem UAV. The UX4 UAV is almost 90 cm in diameter and 30 cm in height without counting the GPS antenna, which is approximately 15 cm. It is controlled by a PixHawk autopilot. The energy source for the UAV system and its sensors is one six-cell battery of 22 volts and one smaller three-cell battery of 11 volts, which allow the UAV to fly for up to 30 min, depending on payload, battery, and weather conditions. A flight speed of 4 m/s was used to limit the maximum gap between the first and last band of the hyperspectral imager to 3.1 m in a single cube.

During the field campaigns, three radiometric reference targets were placed in the area to enable reflectance calibration. Flight campaigns were performed over the study area (Figure 1) on 1 July 2017, 16 June 2018, and 13 July 2019, with an above-ground flight height of approximately 160 m and flight speed of 4 m/s. The flight height was selected so that a GSD of 10 cm was obtained. This ensured a good representation of tree crowns that were predominantly over 3 m in diameter. Table 4 provides more details about the flight time of each campaign and the mean zenith and azimuth angles of the Sun during the image acquisitions.

Table 4. Details of the image acquisition in each flight campaign.

Flight Campaign	Time (UTC-3)	Sun Zenith	Sun Azimuth
1 July 2017	10:14–10:24	56.35°	38.46°
16 June 2018	11:47–11:54	46.75°	12.55°
13 July 2019	14:27–14:34	52.32°	325.61°

Images were geometrically and radiometrically processed to obtain hyperspectral image orthomosaics. First, the images were radiometrically corrected from the dark current and nonuniformity of sensors using a dark image acquired before each flight and laboratory parameters [47,49].

The geometric processing was performed using the Agisoft PhotoScan software (version 1.3) (Agisoft LLC, St. Petersburg, Russia). In the orientation process, for each year, the exterior orientation parameters (EOPs) of four reference bands (band 3: 550.39 nm; band 8: 609.00 nm; band 14: 679.84 nm; and band 22: 769.89 nm) were estimated in the same Agisoft PhotoScan project in order to reduce misregistration between the datasets. The EOPs of the other bands were calculated using the method developed in [49,50]. Positions from the camera GPS were used as initial values and refined using a bundle block adjustment (BBA) and ground control points (GCPs). The number of GCPs varied between datasets, with 3, 3, and 4 used in 2017, 2018, and 2019, respectively. GCPs were placed outside the forest since it was not possible to see the ground from imagery acquired over the forested area. Initially, the base station was defined near the study area, and the global navigation satellite system (GNSS) observations from GCPs were collected and processed in differential mode.

A self-calibrating bundle adjustment was used to estimate the interior orientation parameters (IOPs) of each sensor and for each year of the dataset. After initial image alignment, parameter estimation was optimized with automatic outlier removal using a gradual selection of tie points based on reconstruction uncertainty and reprojection error, together with the manual removal of points. The final products of this step were the calibrated IOPs, EOPs, sparse and dense point clouds, and digital surface model (DSM) of the area with a GSD of 10 cm. These were used in the following radiometric block adjustment and mosaic generation.

Radiometric adjustment processing aims to correct the digital number (DN) of pixels of images from the bidirectional reflectance distribution function (BRDF) effects and differences caused by the different geometries of acquisition due to the UAV and Sun movements. Thus, nonuniformities among images were compensated for by the radBA software, developed at the Finnish Geospatial Research Institute (FGI) [49,50]. Equation (1) shows the model used in the software to extract the reflectance value from the DN of each pixel.

$$DN_{jk} = a_{relj}\left(a_{abs}\cdot R_{jk}(\theta_i, \theta_r, \varphi) + b_{abs}\right), \quad (1)$$

where DN_{jk} is the digital number of pixel k in image j; $R_{jk}(\theta_i, \theta_r, \varphi)$ is the corresponding reflectance factor with respect to the zenithal angle θ of the incident and reflected light, i and r, respectively, and with the relative azimuthal angle φ $(\varphi_r - \varphi_i)$, where φ_r refers to the reflected azimuthal angle, and φ_i

denotes the incident azimuthal angle; a_{relj} is the relative correction factor of illumination differences with respect to the reference image; and a_{abs} and b_{abs} are the empirical line parameters for the linear transformation between reflectance and DNs.

According to a previous study by Miyoshi et al. [47], for the study area, the best initial relative correction factor (a_{relj}) was one (1), with a standard deviation equal to 0.05. It is worth noting that an exception was necessary for the dataset from 2018 because of higher density differences in cloud covering. The 2017 and 2019 flights were carried out in almost blue-sky conditions, with slight differences compensated for by the radiometric block adjustment. The radiometric block adjustment was performed in two steps for the 2018 dataset. First, an initial radiometric block adjustment was performed using initial values of $a_{relj} = 1$. In sequence, the final values of a_{relj} were used as the initial values for the second radiometric block adjustment. Then, the reflectance factor values were estimated using the empirical line method [51]. The empirical line parameters (a_{abs} and b_{abs}) were estimated from the linear relationship between the DN values of three radiometric reference targets with a mean reflectance of 4%, 11%, and 37%. Radiometric reference targets were 90 cm × 90 cm and composed of light-gray, gray, and black synthetic material. Thus, the mosaics of hyperspectral images for each dataset representing the reflectance factor values were obtained.

Additionally, point cloud data from airborne laser scanning (ALS) were provided by the company Fototerra. ALS data were acquired in November 2017 using a Riegl LMS-Q680i laser scanner (RIEGL, Horn, Austria) onboard a manned aircraft at a flight height of 400 m, which resulted in an average density of 8.4 points/m^2. The canopy height model (CHM) was obtained by extracting the digital terrain model (DTM) from the DSM. The processing was performed using the LAStools software (Martin Isenburg, LAStools—efficient tools for LiDAR processing) [52]. First, the lasnoise tool was applied to withdraw possible noises in the point cloud. Then, the CHM was extracted using the lasheight tool to obtain tree heights in the study area.

Figure 5 shows the mean height of each tree sample recognized in the field; most of the observed samples fell within a similar height range. Additionally, taller trees were found in the more developed successional stage of the area. Trees of the same species varied in age and were found in regions of different successional stages. For example, PP trees had crown areas of around 25 m^2 and mean heights of 10–20 m. Similarly, HC samples had a mean height of almost 14 m, with tree crown areas ranging from 16 to 90 m^2. It is important to highlight that the ALS data were not used in the classification step, since the objective of this research was to evaluate hyperspectral multitemporal data to improve tree species identification.

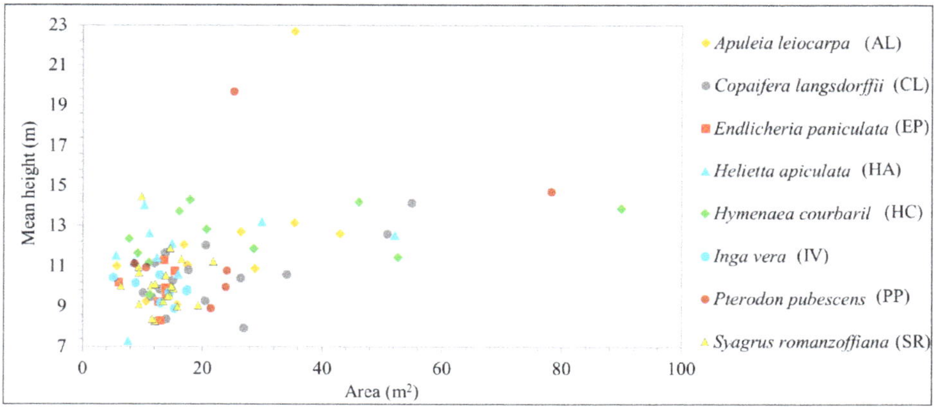

Figure 5. Mean tree height versus the tree crown area for all samples identified in the field. Data are from the 2017 dataset.

2.4. Extraction of Spectral Features

Spectral features were extracted using manually delineated crown polygons. We used spectral features taken from the reflectance mosaics and normalized features extracted after pixel normalization. The normalization reduced the differences between the sunlit and shadowed pixels, assuming a uniform distribution across the crown [13,53]. The normalization process was performed to reduce the spectral variability of spectra from the same tree species (within-species). The normalized pixel value was calculated by dividing the pixel value of a band by the sum of values of this pixel in all bands [53]. The mean values of normalized and nonnormalized pixels in each polygon were extracted for use in the region-based classification method. These values are referred to as MeanNorm and Mean, respectively.

Despite performing joint geometric processing, there were differences in the spatial position of trees, especially when using very high spatial resolution imagery. These differences are mainly caused by tree growth, changes in leaves with changing seasons, and weather conditions or projection differences due to the characteristics of the surface used. Figure 6 shows the slight difference in the spatial distribution of the leaves of SR trees.

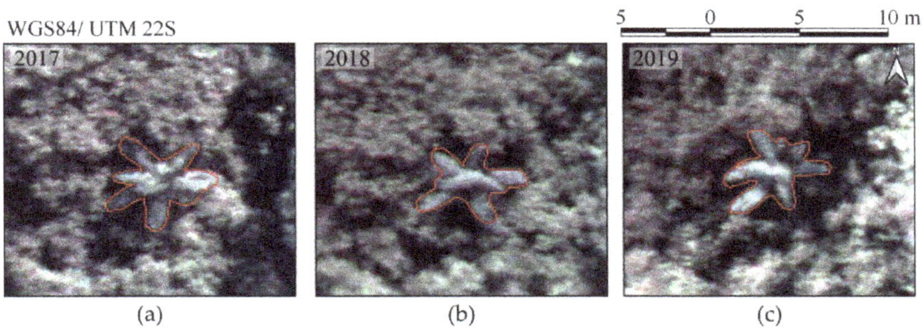

Figure 6. The spatial difference in the leaves of one sample of SR tree in each dataset (R: 690.28 nm; G: 565.10 nm; B: 519.94 nm; automatic contrast from QGIS software, version 3.0.0): (**a**) 2017, (**b**) 2018, and (**c**) 2019.

2.5. Tree Species Identification with RF

The RF method was used for tree species identification using only the spectral information, since the objective was to verify whether the use of temporal information could improve tree species detection. This method is based on multiple decision trees, and the class is determined by the most popular vote [6]. Decision trees are composed of different samples, which are drawn with replacement, i.e., one sample can belong to more than one tree [54]. The RF has been successfully applied for image classification when working with high-dimensional data [8,13], and it is less sensitive to feature selection [54]. The RF was applied using the default parameters of the Weka software version 3.8.3 (The University of Waikato, Hamilton, New Zealand) [55].

The classification process was carried out five times with four different datasets: (i) the 2017 spectral information (D17); (ii) the 2018 spectral information (D18); (iii) the 2019 spectral information (D19); (iv) the combination of the 2017, 2018, and 2019 spectral information (Dall). For the D17, D18, and D19 datasets, we used the normalized pixel values to extract the spectral features, which are referred to as cases D17_MeanNorm, D18_MeanNorm, and D19_MeanNorm, respectively. Additionally, in the case of the combined dataset (item (iv) in the previously described datasets), the classification was performed using both the normalized and nominal values, referred to as Dall_MeanNorm and Dall_Mean, respectively. Table 5 summarizes the number of features used in each case.

The number of samples of different species is relatively low and also unbalanced. The leave-one-out cross-validation (LOOCV) method was used to circumvent this problem. LOOCV is a particular case of k-fold cross-validation, where k is equal to the total number of samples of the dataset [13,56]. The classification model is trained k times, followed by testing with one subset and training with the remaining subsets. In each iteration, the model is trained using k − 1 samples and tested with the remaining sample. The final accuracy values are obtained by averaging the accuracy values of each iteration [56]. LOOCV has been successfully applied in tree species classification studies with a small sample size (e.g., less than 10 samples per class [14]) or an unbalanced number of samples per class [13].

Table 5. The number of features used in each test.

Cases	Spectral Data From			Number of Features
	2017	2018	2019	
D17_MeanNorm	X			25
D18_MeanNorm		X		25
D19_MeanNorm			X	25
Dall_Mean	X	X	X	75
Dall_MeanNorm	X	X	X	75

The results were evaluated through the area under the receiver operating characteristic curve, known as AUC (area under the curve) ROC (receiver operating characteristics) or AUCROC [57–59]. ROC is the relationship between the false positive rate (FPR), or "1-specificity", and the true positive rate (TPR), or sensitivity, and it is useful when working with unbalanced classes because it is independent of the class distribution [59,60]. When using classifiers such as RF that provide probabilities or scores, thresholds can be applied to acquire different points in the ROC space to form an ROC curve [60]. AUCROC is the area under the ROC curve, and represents the probability of the classification model correctly classifying a random sample in a specific class. AUCROC varies from 0 to 1 for each class, where a value of 0.5 indicates that the specific classification model is no better than a random assignment, and a value of 1 represents perfect discrimination of a class from the remaining ones [59]. To the best classification (i.e., the one with highest value of average AUCROC value), the overall accuracy (i.e., the percentage of correctly classified instances of the total number of samples), the user accuracy, and the producer accuracy [61] were calculated as well.

3. Results

3.1. Spectral Response of Each Tree Species Recognized in the Field

The spectral variability within samples of the same tree species was verified through the mean, minimum, and maximum values of Mean and MeanNorm (defined in Section 2.3). The mean values are presented in Figure 7, showing similar spectral responses in the VIS region for both Mean and MeanNorm. In the NIR region, the Mean spectra are visually similar between IV, HA, HC, and AL. Despite smaller differences among the MeanNorm spectra, which may lead to higher classification confusion, the spectral variability within the samples of Mean had a higher range (Figure 8). In Figure 8, the range between the minimum and maximum values is visually the same for both the Mean and MeanNorm spectra of all tree species in the VIS part of the electromagnetic spectrum. It is noted that the number of samples of each tree species can affect this range of variation, as observed for SR with 20 samples. However, this behavior was not observed for AL (10 samples) and HC (11 samples). The range variation in the Mean values from the red-edge (700 nm) to near-infrared region (820 nm) had a higher variability when compared with the MeanNorm values, leading to the conclusion that a higher variability may influence classifier performance. Moreover, in Figure 7, an unusual peak may be noticed at the spectral response at 650 nm, probably due to the fact that this spectral band is located near the edge of the first sensor from the FPI, which acquires information from 647 nm to 900 nm.

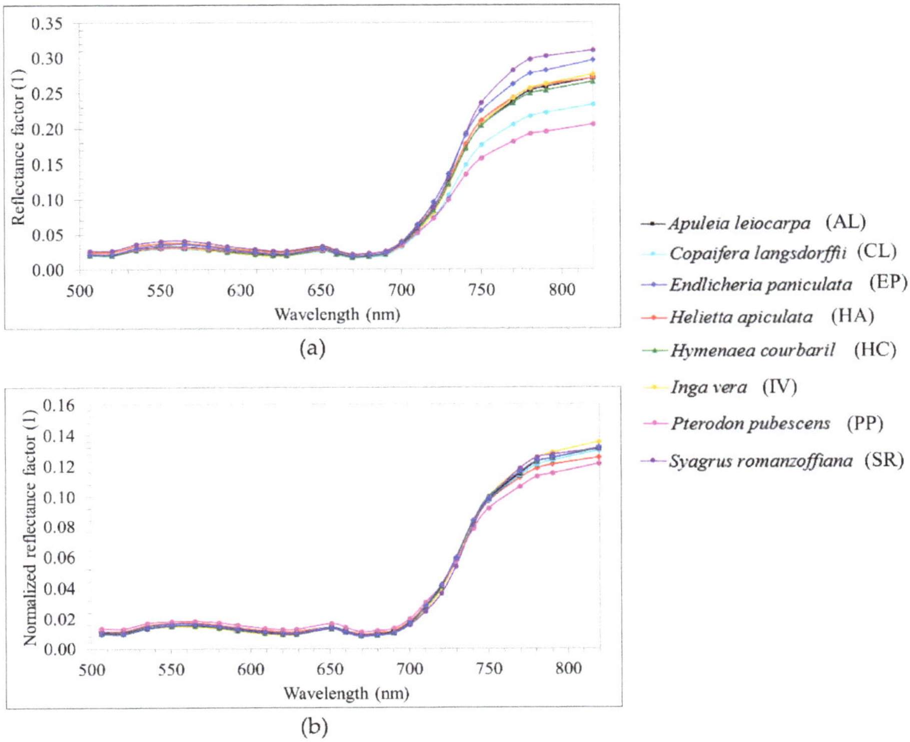

Figure 7. Mean spectra of each tree species considering the: (**a**) Mean, and the (**b**) MeanNorm.

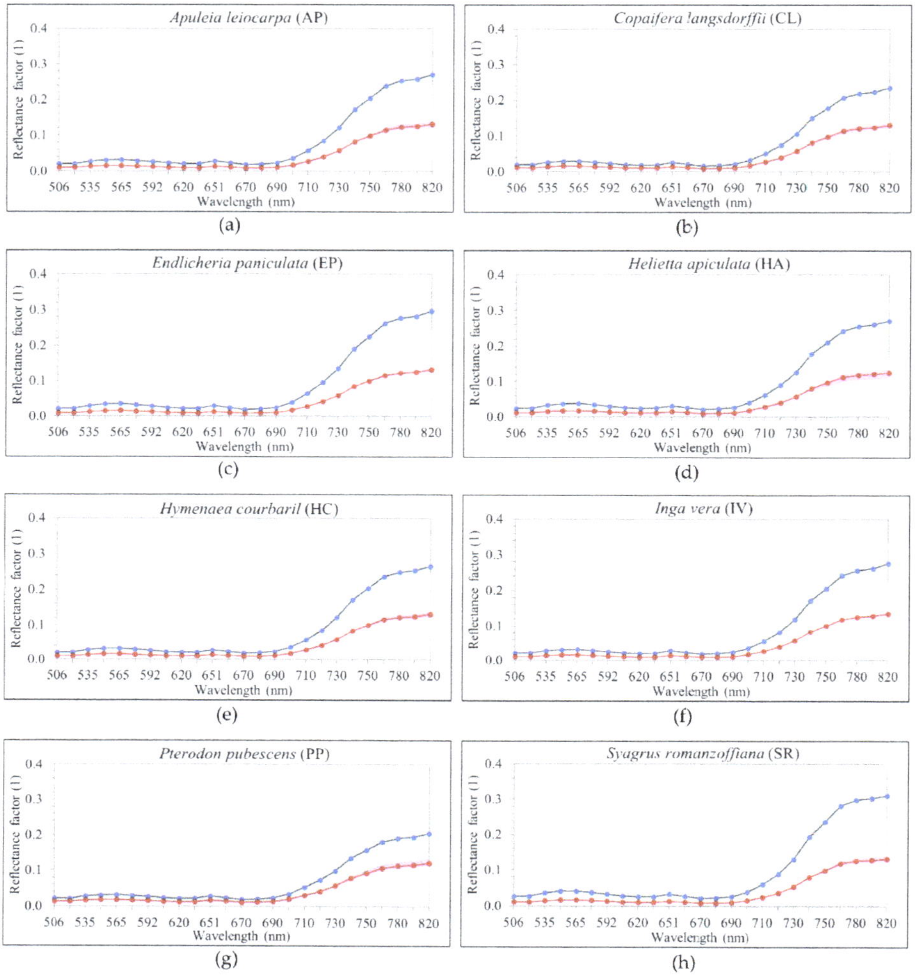

Figure 8. Values for each tree species considering the mean reflectance factor spectra (Mean) and the mean normalized spectra (MeanNorm) calculated using pixels in the delineated crown polygons. The blue line represents the Mean, the red line represents the MeanNorm, and the shaded area represents the minimum and maximum values. (**a**) *Apuleia leiocarpa* (AP), (**b**) *Copaifera langsdorffii* (CL), (**c**) *Endlicheria paniculata* (EP), (**d**) *Helietta apiculata* (HA), (**e**) *Hymenaea courbaril* (HC), (**f**) *Inga vera* (IV), (**g**) *Pterodon pubescens* (PP), (**h**) *Syagrus romanzoffiana* (SR).

3.2. Identification of Tree Species Results

Table 6 provides the AUCROC values after applying the RF with LOOCV to each dataset. Dall_MeanNorm presented the highest average AUCROC value (0.807), and thus, it can be considered the best dataset with which to identify the tree species. Average AUCROC values for Dall_Mean, D17_MeanNorm, D18_MeanNorm, and D19_MeanNorm were 0.783, 0.746, 0.754, and 0.682, respectively. Next, a more detailed analysis was performed using AUCROC values of Dall_MeanNorm.

Table 6. AUCROC values for each tree species identified in each dataset. AUCROC values are from imagery data of (i) only 2017 (D17_MeanNorm); (ii) only 2018 (D18_MeanNorm); (iii) only 2019 (D19_MeanNorm); (iv) all years and the mean spectral values (Dall_Mean); and (v) all years and the mean normalized values (Dall_MeanNorm).

Tree Species [1]	AUCROC				
	D17_MeanNorm	D18_MeanNorm	D19_MeanNorm	Dall_Mean	Dall_MeanNorm
AL	0.608	0.438	0.313	0.754	0.613
CL	0.821	0.678	0.517	0.742	0.768
EP	0.818	0.827	0.664	0.743	0.836
HA	0.594	0.576	0.899	0.798	0.846
HC	0.800	0.809	0.847	0.699	0.847
IV	0.627	0.886	0.622	0.837	0.824
PP	0.713	0.817	0.680	0.758	0.723
SR	0.986	0.997	0.915	0.936	0.999
Average AUCROC	0.746	0.754	0.682	0.783	0.807

[1] AL: Apuleia Leiocarpa; CL: Copaifera Langsdorffii; EP: Endlicheria Paniculata; HA: Helietta Apiculata; HC: Hymenaea Courbaril; IV: Inga Vera; PP: Pterodon Pubescens; SR: Syagrus Romanzoffiana.

Compared with the other datasets, Dall_MeanNorm had the highest AUCROC values for three of the eight tree species, namely, EP, HC, and SR. HA was better modeled in the D19_MeanNorm dataset, with an AUCROC value of 0.899, and it was worst modeled in the D18_MeanNorm dataset (AUCROC = 0.576). In contrast, IV was best and worst identified in D18_MeanNorm and D19_MeanNorm, respectively. Additionally, no significant differences were obtained when using normalized pixels compared with unnormalized ones for this tree species since the AUCROC values were 0.837 for Dall_Mean and 0.824 for Dall_MeanNorm. The identification of CL was similar between Dall_Mean (AUCROC = 0.742) and Dall_MeanNorm (AUCROC = 0.768), and it was best modeled in D17_MeanNorm (AUCROC = 0.821). AL had the lowest AUCROC value in D19_MeanNorm (0.313), which probably affected its identification in the Dall_MeanNorm dataset, in which its AUCROC was 0.613.

Since Dall_MeanNorm generated the best results in general, its ROC curves are shown in Figure 9, and its confusion matrix and user and producer accuracies are presented in Table 7. Figure 9 reveals different threshold values for each tree species, which are related to predictive probabilities [62]. For AL, which had the lowest AUCROC value (0.613), the FPR was higher than 0 (0.088), even when the TPR was equal 0, which indicates that the RF performed poorly in identifying this tree species, as confirmed by the confusion matrix, since none of AL were correctly identified. PP had the second lowest AUCROC value (0.723), and its threshold varied from 0 to 0.46, i.e., similar to AL. As shown in Figure 9 and in the confusion matrix of Table 7, only one tree species was correctly identified, and the TPR was only higher than 0 (TPR = 0.143) when the FPR was 0.024 to a threshold of 0.4. The highest AUCROC value, 0.999 for SR, corresponded to the tree species with the fewest false positives; that is, it was less frequently confused with the other tree species. The ROC curve of SR in Figure 9 shows that a TPR of 1 was obtained when the FPR was 0.014. This fact indicates that samples of this tree species will always be correctly identified; however, even with a low degree of variation, they could be confused with other tree samples, even in small proportions. Interestingly, for IV, that was not among the highest AUCROC values, the FPR is equal to 0 until a threshold of 0.44, when the TPR is 0.375. This fact is associated with the confusion matrix of Table 7, which has few false positives for this tree species.

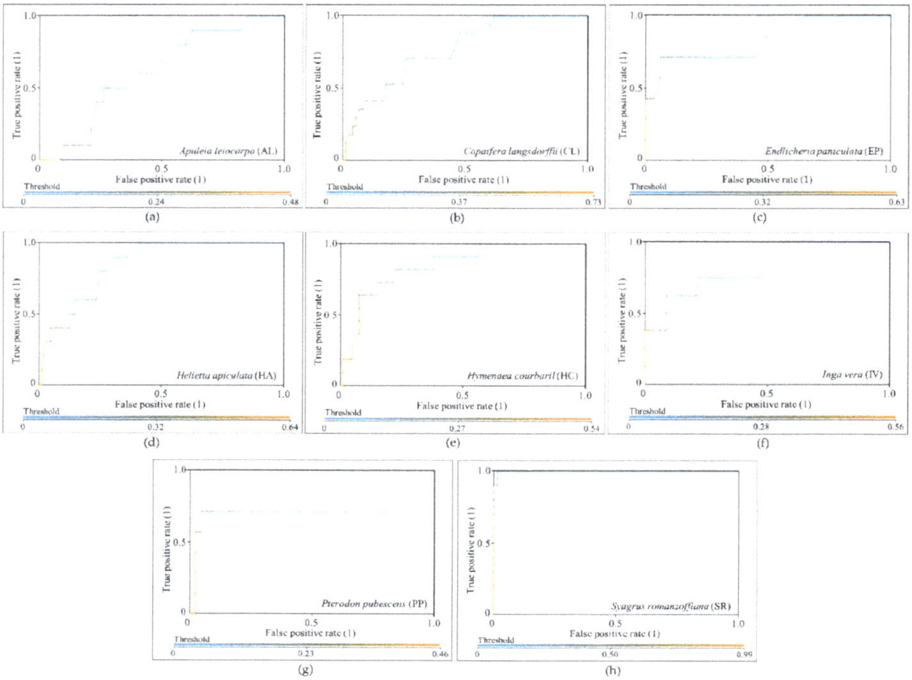

Figure 9. ROC curves of the identification of each tree species from the application of RF to all imagery datasets (Dall_MeanNorm). (**a**) *Apuleia leiocarpa* (AP), (**b**) *Copaifera langsdorffii* (CL), (**c**) *Endlicheria paniculata* (EP), (**d**) *Helietta apiculata* (HA), (**e**) *Hymenaea courbaril* (HC), (**f**) *Inga vera* (IV), (**g**) *Pterodon pubescens* (PP), (**h**) *Syagrus romanzoffiana* (SR).

Table 7. Confusion matrix of the classification of 8 tree species and all datasets (Dall_MeanNorm) and its user accuracy and producer accuracy.

Tree Species [1]	AL	CL	EP	HA	HC	IV	PP	SR	User Accuracy (%)
AL	0	2	0	1	3	0	0	0	0
CL	3	8	2	3	0	2	2	0	40
EP	0	1	4	1	1	0	0	0	57.1
HA	2	2	0	3	0	0	2	0	33.3
HC	4	2	1	0	6	2	2	0	35.3
IV	0	2	0	0	0	3	0	0	60
PP	0	0	0	2	0	0	1	0	33.3
SR	1	0	0	0	1	1	0	20	87
Producer accuracy (%)	0	47.1	57.1	30	54.5	37.5	14.3	100	**Overall Accuracy = 50%**

[1] AL: Apuleia Leiocarpa; CL: Copaifera Langsdorffii; EP: Endlicheria Paniculata; HA: Helietta Apiculata; HC: Hymenaea Courbaril; IV: Inga Vera; PP: Pterodon Pubescens; SR: Syagrus Romanzoffiana.

Spectral Feature Importance

The feature importance in the Dall_MeanNorm dataset, which had the best classification results, is given in Figure 10. The feature importance was scaled from 0 to 1, where 0 represents the least important feature, and 1 represents the most important feature. The least important feature was band 21, centered at 750.16 nm, in the 2019 dataset, and the most important was band 10, centered at 628.73 nm, in the 2017 dataset. Additionally, in general, the most important features in the 2017 dataset were from the VIS part to the beginning of the red-edge part of the electromagnetic spectrum. In the 2018 dataset, an exception in feature importance may be observed at bands 15, 19, and 23, centered at 690.28 nm, 729.57 nm, and 780.49 nm. These bands were more important than most of the NIR bands in the 2018 dataset. In the 2019 dataset, bands 3, 6, and 12, centered at 535.09 nm, 580.16 nm, and 659.72 nm, respectively, were highlighted because of the peak in the feature importance value when compared with the other bands from 2019. These bands are in the VIS part of the electromagnetic spectrum; this is related to the leaves' pigment, e.g., chlorophyll and carotenoids, content.

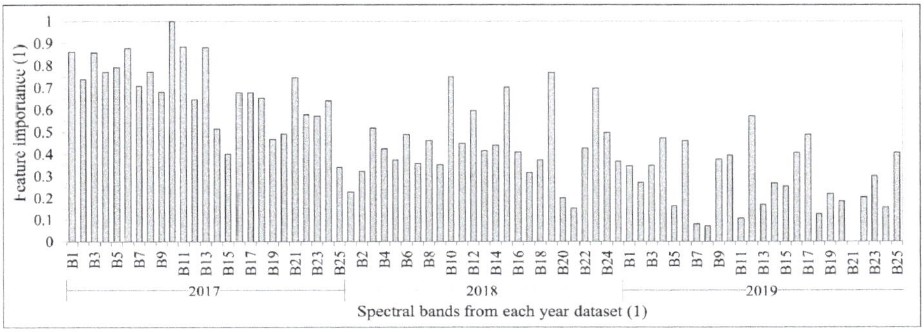

B1 = 506.22 B2 = 519.94 B3 = 535.09 B4 = 550.39 B5 = 565.10 B6 = 580.16 B7 = 591.90 B8 = 609.00 B9 = 620.22
B10 = 628.73 B11 = 650.96 B12 = 659.72 B13 = 669.75 B14 = 679.84 B15 = 690.28 B16 = 700.28 B17 = 710.06
B18 = 720.17 B19 = 729.57 B20 = 740.42 B21 = 750.16 B22 = 769.89 B23 = 780.49 B24 = 790.30 B25 = 819.66

Figure 10. Feature importance when using the RF for 8 tree species and all datasets (Dall_MeanNorm). The x-axis represents the spectral bands in each year. Wavelengths in nm.

4. Discussion

Our study area is located in the highly diverse Brazilian Atlantic forest and comprises a great number of tree species at varying successional stages, which makes tree species identification in the field challenging. Most of the study area is in the initial stage of regeneration [39], which indicates that the tree heights have a low degree of variation, as verified in Figure 5. The southernmost part of the study area is more preserved and exhibits a medium degree of regeneration [39], as supported by the existence of taller trees in Figure 5. Monitoring this type of vegetation is a feasible way to increase the knowledge of forest composition and development, especially when the structural data do not have enough variation among classes to identify the tree species. The importance of regenerating forest paths is directly related to the maintenance of biodiversity.

The use of joint spectral normalized features (i.e., Dall_MeanNorm) increased the AUCROC values of three tree species (EP, HC, and SR). In general, when using the mean spectral features together (i.e., Dall_Mean), variations in the AUCROC values were more apparent compared with the use of spectral information from each dataset separately. The exception in the Dall_Mean results is to AL, whose AUCROC value increased with the use of temporal spectral information without normalization. Moreover, it was observed that weather conditions directly affected most of the trees' phenology and, consequently, their spectral response, thus affecting RF performance on each dataset.

All the AUCROC values for SR were higher than 0.90, leading to the conclusion that the identification of this tree species did not depend on multitemporal information or the use of normalized spectra. A similar analysis can be applied to HC, which had similar AUCROC values in all tests, without counting the normalized spectra. CL had similar AUCROC values in Dall_Mean and Dall_MeanNorm and was better identified in D17_MeanNorm. Therefore, the weather pattern in 2017 was related to the identification accuracy of CL. There was a higher volume of rain before the 2017 flight campaign (Figure 2). Similarly, the weather influenced the detection of other tree species when using a single spectral dataset. The dry weather before the 2018 and 2019 flight campaigns hindered the ability to identify the AL tree species when using the spectral data of these years.

These tree species have different structures, such as the leaf format, and they have different blossom and fruit sets [41,42]. SR is a palm tree with leaves that are 2–3 m in length and spadices that are 80–120 cm in length [41]. HC has pinnate leaves and requires sunlight to grow and emerge from the canopy; blooming occurs in the dry season, and fruit appears after 3–4 months [42]. AL blooms without leaves, usually in September, and its flowers are white [42]. Thus, the use of multitemporal data influenced the detection of tree species. Of the previous works in the literature, the research of Ferreira et al. [28] is highlighted. They acquired WorldView images during the wet and dry seasons of a well-developed Brazilian semideciduous forest to classify tree species; no improvement in the classification results was observed when using the combined data. On the other hand, Somers and Asner [24], Deventer et al. [25], and Hill et al. [26] found that tree species classification improved when using multitemporal data because of the different spectral changes in the data.

As supported by previous research, VIS bands were among the most important features in tree species classification at the crown scale [10,13,63,64]. Vegetation spectra are characterized by the peak and absorption in the green and red parts of the electromagnetic spectrum, which is helpful for differentiating tree species. Similarly, the use of CHM has been shown to improve classification accuracies [13,14,65]. CHM data were not applied in our study since the focus was on the usefulness of temporal spectral information. The use of CHM data, combined with the use of tree crown segments, is highly recommended for future studies on the classification of tree species in the area.

This study was the first to investigate tree species classification using multitemporal hyperspectral UAV data acquired over the Atlantic forest. The previous studies that used multitemporal data acquired datasets from different seasons, and they did not use UAVs or consider a semideciduous forest with different development stages. Deventer et al. [25] simulated WorldView and RapidEye data from the leaf spectra of a subtropical forest in South Africa. Hill et al. [26] used the Daedalus 1268 AirborneThematic Mapper (ATM) sensor to acquire data over a deciduous forest in England. Using WorldView images, Li et al. [7] studied the multitemporal information of tree species in urban environments. When using UAVs, image acquisition depends on several factors (such as wind conditions since a UAV is a lightweight platform), and there are safety requirements to fulfill. During the spring and summer, when some trees may be blooming, the rainfall is higher; for example, summer rains may occur every day. Although images were acquired in the same season in this study, annual differences in tree phenology provided additional information and enhanced the classification accuracy.

The utilization of multitemporal data introduces some challenges to the data processing and classification processes. As shown in Figure 6, there are small differences in tree positions due to tree growth and probably also due to geometric projection characteristics; thus, trees were delineated separately in each dataset. When using structural features, the use of different polygons in the same point cloud might affect the classifier. Furthermore, these variations are challenging when working with very high spatial resolution imagery. Ferreira et al. [28] used resampled WorldView images at 0.30 and 1.2 m and needed to adjust the polygons of each ITC. Special attention must be paid to the radiometric processing of multitemporal spectral datasets. In this study, the datasets from each year were first processed to ensure that reflectance mosaics were uniform using the radiometric block adjustment, and further normalization of the shadows was shown to be advantageous.

Classification accuracies are always affected by the forest characteristics, the existence of several classes, and dataset characteristics, which should be considered for a reliable comparison of studies. Tuominen et al. [65] used multisource data to classify 26 different tree species of a Finnish forest into species and genus. They had more than 650 samples and achieved accuracies from 59.9% (when classifying tree species using the RF classifier and DN values of the shortwave infrared range) to 86.9% (when using selected features and the k-NN algorithm to classify the genus). Dalponte et al. [66] classified three types of trees in a boreal forest with more than 2300 samples and obtained an overall accuracy of 93.5% using manually delimited ITCs. Sothe et al. [14] used hyperspectral imagery and structural features to classify 12 tree species of a mixed ombrophilous forest, and achieved a maximum overall accuracy of 72.4%.

The number of samples affects the classification results and, thus, the analysis results, especially when using an unbalanced number of samples and statistics that consider the overall accuracy rather than the class accuracy. Therefore, the use of LOOCV followed by AUCROC analysis is extremely important because AUCROC values are specific to each class. In this study, the number of samples was quite low because of challenges in detecting the training data of a complex forest.

The delineation of the tree crown is equally important. When working with different successional stages and, thus, trees of potentially similar heights, correct spatial identification and tree crown delineation are crucial to the performance of any classifier. Since the region-based classification method provides better classification accuracy of tree species, future studies will be performed to fulfill this need. Approaches using ALS data have been successfully applied in boreal forests or pulpwood forests [67]. Wagner et al. [11] segmented the semideciduous Atlantic forest in a well-developed stage using imagery information only. Automating the delineation of the tree crown with acceptable accuracy would enhance the production of tree species maps to supplement forest monitoring.

5. Conclusions

The objective of this investigation was to develop hyperspectral unmanned aerial vehicle (UAV) imaging-based methods for tree species classification in an area of the Brazilian Atlantic forest that has great species diversity and a multitude of successional stages. An objective further was to assess the contribution of temporal spectral information to classification accuracy.

Temporal spectral information improved random forest performance for three of the eight tree species analyzed, indicating that promising accuracy could be obtained when using temporal spectral information. Separately analyzing single-date datasets showed that the weather patterns directly influenced the classification performance of some of the tree species. The analysis of datasets from several years of the same season showed that differences in weather conditions in different years resulted in some changes in the species spectra; these changes were useful for differentiating between tree species. The use of multitemporal spectra did not improve the identification of *Inga vera*, which had the highest area under the receiver operating characteristic curve (AUCROC) value when using only the dataset from 2018, and for *Copaifera langsdorffii*, which was better identified when using only the dataset from 2017. However, it is important to note that the AUCROC value for *Apuleia leiocarpa* resulting from the use of all the datasets (i.e., Dall_MeanNorm) might have been affected by the poor identification resulting from the use of the 2018 and 2019 datasets (AUCROC values equal 0.438 and 0.313, respectively). Weather conditions were observed to directly affect the tree species bloom because some species, such as *Apuleia leiocarpa*, bloom only in the dry season in trees that are completely without leaves. The normalization of spectra was necessary.

To the authors' knowledge, this is the first work to use hyperspectral UAV images acquired over several years to classify the highly diverse Atlantic Forest. Improvements should be applied regarding the number of samples per class and the automated segmentation of individual tree crowns to enhance the applicability of the methodology, in addition to the use of tree height information. Moreover, care should be taken when using very high spatial resolution and automatic tree crown segmentation because of the slightly different positions of tree leaves caused by tree development.

Author Contributions: Conceptualization, G.T.M., N.N.I., and A.M.G.T.; data collection, G.T.M., N.N.I., A.M.G.T., M.V.A.d.M.; methodology, G.T.M., N.N.I., A.M.G.T., M.V.A.d.M., E.H.; writing—original draft preparation, G.T.M.; writing—review and editing, G.T.M., N.N.I., A.M.G.T., M.V.A.d.M., E.H.; supervision: N.N.I.; funding acquisition, N.N.I., A.M.G.T., E.H. All authors have read and agreed to the published version of the manuscript.

Funding: This study was financed in part by the Coordenação de Aperfeiçoamento de Pessoal de Nível Superior—Brasil (CAPES)—Finance Code 001 (process number 88881.187406/2018-01); in part by the National Council for Scientific and Technological Development (CNPq), grant number 153854/2016-2; in part by the São Paulo Research Foundation (FAPESP), grant number 2013/50426-4; and in part by the Academy of Finland, grant number 327861.

Acknowledgments: The authors would like to thank the company Fototerra S.A., through the individual César Francisco de Paula, for providing the ALS point cloud from the study area, as well as Valter Ribeiro Campos and the members of the Photogrammetry group from São Paulo State University (UNESP) who helped us during the fieldwork.

Conflicts of Interest: The authors declare no conflict of interest.

References

1. Immitzer, M.; Atzberger, C.; Koukal, T. Tree Species Classification with Random Forest Using Very High Spatial Resolution 8-Band WorldView-2 Satellite Data. *Remote Sens.* **2012**, *4*, 2661–2693. [CrossRef]
2. Colgan, M.S.; Baldeck, C.A.; Féret, J.-B.; Asner, G.P. Mapping Savanna Tree Species at Ecosystem Scales Using Support Vector Machine Classification and BRDF Correction on Airborne Hyperspectral and LiDAR Data. *Remote Sens.* **2012**, *4*, 3462–3480. [CrossRef]
3. Heinzel, J.; Koch, B. Investigating multiple data sources for tree species classification in temperate forest and use for single tree delineation. *Int. J. Appl. Earth Obs. Geoinf.* **2012**, *18*, 101–110. [CrossRef]
4. Zhang, C.; Chen, K.; Liu, Y.; Kovacs, J.M.; Flores-Verdugo, F.; de Santiago, F.J.F. Spectral response to varying levels of leaf pigments collected from a degraded mangrove forest. *J. Appl. Remote Sens.* **2012**, *6*, 063501.
5. Melgani, F.; Bruzzone, L. Classification of hyperspectral remote sensing images with support vector machines. *IEEE Trans. Geosci. Remote Sens.* **2004**, *42*, 1778–1790. [CrossRef]
6. Breiman, L. Random forests. *Mach. Learn.* **2001**, *45*, 5–32. [CrossRef]
7. Li, D.; Ke, Y.; Gong, H.; Li, X. Object-Based Urban Tree Species Classification Using Bi-Temporal WorldView-2 and WorldView-3 Images. *Remote Sens.* **2015**, *7*, 16917–16937. [CrossRef]
8. Maschler, J.; Atzberger, C.; Immitzer, M. Individual Tree Crown Segmentation and Classification of 13 Tree Species Using Airborne Hyperspectral Data. *Remote Sens.* **2018**, *10*, 1218. [CrossRef]
9. Matsuki, T.; Yokoya, N.; Iwasaki, A. Hyperspectral Tree Species Classification of Japanese Complex Mixed Forest with the Aid of Lidar Data. *IEEE J. Sel. Top. Appl. Earth Obs. Remote Sens.* **2015**, *8*, 2177–2187. [CrossRef]
10. Ferreira, M.P.; Zortea, M.; Zanotta, D.C.; Shimabukuro, Y.E.; de Souza Filho, C.R. Mapping tree species in tropical seasonal semi-deciduous forests with hyperspectral and multispectral data. *Remote Sens. Environ.* **2016**, *179*, 66–78. [CrossRef]
11. Wagner, F.H.; Ferreira, M.P.; Sanchez, A.; Hirye, M.C.M.; Zortea, M.; Gloor, E.; Phillips, O.L.; Filho, C.R.; de Souza Filo, C.; Shimabukuro, Y.E.; et al. Individual tree crown delineation in a highly diverse tropical forest using very high resolution satellite images. *ISPRS J. Photogramm. Remote Sens.* **2018**, *145*, 362–377. [CrossRef]
12. Feret, J.; Asner, G.P. Tree Species Discrimination in Tropical Forests Using Airborne Imaging Spectroscopy. *IEEE Trans. Geosci. Remote Sens.* **2013**, *51*, 73–84. [CrossRef]
13. Nevalainen, O.; Honkavaara, E.; Tuominen, S.; Viljanen, N.; Hakala, T.; Yu, X.; Hyyppä, J.; Saari, H.; Pölönen, I.; Imai, N.N.; et al. Individual Tree Detection and Classification with UAV-Based Photogrammetric Point Clouds and Hyperspectral Imaging. *Remote Sens.* **2017**, *9*, 185. [CrossRef]
14. Sothe, C.; Dalponte, M.; de Almeida, C.M.; Schimalski, M.B.; Lima, C.L.; Liesenberg, V.; Miyoshi, G.T.; Tommaselli, A.M.G. Tree Species Classification in a Highly Diverse Subtropical Forest Integrating UAV-Based Photogrammetric Point Cloud and Hyperspectral Data. *Remote Sens.* **2019**, *11*, 1338. [CrossRef]
15. Otero, V.; Kerchove, R.V.D.; Satyanarayana, B.; Martínez-Espinosa, C.; Fisol, M.A.B.; Ibrahim, M.R.B.; Sulong, I.; Mohd-Lokman, H.; Lucas, R.; Dahdouh-Guebas, F. Managing mangrove forests from the sky: Forest inventory using field data and Unmanned Aerial Vehicle (UAV) imagery in the Matang Mangrove Forest Reserve, peninsular Malaysia. *For. Ecol. Manag.* **2018**, *411*, 35–45. [CrossRef]
16. Nuvem UAV Batmap. Available online: http://nuvemuav.com/batmap (accessed on 5 December 2019).
17. Aasen, H.; Honkavaara, E.; Lucieer, A.; Zarco-Tejada, P.J. Quantitative Remote Sensing at Ultra-High Resolution with UAV Spectroscopy: A Review of Sensor Technology, Measurement Procedures, and Data Correction Workflows. *Remote Sens.* **2018**, *10*, 1091. [CrossRef]
18. Sanchez-Azofeifa, A.; Antonio Guzmán, J.; Campos, C.A.; Castro, S.; Garcia-Millan, V.; Nightingale, J.; Rankine, C. Twenty-first century remote sensing technologies are revolutionizing the study of tropical forests. *Biotropica* **2017**, *49*, 604–619. [CrossRef]

19. Colomina, I.; Molina, P. Unmanned aerial systems for photogrammetry and remote sensing: A review. *ISPRS J. Photogramm. Remote Sens.* **2014**, *92*, 79–97. [CrossRef]
20. Paneque-Gálvez, J.; McCall, M.K.; Napoletano, B.M.; Wich, S.A.; Koh, L.P. Small Drones for Community-Based Forest Monitoring: An Assessment of Their Feasibility and Potential in Tropical Areas. *Forests* **2014**, *5*, 1481–1507. [CrossRef]
21. Cubert UAV Mapping with FireflEYE—Cubert. Available online: https://cubert-gmbh.com/applications/uav-mapping-with-firefleye/ (accessed on 25 October 2019).
22. Senop Datasheet. Available online: https://senop.fi/files/rikola/pdf/Hyperspectral+Camera_Datasheet.pdf (accessed on 25 October 2019).
23. Key, T.; Warner, T.A.; McGraw, J.B.; Fajvan, M.A. A Comparison of Multispectral and Multitemporal Information in High Spatial Resolution Imagery for Classification of Individual Tree Species in a Temperate Hardwood Forest. *Remote Sens. Environ.* **2001**, *75*, 100–112. [CrossRef]
24. Somers, B.; Asner, G.P. Tree species mapping in tropical forests using multi-temporal imaging spectroscopy: Wavelength adaptive spectral mixture analysis. *Int. J. Appl. Earth Obs. Geoinf.* **2014**, *31*, 57–66. [CrossRef]
25. Van Deventer, H.; Cho, M.A.; Mutanga, O. Improving the classification of six evergreen subtropical tree species with multi-season data from leaf spectra simulated to WorldView-2 and RapidEye. *Int. J. Remote Sens.* **2017**, *38*, 4804–4830. [CrossRef]
26. Hill, R.A.; Wilson, A.K.; George, M.; Hinsley, S.A. Mapping tree species in temperate deciduous woodland using time-series multi-spectral data. *Appl. Veg. Sci.* **2010**, *13*, 86–99. [CrossRef]
27. Castro-Esau, K.L.; Sánchez-Azofeifa, G.A.; Rivard, B.; Wright, S.J.; Quesada, M. Variability in leaf optical properties of Mesoamerican trees and the potential for species classification. *Am. J. Bot.* **2006**, *93*, 517–530. [CrossRef]
28. Ferreira, M.P.; Wagner, F.H.; Aragão, L.E.O.C.; Shimabukuro, Y.E.; de Filho, C.R.S. Tree species classification in tropical forests using visible to shortwave infrared WorldView-3 images and texture analysis. *ISPRS J. Photogramm. Remote Sens.* **2019**, *149*, 119–131. [CrossRef]
29. Karasiak, N.; Dejoux, J.-F.; Fauvel, M.; Willm, J.; Monteil, C.; Sheeren, D. Statistical Stability and Spatial Instability in Mapping Forest Tree Species by Comparing 9 Years of Satellite Image Time Series. *Remote Sens.* **2019**, *11*, 2512. [CrossRef]
30. Immitzer, M.; Neuwirth, M.; Böck, S.; Brenner, H.; Vuolo, F.; Atzberger, C. Optimal Input Features for Tree Species Classification in Central Europe Based on Multi-Temporal Sentinel-2 Data. *Remote Sens.* **2019**, *11*, 2599. [CrossRef]
31. Fassnacht, F.E.; Latifi, H.; Stereńczak, K.; Modzelewska, A.; Lefsky, M.; Waser, L.T.; Straub, C.; Ghosh, A. Review of studies on tree species classification from remotely sensed data. *Remote Sens. Environ.* **2016**, *186*, 64–87. [CrossRef]
32. Available online: http://www.planalto.gov.br/ccivil_03/_Ato2004-2006/2006/Lei/L11428.htm (accessed on 10 September 2019).
33. Ribeiro, M.C.; Metzger, J.P.; Martensen, A.C.; Ponzoni, F.J.; Hirota, M.M. The Brazilian Atlantic Forest: How much is left, and how is the remaining forest distributed? Implications for conservation. *Biol. Conserv.* **2009**, *142*, 1141–1153. [CrossRef]
34. Lira, P.K.; Tambosi, L.R.; Ewers, R.M.; Metzger, J.P. Land-use and land-cover change in Atlantic Forest landscapes. *For. Ecol. Manag.* **2012**, *278*, 80–89. [CrossRef]
35. Ibge Manual técnico da vegetação brasileira. Available online: https://biblioteca.ibge.gov.br/visualizacao/livros/liv63011.pdf (accessed on 25 May 2019).
36. Alvares, C.A.; Stape, J.L.; Sentelhas, P.C.; de Moraes Gonçalves, J.L.; Sparovek, G. Köppen's climate classification map for Brazil. *Meteorol. Z.* **2013**, *22*, 711–728. [CrossRef]
37. INMET—Instituto Nacional de Meteorologia Estações Automáticas—Gráficos. Available online: http://www.inmet.gov.br/portal/index.php?r=home/page&page=rede_estacoes_auto_graf (accessed on 8 November 2019).
38. Berveglieri, A.; Tommaselli, A.M.G.; Imai, N.N.; Ribeiro, E.A.W.; Guimarães, R.B.; Honkavaara, E. Identification of Successional Stages and Cover Changes of Tropical Forest Based on Digital Surface Model Analysis. *IEEE J. Sel. Top. Appl. Earth Obs. Remote Sens.* **2016**, *9*, 5385–5397. [CrossRef]
39. Berveglieri, A.; Imai, N.N.; Tommaselli, A.M.G.; Casagrande, B.; Honkavaara, E. Successional stages and their evolution in tropical forests using multi-temporal photogrammetric surface models and superpixels. *ISPRS J. Photogramm. Remote Sens.* **2018**, *146*, 548–558. [CrossRef]

40. Da Silva, F.R.; Begnini, R.M.; Lopes, B.C.; Castellani, T.T. Seed dispersal and predation in the palm Syagrus romanzoffiana on two islands with different faunal richness, southern Brazil. *Stud. Neotropical Fauna Environ.* **2011**, *46*, 163–171. [CrossRef]
41. Lorenzi, H. Árvores brasileiras. In *Plant. Nova Odessa*, 1st ed.; Editora Plantarum: Nova Odessa, SP, Brazil, 1992; Volumes 1–2.
42. Lorenzi, H. Árvores brasileiras. In *Plant. Nova Odessa*, 2nd ed.; Editora Plantarum: Nova Odessa, SP, Brazil, 1992; Volume 1.
43. Da Scaranello, M.A.S.; Alves, L.F.; Vieira, S.A.; de Camargo, P.B.; Joly, C.A.; Martinelli, L.A. Height-diameter relationships of tropical Atlantic moist forest trees in southeastern Brazil. *Sci. Agric.* **2012**, *69*, 26–37. [CrossRef]
44. Lima, R.B.; Bufalino, L.; Alves Junior, F.T.; da Silva, J.A.A.; Ferreira, R.L.C. Diameter distribution in a Brazilian tropical dry forest domain: Predictions for the stand and species. *Anais Acad. Bras. Ciênc.* **2017**, *89*, 1189–1203. [CrossRef]
45. D'Oliveira, M.V.N.; Alvarado, E.C.; Santos, J.C.; Carvalho, J.A. Forest natural regeneration and biomass production after slash and burn in a seasonally dry forest in the Southern Brazilian Amazon. *For. Ecol. Manag.* **2011**, *261*, 1490–1498. [CrossRef]
46. Miyoshi, G.; Imai, N.; Tommaselli, A.; Honkavaara, E. Comparison of Pixel and Region-Based Approaches for Tree Species Mapping in Atlantic Forest Using Hyperspectral Images Acquired by Uav. *ISPRS-Int. Arch. Photogramm. Remote Sens. Spat. Inf. Sci.* **2019**, *4213*, 1875–1880. [CrossRef]
47. Miyoshi, G.T.; Imai, N.N.; Tommaselli, A.M.G.; Honkavaara, E.; Näsi, R.; Moriya, É.A.S. Radiometric block adjustment of hyperspectral image blocks in the Brazilian environment. *Int. J. Remote Sens.* **2018**, *39*, 4910–4930. [CrossRef]
48. De Oliveira, R.A.; Tommaselli, A.M.G.; Honkavaara, E. Geometric Calibration of a Hyperspectral Frame Camera. *Photogramm. Rec.* **2016**, *31*, 325–347. [CrossRef]
49. Honkavaara, E.; Saari, H.; Kaivosoja, J.; Pölönen, I.; Hakala, T.; Litkey, P.; Mäkynen, J.; Pesonen, L. Processing and Assessment of Spectrometric, Stereoscopic Imagery Collected Using a Lightweight UAV Spectral Camera for Precision Agriculture. *Remote Sens.* **2013**, *5*, 5006–5039. [CrossRef]
50. Honkavaara, E.; Rosnell, T.; Oliveira, R.; Tommaselli, A. Band registration of tuneable frame format hyperspectral UAV imagers in complex scenes. *ISPRS J. Photogramm. Remote Sens.* **2017**, *134*, 96–109. [CrossRef]
51. Smith, G.M.; Milton, E.J. The use of the empirical line method to calibrate remotely sensed data to reflectance. *Int. J. Remote Sens.* **1999**, *20*, 2653–2662. [CrossRef]
52. Isenburg, M. Available online: http://lastools.org/ (accessed on 21 November 2018).
53. Dalponte, M.; Frizzera, L.; Gianelle, D. Individual tree crown delineation and tree species classification with hyperspectral and LiDAR data. *PeerJ* **2019**, *6*, e6227. [CrossRef] [PubMed]
54. Belgiu, M.; Drăguţ, L. Random forest in remote sensing: A review of applications and future directions. *ISPRS J. Photogramm. Remote Sens.* **2016**, *114*, 24–31. [CrossRef]
55. Hall, M.; Frank, E.; Holmes, G.; Pfahringer, B.; Reutemann, P.; Witten, I.H. The WEKA data mining software: An update. *SIGKDD Explor.* **2009**, *11*, 10–18. [CrossRef]
56. Brovelli, M.A.; Crespi, M.; Fratarcangeli, F.; Giannone, F.; Realini, E. Accuracy assessment of high resolution satellite imagery orientation by leave-one-out method. *ISPRS J. Photogramm. Remote Sens.* **2008**, *63*, 427–440. [CrossRef]
57. Fan, J.; Upadhye, S.; Worster, A. Understanding receiver operating characteristic (ROC) curves. *Can. J. Emerg. Med.* **2006**, *8*, 19–20. [CrossRef]
58. Bradley, A.P.; Duin, R.P.W.; Paclik, P.; Landgrebe, T.C.W.; Bradley, A.P.; Duin, R.P.W.; Paclik, P.; Landgrebe, T.C.W. Precision-recall operating characteristic (P-ROC) curves in imprecise environments. In Proceedings of the 18th International Conference on Pattern Recognition (ICPR'06), Hong Kong, China, 20–24 August 2006; Volume 4, pp. 123–127.
59. Evangelista, P.H.; Stohlgren, T.J.; Morisette, J.T.; Kumar, S. Mapping Invasive Tamarisk (Tamarix): A Comparison of Single-Scene and Time-Series Analyses of Remotely Sensed Data. *Remote Sens.* **2009**, *1*, 519–533. [CrossRef]
60. Fawcett, T. An introduction to ROC analysis. *Pattern Recognit. Lett.* **2006**, *27*, 861–874. [CrossRef]

61. Li, W.; Guo, Q.; Jakubowski, M.K.; Kelly, M. A New Method for Segmenting Individual Trees from the Lidar Point Cloud. *Photogramm. Eng. Remote Sens.* **2012**, *78*, 75–84. [CrossRef]
62. Witten, I.H.; Frank, E. *Data Mining: Practical Machine Learning Tools and Techniques*, 2nd ed.; Morgan Kaufmann Series in Data Management Systems; Morgan Kaufmann: San Francisco, CA, USA, 2005.
63. Michez, A.; Piégay, H.; Lisein, J.; Claessens, H.; Lejeune, P. Classification of riparian forest species and health condition using multi-temporal and hyperspatial imagery from unmanned aerial system. *Environ. Monit. Assess.* **2016**, *188*, 146. [CrossRef] [PubMed]
64. Clark, M.L.; Roberts, D.A.; Clark, D.B. Hyperspectral discrimination of tropical rain forest tree species at leaf to crown scales. *Remote Sens. Environ.* **2005**, *96*, 375–398. [CrossRef]
65. Tuominen, S.; Näsi, R.; Honkavaara, E.; Balazs, A.; Hakala, T.; Viljanen, N.; Pölönen, I.; Saari, H.; Ojanen, H. Assessment of Classifiers and Remote Sensing Features of Hyperspectral Imagery and Stereo-Photogrammetric Point Clouds for Recognition of Tree Species in a Forest Area of High Species Diversity. *Remote Sens.* **2018**, *10*, 714. [CrossRef]
66. Dalponte, M.; Ørka, H.O.; Ene, L.T.; Gobakken, T.; Næsset, E. Tree crown delineation and tree species classification in boreal forests using hyperspectral and ALS data. *Remote Sens. Environ.* **2014**, *140*, 306–317. [CrossRef]
67. Kaartinen, H.; Hyyppä, J.; Yu, X.; Vastaranta, M.; Hyyppä, H.; Kukko, A.; Holopainen, M.; Heipke, C.; Hirschmugl, M.; Morsdorf, F.; et al. An International Comparison of Individual Tree Detection and Extraction Using Airborne Laser Scanning. *Remote Sens.* **2012**, *4*, 950–974. [CrossRef]

remote sensing

MDPI

Article

Machine Learning Classification of Endangered Tree Species in a Tropical Submontane Forest Using WorldView-2 Multispectral Satellite Imagery and Imbalanced Dataset

Colbert M. Jackson * and Elhadi Adam

School of Geography, Archaeology and Environmental Studies, University of the Witwatersrand, Johannesburg 2050, South Africa; elhadi.adam@wits.ac.za
* Correspondence: 1615228@students.wits.ac.za; Tel.: +254-71-413-7957

Abstract: Accurate maps of the spatial distribution of tropical tree species provide valuable insights for ecologists and forest management. The discrimination of tree species for economic, ecological, and technical reasons is usually necessary for achieving promising results in tree species mapping. Most of the data used in tree species mapping normally have some degree of imbalance. This study aimed to assess the effects of imbalanced data in identifying and mapping trees species under threat in a selectively logged sub-montane heterogeneous tropical forest using random forest (RF) and support vector machine with radial basis function (RBF-SVM) kernel classifiers and WorldView-2 multispectral imagery. For comparison purposes, the original imbalanced dataset was standardized using three data sampling techniques: oversampling, undersampling, and combined oversampling and undersampling techniques in R. The combined oversampling and undersampling technique produced the best results: F1-scores of 68.56 ± 2.6% for RF and 64.64 ± 3.4% for SVM. The balanced dataset recorded improved classification accuracy compared to the original imbalanced dataset. This research observed that more separable classes recorded higher F1-scores. Among the species, *Syzygium guineense* and *Zanthoxylum gilletii* were the most accurately mapped whereas *Newtonia buchananii* was the least accurately mapped. The most important spectral bands with the ability to detect and distinguish between tree species as measured by random forest classifier, were the Red, Red Edge, Near Infrared 1, and Near Infrared 2.

Keywords: tropical forests; endangered tree species; selective logging; imbalanced data; pixel-based classification; machine learning algorithm

Citation: Jackson, C.M.; Adam, E. Machine Learning Classification of Endangered Tree Species in a Tropical Submontane Forest Using WorldView-2 Multispectral Satellite Imagery and Imbalanced Dataset. *Remote Sens.* **2021**, *13*, 4970. https://doi.org/10.3390/rs13244970

Academic Editors: Mutanga Immitzer, Onisimo Mutanga and Clement Atzberger

Received: 10 October 2021
Accepted: 3 December 2021
Published: 7 December 2021

1. Introduction

Tropical forests comprise woody, evergreen vegetation, cover 47% of the world's total forest area [1], and host the highest proportion of global tree diversity, i.e., with more than 53,000 tree species, compared to approximately 124 in temperate Europe [2]. The height of the tree crowns forming the forest canopy is in the range of 30 to 50 m, but emergent trees may attain heights of approximately 70 m. Tropical forests cover approximately 7% of the globe, and they are home to more than half of all earth's biodiversity [1]. Vital environmental processes such as the water cycle, soil conservation, carbon sequestration, and habitat protection are immensely regulated by tropical tree species [1], therefore, those forests maintain the ecosystem services and mitigate climate change [3]. Information on key parameters such as tree species, tree diameter, and height, crown size, and location are important for resource management, biodiversity assessment, ecosystem services assessment, and conservation [3]. Ecologists have long been interested in explaining species' distribution in ecosystems [2], as it can drive the exploitation and management policies of forests. In tropical forests, the maintenance of a canopy composed of emergent trees of fundamental species has been shown to provide conditions favorable to ecological processes, playing an essential role in the forest community's resilience and perpetuation [4].

Traditionally, detailed tree species identification is obtained in relatively small areas with time-consuming, high levels of manpower and the associated high costs and often operationally prohibitive field inventories [4,5]. Tropical rainforests are characterized by very high species richness and lack of access to some parts of the forests [5]. Therefore, obtaining exact information about the occurring species using field inventories is almost impossible [5], thus, they have difficulty resolving large geographic patterns. Remote sensing captures information over extensive areas in fine detail [6]. Plant species mapping with remote sensing is linked to an understanding that species have unique spectral signatures associated with characteristic biochemical and biophysical properties [7].

Several studies have assessed the potential of multispectral data in tree species mapping in tropical forests. The earliest study by Clark et al. [7] attempted a classification of seven species of emergent trees in a tropical rain forest in Costa Rica using hyperspectral digital collection experiment (HYDICE) imagery. Spectral-based species classification was performed using linear discriminant analysis (LDA), maximum likelihood (ML), and spectral angle mapper (SAM) classifiers applied to combinations of bands from a stepwise-selection procedure. Zhang et al. [8] selected five species to assess intra- and inter-class variability of tree species using a high spectral and spatial resolution imagery acquired using the airborne sensor HYDICE data. Trichon and Julien [9] used two sets of aerial photographs to identify tree species through air photo interpretation. Somers and Asner [10] performed hyperspectral time series analysis of two native and two invasive species in Hawaiian rainforests, using the separability index (SI). Féret and Asner [11] applied semi-supervised support vector machine classification using tensor summation kernel to identify individual crowns using imaging spectroscopy and light detection and ranging (LiDAR). Clark and Roberts [12] mapped seven tropical rainforest tree species using hyperspectral data. Narrowband indices, derivative- and absorption-based techniques, and spectral mixture analysis were used to derive metrics that respond to vegetation chemistry and structure. The random forest (RF) classifier was used to discriminate species with minimally-correlated, importance-ranked metrics. Papeş et al. [13] used Earth Observing-1 Hyperion satellite hyperspectral imagery to spectrally discriminate between crowns of 42 individual trees of 5 taxa using linear discriminant analysis, and they also evaluated seasonal variation in species discriminations related to phenology. Féret and Asner [14] applied supervised classification to identify canopy species using airborne hyperspectral imagery acquired with the Carnegie Airborne Observatory-Alpha system. Singh et al. [15] mapped and characterized selected tree species using aerial data. To delineate individual tree crowns (ITCs) from very-high-resolution (VHR) aerial imagery and LiDAR data, the study used object-based image analysis (OBIA). Both maximum likelihood (ML) and spectral angle mapper (SAM) classifiers were applied to the aerial imagery. Other studies have combined hyperspectral and LiDAR sensors. For example, Baldeck et al. [16] used airborne imaging spectroscopy to identify individuals of three focal canopy tree species in a diverse tropical forest on Barro Colorado Island, Panama. The addition of co-registered LiDAR data further improved performance by identifying intra- and inter-canopy shadows that alter species signatures [16]. Ferreira et al. [17] used LDA, RF, and SVM classifiers on hyperspectral and multispectral data to discriminate and map tree species, at the pixel level. Simulated WorldView-3 data were used to assess the role of SWIR bands in species classification. A tree crown segmentation approach on the hyperspectral data was used to map tree species. Graves et al. [18] assessed the accuracy of a support vector machine (SVM) model with a highly imbalanced dataset using a hyperspectral image mosaic. Wagner et al. [3] applied automatic ITC delineation in a highly diverse tropical forest using WorldView-2 satellite images. Ferreira et al. [4] applied visible to shortwave infrared WorldView-3 images and texture analysis to classify tropical tree species in a semi-deciduous forest in different seasons.

Conventional multispectral sensors such as Landsat or MODIS lack both the spatial and spectral resolution to detect changes in tree species composition [5,6]. Other satellite-borne multispectral sensors such as IKONOS or QuickBird have a high spatial resolution but they lack the spectral resolution to map tree species in tropical forests [6]. High spatial resolution sensors are not suited for species-level mapping, but rather adequately suited for mapping fine targets, e.g., tree canopies or canopy gaps [6]. Air-borne cameras that provide the highest spatial resolution do not offer the spectral resolution required [5]. Moreover, aerial photos' wide field of view results in strong effects from bi-directional reflectance characteristics of most land cover types [5]. Therefore, the spectral signature of an object can differ significantly. Tropical forests consist of trees of different species and ages, growing close to each other, with their crowns intertwined, and this has made tree species mapping a challenge. Their mapping requires remote sensing systems that can provide high spatial resolution as well as high spectral resolution. Airborne hyperspectral and LiDAR sensors enable mapping at very fine scales, but the high cost related to hyperspectral and LiDAR data acquisition and processing has hindered their application in mapping tree species in tropical forests [6]. However, some of the inherent features of hyperspectral data such as the carotenoids and chlorophyll sensitive bands are preserved in WorldView-2 multispectral data [19]. Thus, the high-resolution multispectral sensor, Worldview-2, has shown great potential to generate the information required in the identification of tree species and canopy attributes in complex tropical forest environments [17].

Kenya's rainforest cover, mostly montane forests, are scattered patches that are being further degraded [20]. The intense growth of population around Mount Kenya Forest Reserve (MKFR) since the early 1970s has led to degradation of the indigenous forests due to illegal logging of important timber trees [21]. This greatly reduced plant diversity and the regenerative capacity of such tree species [21–24]. More demand than supply for wood in Kenya has led to over-exploitation of high valued tree species [25]. Examples of the targeted tree species are African pencil-cedar (*Juniperus procera*), Wild olive (*Olea europaea*), East African rosewood (*Hagenia abyssinica*), East African camphor (*Ocotea usambarensis*), red stinkwood (*Prunus africana*), East African newtonia (*Newtonia buchananii*), East African yellow-wood (*Podocarpus spp*), East African olive (*Olea capensis*), Meru oak (*Vitex keniensis*), Peacock flower (*Albizzia gummifera*), and others [21–24].

In mapping tree species in closed-canopy tropical forests, the performance of classification algorithms may be affected by class imbalance [26]. Both deliberative and purely random sampling may attract instances of imbalanced data [26]. In simple random sampling, the chance of choosing a class is related to the areal coverage of the class, thus relatively rare classes will consist of smaller proportions of the training set. Therefore, this study aims to assess the effect of imbalanced data in identifying and mapping trees species under threat in a selectively logged sub-montane heterogeneous tropical forest using RF and SVM classifiers and WorldView-2 multispectral imagery. In assessing the significance of imbalanced data in mapping endangered tree species in the study site, different training class sizes comprising imbalanced and balanced datasets were used. To standardize the original imbalanced data, the oversampling, undersampling, and combined oversampling and undersampling techniques were applied. In addition,, the explanatory power of the WorldView-2 spectral bands in discriminating the tree species was evaluated. The maps showing the spatial distribution and abundance of the endangered tree species in the study area will form the basis by which efforts can be made for the restoration of the tree species, among other ecological applications.

2. Materials and Methods

2.1. Study Area

This total study area is 130 ha in MKFR, located between 0°21′5″ S t 0°20′5″ S latitude and 37°31′18″ E to 37°32′43″ E longitude (Figure 1). Mount Kenya is an ice-capped mountain and the second-highest in Africa, at approximately 5199 m asl. It is an extinct strato-volcano developed in the Late Pliocene–Quaternary periods [27]. The mountain is located on the Equator in Kenya, East Africa. The climate of the region is characterized by large diurnal temperature fluctuations and small monthly disparities. The rainfall pattern comprises short rains from October to November and long rains from March to

June [21,27]. The mean annual rainfall values are 1015 mm at the foothills to over 2000 mm in the montane forest and declining to 1015 mm per year in the alpine zone [28]. The annual-mean maximum temperatures are 26 °C at the foothills, decreasing to 2 °C at the nival zone [21].

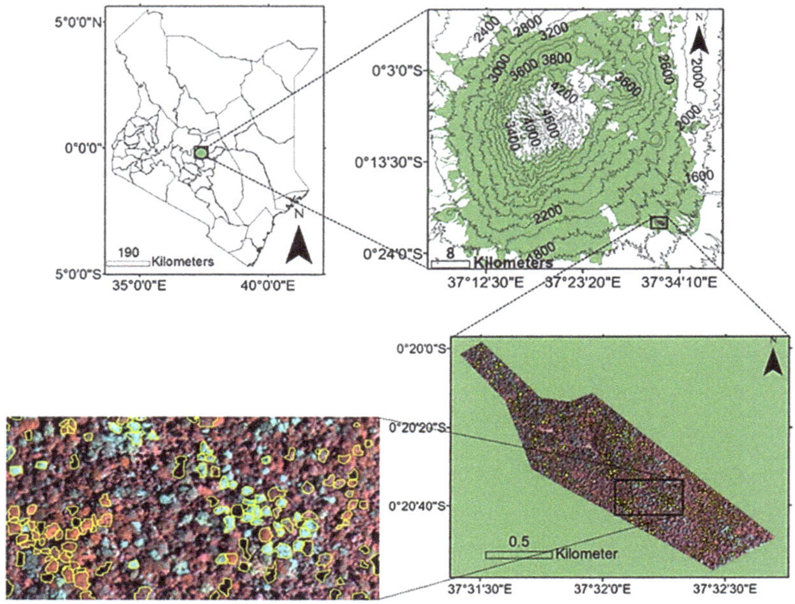

Figure 1. WorldView-2 false-color composite (near-infrared 2: yellow: coastal) of the study area showing tree crowns identified to the species level and other forest landscape features in the field.

Mount Kenya's abrupt changes in altitude within short distances result in a variety of plant species in a quite small area [21]. MKFR's highly heterogeneous canopy consists of deciduous and evergreen tree species. Anthropogenic activities determine vegetation types and their distribution at the lower altitudes [21,29]. The montane forest zone lies between 1980 and 3000 m asl on the western, eastern, and southern sides of the mountain [30]. The northern side is covered by grass, scattered trees, and Ericaceae and Protea scrub [30]. Nyayo Tea Zone, established by Legal Notice No. 265 of 1986 provides a buffer zone to check against human encroachment into MKFR.

2.2. Acquisition and Pre-Processing of WorldView-2 Satellite Data

The study site was covered by one WorldView-2 scene, acquired on 30 January 2019. The image utilized in this study was pre-processed and orthorectified by the image distributor [31]. It was geo-referenced to World Geodetic System (WGS) 1984 datum and the Universal Transverse Mercator (UTM) zone 37S projection. Theoretically, the radiometric correction and calibration of image data are necessary for detailed feature extraction. Atmospheric effects can complicate the spectral separability between landscape features with fine spectral differences [32]. Therefore, for optimal feature extraction, the WorldView-2 image was atmospherically calibrated by converting the digital numbers (DN) to the top-of-atmosphere reflectance. The conversion was done using the ENVI module (ENVI 5.3) FLAASH.

2.3. Field Data Collection

Because a published inventory of threatened plant taxa for Kenya does not exist, most botanists discover a rising number of species requiring special protection [33]. Different sources have tried to classify tree species considered to be threatened or endangered. For example, a State of the Environment report by the National Environment Management Authority (NEMA) terms East African camphor (*Ocotea usambarensis*), Red stinkwood (*Prunus Africana*), African satinwood (*Zanthoxylum gilletii*), East African sandalwood (*Osyris lanceolata*), and Meru oak (*Vitex keniensis*) as endangered [20]. Ng'eno [24] listed *Prunus africana* as an endangered tree species in Kenya, and *Podocarpus spp.*, *Vetex keniensis*, *Newtonia buchananii*, and *Albizia gummifera* are threatened species. *Hagenia abyssinica*, *Juniperus procera*, and *Ocotea usambaraensis* are listed as restricted tree species. According to KWS [21], threatened plant species in Kenya include *Prunus africana*, *Vitex keniensis*, *Ocotea usambarensis*, among others. All these tree species are commercial indigenous tree species of Kenya [34]. Therefore, field data collection involved the following endangered tree species: *Prunus Africana*, *Zanthoxylum gilletii*, *Albizia gummifera*, and *Newtonia buchananii*. Other species covered in this study were *Anthocleista grandiflora*, *Syzygium guineense*, and *Macaranga kilimandscharica*; they are not currently under any threat, however, an increase in the volume of extraction could endanger them. Rampant selective logging targeting especially *Ocotea usambarensis* has made the species to be very rare in the study site, and only eight samples were identified in the field. Other landscape classes, i.e., other woody vegetation and shadow, were also sampled. Other woody vegetation class consists of all species with very few samples, brought together to form a single mixed-species class. This allowed for their inclusion in the model so that they could be mapped, although individual species distinctions could not be made.

The ground truth points were collected in January and February 2020. Due to very steep mountainous terrain and dense forest cover, only a few areas were accessible, therefore, the field campaign randomly selected tree species along trails traversing the study area. The field measurement involved identification of tree species, tree height, diameter at breast height (DBH), tree dominance, and size, which varied among stands. A handheld global positioning system (Garmin eTrex® 20 GPS Receiver) and an RGB false-color composite (Near Infrared 2: Yellow: Coastal) of WorldView-2 image (pixel = 1.89 m), aided in locating tree crowns in the field. Using a GIS (ArcGIS v. 10.3®, ESRI, Redlands, CA, USA), points were set on all relevant crown pixels on the WorldView-2 imagery. By following the edges of the pixels, the points were made into polygons. A lot of care was taken to ensure that the pixels extracted were as pure as possible, without contamination from lianas or neighboring trees [7]. Any additional valuable information, such as tree flowering and leaves from identified crowns, was noted. Tree species information was assigned to each tree crown. The collected field data resulted in an imbalanced dataset, i.e., with common species having many samples and less common species containing few samples (Table 1). The dataset was randomly partitioned into 70% for training and 30% for testing.

Table 1. List of tree species, codes, family, leafy phenology, diameter at breast height (DBH), number of individual tree crowns, and composition of training and test data for each species used to map trees in Mt. Kenya Forest Reserve.

Botanical Name	Code	Family	Leaf Phenology	DBH (m)	Train Data	Test Data	Total
Macaranga k.	MK	Euphorbiaceae	Semi-deciduous	0.30–0.57	56	24	80
Zanthoxylum g.	ZG	Rutaceae	Semi-deciduous	0.97–3.00	56	24	80
Syzygium g.	SG	Myrtaceae	Evergreen	1.00–2.50	56	24	80
Newtonia b.	NB	Fabaceae	Deciduous	1.50–4.50	56	24	80
Anthocleista g.	AnG	Gentianaceae	Evergreen	1.01–2.50	36	15	51
Prunus a.	PA	Rosaceae	Evergreen	1.24–3.50	22	10	32
Albizzia g.	AlG	Fabaceae	Deciduous	1.37–4.00	20	8	28
Other woody vegetation	OWV	—	—	—	56	24	80
Shadow	SD	—	—	—	56	24	80

Shown in Figure 2 are samples of the tree species in the study area.

Figure 2. Samples of endangered tree species in the study area.

2.4. Spectral Separability

On-screen digitizing of samples from the display of the WorldView-2 false-color composite was implemented to generate the respective signature files. An assessment of the signatures was done to examine the spectral properties of the training sample classes and their separability over others. The spectral information was extracted using the central pixel within the crown polygon. Jeffries–Matusita (JM) distance was chosen as a separability measure. The JM distance among the distributions of two classes ω_i and ω_j has been defined as follows [35]:

$$JM_{ij} = 2\left(1 - e^{-B_{ij}}\right), \tag{1}$$

where B_{ij} is the Bhattacharyya distance calculated as [36]:

$$B_{ij} = \frac{1}{8}(\mu_i - \mu_j)^T \left(\frac{\Sigma_i + \Sigma_j}{2}\right)^{-1} (\mu_i - \mu_j) + \frac{1}{2} In \left(\frac{1}{2} \frac{\left|\Sigma_i + \Sigma_j\right|}{\sqrt{\left|\Sigma_i + \Sigma_j\right|}} \right) \tag{2}$$

where μ_i and μ_j are the mean reflectances of species i and j, and Σ_i and Σ_j correspond to their covariance matrices, whereas $|\Sigma_i|$ and $|\Sigma_j|$ are the determinants of Σ_i and Σ_j, respectively. ln is the natural logarithm function, and T is the transposition function.

The JM distance was calculated for all pairwise combinations between the mean spectral reflectance of the samples of a species and those of all other species, one at a time. The VNIR (450–1040 nm) WorldView-2 bands were used for computing the JM distance.

2.5. Training of Random Forest and Support Vector Machine Classifiers

The RF and SVM algorithms were applied using RStoolbox in R software in a pixel-based classification setting approach. The RF algorithm groups decision trees and the splitting variables are randomly chosen feature subsets with bagged samples [37]. Normally, the number of decision trees (*ntree*) in the ensemble and the number of predictor variables (*mtry*) randomly selected at each node need to be defined before applying the RF model. To extract the best performing parameters to be utilized in training the algorithm, a 10-fold grid search method was applied. RF brings together many weak learners, i.e., decision trees, into a stronger predictor by aggregating the predictions from all decision trees [37]. The majority 'vote' of all the trees is used to allocate a final class for unknown features. The mean decrease in accuracy (MDA) was used to evaluate the explanatory power of the input variables [37]. The MDA shows how much accuracy the model loses by excluding each variable [37]. The higher the MDA value, the more important the variable in the model.

The SVM finds the position of the optimal separating hyperplane, i.e., decision boundary, which meets two ultimate goals at a go: (i) separate the original data while maximizing the margin between classes, and (ii) minimize the misclassification error [38]. In the present study, the radial basis function (RBF) was selected as the kernel function because it has fewer parameter values to predefine, and it is as robust as other kernel types [26]. The success of an SVM model depends on the C (penalty) and γ (Gamma) parameters in the kernel function. The goal is to identify the best C and γ for the tree species mapping problem so that the classifier can accurately predict unknown data [39].

2.6. Class Imbalance

A dataset's imbalanced class distribution is characterized by cases where some classes have more samples than others. Most classifier learning algorithms that assume a fairly balanced distribution have found it challenging [26,40]. Previous studies have shown that the *k*-NN (*k*-nearest neighbor), RF, and SVM algorithms are affected by imbalanced training data [26].

To evaluate the impact of training data imbalance in mapping endangered tree species, this study applied data sampling techniques that aimed to modify the imbalanced dataset into balanced distribution, i.e., by altering the overall number of samples used in the training [26]. In choosing the best model in the classification of tree species in the study area, four datasets were compared: category 1 was composed of the original imbalanced dataset (Table 1) from the field. In category 2, the training data for all classes were randomly oversampled to match that of the majority classes, i.e., 56 data points. Category 3 consisted of the same number of samples for all classes, which was attained by randomly undersampling the training data for all classes to match that of the smallest class, i.e., 20. Finally, category 4 comprised a dataset created by combining both downsampling and oversampling techniques. First, the oversampling technique is applied to create duplicate and artificial data points, then the undersampling technique was used to eradicate noise data points, thus creating a robust balanced dataset suitable for model training. Therefore, the last three categories are subsets of the original field acquired dataset.

In all cases, the datasets were split into test and training data before running the models. Running models before splitting datasets can allow identical samples to be present in both the test and training data, leading to the models overfitting the training data. In all models, the 10-fold cross-validation method was repeated 10 times.

2.7. Measures of Model Performance

A multidimensional scaling (MDS) technique was applied in the analysis of (dis)similarities in the tree species datasets. MDS calculates a (dis)similarity matrix among pairs of tree species and then displays the data in a low-dimensional representation [41]. Therefore, MDS shows (dis)similarities among pairs of tree species as distances between points in a low-dimensional space. The Euclidean distance formula can be used to calculate distance between two points, e.g., i and j, as follows [42]:

$$d_{ij} = \sqrt{\sum_{k=1}^{p} \left(x_{ik} - x_{jk}\right)^2} \quad (3)$$

where p is the number of dimensions, d_{ij} is the distance, and x_{ik} is the data value of the *ith* row and *kth* column. The dissimilarity between points i and j is denoted d_{ij} and s_{ij} for similarity. Small d_{ij} values indicate points that are close together, thus, they belong to the same group and vice versa. The similarity values are the opposite, i.e., small s_{ij} values indicate points that are farther apart, hence they are not in the same group, and vice versa [43]. The goodness-of-fit statistic, the stress measure, which is based on the differences between the observed data and their predicted values, was used to express how well the datasets are represented by the model [42].

$$stress = \sqrt{\frac{\sum(d_{ij} - \hat{d}_{ij})^2}{\sum d_{ij}^{\,2}}} \quad (4)$$

where $\hat{d}_{ij}$ is the predicted distance based on the MDS model.

To fully evaluate model effectiveness, precision and recall performance metrics were used. Precision measures how accurate is the classifier's prediction of a class [18]. Low precision indicates a high number of false positives. Recall, also known as sensitivity, measures the classifier's ability to identify a class [18]. Low recall indicates a high number of false negatives. For an imbalanced dataset, F1-score is a more appropriate metric. It is the harmonic mean of the precision and recall, and the expression is [18]:

$$F1 - score = 2 \cdot \frac{Precision \cdot Recall}{Precision + Recall} \quad (5)$$

To evaluate the performance of RF and SVM algorithm classifiers on mapping the selected tree species using WorldView-2 data, 30% of the ground truth data were used to generate confusion matrices from which precision, recall, and F1-score were calculated. The overall, producer's, and user's accuracies were also calculated. The producer's accuracy (PA) is the ratio of the correctly detected trees to all the positive ground truth tree samples, whereas the user's accuracy (UA) is the ratio of the correctly detected trees to all the positive model-predicted tree samples [18].

The McNemar test, a non-parametric test focused on the binary distinction between correctly classified and misclassified class allocations of two classification outputs, was used to compare and indicate the statistical significance of any difference in results [44]:

$$Z = \frac{f_{12} - f_{21}}{\sqrt{f_{12} + f_{21}}} \quad (6)$$

where f_{12} is the total number of samples classified correctly by the first classification, but misclassified by the second, and f_{21} is the total samples classified correctly by the second classification but misclassified by the first one. A difference in accuracy between the confusion matrices of the different models is regarded statistically significant ($p \leq 0.05$) if $Z \geq 1.96$ (44).

3. Results

3.1. Spectral Separability between the Tree Species

The first step in mapping tree species is to figure out if the classes are spectrally different. Figure 3 shows the mean spectral signatures for the eight tree species in the study area. In the visible bands, *Syzygium guineense* has higher reflectance values compared to the other seven tree species which reflect almost the same. The Coastal band shows the highest reflectance values for all tree species. The difference in reflectance between *Syzygium guineense* and the other tree species keeps on increasing from the Coastal band towards the Red band.

The reflectance values in the Red Edge, Near Infrared 1, and Near Infrared 2 bands are generally higher for the deciduous and semi-deciduous tree species than for the evergreen tree species. *Zanthoxylum gilletii* and *Albizzia gummifera* showed the highest reflectance values and differ significantly from the others. Among the evergreen tree species, *Prunus africana* shows the highest reflectance values, followed by *Anthocleista grandiflora*. *Syzygium guineense* has notably low reflectance values. Among the three bands, Near Infrared 1 shows the highest reflectance values by all tree species. The Near Infrared 1 and Near Infrared 2 bands show the largest differences in reflectance between the seven tree species in the study area.

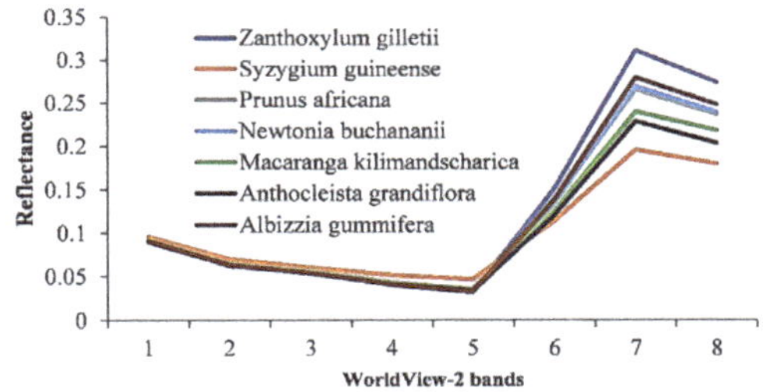

Figure 3. Tree species mean spectral signatures derived from the WorldView-2 data.

Figure 4 shows significant spectral overlaps between tree species. Band-specific within-species variance is evident, e.g., *Syzygium guineense* within-species variance is bigger in Coastal and Blue bands, and quite small in the Near Infrared 2. *Zanthoxylum gilletii* has the smallest within-species variance in the Red band, whereas the Coastal and Green bands show larger variances.

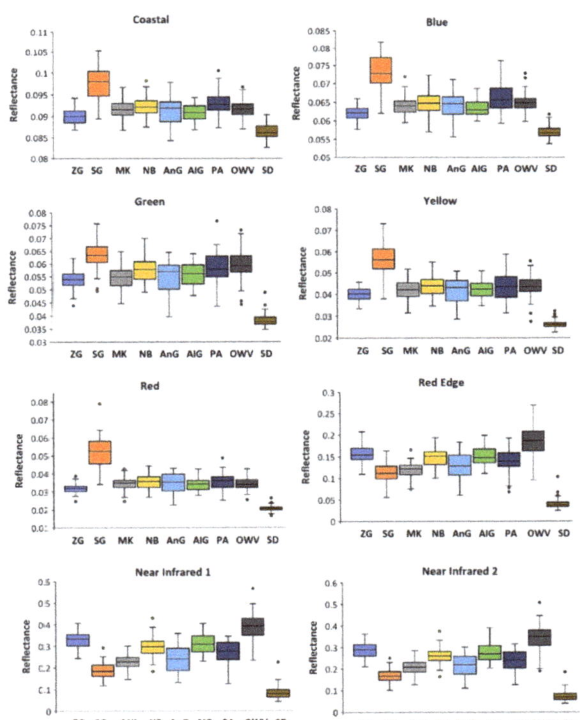

Figure 4. Box-whisker plots showing tree species mean reflectance values derived from WorldView-2 data. The central lines within each box represent the medians, while the top and bottom edges of the boxes are the upper and lower quartiles, respectively. The dots represent the outlier values within tree species. Full species names for each species code are given in Table 1.

Table 2 is a matrix showing the inter-specific spectral separability of the tree species calculated from the WorldView-2 VNIR bands using the JM distance. The highest JM distances were reported for *Syzygium guineense*, i.e., values greater than 1. The lowest values were 0.42 for separability between *Newtonia buchananii* and *Albizzia gummifera*, and

0.43 for *Newtonia buchananii* and *Anthocleista grandiflora*. The highest values are shown in bold.

Table 2. Tree species' inter-specific spectral separability as calculated by the Jeffries–Matusita distance (Equation (1)). The bold shows those species which are separable and those not separable (not in bold).

	SG	ZG	AnG	AlG	PA	NB	MK
SG		**1.37**	**1.21**	**1.23**	**1.27**	**1.25**	**1.16**
ZG			**1.01**	0.75	**1.05**	**0.89**	**1.22**
AnG				0.51	0.59	0.43	0.66
AlG					0.65	0.42	**0.87**
PA						0.58	**0.90**
NB							0.74
MK							

3.2. Optimization of Random Forest and Support Vector Machine

In all models, extracting the best performing *ntree* and *mtry* parameters to be used in training of the RF algorithm, repeated 10-fold cross-validation (CV) technique, dependent on the out-of-bag (OOB) error was applied. Using the same approach, the SVM parameters were optimized for the WorldView-2 dataset. The best parameters combinations under each of the four models are summarized in Table 3. As expected, the dataset produced by the combined oversampling and then downsampling techniques performed better than the others. For the R model, the default *mtry* value of 2 and *ntree* value of 3500 produced the lowest OOB error rate, 0.271 (Figure 5a). The SVM parameters, i.e., gamma and cost, were optimized for the WorldView-2 dataset, also using a 10-fold grid search approach. As seen in Figure 5b, a gamma value of 1 and a cost value of 100 yielded the best performance with a cross-validation error of 0.326. For the undersampled dataset, the best performance was an OOB error rate of 0.476 produced by *mtry* value of 2 and *ntree* value of 3500. The SVM yielded a gamma value of 0.01 and a cost value of 10, with a cross-validation error of 0.422.

Table 3. The RF and SVM model optimization parameters.

Model	Train Data per Species	RF			SVM		
		mtry	ntree	OOB Error	Gamma	Cost	CV Error
Original dataset	Refer to Table 1	3	1500	0.377	0.01	1000	0.387
Oversampling technique	56	4	4500	0.279	0.01	1000	0.327
Undersampling technique	20	2	3500	0.476	0.01	10	0.422
Combined technique	varied	2	3500	0.271	1	100	0.326

The sampling techniques were only applied to training data to avoid model overfitting.

3.3. Relative Importance of Variables

After running the classification models, the MDA provided the relative importance of each band (Figure 6). The most important spectral bands in all models, i.e., bands depicted by highest MDA were the Red, Red Edge, Near Infrared 1, and Near Infrared 2.

The classification models also evaluated the ability of the WorldView-2 spectral bands to detect the tree species. Figure 7 shows that the Red band was critical in the identification of especially *Syzygium guineense*, and also involved to a greater extent in mapping *Albizzia gummifera*, *Anthocleista grandiflora*, *Prunus africana*, and *Zanthoxylum gilletii*. The Near Infrared 1 band contributed significantly to the mapping of *Albizzia gummifera* and *Prunus africana*, as well as *Anthocleista grandiflora*, *Macaranga kilimandscharica*, *Newtonia buchananii*, and *Zanthoxylum gilletii*. The Red Edge band majorly contributed to the identification of *Macaranga kilimandscharica* and other woody vegetation, and, to a greater extent, shadow. The Near Infrared 2 band was helpful in the mapping of all tree species at different proportions, but it was very vital in the identification of *Albizzia gummifera*, *Anthocleista grandiflora*, and *Prunus Africana*.

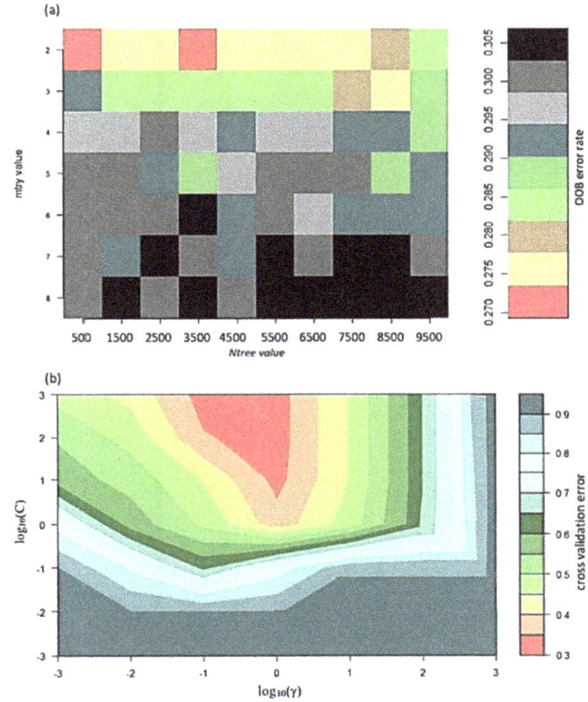

Figure 5. Optimization by 10—fold cross—validation: (**a**) random forest parameters (*mtry* and *ntree*), and (**b**) support vector machine parameters (gamma and cost).

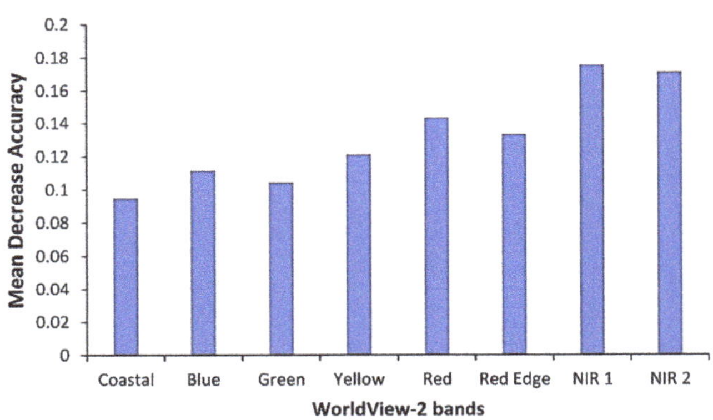

Figure 6. The relative importance of WorldView-2 bands in discriminating between different tree species as measured by RF classifier.

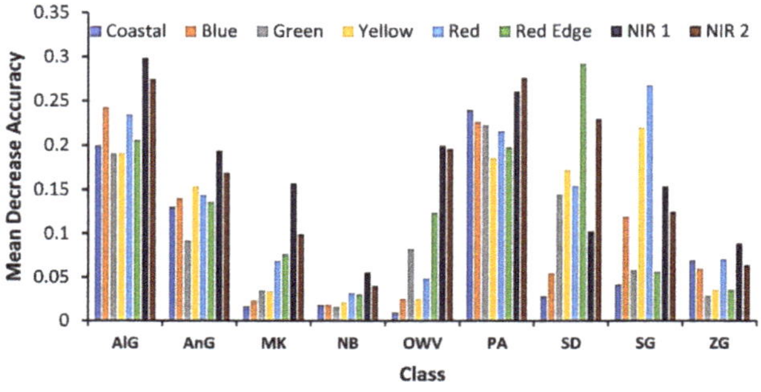

Figure 7. The mean decrease accuracy (MDA) values show the relationship between each tree species and WorldView-2 spectral bands as measured by the RF classifier. Full species names for each species code are given in Table 1.

3.4. Model Performance

Figure 8a shows RF multidimensional scaling (MDS) is a nice way to visualize (dis)similarity within and among tree species, using the proximity matrix calculated from the variables in the training dataset. SVM multidimensional scaling also visualizes (dis)similarity within and among tree species, using the test data, as seen in Figure 8b. The undersampled dataset performed poorly by attaining higher stress values, i.e., 2.41% for RF and 2.09% for SVM. The rest of the datasets performed fairly as their stress values were in the range of 1.41–1.89% for RF and SVM classifiers.

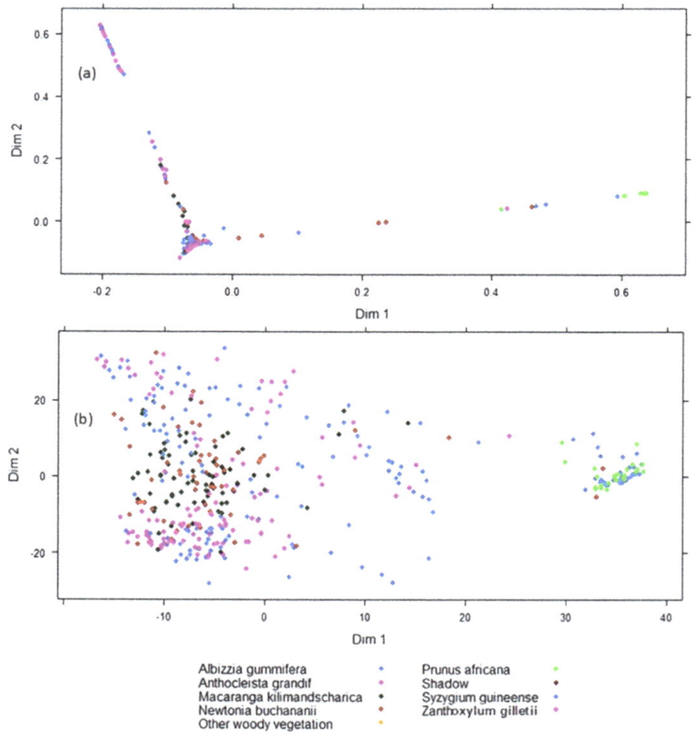

Figure 8. Visualizing tree species (dis)similarity using MDS scaling: (**a**) RF; (**b**) SVM.

The confusion matrices in Table 4 are for RF and SVM classifiers, for the combined technique, and F1-scores were closest to their model mean. They were computed using the test data, which comprised 30% of the ground truth data.

Table 4. Confusion matrices for (a) RF algorithm and (b) SVM algorithm, for the combined oversampling and undersampling technique. Full species names for each species code are given in Table 1.

(a) Random Forest

	AlG	AnG	MK	NB	OWV	PA	SD	SG	ZG	Total	UA (%)
AlG	4	1	1	3	0	0	1	0	0	10	40.0
AnG	0	9	4	1	2	2	0	0	0	18	50.0
MK	0	0	16	2	0	2	0	0	0	20	80.0
NB	3	4	2	9	1	0	0	0	1	20	45.0
OWV	0	0	0	2	17	0	0	0	1	20	85.0
PA	0	0	1	5	1	6	0	0	0	13	46.2
SD	0	0	0	0	0	0	23	0	0	23	100.0
SG	0	0	0	0	0	0	0	23	0	23	100.0
ZG	1	1	0	2	3	0	0	1	22	30	73.3
Total	8	15	24	24	24	10	24	24	24	177	
PA (%)	50.0	60.0	66.7	37.5	70.8	60.0	90.0	90.0	80.0		

Overall accuracy = 72.9% F1-score = 68.0%

(b) Support Vector Machine algorithm

	AlG	AnG	MK	NB	OWV	PA	SD	SG	ZG	Total	UA (%)
AlG	3	1	1	2	1	0	0	0	0	8	37.5
AnG	1	8	3	2	2	2	1	0	1	20	40.0
MK	0	0	15	3	0	2	1	0	0	21	71.4
NB	3	4	2	7	1	0	0	0	1	18	38.9
OWV	0	0	0	2	17	0	0	0	0	19	89.5
PA	1	1	2	6	2	6	0	0	0	18	33.3
SD	0	0	0	0	0	0	22	0	0	22	100.0
SG	0	0	0	0	0	0	0	24	0	24	100.0
ZG	0	1	1	2	1	0	0	0	22	27	81.5
Total	8	15	24	24	24	10	24	24	24	177	
PA (%)	37.5	53.3	62.5	29.2	70.8	60.0	80.0	100.0	70.8		

Overall accuracy = 70.1% F1-score = 64.2%

The combined technique dataset achieved the highest average overall accuracy on both RF and SVM models, i.e., 73.2 ± 2.5% and 70.9 ± 2.7%, respectively. The average species F1-score was 68.56 ± 2.6% for RF and 64.64 ± 3.4% for SVM. The F1-score ranged between 18 and 100%. *Syzygium guineense* and *Zanthoxylum gilletii* are characterized with high accuracy and low variability, unlike *Albizzia gummifera*, *Anthocleista grandiflora*, and *Newtonia buchananii*, which had low variability across iterations and also low F1-score (Figure 9).

The undersampled data reported the lowest F1-score, i.e., 45.8 ± 5% for RF, compared to SVM's 48 ± 6%, unlike the other datasets where RF performed better than SVM. The original dataset's F1-score were 63.8 ± 3.9% for RF and 62.7 ± 4.3% for SVM. The oversampled dataset achieved F1-score values of 67.8 ± 4.1% and 63.6 ± 3.7% for RF and SVM, respectively (Figure 10).

The McNemar's test, applied to test whether there was a significant difference between the two best maps produced by the RF and SVM models, using the combined oversampling and undersampling techniques, returned a Z value of 0.96 (Table 5). Thus, there were no significant differences ($Z \geq 1.96$) at 95% confidence level, existing amongst the confusion matrices of the two classifiers.

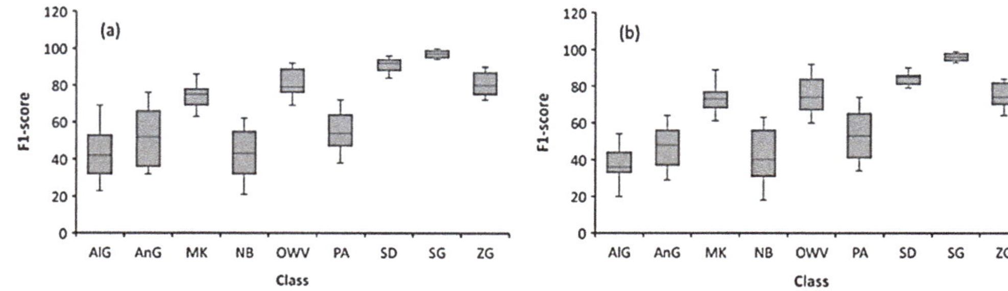

Figure 9. Species-level F1-score for the combined oversampling and undersampling technique: (**a**) RF; (**b**) SVM.

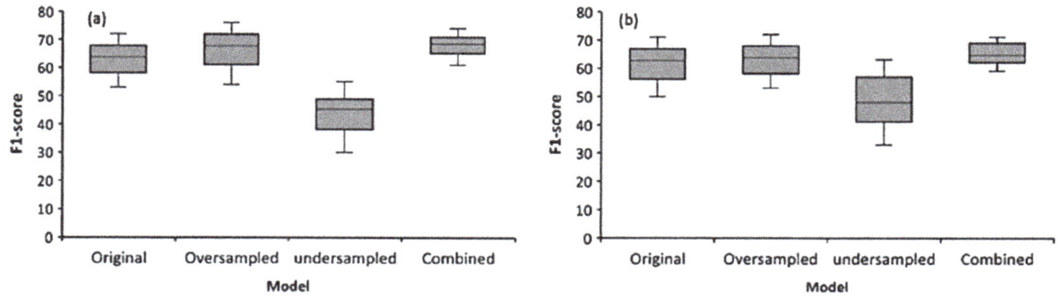

Figure 10. Model-level F1-score for the datasets: (**a**) RF; (**b**) SVM.

Table 5. McNemar's test results to compare RF and SVM classification models using the combined oversampling and undersampling technique.

Classifier		RF		
		CC	IC	Total
SVM	CC	98	11	111
	IC	16	52	66
	Total	114	63	177

CC stands for correctly classified and IC is for incorrectly classified samples.

3.5. The Spatial Distribution of the Endangered Tree Species

In showing the distribution of tree species in the study area, two classified maps, i.e., one by RF and the other by SVM, produced using the combined technique dataset, whose F1-score was closest to their model mean are used (Figure 11). These maps show that tree species distribution is not haphazard. In the northwestern part of the study area is the *Macaranga kilimandscharica*. *Syzygium guineense* is mostly occupying the lower and central parts of the study area. *Zanthoxylum gilletii* covers the central-eastern and western parts. A combination of other vegetation types can be seen dominating mostly the lower parts of the study area. The other tree species, i.e., *Albizzia gummifera*, *Anthocleista grandiflora*, *Prunus africana*, and *Newtonia buchananii* are found throughout the study area.

The RF and SVM classifiers show differences in the size of each class. *Macaranga kilimandscharica* shows an area of 45.2 ha and 44.1 ha for RF and SVM classifier, respectively (Table 6). *Zanthoxylum gilletii* reported 7.1 ha for RF and 7.2 ha for SVM, a negligible difference of 0.1% between them. For *Syzygium guineense*, there was a 0.8% difference between the two classifiers: RF recorded 9.6 ha and SVM recorded 10.6 ha. *Newtonia buchananii* reported a difference of 0.4 ha between the two classifiers. *Anthocleista grandiflora* had the second largest difference between the two classifiers, 1.8 ha, i.e., 13.4 ha for RF and

15.2 ha for SVM. Other woody vegetation showed a difference of 1.9 ha between RF and SVM classifiers. Shadow recorded 10.6 ha and 11.1 ha for RF classifier and SVM classifier, respectively. The areal coverage of *Prunus Africana* and *Albizzia gummifera* was insignificant compared to that of the other classes.

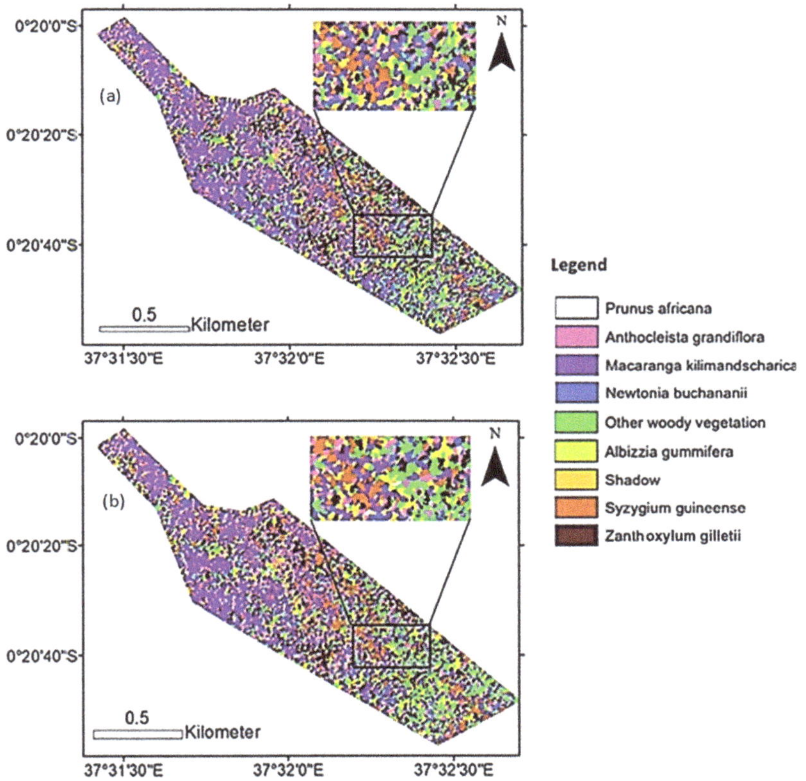

Figure 11. Classification maps of tree species in Mt. Kenya Forest, obtained using (**a**) random forest and (**b**) support vector machine.

Table 6. Area coverage in hectares and percentage of the classes using RF and SVM classifiers.

Classes	RF Classified Area (ha)	Percentage (%)	SVM Classified Area (ha)	Percentage (%)
Macaranga kilimandscharica	45.2	34.8	44.1	33.9
Zanthoxylum gilletii	7.1	5.5	7.2	5.6
Syzygium guineense	9.6	7.4	10.6	8.2
Newtonia buchananii	23.6	18.1	23.2	17.8
Anthocleista grandiflora	13.4	10.3	15.2	11.7
Prunus africana	0.0	0.0	0.0	0.0
Albizzia gummifera	0.0	0.0	0.0	0.0
Other woody vegetation	20.5	15.8	18.6	14.3
Shadow	10.6	8.1	11.1	8.5
Total	130	100	130	100

4. Discussion

4.1. Spectral Separability between the Tree Species

This study aimed to evaluate the impact of imbalanced data in mapping endangered trees species in a selectively logged sub-montane heterogeneous tropical forest using random forest and support vector machine classifiers, and WorldView-2 multispectral imagery. As seen in Figure 4, there are spectral overlaps within the tree species classes. The overlaps could be due to, among other reasons, the study area's complex forest structure, because some stands were multi-storied and also some tree crowns were relatively small, thus the presence of mixed pixels [5]. This led to some of the misclassifications between species, which was in agreement with their inter-specific separability (Table 2). *Newtonia buchananii*, *Anthocleista grandiflora*, and *Albizzia gummifera* are examples of tree species with significantly low F1-scores. The J–M distance values showed that the species were not as separable as *Syzygium guineense*, *Zanthoxylum gilletii*, and *Macaranga kilimandscharica*, which indicated higher spectral separability values and were more separable, and therefore had higher F1-score values. In addition, very steep mountainous terrain and dense forest made it difficult to collect data in the field. This may have interfered with the quality of ground data used in the validation process. This research observed that higher spectral similarity within a class and higher spectral variability among classes led to higher classification accuracy. However, the contribution of the within- and among-species spectral variability on the classification accuracy of tropical tree species is poorly understood [17].

4.2. Relative Importance of Variables

In the classification process, the study was able to single out the most valuable bands, by making use of the variable importance (VI) measurement in the RF algorithm. RF uses the mean decrease in accuracy (MDA) and mean decrease in Gini (MDG) to evaluate the explanatory power of the input variables [37]. Generally, the best four bands in the classification process were the Red, Red Edge, Near Infrared 1, and Near Infrared 2 bands. This shows that both the visible and the infrared portions of the electromagnetic spectrum are quite relevant in the mapping of endangered tree species in the study area. The MDA was also used to show bands that were crucial in detecting particular tree species (Figure 7). However, both MDA and MDG are biased in adapting to the variables in the tree structure, hence larger values are provided than the actual value. Strobl et al. [45] point out that two indicators are unable to determine the explanatory power of variables in the classification process because they cannot distinguish false correlations due to data characteristics. Strobl et al. [45] developed a technique to evaluate the influence of conditional variable classification to solve this problem, but the technique is inconsistent in grasping the explanatory power variables [46]. Hur et al. [46] developed another technique based on the Shapley Value method on random forest regression. The determination of the relative importance of variables remains an active area of research.

4.3. Class Imbalance

The effect of unequal training class sizes for species classification from WorldView-2 imagery was reported in Section 3.4. The original dataset was imbalanced despite efforts to add samples of rare species, such as *Prunus Africana* and *Albizzia gummifera*, whose density has diminished in MKFR. The variation in the sample size across classes in the original dataset increased the rate of classification errors [26]. As expected, tree species with larger sample sizes had positive prediction bias, i.e., commission errors were greater than omission errors, and vice versa [18,40]. In addition,, species with small samples recorded high F1-score variability across iterations (Figure 9). The degree of bias in the model was minimized when the training data size was standardized across classes. The model that used the original dataset reported a reduced overall accuracy. In recent times, research shows that balanced data offers better overall classification performance [26,40]. The F1-score for the combined technique model was 68.56 $\pm$ 2.6% and 64.64 $\pm$ 3.4% for RF and SVM, respectively, whereas the unbalanced, original dataset reported F1-scores of 63.8 $\pm$ 3.9%

for RF and 62.7 ± 4.3% for SVM. Previous studies have shown that the *k*-NN (*k*-nearest neighbor), RF, and SVM algorithms are affected by imbalanced training data [26]. Although these results are consistent with other studies using similar classification algorithms, they may not be directly comparable because of factors such as differences in the topography of the study area, size of samples, species diversity in the study area, a continuous closed canopy, size of the study area, and spectral and spatial resolutions of the images used.

The undersampling techniques reduce observation numbers from the majority class to make the dataset balanced. This technique may lead to the loss of important information in the training data. According to Graves et al. [18], lowering the number of the sampled common tree species while maintaining the species' full range spectral variability solves the reduced overall accuracy problem. Spectral variability in data can be retained when using SVM algorithms by selecting data points on the border between classes to delineate the separation between them [44], or to iteratively prune the support vectors to attain the best separation between classes [47]. Another way is the manipulation of the SVM algorithm by adjusting margin variable misclassification costs, e.g., the cost of misclassifying a feature in the minority class is set higher than that of misclassifying a feature in the majority class [26]. However, the effects of these selection processes on bias in models and successive application to larger areas with many classes have not been quantified [18].

Apart from the techniques applied in this study to reduce class imbalance, other methods, e.g., balanced class weight, and generating synthetic data of the minority class that are similar to the original minority examples in the feature space using methods such as the synthetic minority over-sampling (SMOTE) technique [40], among others, can be used. In this study, the idea was to combine the undersampling and oversampling techniques to create a robust balanced dataset fit for model training.

According to Krawczyk [48], the imbalanced classification problem is not solved; at a time when we have such terms as big data, large neural network models, deep learning, and models such as the xgboost, solutions should be identified and addressed specifically for each training dataset [48].

4.4. Model Performance

In using an imbalanced dataset, the final classification under-predicts the classes with fewer samples, thus minority classes will have less effect on the accuracy compared to larger classes [49]. Therefore, test samples belonging to smaller classes are more often misclassified than those belonging to the dominant classes [26,40]. In such a case, a model may report a high accuracy level, but the map would not be useful. Accuracy is appropriate for balanced datasets but not good for imbalanced ones [18]. In some cases, the accuracy of samples in the smaller classes can be of greater importance than the contrary case [26,40]. This is the case in this research because the mapping of sparse tree species, e.g., *Prunus Africana* and *Albizzia gummifera* is crucial.

The balancing of the original dataset slightly improved the F1-scores for RF and SVM classifiers. Omission errors were recorded for each species, but *Newtonia buchananii* had the highest omission errors in all models. *Newtonia buchananii* and *Anthocleista grandiflora* exhibited a higher level of spectral confusion. This further degraded the F1-scores in all models, because tree species with low separability tend to have high misclassification rates. Furthermore, these tree species exhibited different crown color and density [50]. In addition,, because random sampling was used to collect field data, tree crowns used to train and validate the classifications of tree species are not distributed equally over the forest. The dense forest and rough mountainous terrain limited the places we could access within the forest. This may as well have influenced the outcome of the classification.

Finally, although error matrices are very crucial in comparing classification results, they only give an estimate of the accuracy of the classification, determined by the samples collected from the field. Thus, only biased conclusions can be made from such data [44]. Other metrics of model performance should be tried, e.g., balanced accuracy, bias score, or the F-score, among others [16].

4.5. The Spatial Distribution of Endangered Tree Species

This study found that, *Macaranga kilimandscharica*, an invasive species, mostly occupies the northwestern part of the study area. This tree species is found in areas that have experienced heavy intensive logging and forest disturbance. Logging activities have taken place in this area in the recent past. The RF and SVM classified maps of the study area have shown that *Newtonia buchananii* is more dominant than *Syzygium guineense*, *Zanthoxylum gilletii*, *Anthocleista grandiflora*, *Albizzia gummifera*, and *Prunus africana*. Both *Prunus africana* and *Albizzia gummifera* are hard to find in the study area. The same applies to *Occtea usambarensis*. Once dominant in the wet forests of Mount Kenya, the endangered tree species are now rare.

In cases where differences in accuracy are marginal, i.e., a few percentage points apart, Janssen and van der Wel [51] propose that a statistically rigorous way must be used to compare the accuracies, and the results should be expressed with confidence limits. In this study, the McNemar test was used to evaluate the RF and SVM classification outputs of the combined technique dataset. The McNemar test was meant to indicate whether a difference of 3.9% was statistically important. A difference in accuracy between the confusion matrices of different WorldView-2 spectral subsets is statistically significant ($p \leq 0.05$) if the Z value is more than 1.96 [44] The Z value was 0.96, meaning there is no significant difference between the two maps. Therefore, either of the two maps can be considered for conservation purposes.

5. Conclusions

This study aimed to assess the effects of imbalanced data on identifying and mapping trees species under threat in a selectively logged sub-montane heterogeneous tropical forest using RF and SVM classifiers and WorldView-2 multispectral imagery. The study obtained average F1-scores of 68.56 ± 2.6% and 64.64 ± 3.4% for RF and SVM, respectively, for the best model, the combined oversampling and undersampling technique. This was an improvement from the original imbalanced dataset. The F1-scores reported were directly related to the differences between the spectral variability within and among species. The most important spectral bands identified to have played a major role in mapping the endangered tree species in the study area were the Red, Red Edge, Near Infrared 1, and Near Infrared 2 bands. The tree species portrayed significant spectral overlaps, and this may have led to misclassification errors between classes. As well, difficult mountainous terrain and dense forest made it hard to collect data in the field, and this may have interfered with the quality of data, and thus contributed to increased classification errors.

Given the results presented here, the approach used in this study may serve as the basis for forest recovery initiatives in MKFR, an ecosystem that is considered a biodiversity hot-spot for conservation priorities. However, these applications are based on models of species classification that are not perfect [18]; therefore, more methods need to be developed to overcome the challenges caused by imbalanced data. Further research will target a larger study area and a higher number of tree species using VHR satellite data and object-based image analysis techniques.

Author Contributions: C.M.J.: conceptualization and methodology, C.M.J. & E.A.: data curation, E.A.: supervision, C.M.J.: validation, C.M.J. & E.A.: writing—review and editing. All authors have read and agreed to the published version of the manuscript.

Funding: This research did not receive any specific grant from funding agencies in the public, commercial, or not-for-profit sectors.

Data Availability Statement: Not applicable.

Conflicts of Interest: The authors have declared that no competing interests exist.

References

1. Solberg, R.; Malnes, E.; Amlien, J.; Danks, F.; Haarpaintner, J.; Høgda, K.-A.; Johansen, B.E.; Karlsen, S.R.; Koren, H. State of the art for tropical forest monitoring by remote sensing. In *A Review Carried out for the Ministry for the Environment of Norway and the Norwegian Space Centre*; Norwegian Computing Centre: Oslo, Norway, 2008; pp. 1–76.
2. Slik, J.W.F.; Arroyo-Rodrguez, V.; Aiba, S.-I.; Alvarez-Loayza, P.; Alves, L.F.; Ashton, P.; Balvanera, P.; Bastian, M.L.; Bellingham, P.J.; van den Berg, E.; et al. An estimate of the number of tropical tree species. *Proc. Natl. Acad. Sci. USA* **2015**, *112*, 7472–7477. [CrossRef] [PubMed]
3. Wagner, F.H.; Ferreira, M.P.; Sanchez, A.; Hirye, M.C.M.; Zortea, M.; Gloor, E.; Phillips, O.L.; Filho, C.R.S.; Shimabukuro, Y.E.; Aragão, L.E.O.C. Individual tree crown delineation in a highly diverse tropical forest using very high-resolution satellite images. *ISPRS J. Photogramm. Remote Sens.* **2018**, *145 Pt B*, 362–377. [CrossRef]
4. Ferreira, M.P.; Wagner, F.H.; Aragão. L.E.O.C.; Shimabukuro, Y.E.; de Souza Filho, C.R.S. Tree species classification in tropical forests using visible to shortwave infrared WorldView-3 images and texture analysis. *ISPRS J. Photogramm. Remote Sens.* **2019**, *149*, 119–131. [CrossRef]
5. Immitzer, M.; Atzberger, C.; Koukal, T. Tree species classification with random forest using very high spatial resolution 8-band WorldView-2 satellite data. *Remote Sens.* **2012**, *4*, 2661–2693. [CrossRef]
6. Nagendra, H.; Rocchini, D. High resolution satellite imagery for tropical biodiversity studies: The devil is in the detail. *Biodiv. Conserv.* **2008**, *17*, 3431–3442. [CrossRef]
7. Clark, M.L.; Roberts, D.A.; Clark, D.B. Hyperspectral discrimination of tropical rain forest tree species at leaf to crown scales. *Remote Sens. Environ.* **2005**, *96*, 375–398. [CrossRef]
8. Zhang, J.; Rivard, B.; Sánchez-Azofeifa, A.; Castro-Esau, K. Intra- and inter-class spectral variability of tropical tree species at La Selva, Costa Rica: Implications for species identification using HYDICE imagery. *Remote Sens. Environ.* **2006**, *105*, 129–141. [CrossRef]
9. Trichon, V.; Julien, M.-P. Tree species identification on large-scale aerial photographs in a tropical rain forest, French Guiana-application for management and conservation. *For. Ecol. Manag.* **2006**, *225*, 51–61. [CrossRef]
10. Somers, B.; Asner, G.P. Hyperspectral time series analysis of native and invasive species in Hawaiian rainforests. *Remote Sens.* **2012**, *4*, 2510–2529. [CrossRef]
11. Féret, J.-B.; Asner, G.P. Semi-supervised methods to identify individual crowns of lowland tropical canopy species using imaging spectroscopy and LiDAR. *Remote Sens.* **2012**, *4*, 2457–2476. [CrossRef]
12. Clark, M.L.; Roberts, D.A.; Clark, D.B. Species-Level Differences in Hyperspectral Metrics among Tropical Rainforest Trees as Determined by a Tree-Based Classifier. *Remote Sens.* **2012**, *4*, 1820–1855. [CrossRef]
13. Papeş, M.; Tupayachi, R.; Martínez, P.; Peterson, A.T.; Asner, G.P.; Powell, G.V.N. Seasonal variation in spectral signatures of five genera of rainforest trees. *IEEE J. Sel. Top. Appl. Earth Obs. Remote Sens.* 2013; 6, 339–350. [CrossRef]
14. Feret, J.B.; Asner, G.P. Tree species discrimination in tropical forests using airborne imaging spectroscopy. *IEEE Trans. Geosci. Remote Sens.* **2013**, *51*, 73–84. [CrossRef]
15. Singh, M.; Evans, D.; Tan, B.S.; Nin, C.S. Mapping and Characterizing Selected Canopy Tree Species at the Angkor World Heritage Site in Cambodia Using Aerial Data. *PLoS ONE* **2015**, *10*, e0121558. [CrossRef]
16. Baldeck, C.A.; Asner, G.P.; Martin, R.E.; Anderson, C.B.; Knapp, D.E.; Kellner, J.R.; Wright, S.J. Operational tree species mapping in a diverse tropical forest with airborne imaging spectroscopy. *PLoS ONE* **2015**, *10*, e0118403. [CrossRef]
17. Ferreira, M.P.; Zortea, M.; Zanotta, D.C.; Shimabukuro, Y.E.; de Souza Filho, C.R. Mapping tree species in tropical seasonal semi-deciduous forests with hyperspectral and multispectral data. *Remote Sens. Environ.* **2016**, *179*, 66–78. [CrossRef]
18. Graves, S.J.; Asner, G.P.; Martin, R.E.; Anderson, C.B.; Colgan, M.S.; Kalantari, L.; Bohlman, S.A. Tree species abundance predictions in a tropical agricultural landscape with a supervised classification model and imbalanced data. *Remote Sens.* **2016**, *8*, 161. [CrossRef]
19. Mutanga, O.; Adam, E.; Cho, M.A. High density biomass estimation for wetland vegetation using WorldView-2 imagery and random forest regression algorithm. *Int. J. Appl. Earth Obs. Geoinf.* **2012**, *18*, 399–406. [CrossRef]
20. NEMA (National Environment Management Authority). Kenya State of the Environment and Outlook 2010; Supporting the Delivery of Vision 2030. Available online: http://www.enviropulse.org/documents/Kenya_SOE.pdf (accessed on 3 January 2019).
21. KWS (Kenya Wildlife Service). Mt Kenya Ecosystem Management Plan 2010–2020. Available online: http://www.kws.go.ke/file/1470/download?token=1lO6G3zI (accessed on 16 February 2019).
22. Wass, P. *Kenya's Indigenous Forests: Status, Management and Conservation*; IUCN: Gland, Switzerland; Cambridge, UK, 1995; 205p.
23. Bussmann, R.W. Destruction and management of Mount Kenya's forests. *Ambio* **1996**, *25*, 314–317.
24. Ng'eno, J.K. Kenya. Proceedings of Country report to the FAO International Technical Conference on Plant Genetic Resources, Leipzig, Germany, 17–23 June 1996.
25. KFS (Kenya Forest Service). Mt. Kenya Forest Reserve Management Plan 2010–2019. Available online: http://www.kenyaforestservice.org/documents/MtKenya.pdf (accessed on 3 January 2019).
26. Maxwell, A.E.; Warner, T.A.; Fang, F. Implementation of machine-learning classification in remote sensing: An applied review. *Int. J. Remote Sens.* **2018**, *39*, 2784–2817. [CrossRef]

27. Bussmann, R.W. Stand structure and regeneration of the subalpine *Hagenia abyssinica* forest of Mt. Kenya. *Bot. Act.* **1997**, *110*, 473–480. [CrossRef]
28. Baker, B.H. Geology of the Mount Kenya Area. *Geol. Surv. Kenya Rep.* **1967**, *79*, 464–465.
29. Nyongesa, K.W.; Vacik, H. Evaluating management strategies for Mount Kenya Forest Reserve and National Park to reduce fire danger and address interests of various stakeholders. *Forests* **2019**, *10*, 426. [CrossRef]
30. Ogondo, J.A. Geomorphological formation of Mount Kenya. Proceedings of Kenya National Commission for UNESCO, Stakeholders sensitization Workshop on the extension of Mt. Kenya World Heritage Site, at Sportsman's Arm Hotel, Nanyuki, Kenya, 27–28 August 2009. [CrossRef]
31. DigitalGlobe. The Benefits of the 8 Spectral Bands of WorldView-2. Available online: https://dg-cms-uploads-production.s3.amazonaws.com/uploads/document/file/35/DG-8SPECTRAL-WP_0.pdf (accessed on 2 February 2019).
32. Jensen, J.R. *Introductory Digital Image Processing: A Remote Sensing Perspective*, 3rd ed.; Prentice Hall: Upper Saddle River, NJ, USA, 2005; pp. 210–215.
33. GoK (Government of Kenya). Fifth National Report to the Conference of Parties to the Convention on Biological Diversity. Available online: https://www.cbd.int/doc/world/ke/ke-nr-05-en.pdf (accessed on 16 June 2019).
34. GoK (Government of Kenya). *The Wildlife Conservation and Management Act, 2013*; No. 47 of 2013; GoK: Nairobi, Kenya, 2013.
35. Richards, J.A.; Jia, X. *Remote Sensing Digital Image Analysis: An Introduction*, 3rd ed.; Springer-Verlag: Berlin/Heidelberg, Germany, 1999.
36. Kailath, T. The divergence and bhattacharyya distance measures in signal selection. *IEEE Trans. Commun. Technol.* **1967**, *15*, 52–60. [CrossRef]
37. Breiman, L. Random forests. *Machin. Learn.* **2001**, *45*, 5–32. [CrossRef]
38. Vapnik, V. *The Nature of Statistical Learning Theory*, 2nd ed.; Springer-Verlag: New York, NY, USA, 2000.
39. Kuter, S. Completing the machine learning saga in fractional snow cover estimation from MODIS Terra reflectance data: Random forests versus support vector regression. *Remote Sens. Environ.* **2021**, *255*, 112294. [CrossRef]
40. Chawla, N.V.; Japkowicz, N.; Kolcz, A. Editorial: Special issue on learning from imbalanced data sets. *ACM SIGKDD Explor. Newsl.* **2004**, *6*, 1–6. [CrossRef]
41. Borg, I.; Groenen, P.J.F. *Modern Multidimensional Scaling: Theory and Applications*, 2nd ed.; Springer Science + Business Media: Berlin, Germany, 2005.
42. NCSS. Chapter 435. Multidimensional Scaling. Available online: https://ncss-wpengine.netdna-ssl.com/wp-content/themes/ncss/pdf/Procedures/NCSS/Multidimensional_Scaling.pdf (accessed on 18 November 2021).
43. Buja, A.; Swayne, D.F.; Littman, M.L.; Dean, N.; Hofmann, H.; Chen, L. Data Visualization with Multidimensional Scaling. *J. Comput. Graph. Stat.* **2007**, *17*, 444–472. [CrossRef]
44. Foody, G.M.; Mathur, A. Toward intelligent training of supervised image classifications: Directing training data acquisition for SVM classification. *Remote Sens. Environ.* **2004**, *93*, 107–117. [CrossRef]
45. Strobl, C.; Boulesteix, A.-L.; Kneib, T.; Augustin, T.; Zeileis, A. Conditional variable importance for random forests. *BMC Bioinform.* **2008**, *9*, 307. [CrossRef]
46. Hur, J.-H.; Ihm, S.-Y.; Park, Y.-H. A variable impacts measurement in random forest for mobile cloud computing. *Wirel. Commun. Mob. Comput.* **2017**, *2017*, 1–13. [CrossRef]
47. Chen, X.; Gerlach, B.; Casasent, D. Pruning support vectors for imbalanced data classification. Proceedings of International Joint Conference on Neural Networks, Montreal, QC, Canada, 31 July–4 August 2005; pp. 1883–1888. [CrossRef]
48. Krawczyk, B. Learning from imbalanced data: Open challenges and future directions. *Prog. Artif. Intell.* **2016**, *5*, 221–232. [CrossRef]
49. He, H.; Garcia, E.A. Learning Form Imbalanced Data. *IEEE Trans. Knowl. Data Eng.* **2009**, *21*, 1263–1284. [CrossRef]
50. Adam, E.; Mutanga, O.; Odindi, J.; Abdel-Rahman, E.M. Land-use/cover classification in a heterogeneous coastal landscape using RapidEye imagery: Evaluating the performance of random forest and support vector machines classifiers. *Int. J. Remote Sens.* **2014**, *35*, 3440–3458. [CrossRef]
51. Janssen, L.L.F.; van der Wel, F.J.M. Accuracy assessment of satellite-derived land-cover data—A review. *Photogramm. Eng. Remote Sens.* **1994**, *60*, 419–426.

remote sensing

MDPI

Article

Object-Based Approach Using Very High Spatial Resolution 16-Band WorldView-3 and LiDAR Data for Tree Species Classification in a Broadleaf Forest in Quebec, Canada

Mathieu Varin *, Bilel Chalghaf and Gilles Joanisse

Centre D'enseignement et de Recherche en Foresterie de Sainte-Foy (CERFO), 2440 Ch. Ste-Foy, Quebec City, QC G1V 1T2, Canada; b.chalghaf@cerfo.qc.ca (B.C.); g.joanisse@cerfo.qc.ca (G.J.)
* Correspondence: m.varin@cerfo.qc.ca

Received: 18 August 2020; Accepted: 18 September 2020; Published: 21 September 2020

Abstract: Species identification in Quebec, Canada, is usually performed with photo-interpretation at the stand level, and often results in a lack of precision which affects forest management. Very high spatial resolution imagery, such as WorldView-3 and Light Detection and Ranging have the potential to overcome this issue. The main objective of this study is to map 11 tree species at the tree level using an object-based approach. For modeling, 240 variables were derived from WorldView-3 with pixel-based and arithmetic feature calculation techniques. A global approach (11 species) was compared to a hierarchical approach at two levels: (1) tree type (broadleaf/conifer) and (2) individual broadleaf (five) and conifer (six) species. Five different model techniques were compared: support vector machine, classification and regression tree, random forest (RF), k-nearest neighbors, and linear discriminant analysis. Each model was assessed using 16-band or first 8-band derived variables, with the results indicating higher precision for the RF technique. Higher accuracies were found using 16-band instead of 8-band derived variables for the global approach (overall accuracy (OA): 75% vs. 71%, Kappa index of agreement (KIA): 0.72 vs. 0.67) and tree type level (OA: 99% vs. 97%, KIA: 0.97 vs. 0.95). For broadleaf individual species, higher accuracy was found using first 8-band derived variables (OA: 70% vs. 68%, KIA: 0.63 vs. 0.60). No distinction was found for individual conifer species (OA: 94%, KIA: 0.93). This paper demonstrates that a hierarchical classification approach gives better results for conifer species and that using an 8-band WorldView-3 instead of a 16-band is sufficient.

Keywords: tree species; object-based; classification; mapping; WorldView-3; LiDAR; machine learning

1. Introduction

Forest characterization in Quebec, Canada, is usually assessed based on photo-interpretation using three-dimensional appearance. This approach has been used since the last century and is still in use for forest planning and forest composition analysis [1]. New techniques, such as image enhancement, have been developed over the years using aerial imagery and user-friendly software, and the information provided has been well accepted by and proven useful for foresters [2,3]. However, species identification with these newer methods still lacks precision, and varies among photo-interpreters, mainly because this characterization is made at the stand level, as species identification at the tree level would be time consuming and expensive [3,4]. Recently, very high spatial resolution satellite imagery has become more available and could be used to classify tree species at tree level across different biomes [5–7]. In addition, with an airborne laser scanner or "LiDAR" (light detection and ranging), an infrared laser can scan the surface of the earth, generating a 3D point cloud that can be used to analyze the tree structure [8,9].

Furthermore, LiDAR data allows a forest to be characterized at the tree level, which can lead to a better estimation of timber volume and hence better planning by foresters [10]. Individual tree crown (ITC) segmentation is being studied more and more [11–14]. Forest segmentation can be done through two different techniques: (1) point cloud-based and (2) raster-based, using the canopy height model (CHM) [8,15,16]. The first technique generally gives good results, but it is time-consuming, complex and requires advanced LiDAR sensors [17]. The second technique has been studied much more, both at the stand level [18,19] and at the tree level [20–22], as there are a variety of algorithms that provide rapid ITC segmentation, which gives satisfactory results [14,16,23].

Many studies have investigated tree species mapping at the tree level in a forest environment, however they usually process hyperspectral data [24–28]. Fewer studies have tried to map tree species with satellite imagery at tree level scale [6,29–31]. Pham et al. [32] tried to combine imagery, LiDAR and GIS topography indices, which led to better results than using a single data source. In previous projects, we used aerial hyperspectral data fused with LiDAR and GIS data to map individual tree species [4,33,34]. The results showed global precisions of over 93% to classify ash and spruce against 14 other species, and precisions of 62% and higher to classify seven species, in an urban and in a forest environment, respectively. In the latter case, yellow birch and hemlock were the species identified with the best accuracy (mean precisions of 77% and 83%, respectively). Both studies were carried out using an experimental hyperspectral sensor. While using aerial hyperspectral data gives interesting results, the complexity of the processes, as well as the high acquisition costs over large areas must be taken into account [35].

The use of satellite multispectral imagery is relevant for tree species mapping. Indeed, satellite data has been widely used to classify tree species [30,36–39], but since the launch of DigitalGlobe multispectral sensors, none of these efforts has had the ability to capture very high spatial resolution (<2 m) that is as detailed as WorldView-3. Moreover, the eight new bands in the Short-Wave Infrared (SWIR) may improve tree species classification [7,40]. In remote areas, such precise satellite images can become an alternative to aerial photography [41]. These images provide more spectral bands for analysis with a relatively competitive acquisition cost. Some studies also combined satellite imagery with LiDAR and demonstrated that a combination of both can significantly increase the accuracy of classification [19,38,42,43], but they have essentially worked at the stand level. Others have used fused data to classify tree species at the tree level using high spatial resolution sensors [32,43,44]. More recently, Li et al. [45] worked on tree species classification with WorldView-3 and LiDAR at the tree level in an urban context with isolated trees. Nevertheless, the number of tree species in those studies was limited, usually less than ten species.

Recently, the use of machine learning techniques, including the support vector machine (SVM), classification and regression tree (CART), random forest (RF), k-nearest neighbors (k-NN) and linear discriminant analysis (LDA) techniques for classifying forest characteristics have been gaining popularity. These techniques have been widely used in remote sensing for species classification [46,47], vegetation health assessment [48–51], biomass mapping [52–54], wetland mapping [55–57] and landslide risk evaluation [58,59]. He et al. [40] also used RF in a hierarchical approach in order to classify tree species. However, few studies evaluated the use of multiple techniques in a hierarchical approach at the tree level [60].

The SVM algorithm, initially suggested by Vapnik [61], maximizes the margin around the hyperplane that separates features into different domains [62]. For classes that are not linearly separable, the SVM uses a kernel function, reducing a nonlinear problem to a linear problem based on a radial basis function or Gaussian kernels. The penalty parameter (C) and the kernel parameter gamma (γ) for the radial basis function kernel should be optimized, and can heavily impact the classification accuracy when using SVM models [63]. It is C that determines the trade-off between margin maximization and training error minimization [64], while the γ parameter defines how far the influence of a single training example reaches, with low values meaning 'far' and high values meaning 'close' [65–67].

The CART approach operates by recursively splitting the data until the ending points, or terminal nodes, are achieved using pre-set criteria [68]. The CART therefore begins by analyzing all explanatory variables and determining which binary division of a single explanatory variable best reduces deviance in the response variable [69] The main elements of the CART, and of any decision tree algorithm, are: (1) rules for splitting data at a node based on the value of one variable; (2) stopping rules for deciding when a branch is terminal and cannot be split anymore; and (3) a prediction for the target variable in each terminal node.

Introduced by Breiman [70], RF is a classifier that evolves from decision trees. It actually consists of many CARTs. To classify a new instance, each tree is trained with a randomly selected subset of the training samples and variables based on bootstrap sampling, and then the final classification is conducted based on a majority vote on the trees in the forest [71]. Although the classifier was originally developed for the machine learning community, thanks to its accuracy, interest in RF has grown rapidly in ecology [72] and in the classification of remotely-sensed imagery [73].

The k-NN [74] algorithm is a non-parametric method that assigns to an unseen point the dominant class among its k-nearest neighbors within the training set. Unlike most other methods of classification, k-NN falls under lazy learning, which means that there is no explicit training phase before the classification. The classification with k-NN is carried out by following three steps: (1) compute a distance value between the item to be classified and every other item in the training data set; (2) choose the k-closest data points (the items with the k-lowest distances); and (3) conduct a "majority vote" among those data points to decide the final classification.

LDA has been widely used in various tree species classification studies [75–78]. LDA projects the original features onto a lower dimensional space by means of three steps [79]: (1) calculate the separability between different classes, called the between-class variance; (2) calculate the distance between the mean and sample of each class, called the within-class variance; and (3) construct the lower dimensional space which maximizes the between-class variance and minimizes the within-class variance.

The main objective of this study is to map 11 tree species using an object-based approach with WorldView-3 imagery and LiDAR data. Object-based image analysis brings the capacity to group homogenous pixels into meaningful objects based on their spectral values, which can then be analyzed by their shape, size, texture and contextual information [19], in contrary to the pixel-based approach [80]. We implemented modeling techniques in a global and hierarchical approach. More specifically, this study aims to (1) delineate ITC using fused data (WorldView-3 imagery and LiDAR data); (2) compare models at each classification level (global and hierarchical); (3) evaluate classification improvement using 16-band instead of 8-band WorldView-3; and (4) apply the best models to map tree species over the study areas. This implies the ability to delineate ITCs in order to extract spectral signatures, and to assign a specie class to each object. For ITC segmentation, we used three different techniques. We propose an ITC segmentation using fused data (CHM and satellite imagery) to refine tree species' crown delineation [32]. The classification of species is divided into three parts, on two levels: (1) tree types and (2) broadleaf and conifer tree species. In the present study we used five different models (SVM, CART, RF, k-NN and LDA) to overcome the uncertainty derived from the use of an individual model, given that the results can vary depending on the modeling technique.

2. Data and Methods

2.1. Study Areas and Data

2.1.1. Study Areas

The study was conducted on the Kenauk Nature property, which is located in the south-west region of the province of Quebec, Canada (N 45°52′1″–45°39′36″, W 74°58′22″–74°44′7″) (Figure 1). This private property is over 250 km^2, has a mean elevation of 226 m and is composed of a diverse broadleaf forest including more than 25 tree species [81]. Medium slopes characterize its topography.

For the purposes of the project and due to time limitations, three study areas totaling 26.1 km^2 were selected to collect field data, train and apply the models. Those areas contain mature forest stands composed of dominant tree species with diverse structural stands and topography.

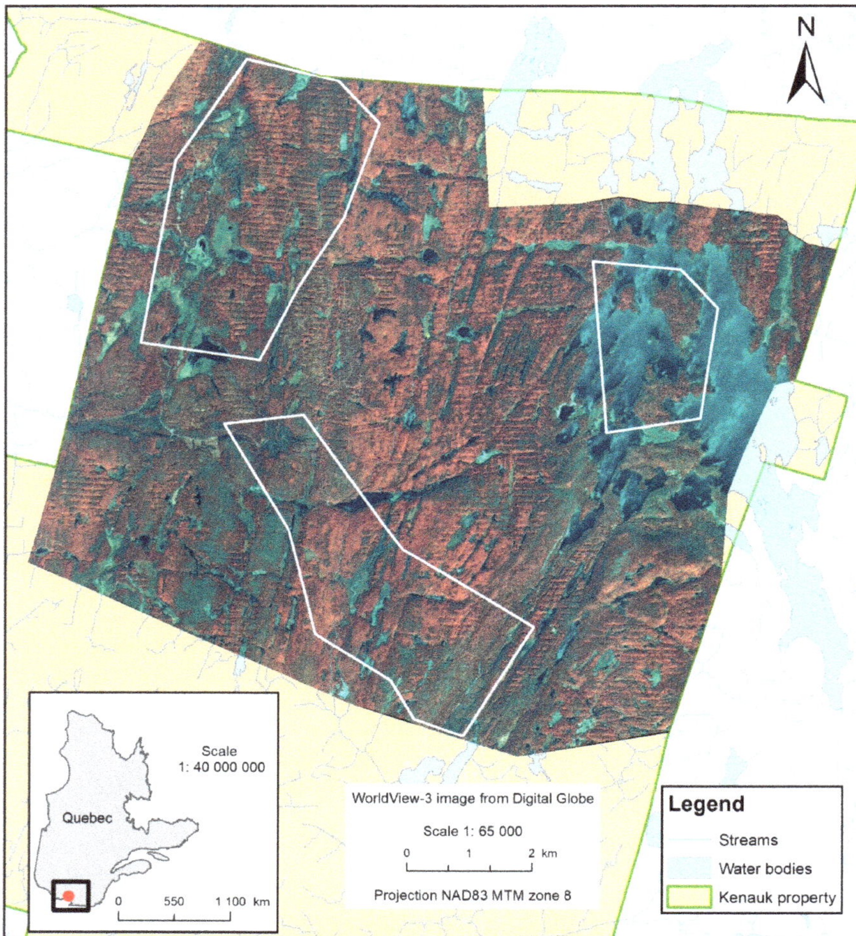

Figure 1. Study areas (delineated in white) at the Kenauk Nature property. The background image displays WorldView-3 in false colors (infrared, green and blue).

2.1.2. Imagery and Airborne Laser Scanner Data

Very high spatial resolution 16-band WorldView-3 imagery was acquired for the study areas on 26 August 2016 with a nadir view angle of 12.9° and a solar elevation of 54.3° (Table 1). Two cloud-free images were collected and preprocessed. The images were geometrically corrected with Rational Polynomial Coefficients (RPCs), radiometrically calibrated (radiance to reflectance) and then pansharpened using the least squares algorithm [82]. No atmospheric correction or dehaze reductions were applied since many artifacts were produced, and this preprocessing has proven to be less effective than pansharpening for tree classification purposes using high spatial resolution imagery [6,30,82]. The orthorectification was done using a 5 m LiDAR Digital Elevation Model with OrthoEngine of PCI Geomatica [6,7]. For the purpose of the study, all sixteen bands were rescaled to 30 cm, similarly to Li et al. [45]. Finally, the images were mosaicked with a bundle adjustment and then fitted with the CHM using 100 tie points to reduce the offset in the canopy [83]. A second-order polynomial regression was used to create the final mosaic with a root mean square error of 0.97 m. Inspired by Zhou and Qiu [84] and Hartling et al. [31], deep shadow was extracted from the mosaic using a maximum likelihood classification with a shadow index [85]. The Bhattacharyya index showed a separability of 1.94 to detect deep shadow against other elements. This result was used to mask the mosaic for subsequent processes. Those preprocesses were carried out using PCI Geomatica (version 2016) and ENVI (version 5.4).

LiDAR data were acquired on 17 June 2015, with a point density of 10 pts/m^2. The sensor used was a Riegl Q-780 system with a pulse repetition frequency of 400 kHz and a laser wavelength of 1064 nm. For the acquisition, a Cessna 172 airplane was flown at a mean height of 1200 m above ground level with a flight speed of 185 km/h. The absolute accuracy in xyz was 30 cm. The point cloud was classified by the provider into five classes: Unassigned, Ground, High vegetation, Building and Water. We used the Ground class (lowest points) to create a 50-cm Digital Elevation Model (DEM) and the High vegetation class (highest points) to produce a 50-cm Digital Surface Model (DSM). We subtracted the DSM from the DEM to obtain the 50-cm CHM. This procedure was performed using the LAS Dataset To Raster function and the Raster Calculator tool in ArcGIS Desktop 10.6.

Table 1. WorldView-3 channels characteristics.

Band	Spectrum	Wavelength Range (nm)	Wavelength Center (nm)	Spatial Resolution (m)
0	Panchromatic	450–800	625	0.31
1	Costal	400–450	425	1.26
2	Blue	450–510	480	
3	Green	510–580	545	
4	Yellow	585–625	605	
5	Red	630–690	660	
6	Red-edge	705–745	725	
7	Near-infrared #1	770–895	832.5	
8	Near-infrared #2	860–1040	950	
9	Short-Wave Infrared #1	1195–1225	1210	3.89
10	Short-Wave Infrared #2	1550–1590	1570	
11	Short-Wave Infrared #3	1640–1680	1660	
12	Short-Wave Infrared #4	1710–1750	1730	
13	Short-Wave Infrared #5	2145–2185	2165	
14	Short-Wave Infrared #6	2185–2225	2205	
15	Short-Wave Infrared #7	2235–2285	2260	
16	Short-Wave Infrared #8	2295–2365	2330	

2.2. Field Survey and Data Collection

A total of 515 dominant trees were identified and positioned in the study areas using a high precision Trimble Pro6H GPS with a mean precision of 1 m after post-processing. We targeted individual trees or groups of trees presenting the same species with a minimum crown diameter of 5 m. Geographical coordinates and physical parameters, such as tree heights and crown sizes, were measured or interpreted from WorldView-3 and aerial images [1]. Based on GPS positions, a manual delineation was made by photo-interpretation to fit the crown to the correct tree on the WorldView-3 images (Figure 2) [6,7,86]. Thus, 185 broadleaf and 153 coniferous tree samples (total of 338) remained after this manual delineating exercise, as only visible and identifiable crowns were kept (Table 2). Mean crown sizes vary between species (22–85 m^2) and the mean height is over 16 m in all cases.

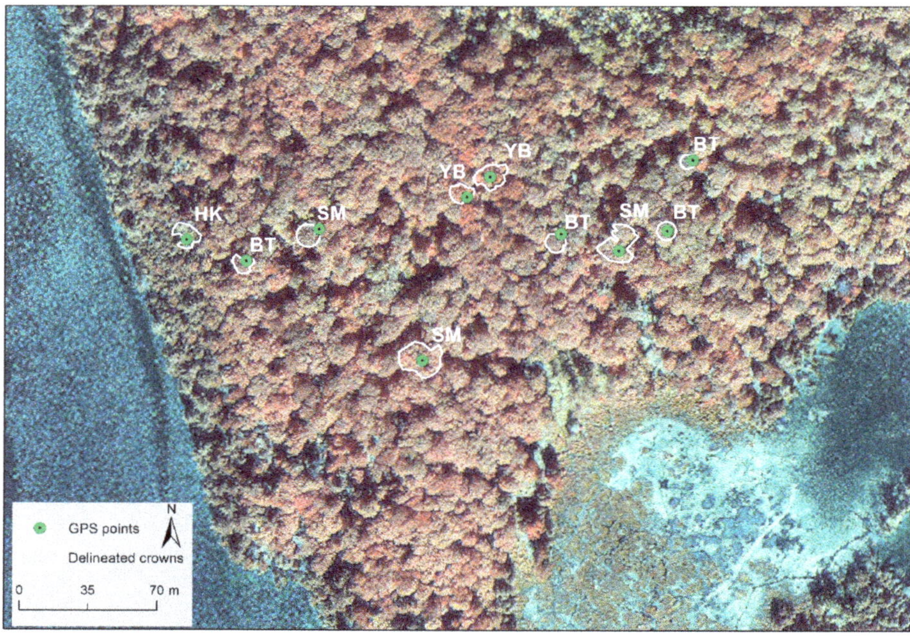

Figure 2. Manually delineated crowns by photo-interpretation and their corresponding GPS points collected during the field campaign. The background image displays WorldView-3 in false colors (infrared, green and blue). The species' abbreviations are described in Table 2.

Table 2. Distribution of the different tree species samples (train and test) and tree crowns statistics of each species. BL: Broadleaf, CF: Conifer.

Species	Acronym	Type	Tree Crowns Statistics			Train	Test	Total
			Mean Size (m^2)	Mean Height (m)	SD Height (m)			
American Beech	AB	BL	32	21	4	31	10	41
Big Tooth Aspen	BT	BL	42	25	4	13	5	18
Red Oak	RO	BL	60	24	3	24	10	34
Sugar Maple	SM	BL	85	24	3	37	9	46
Yellow Birch	YB	BL	63	22	4	36	10	46
Balsam Fir	BF	CN	22	16	3	13	3	16
Eastern White Cedar	EC	CN	31	21	5	16	5	21
Eastern Hemlock	HK	CN	39	23	3	29	9	38
Red Pine	RP	CN	59	28	3	15	5	20
White Pine	WP	CN	64	26	4	38	10	48
White Spruce	WS	CN	35	20	5	7	3	10
Total						259	79	338

2.3. *Derived Variables*

WorldView-3 imagery data were used to extract, calculate and adapt variables based on the available literature. The details of these variables are given in Appendix A. The spectral variables were computed from the 16 bands so as to cover diverse spectral elements (vegetation, wetness, openness etc.). Spectral variables of reflectance values were calculated by pixel [6] and object-based methods using R [87], SAS software [88] and eCognition Developer. Object-based indices consist of a series of customized arithmetic features calculated using the mean of pixel values within an object of specific bands [16,19,80]. Arithmetic features were also calculated for the 95th percentile highest pixels within each object in order to use the brightest (sunlit) parts of each crown. Although they could be correlated with pixel-based indices, arithmetic features are quick to calculate, which is a significant advantage when working with massive data. Textural variables were extracted using eCognition Developer 9.2.

2.4. Tree Crown Segmentation from Fused Data

The first step of the object-based approach is to segment the territory into contextual objects, such as single tree crowns. ITC segmentation using CHM is challenging in complex forest stands [89]. We decided to analyze trees over 17 m in height to reduce the effect of understory on canopy gaps [29,90] and as a trade-off based on the field survey (see Table 2). A 2 m buffer around mature trees (>17 m) was incorporated to keep pixels that are part of the same crown but that have smaller heights; those less than 7 m in height were eliminated and used as a mask for further processing [16,29]. Preprocessed CHM is usually utilized for ITC segmentation [15,21,91–93]. Various spatial filters with different window sizes (two or four pixels) and shapes (rectangular or circular) were tested, and the one that best fit the tree crowns visually was selected. We then combined the original and filtered CHMs with the Jakubowski et al. [16] method to keep both the edges of the original CHM and the smooth central crown of the filtered CHM. Topological operations were also undertaken to merge small crowns [93].

Three CHMs were evaluated in this study: original, filtered and corrected. Similarly to Pham et al. [32] and Koch et al. [15], an inverse watershed segmentation algorithm was used within eCognition Developer. This algorithm uses a CHM to find local maxima to grow a region according to the heights of neighboring pixels until they touch another object [94]. Several parameters were tested in order to select the optimal values for the height differences between local maxima and minima and object areas, using trial and error with various combinations. A neighborhood of eight pixels was used in order to produce disconnected objects. Three thresholds were used; considering the high resolution of the CHM (50 cm), these were: (1) an overflow area of 25 pixels (6.25 m^2); (2) an overflow depth (difference between maximum and minimum height) of 0.5 m; and (3) an overflow volume of 15 m^3 were selected. Objects that were below those thresholds were merged to their neighboring objects.

The approach based on the CHM depends on height variation detection and is therefore related to the shape of the crowns. However, different neighbor tree species may have overlapping branches and could produce confusion when using imagery data to classify tree species. We added a multiresolution segmentation to make the objects generated with CHM more precise by using imagery at a sublevel [5,6]. This segmentation is a bottom-up region merging technique. Combining CHM and imagery to make ITC segmentation allows pure species objects to be delineated. The multiresolution segmentation algorithm with the most significant bands identified with statistical analysis was used, as suggested by Koch et al. [15] (Section 2.5.1). This algorithm requires several empirical parameters, including: (1) the weights of each selected bands; (2) a color/shape weight associated with the spectral/shape homogeneity; (3) a compactness/smoothness weight according to the object shape; and (4) a scale parameter referring to the average size of objects. In this study, to integrate non-correlated visible and infrared information, and by trial and error, four bands (3, 5, 6 and 7) were selected and assigned the same weight (1). The weights of color and compactness were set as 0.3 and 0.7, respectively, in order to insist on spectral variations while retaining more compact objects such as tree crowns. A scale parameter of 100 was selected with visual assessments in order to limit under-segmentation. It has been shown that under-segmentation affects the classification accuracy more than over-segmentation [95]. Those parameters were empirically chosen to ensure best results for ITC segmentation [19]. The performance of ITC segmentation based on CHM alone and then with imagery was assessed using 30 random non-isolated crowns with photo-interpretation.

We first evaluated if a segment represented a single crown. Considering the offset between the data and the complexity of the forest structure, a single crown was identified when the segment contained at least 75% of the corresponding tree [15]. Then, we evaluated the number of species encompassed within the segment. The ITC segmentation technique which gave the best accuracy was retained and used for classification over the study areas.

2.5. Classification Models

2.5.1. Variable Selection

In accordance with the literature, a total of 240 variables were used as predictor variables, including 64 band statistics (mean, skewness, standard deviation), 112 spectral indices and 64 textures (detailed in Appendix A). Variable selection is a crucial procedure before modeling. Using a large number of predictor variables is time consuming, requires high computing capacity, reduces the reproducibility, and the results cannot be easily interpretable. Furthermore, the use of a large number of predictor variables does not necessarily produce the best results, since the model performance can vary widely according to the variables utilized.

As a first step, outliers were removed from the dataset, identified based on the work of Brillinger [96]. We first calculated the interquartile range for our data. Using that range, only the values falling between the median value plus 2.5 times the interquartile, and the median value minus 2.5 times the interquartile were conserved. A datum was removed after it had been attributed to an outlier for more than eight distributed variables.

Secondly, to reduce the dimensionality of the data we proceeded with a correlation analysis to avoid redundant information. Pairs of variables with correlation coefficients greater than 0.85 were considered to be highly correlated, and the variable presenting the highest mean correlation with the remaining predictor variables was discarded from further analysis [97]. To simplify the analysis, those procedures were applied to parametric and non-parametric models. For all models, except LDA, further procedures were applied. At the end of the correlation analysis, this limited number of variables were introduced as input variables in the Boruta algorithm [98]. Boruta is a wrapper algorithm that seeks to capture all the important, interesting features in a dataset with respect to a variable outcome. The 15 most significant variables were selected. Finally, as machine learning algorithms do not always retain the relevant variables [99,100], we created a loop among the 15 selected variables from Boruta to evaluate the performance of models created by all possible combinations of a number of variables (3, 4, 5, 6, 7, 8, 9 and 10). For the LDA, variable selection was done with the Stepdisc procedure in SAS using the stepwise selection method and a variable entry and staying significance level of 0.005. Those iterations were processed within the training dataset. We then retained the combination that gave the best performance for each classification technique. Those procedures were independently done for 16-band and 8-band WorldView-3 derived variables.

2.5.2. Modeling Process

We began by attempting to model all 11 tree species, five broadleaf and six coniferous, in a global approach. Next, we used a hierarchical approach with group classifications, similarly to Wessel et al. [60]. In the first step, we attempted to separate the two tree types (broadleaf and conifer trees). In the second step, we proceeded to the modeling of individual tree species belonging to each type. For the classification, four modeling procedures were implemented under R: RF [101], SVM [62], k-NN [102], CART [102,103] and one under SAS: LDA [76,77].

To avoid overfitting of the classification models, independent validation was conducted by dividing the available reference data into two sets, where 80% of the total samples were used for model calibration [25] and the remaining samples were used as test set (Table 2). Tuning was applied differently on each model.

To find the best gamma and penalty parameters for SVM we used a grid search over a supplied parameter range, and the combination of parameters that maximized model performance was retained. The CART model was tuned with different complexity parameter (cp) values that were estimated by testing cp values using cross-validation approaches. The best cp was then defined as the one that maximized the cross-validation accuracy. For the RF model, we selected the number of trees to grow (ntree) as the number from which the error converged and remained stable based on the out of bag (OOB) error. Careful selection of this parameter is key, as we want to have enough trees to stabilize the error but not so many that we over-correlate the ensemble. On the other hand, the number of variables randomly sampled as candidates at each split (mtry) was selected based on the value that minimised the OOB error and maximised the model performance. The neighborhood parameter (k) for the k-NN algorithm was selected based on the model achieving the best accuracy when varying k. For the LDA, selected variables were included using the Discrim procedure in SAS with a parametric method based on multivariate normal distribution within each class to derive a linear discriminant function [88]. Five models were developed in order to compare their performance using various combinations of predictor variables and tuning parameters. To determine if the eight extra bands of WorldView-3 imagery (SWIR 1 to 8) would improve classification accuracy, those procedures were first applied with variables derived from 8-band WorldView-3 and then repeated with variables derived from 16-band WorldView-3. Once the best-performing model was identified it was used to map the tree species' distribution throughout the study areas.

2.6. Model Performance

The classification performance of each model was assessed based on confusion matrices computed using reference data (20%) selected as a test dataset. Comparing the observed and the predicted data allowed us to assess producer and user accuracies. Overall accuracy (OA) was calculated by averaging accuracies among all classes [104]. We calculated Cohen's Kappa index of agreement (KIA) to evaluate the possibility of an agreement occurring simply by chance [105]. The KIA is a robust statistic useful for reliability testing. Similar to correlation coefficients, it can range from −1 to +1, where 0 represents the amount of agreement that can be expected from random chance, and 1 represents perfect agreement [106,107]. We compared all the models and selected the optimal ones offering the highest OA and KIA.

3. Results

3.1. Individual Tree Crown Segmentation and Assessment

Three CHMs were used in the ITC segmentation: the original CHM, a filtered CHM and a corrected CHM. Figure 3 presents a 3D profile for each CHM. It shows that the original CHM has a high z range and the filtered CHM smooths the z variation. The corrected CHM is composed of both original local variety and filtered CHM smoothness.

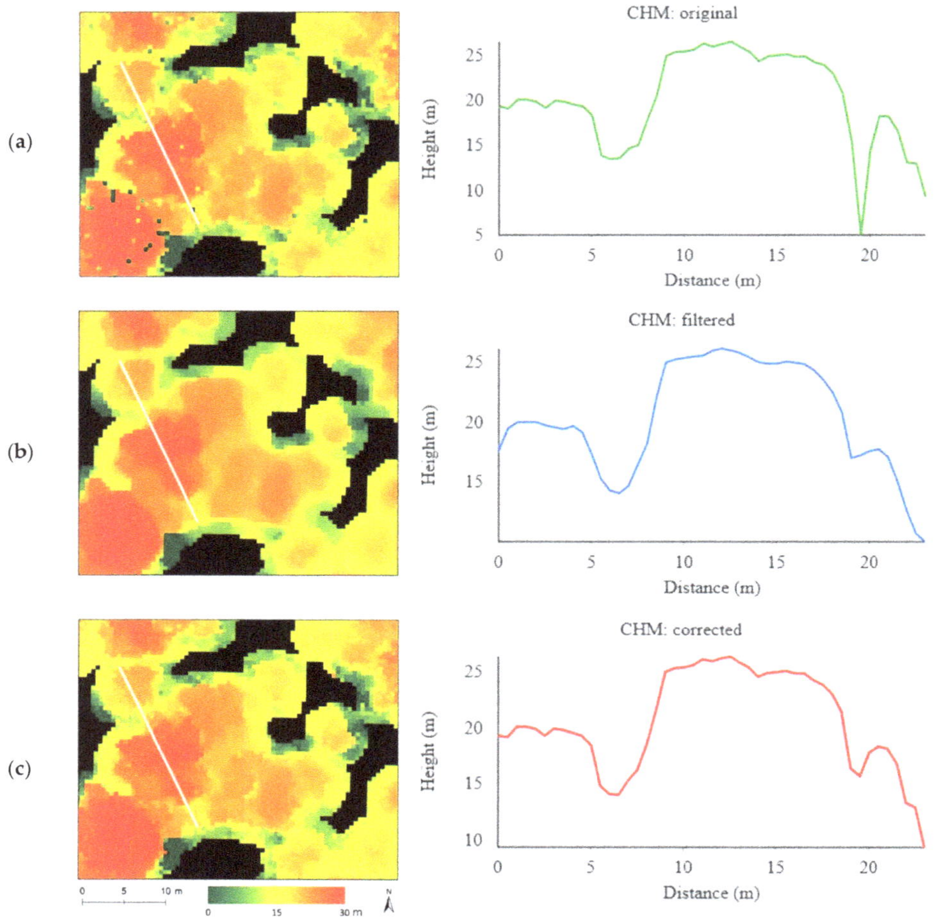

Figure 3. The three canopy height models (CHMs) used to make individual tree crown (ITC) segmentation. (**a**) original CHM; (**b**) filtered CHM; and (**c**) corrected CHM.

As a second step, imagery was added to the ITC segmentation. The assessment shows similar results for filtered and corrected CHMs (Table 3). They produced better delineation than the original CHM for single crown (63% vs. 40%) and single species (73% vs. 70%) segmentation. Our results indicate that the use of CHMs alone for ITC segmentation can lack precision, especially when different species' crowns can be interlaced or when their neighbors are at the same height. The combination of filtered CHM and imagery showed the best result for single crown delineation (68%). Combinations of original and corrected CHMs with imagery indicated that 56% and 64% of the objects fitted a single crown, respectively, showing over-segmented crowns. ITC segmentation assessment showed that the best results for single species were found by combining a corrected CHM with imagery. For this combination, 82% of the objects represented a single species, in contrast to original (74%) and filtered (75%) combinations. For the residual objects (18%), a majority showed a single species for 50% of their area. Figure 4 shows an example of ITC segmentation using a corrected CHM combined with imagery, the combination chosen for tree mapping.

Table 3. Individual tree crown assessment comparing segmented objects produced with original, filtered or corrected canopy height model (CHM) in combination with imagery.

	Original		Filtered		Corrected	
	CHM	CHM+Imagery	CHM	CHM+Imagery	CHM	CHM+Imagery
Single crown	40%	56%	60%	68%	63%	64%
Single species	70%	74%	73%	75%	73%	82%

Figure 4. Individual tree crown segmentation based on LiDAR (canopy height model) and WorldView-3 imagery. The background image displays WorldView-3 in true colors.

3.2. Classification and Assessment

For a visual analysis, the mean spectral values were calculated for each tree species of the training dataset (Figure 5). The species are most discriminated in the NIR region. Conifers (Figure 5B) are more separable than broadleaf species (Figure 5A), as their curves are more distinguishable.

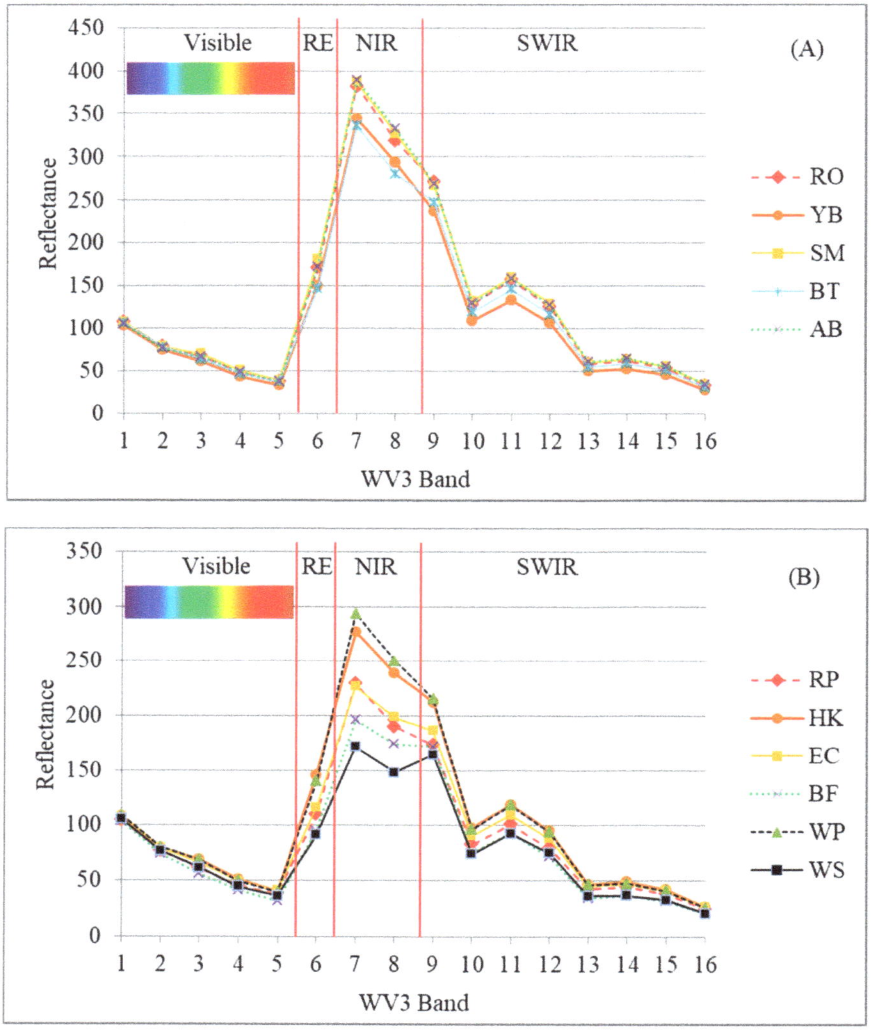

Figure 5. Spectral signatures for selected broadleaf (**A**) and conifer (**B**) species with 16-bands WorldView-3 (WV3). The points are the mean spectral values of the training samples. The spectral signatures are separated into four parts: visible, Red-edge (RE), Near-infrared (NIR) and Short-Wave Infrared (SWIR). The species' abbreviations and the number of training samples used to calculate the mean are described in Table 2.

Variable selection was conducted prior to the classification. From the 240 original derived variables, 50 and 75 were not correlated using the first eight bands and using all the bands, respectively, and were processed in the Boruta algorithm. The 15 most significant variables were then selected and utilized to run all possible combinations for all models except LDA. Among those 15 variables, three were especially notable: (A) the ARI_mean_95pc_higher index (Anthocyanin Reflectance Index); (B) the GLCM_Entropy_Band_7; and (C) the GMI2_mean index (Simple NIR/Red-edge Ratio) (Figure 6). The ARI (Anthocyanin Reflectance Index) was calculated with the 95th percentile highest pixels and allows an estimation of the anthocyanin (water-soluble vacuolar pigments) accumulation in intact stressed and senescing leaves [108]. Broadleaf trees have higher values than conifers, which is consistent with the fact that leaves, in contrary to needles, accumulate more anthocyanin to protect them from sunlight [109]. The GLCM (Grey Level Co-occurrence Matrix) entropy texture index was calculated using band 7 (832.5 nm); this index has a high value when all pixels are of similar magnitude [110,111]. Its values show that Sugar Maple (SM) species presents the most uniform pattern. A crown with small bumps and shallow cavities, with an appearance similar to that of a broccoli, are indeed characteristics used in photo-interpretation to identify this species [112]. The GMI2 (Gitelson and Merzylak Index) index is a simple ratio that allows chlorophyll content estimation in a wide range of pigment variation using insensitive (B7: near-infrared) and sensitive (B6: red-edge) bands and could be considered as an improved Normalized Difference Vegetation Index (NDVI) [113]. This index shows values that vary between conifer species. White pine (WP) is the species with the highest value, indicating a higher chlorophyll content than the other conifer species [114].

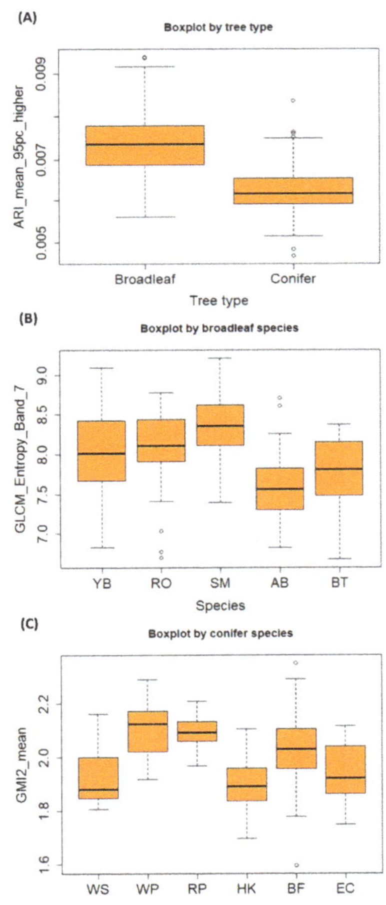

Figure 6. Boxplots of the (**A**) ARI_mean_95pc_higher index; (**B**) GLCM_Entropy_Band_7; and (**C**) GMI2_mean index for tree types, broadleaves and conifers, respectively. The species' abbreviations and the number of training samples used to calculate the mean are described in Table 2. The variables' abbreviations are described in Appendix A.

For the global modeling approach, RF (ntree: 2000; mtry: 3) was selected as the best model based on the performance assessment, as it gives an OA of 75% and a KIA of 0.72 when using 16-band derived variables (Table 4). This performance was achieved using nine variables. SVM also performed well, with 71% OA and a KIA value of 0.68. The k-NN, CART and LDA approaches offered less-precise classification, with OAs below 61%. All the performance models declined or were stable when using 8-band instead of 16-band derived variables, except for LDA, which increased by 5% (61% to 66%).

Table 4. Summary of the performance assessment using training and test sets with five modelling techniques (RF, SVM, k-NN, CART and LDA) for global modeling. The modeling technique with the highest overall accuracy (OA) is highlighted in dark. KIA: Kappa index of agreement.

Model	Technique	Based on 8-Band WorldView-3					Based on 16-Band WorldView-3				
		No of Variables	Training		Test		No of Variables	Training		Test	
			OA	KIA	OA	KIA		OA	KIA	OA	KIA
Global	RF	8	100%	1.00	71%	0.67	9	100%	1.00	75%	0.72
	SVM	10	93%	0.93	70%	0.66	10	98%	0.98	71%	0.68
	k-NN	9	72%	0.68	41%	0.34	10	78%	0.76	48%	0.42
	CART	8	74%	0.70	53%	0.48	10	71%	0.68	53%	0.48
	LDA	11	96%	0.95	66%	0.61	11	95%	0.94	61%	0.56

For the hierarchical modeling approach, RF (ntree: 2000; mtry: 2) presented the best performance with OA of 99% and KIA of 0.97 for tree type (broadleaf/conifer) modeling (Table 5). This performance was achieved using four variables derived from 16-band WorldView-3. If only 8-band instead of 16-band derived variables were used, RF and k-NN achieved the best performances (OA: 97%, KIA: 0.95). All the performance models declined or were stable when using 8-band derived variables. For broadleaf and conifer species modeling, RF (ntree: 2000; mtry: 2) gave the best performances using 8-band derived variables, with OAs of 70% and 94% and using six and seven variables, respectively (Table 6). For broadleaf modeling with RF, the KIA value (0.63) indicated a moderate agreement, which is considered as a substantial agreement as presented by McHugh [106]. The KIA was greater than 0.90 for conifer species, which is considered as an almost perfect agreement [106] with a value of 0.93. For RF, using 16-band instead of 8-band derived variables did not increase OAs for broadleaves (70% vs. 68%) or conifers (94% in both cases).

Table 5. Summary of the performance assessment using training and test sets for the five modelling techniques (RF, SVM, k-NN, CART and LDA) for the tree type (broadleaf/conifer) modeling approach. For tree mapping, the selected modeling technique with the highest overall accuracy (OA) is highlighted in dark. KIA: Kappa Index of agreement.

Model	Technique	Based on 8-Band WorldView-3					Based on 16-Band WorldView-3				
		No of Variables	Training		Test		No of Variables	Training		Test	
			OA	KIA	OA	KIA		OA	KIA	OA	KIA
Tree type	RF	4	100%	1.00	97%	0.95	4	100%	1.00	99%	0.97
	SVM	10	100%	1.00	94%	0.87	6	100%	1.00	97%	0.95
	k-NN	6	100%	1.00	97%	0.95	4	100%	1.00	97%	0.95
	CART	2	97%	0.93	92%	0.85	4	98%	0.96	92%	0.85
	LDA	3	97%	0.93	96%	0.92	4	100%	0.99	97%	0.95

Table 6. Summary of the performance assessment using training and test sets for the five modelling techniques (RF, SVM, k-NN, CART and LDA) for broadleaf and conifer species modeling. For tree mapping, the selected modeling techniques with the highest overall accuracy (OA) for each model are highlighted in dark. KIA: Kappa Index of agreement.

Model	Technique	Based on 8-Band WorldView-3					Based on 16-Band WorldView-3				
		No of Variables	Training		Test		No of Variables	Training		Test	
			OA	KIA	OA	KIA		OA	KIA	OA	KIA
Broadleaf	RF	6	100%	1.00	70%	0.63	6	100%	1.00	68%	0.60
	SVM	10	96%	0.95	59%	0.49	10	95%	0.94	68%	0.60
	k-NN	7	79%	0.73	52%	0.39	6	83%	0.78	36%	0.03
	CART	6	75%	0.68	45%	0.31	6	77%	0.70	59%	0.49
	LDA	10	94%	0.93	64%	0.53	9	93%	0.91	61%	0.51
Conifer	RF	7	100%	1.00	94%	0.93	7	100%	1.00	94%	0.93
	SVM	10	96%	0.95	89%	0.85	9	100%	1.00	83%	0.78
	k-NN	9	89%	0.86	83%	0.79	9	88%	0.84	89%	0.85
	CART	5	79%	0.73	77%	0.71	7	81%	0.75	69%	0.60
	LDA	8	99%	0.99	80%	0.74	9	100%	1.00	71%	0.63

Without considering the global approach, the three models selected to classify tree species used a total of 16 variables: nine spectral indices, three simple bands, one standard deviation and three textures (Table 7). Out of the 16 WorldView-3 bands, a total of nine bands were used in our selected models (Table 8).

The following error matrices come from the best models (RF) using 16-band and 8-band derived variables for the global approach and tree type, and for individual species, respectively. Although the RF model presented an OA of 75% for the global approach, its precision per species fluctuated highly and varied between 58% for SM and 100% for Big Tooth Aspen (BT), Eastern Hemlock (HK) and White Spruce (WS) for the user's accuracy in the confusion matrix, while its producer's accuracy varied between 33% for WS and 100% for HK and WP (Table 9). The tree type model (Table 10) had almost perfect results (OA: 99%), with one error: one conifer was classified as a broadleaf. For broadleaf species identification (OA: 70%) (Table 11), all species were classified over 70%, except Red Oak (RO) and Yellow Birch (YB) (67%) for user's accuracy. For producer's accuracy, all species were classified over 60%. For conifer species (OA: 94%) (Table 12), all species were perfectly classified (100%), except WP with (83%) for user's accuracy and Eastern White Cedar (EC) and Red Pine (RP) (80%) for producer's accuracy.

The RF model was used to map tree types and individual broadleaf and conifer tree species over the study areas. Figure 7 illustrates the classification map in an island composed of both tree types. Tall WP trees are especially visible on the border of the island. The inland is mainly made up of SM and YB. Smaller trees (<17 m) were not mapped.

Table 7. Description of the sixteen variables used in the selected models for the hierarchical approach. For spectral indices calculated with 95th percentile highest pixels, the abbreviation is "variable_mean_95pc_higher", for arithmetic feature calculated for spectral indices, the abbreviation is "variable_mean". Band numbers are described in Table 1.

Abbreviation	Vegetation Index	Adapted Formula	Models	References
ARI_mean	Anthocyanin Reflectance Index	1/B3_mean − 1/B6_mean	Conifer	[115]
ARI_mean_95pc_higher	Anthocyanin Reflectance Index	Arithmetic mean of the 5% higher pixel value of the object with ARI	Tree type	[115]
Band_1_mean	Layer values	Mean value of band 1 of the pixels forming the object	Broadleaf	[6]
Band_12_95pc_higher	Layer values	Arithmetic mean of the 5% higher pixel value of the object using band 12	Tree type	[6]
Band_5_95pc_higher	Layer values	Arithmetic mean of the 5% higher pixel value of the object using band 5	Broadleaf	[6]
GLCM_Entropy_Band_7	Texture values	Entropy calculated with the value of band 7 of the pixels forming an object	Broadleaf; Conifer	[110,111]
GLCM_Homogeneity_Band_3	Texture values	Homogeneity calculated with the value of band 3 of the pixels forming an object	Conifer	[110,111]
GLCM_Homogeneity_Band_4	Texture values	Homogeneity calculated with the value of band 4 of the pixels forming an object	Conifer	[110,111]
GMI2_mean	Simple NIR/Red-edge Ratio	B7_mean/B6_mean	Conifer	[113]
IHS_Hue_Band_7_3_2	Intensity, hue, saturation (HIS) transformation	Hue calculated with B7, B3 and B2 as red, green and blue	Conifer	[94,110]
PRI2_mean	Normalized difference Physiological Reflectance Index	(B3_mean − B4_mean)/(B3_mean + B4_mean)	Broadleaf	[116]
PRI2_mean_95pc_higher	Normalized difference Physiological Reflectance Index	Arithmetic mean of the 5% higher pixel value of the object with PRI2	Broadleaf	[116]
Sredgreen_mean	Simple Red/Green ratio	B5_mean/B3_mean	Conifer	[117]
Sredgreen_mean_95pc_higher	Simple Red/Green ratio	Arithmetic mean of the 5% higher pixel value of the object with Sredgreen	Tree type	[117]
Standard_deviation_Band_3	Layer values	Standard deviation of band 3 of the pixels forming the object	Broadleaf	[6]
TCP_greeness_mean	Tasselled Cap—Green Vegetation Index	(B2_mean * −0.2941)+(B3_mean * −0.243)+(B5_mean * −0.5424)+(B7_mean * 0.7276)+(B10_mean * 0.0713)+(B14_mean * −0.1608)	Tree type	[118]

Table 8. The number of times WorldView-3 bands were used with the hierarchical approach for tree mapping.

Band	B1	B2	B3	B4	B5	B6	B7	B12	B14
Times used	1	2	10	3	4	3	5	1	1

Table 9. Confusion matrix of the RF model for the global approach. AB: American Beech, BT: Big Tooth Aspen, RO: Red Oak, SM: Sugar Maple, YB: Yellow Birch, BF: Balsam Fir, EC: Eastern White Cedar, HK: Eastern Hemlock, RP: Red Pine, WP: White Pine, WS: White Spruce. OA: Overall accuracy, KIA: Kappa Index of agreement.

		Reference											
		AB	BT	RO	SM	YB	BF	EC	HK	RP	WP	WS	User's Accuracy (%)
Prediction	AB	**7**	0	0	0	2	0	0	0	0	0	0	78%
	BT	0	**4**	0	0	0	0	0	0	0	0	0	100%
	RO	0	0	**5**	1	0	0	0	0	0	0	0	83%
	SM	1	0	3	**7**	1	0	0	0	0	0	0	58%
	YB	1	0	2	1	**7**	0	0	0	0	0	0	64%
	BF	0	0	0	0	0	**2**	0	0	0	0	1	67%
	EC	1	1	0	0	0	0	**4**	0	0	0	0	67%
	HK	0	0	0	0	0	0	0	**9**	0	0	0	100%
	RP	0	0	0	0	0	0	1	0	**3**	0	1	60%
	WP	0	0	0	0	0	1	0	0	2	**10**	0	77%
	WS	0	0	0	0	0	0	0	0	0	0	**1**	100%
Producer's accuracy (%)		70%	80%	50%	78%	70%	67%	80%	100%	60%	100%	33%	**OA: 75%** **KIA: 0.72**

Table 10. Confusion matrix of the RF model for the hierarchical approach: binary classification of tree type (broadleaf/conifer). OA: Overall accuracy, KIA: Kappa Index of agreement.

		Reference		
		Broadleaf	Conifer	User's Accuracy (%)
Prediction	Broadleaf	**44**	1	98
	Conifer	0	**34**	100
Producer's accuracy (%)		100	97	**OA: 99%** **KIA: 0.97**

Table 11. Confusion matrix of the RF model for the hierarchical approach: broadleaf species classification. AB: American Beech, BT: Big Tooth Aspen, RO: Red Oak, SM: Sugar Maple, YB: Yellow Birch. OA: Overall accuracy, KIA: Kappa Index of agreement.

		Reference					
		AB	BT	RO	SM	YB	User's Accuracy (%)
Prediction	AB	**6**	0	1	0	1	75%
	BT	1	**5**	0	0	1	71%
	RO	0	0	**6**	3	0	67%
	SM	1	0	1	**6**	0	75%
	YB	2	0	2	0	**8**	67%
Producer's accuracy (%)		70%	60%	100%	60%	67%	**OA: 70%** **KIA: 0.63**

Table 12. Confusion matrix of the RF model for the hierarchical approach: conifer species classification. BF: Balsam Fir, EC: Eastern White Cedar, HK: Eastern Hemlock, RP: Red Pine, WP: White Pine, WS: White Spruce. OA: Overall accuracy, KIA: Kappa Index of agreement.

		Reference						
		BF	**EC**	**HK**	**RP**	**WP**	**WS**	**User's Accuracy (%)**
Prediction	**BF**	**3**	0	0	0	0	0	100%
	EC	0	**4**	0	0	0	0	100%
	HK	0	0	**9**	0	0	0	100%
	RP	0	0	0	**4**	0	0	100%
	WP	0	1	0	1	**10**	0	83%
	WS	0	0	0	0	0	**3**	100%
Producer's accuracy (%)		100%	80%	100%	80%	100%	100%	**OA: 94%** **KIA: 0.93**

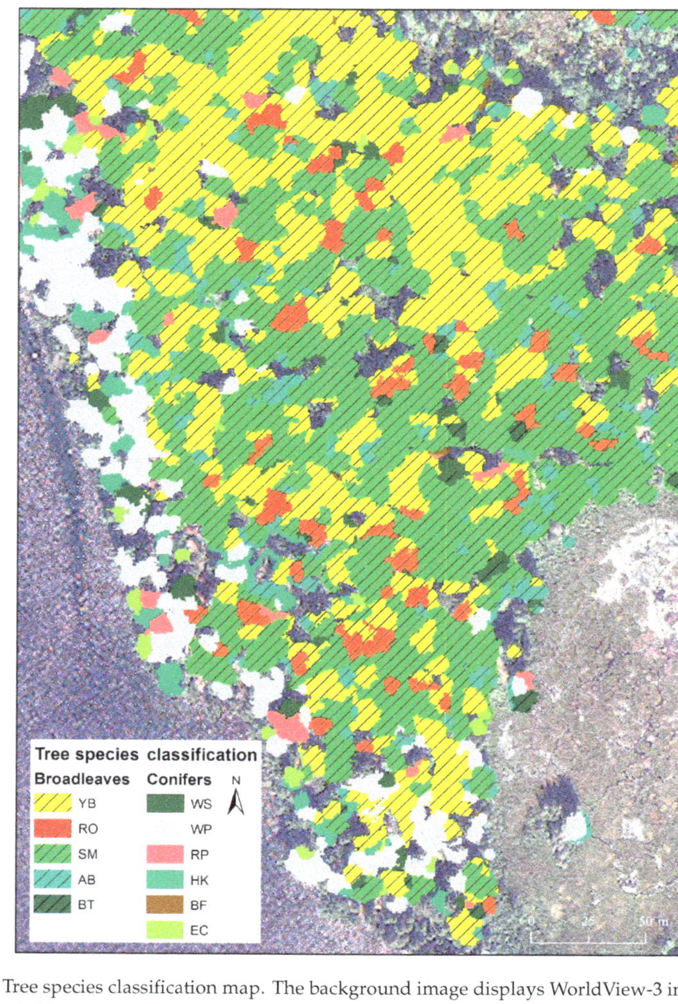

Figure 7. Tree species classification map. The background image displays WorldView-3 in true colors. The species' abbreviations are described in Table 2.

4. Discussion

This study compares five different models to successfully map 11 tree species in a natural North American forest based on WorldView-3 imagery and LiDAR data. The proposed method is highlighted by three main aspects: (1) an object-based segmentation technique using imagery and LiDAR; (2) a hierarchical classification approach with more than ten species; and (3) model iterations for optimal selection.

ITC segmentation is usually implemented when mapping species at the tree level, and studies have often used LiDAR data [13,119] or imagery [6,11,120]. Using only LiDAR or imagery at the tree level results in objects with merged tree crowns [121], especially in a mature broadleaf forest like the one in the present study. Both data types could be used together to limit this effect. As an example, Heinzel and Koch [121] delineated ITCs using a pixel-based classification within the objects to avoid neighbor tree errors. While Alonso-Benito et al. [39] used LiDAR and imagery for segmentation, they did not classify at the tree level. Koukoulas and Blackburn [83] also used both data types, but with a succession of complex GIS procedures to find treetops. The ITC segmentation proposed here follows a watershed algorithm [122] from LiDAR data similarly to Weinacker et al. [93] and Koch et al. [26]. Significant bands for tree types (broadleaf and conifer) were then used to refine the segmentation using a multiresolution algorithm as suggested by Pham et al. [32] and Koukoulas and Blackburn [83]. This approach has similarities with multiscale approaches to separate species in a dense and complex forest. Indeed, raster-based ITC segmentation approaches do not allow object overlaps yet offer a more realistic representation for a broadleaf natural forest [16]. As shown in Table 3, the results indicate that using a filtered or corrected CHM delineates single crowns and species better than using an original CHM (increased accuracy of at least 20% for single crowns and 3% for single species). When imagery is added to ITC segmentation it leads to over-segmentation, creating many objects in large crown cases when compared to their corresponding manually-delineated crowns. Single crown delineation accuracy could be reduced. In such a situation, one option would be to merge similar small objects [24] using spectral difference as a second step [80], although over-segmentation is generally preferred to under-segmentation [40,95]. For this assessment, no isolated tree crowns were used. This could be another reason why accuracies were not over 70% for single crown delineation. For single species delineation, its accuracy improved with imagery; up to 9% for the corrected CHM. For filtered and original CHMs, the accuracy slightly improved with imagery (2–4%). This could be related to the fact that ITC segmentation using filtered CHM alone produced bigger objects. Those were then divided into smaller parts that were not entirely covered (at least by 75%) by a single species.

The Kenauk Nature property is composed of a complex mixed forest with more than 25 tree species. A number of studies have used high spatial resolution sensors to map tree species in a natural forest environment at the tree level; those included relatively few species recognition [6,29,32,43,44,93,121]. For example, Immitzer et al. [7] classified 10 species while concentrating on pure stands for reference data, where spectral variability could be limited. Having such a high number of species in our study area forced us to survey only the dominant species (11). Misclassification could therefore be influenced by the complex forest environment that made it difficult to target suitable data for references. For this reason, we manually delineated tree samples by stereo photo-interpretation to have reliable data as suggested by Immitzer et al. [7].

Previous studies generally limited their classification to a global approach without new machine learning techniques such as SVM and RF. For example, Waser et al. [6] classified seven species with an OA of 83% with a global approach using LDA. The present study demonstrates a hierarchical classification approach as a significant procedure in order to classify and map 11 tree species. This approach conserves the integrity of the tested algorithms in a hierarchical perspective by first classifying tree types and then the individual species by their corresponding type. Our results show that the hierarchical approach gives a better performance than using a single global approach, especially for conifers, which is consistent with other studies [123]. However, the hierarchical approach needed more variables (16) than the global approach (nine). Also, the hierarchical approach presented here shows that using multiple modeling techniques at each level allows the best models to be selected, which could vary. Therefore, this approach has the ability to reduce unbalanced accuracies between species as reported by studies working in a global approach [7]. In our case, RF was the best model for all levels, followed by SVM. This observation is in opposition to other studies working with coarser imagery, such as Sentinel-2 [60].

Another interesting element is that this approach allows the selection of relevant variables and specific model techniques for each classification level. The variables selected for each model were not the same for broadleaf and conifer species. For example, broadleaf species are more distinguishable using texture variables because their branch structures are much more varied (Figure 6(B)). A similar technique was used in Krahwinkler and Rossmann [124] to make a binary decision tree hierarchical structure by classifying each single species. Our approach permits a simpler way to classify species by type with satisfying results, and limits the hierarchical structure to two levels. Moreover, instead of using only the SVM, we tried five different models to optimize the accuracy. On the other hand, it is worth noting that SVM and RF are generally the best algorithms according to their OA. For tree species classification, SVM is generally recognized to be more effective when working with a small number and imbalanced distribution of samples [45]. It should be pointed out that ancillary variables (topographic position index, topographic wetness index and water proximity, etc.) could also improve classification accuracy [32,53,125], although it would be important to collect stratified samples evenly distributed among those variables.

Mitigated improvements were observed when using 16-band or 8-band WorldView-3 derived variables. The additional eight bands (SWIR 1 to 8) slightly enhanced the global approach (OA: 75% vs. 71%, KIA: 0.72 vs. 0.67) and tree type classifications (OA: 99% vs. 97%, KIA: 0.97 vs. 0.95), but did not improve individual species classification. This is partially consistent with other studies that observed an improved classification accuracy when adding new bands, especially with a large number of tree species [7,40]. For example, Ferreira et al. [126] simulated WorldView-3 bands for tree species discrimination and found that incorporating SWIR bands significantly increased the average accuracy. Despite the low spatial resolution compared to other multispectral bands (5.25×7.5 m vs. 0.84×1.2 m), the spectral information of SWIR bands was significant for certain inter-species separability, despite the fact that their integration should be made with caution when mapping smaller trees because their crowns could be covered by just a few pixels. Adding the SWIR bands also permitted to integrate spectral indices that were developed within hyperspectral studies [127–130]. Finally, the small accuracy improvement shows that it may be sufficient to use only 8-band derived variables to simplify the method.

The model iterations procedure for optimal selection is an important contribution of this study compared to other similar studies. Studies generally integrate all variables without an oriented variable selection or by using complicated methods such as linear mixed-effects modeling and genetic algorithms [32,45,131–133]. However, this selection aspect is essential to insure reproducibility for operational purposes [14]. Moreover, our results showed that using fewer variables could actually improve the classification. We proposed a simple method using all the variables in order to select the 15 most significant variables provided by the Boruta algorithm [98], and eliminated the inter-correlated variables similarly to Budei et al. [14]. We then computed all combinations to determine the one that obtained the best results using the least possible number of variables.

Spectral variable calculation techniques are also an important aspect of this procedure. A majority of the recent studies use a pixel-based calculation technique to perform spectral variables [6,44,45]. We used two different calculation techniques: pixel-based and arithmetic feature (mean of all pixels or 95th percentile highest pixels within each object). For example, a tree crown could have an NDVI value that differs depending on if it is calculated using the mean of each red and near-infrared band (arithmetic feature) or if the mean of the NDVI calculated by pixel is extracted. The first case allows spectral variables to be calculated rapidly, while the second case makes it possible to calculate textural variables, for example. Indeed, an arithmetic feature has the advantage of creating variables rapidly within R or SAS instead of adding a new raster band each time, which would make massive data management difficult. Additionally, using the 95th percentile of the highest pixel values allowed us to keep the sunlit parts of crowns and thereby limit the shadow effects which could affect classification accuracy [7].

Machala et al. [19] was concerned about using maximum values in features where objects are heterogeneous (e.g., high and low trees), but this is not the case in our study since ITC segmentation is aiming for homogeneous objects. While testing correlations for both calculation techniques, we obtained high coefficients for many variables. For the arithmetic feature of NDVI with the mean of all pixels within each object, we found correlations of 0.99 and 0.93 for pixel-based and 95th percentile of highest pixel values' corresponding variables, respectively. This method allowed more variables to be implemented in the modeling process.

Although the proposed approach is robust to identify 11 tree species, three main limitations were identified. The first limitation was that unevenly distributed samples between the 11 species made it difficult to correctly use machine learning models such as RF. This limitation was also identified by Tao et al. [134] and Farquad and Bose [135]. It is known that using an unbalanced training dataset tends to affect the prediction accuracy of the dominant classes, which implies lower accuracies in the less represented classes [60]. To limit this impact, new samples could balance the dataset, but this simple solution is also the most expensive, involving additional field surveys and photo-interpretation. As suggested by Farquad and Bose [135], another solution could be to automatically over- or under-balance the dataset [136].

The second limitation concerns spatial and spectral resolutions and the georeferencing of imagery. The research presented here was based on 16-band WorldView-3 imagery. Firstly, the spectral quality could have been affected by rescaling. WorldView-3 bands contain various spatial resolutions from 0.21 m for panchromatic up to 7.5 m for shortwave infrared. The panchromatic band ranges from 450 to 800 nm, covering the first seven bands. Despite the fact that the nine other bands were out of range, for methodological purposes, all bands were rescaled and pansharpened. Those last nine bands could have been degraded, which may have affected the modeling and the reproducibility of the method [137]. To limit this impact, the last nine bands should not be used for the texture variables. Secondly, despite preprocessing, an offset between imagery and LiDAR CHM persist (RMS: 0.97 m) and could affect the ITC segmentation and classification modeling. The offset at the ground level was almost perfect, but the misalignment of the crowns was sometimes over 3 m. Tree crowns tilted in the image could be used for stereo-reconstruction when at least two images are used [78,138], but using a single image caused complex situations where segmented LiDAR crowns were not matching their corresponding trees in the WorldView-3 image. A digital surface model derived from LiDAR could also be used to orthorectify the image [25,29]. However, we did several tests and decided not to use this technique because it created many artefacts when high spatial resolution images such as WorldView-3 were used. In this study, where mature trees were present all over the area, manual points were collected to fit the CHM and thereby reduce this offset. To limit this effect, a threshold of 17 m was set as a mask in order to analyze only tall and large trees. The imagery was also integrated in the ITC segmentation to divide objects including more than one species as a solution to eliminate the offset between data sources.

The third limitation of the proposed approach concerns the fact that the territory is composed of more than 11 species. Given that the species modeling does not include the full diversity, a marginal species will be classified into one of the 11 species classes used in the modeling. Also, small trees were not mapped, as a threshold of 17 m was incorporated. It would be interesting to integrate more species classes in the modeling, considering groups of age or height [26]. Although more species will make the model more complex, functional groups could be tested in the hierarchical approach [139], multi-temporal imagery could be used [40,41,45] or more advanced algorithms like deep learning techniques [31]. Li et al. [45] argued that using bi-temporal WorldView imagery could improve the classification on average by 10.7%. He et al. [40] found their best results when combining late-spring, mid-summer and early-fall images. Hartling et al. [31] demonstrated that deep learning techniques could improve broadleaf species classification by at least 30% compared to RF and SVM. Adding other variables such as LiDAR metrics or topological measures could also improve the classification [8,14,16,22,39,131]. Finally, an expert procedure could be implemented to select a maximum number of each categorical variable to limit over-representation [136]. For example, this procedure would avoid the need to automatically select only LiDAR variables and instead allow for a mix of LiDAR, spectral indices, topological variables, etc.

5. Conclusions

This study proposes a method to map individual tree species by using machine learning techniques with very high spatial resolution imagery (WorldView-3) in a complex natural North American forest at the tree level. An object-based approach at multiple scales was conducted. We found that adding spectral information to CHM improved ITC segmentation. We were able to successfully classify five broadleaf species and six conifer species using a hierarchical approach, with OAs of 70% and 94%, respectively. This hierarchical approach had better accuracies for conifers than using a global approach (75%). Only sixteen variables were used with three models (tree type, broadleaf and conifer) corresponding to nine spectral bands. Among the five tested machine learning techniques, RF provided the best results for all cases. This method could also be applied on a large scale with limited manipulations. The resulting maps represent a valuable tool with which to analyze forest composition and to guide forest planners. However, ITC segmentation could be enhanced with automatic evaluation techniques which could allow additional iterations. Ancillary variables such as topographic and hydrographic indices could also improve the classification accuracy. This approach could also be enriched by balancing dataset and expert procedures, by integrating LiDAR metrics and multi-temporal imagery or by combining other sensors such as hyperspectral sensors.

Author Contributions: Conceptualization, M.V.; methodology, M.V., G.J. and B.C.; software, M.V., G.J. and B.C.; validation, M.V.; formal analysis, B.C. and G.J.; writing—original draft preparation, M.V.; writing—review and editing, M.V., G.J. and B.C.; supervision, M.V.; project administration, M.V.; funding acquisition, M.V. All authors have read and agreed to the published version of the manuscript.

Funding: This work was supported by the Canadian Wood Fiber Centre, which is part of the Canadian Forest Service in Natural Resources Canada [CWFC1718-017], in partnership with the Forest Innovation Program, and by the Ministère de l'Éducation et de l'Enseignement Supérieur du Québec.

Acknowledgments: The authors gratefully acknowledge Antoine Cullen, Martin Dupuis, Anne-Marie Dubois and Philippe Bournival for their support during the project.

Conflicts of Interest: The authors declare no conflict of interest.

Appendix A

Table A1. Variables' details. For pixel-based spectral indices, the abbreviation is "variable_mean_pixel", for spectral indices calculated with 95th percentile highest pixels, the abbreviation is "variable_mean_95pc_higher", for arithmetic feature calculated for spectral indices, the abbreviation is "variable_mean". Band numbers are described in Table 1.

Type	Abbreviation	Spectral Variable	Description/Adapted Formula	Reference	Pixel-Based	Higher 95%	Arithmetic Feature	Total Number
Calculated on band	Band_X_mean	Band 1 to 16	Arithmetic mean value of band X of the pixels forming the object				x	16
	Standard_deviation_Band_X	Band 1 to 16	Standard deviation of band X of the pixels forming the object				x	16
	Skewness_Band_X	Band 1 to 16	Skewness of band X of the pixels forming the object				x	16
	Band_X_mean_95pc_highest	Band 1 to 16	Arithmetic mean of the 5% higher pixel value of the object				x	16
Spectral indices	ARI	Anthocyanin Reflectance Index	1/B3 – 1/B6	[115]	x	x	x	3
	ARI2	Anthocyanin Reflectance Index	1/B3 – 1/B5	[115]	x	x	x	3
	CI	Carter Index	B7/B5	[128]	x	x	x	3
	CRI	Carotenoid Reflectance Index	B7 * (1/B2 – 1/B3)	[140]	x	x	x	3
	CRI2	Carotenoid Reflectance Index	1/B2 – 1/B5	[140]	x	x	x	3
	DRI	Datt Reflectance Index	(B7 – B14)/(B7 + B14)	[141]	x	x	x	3
	DWSI	Disease Water Stress Index	B7/B10	[130]	x	x	x	3
	GMI1	Simple NIR/Red-edge Ratio	B8/B6	[113]	x	x	x	3
	GMI2	Simple NIR/Red-edge Ratio	B7/B6	[113]	x	x	x	3
	MSI	Moisture Stress Index	B10/B7	[142]	x	x	x	3
	MSISR	Ratio MSI/simple ratio	(B10/B7)/(B8/B5)	[143]	x	x	x	3
	NDII	Normalized Difference Infrared Index	(B7 – B11)/(B7 + B11)	[144]	x	x	x	3
	NDLI	Normalized Difference Lignin Index	[log(1/B12) – log(1/B11)]/[log(1/B12) + log(1/B11)]	[145]	x	x	x	3
	NDNI	Normalized Difference Nitrogen Index	[log(1/B10) – log(1/B11)]/[log(1/B10) + log(1/B11)]	[145]	x	x	x	3
	NDVI1	Normalized Difference Vegetation Index	(B7 – B5)/(B7 + B5)	[128]	x	x	x	3
	NDVI2	Normalized Difference Vegetation Index	(B8 – B5)/(B8 + B5)	[127]	x	x	x	3

Table A1. *Cont.*

Type	Abbreviation	Spectral Variable	Description/Adapted Formula	Reference	Pixel-Based	Higher 95%	Arithmetic Feature	Total Number
	NDWI	Normalized Difference Water Index	(B7 − B9)/(B7 + B9)	[146]	x	x	x	3
	NDWI2130	Normalized Difference Water Index	(B7 − B13)/(B7 + B13)	[147]	x	x	x	3
	NMDI	Normalized Multi-Band Drought Index	[RB7 − (B11 − B13)]/[B7 + (B11 − B13)]	[148]	x	x	x	3
	PBI	Plant Biochemical Index	B7/B3	[129]	x	x	x	3
	PRI1	Normalized difference Physiological Reflectance Index	(B2 − B3)/(B2 + B3)	[149,150]	x	x	x	3
	PRI2	Normalized difference Physiological Reflectance Index	(B3 − B4)/(B3 + B4)	[149,150]	x	x	x	3
	PSRI1	Plant Senescence Reflectance Index	(B5 − B2)/B7	[150]	x	x	x	3
	PSRI2	Plant Senescence Reflectance Index	(B5 − B2)/B6	[150]	x	x	x	3
	R5R7	Ratio of Landsat TM band 5 to band 7	B11/B14	[151]	x	x	x	3
	RENDVI	Red-edge Normalized Difference Vegetation Index	(B7 − B6)/(B7 + B6)	[127]	x	x	x	3
	RGR1	Simple Red/Green ratio	B5/B2	[150]	x	x	x	3
	SIPI	Structure Insensitive Pigment Index	(B7 − B2)/(B7 + B5)	[116,150]	x	x	x	3
	Sredgreen	Simple Red/Green ratio	B5/B3	[117]	x	x	x	3
	SRWI	Simple Ratio Water Index	B7/B9	[152]	x	x	x	3
	TCP_brightness	Tasseled Cap—Brightness	(B2 * 0.3029)+(B3 * 0.2786)+(B5 * 0.4733)+ (B7 * 0.5599)+(B10 * 0.508)+(B14 * 0.1872)	[118]	x	x	x	3
	TCP_greeness	Tasseled Cap—Green Vegetation Index	(B2 * −0.2941)+(B3 * −0.243)+(B5 * −0.5424)+ (B7 * 0.7276)+(B10 * 0.0713)+(B14 * −0.1608)	[118]	x	x	x	3
	TCP_wetness	Tasseled Cap—Wetness	(B2 * 0.1511)+(B3 * 0.1973)+(B5 * 0.3283)+ (B7 * 0.3407)+(B10 * −0.7117)+(B14 * −0.4559)	[118]	x	x	x	3
	VARI	Visible Atmospherically Resistant Index	(B3 − B5)/(B5 + B3 − B2)	[140]	x	x	x	3

Table A1. *Cont.*

Type	Abbreviation	Spectral Variable	Description/Adapted Formula	Reference	Pixel-Based	Higher 95%	Arithmetic Feature	Total Number
	Vigreen	Visible Atmospherically Resistant Indices Green	(B3 − B5)/(B5 + B3)	[140]	x	x	x	3
	WBI	Water Band Index	B7/B8	[153]	x	x	x	3
	IHS_Hue_Band_5_3_2	Intensity, hue, saturation (HIS) transformation	Hue calculated with B5, B3 and B2 as red, green and blue	[94]			x	1
	IHS_Hue_Band_7_3_2	Intensity, hue, saturation (HIS) transformation	Hue calculated with B7, B3 and B2 as red, green and blue	[94]			x	1
	IHS_Sat_Band_5_3_2	Intensity, hue, saturation (HIS) transformation	Saturation calculated with B5, B3 and B2 as red, green and blue	[94]			x	1
	IHS_Sat_Band_7_3_2	Intensity, hue, saturation (HIS) transformation	Saturation calculated with B7, B3 and B2 as red, green and blue	[94]			x	1
Textures	GLCM_Contrast_ Band_X	Band 1 to 16	Contrast calculated with the pixels forming an object	[111]	x			16
	GLCM_Dissimilarity_ Band_X	Band 1 to 16	Dissimilarity calculated with the pixels forming an object	[111]	x			16
	GLCM_Entropy_ Band_X	Band 1 to 16	Entropy calculated with the pixels forming an object	[111]	x			16
	GLCM_Homogeneity_ Band_X	Band 1 to 16	Homogeneity calculated with the pixels forming an object	[111]	x			16
Total								232

References

1. Leboeuf, A.; Vaillancourt, É. *Guide de Photo-Interprétation des Essences Forestières du Québec Méridional—Édition 2015*; Direction des Inventaires Forestiers du MFFP: Québec, QC, Canada, 2015; p. 72.
2. Berger, J.-P. *Norme de Stratification Écoforestière-Quatrième Inventaire Écoforestier*; Comité Permanent de la Stratification Forestière de la Direction des Inventaires Forestiers du MRNFQ et Forêt Québec: Québec, QC, Canada, 2008; p. 64, ISBN 978-2-550-73857-2.
3. Wulder, M.A.; White, J.C.; Hay, G.J.; Castilla, G. Towards automated segmentation of forest inventory polygons on high spatial resolution satellite imagery. *For. Chron.* **2008**, *84*, 221–230. [CrossRef]
4. Varin, M.; Joanisse, G.; Ménard, P.; Perrot, Y.; Lessard, G.; Dupuis, M. *Utilisation D'images Hyperspectrales en Vue de Générer une Cartographie des Espèces Forestières de Façon Automatisée*; Centre D'enseignement et de Recherche en Foresterie de Sainte-Foy Inc. (CERFO): Québec, QC, Canada, 2016; p. 68.
5. Cho, M.A.; Malahlela, O.; Ramoelo, A. Assessing the utility WorldView-2 imagery for tree species mapping in South African subtropical humid forest and the conservation implications: Dukuduku forest patch as case study. *Int. J. Appl. Earth Obs. Geoinf.* **2015**, *38*, 349–357. [CrossRef]
6. Waser, L.T.; Küchler, M.; Jütte, K.; Stampfer, T. Evaluating the potential of worldview-2 data to classify tree species and different levels of ash mortality. *Remote Sens.* **2014**, *6*, 4515–4545. [CrossRef]
7. Immitzer, M.; Atzberger, C.; Koukal, T. Tree species classification with Random forest using very high spatial resolution 8-band worldView-2 satellite data. *Remote Sens.* **2012**, *4*, 2661–2693. [CrossRef]
8. Maltamo, M.; Vauhkonen, J.; Næsset, E. *Forestry Applications of Airborne Laser Scanning-Concepts and Case Studies*; Springer: Dordrecht, The Nederlands, 2014; Volume 32, ISBN 978-94-017-8663-8.
9. Eid, T.; Gobakken, T.; Næsset, E. Comparing stand inventories for large areas based on photo-interpretation and laser scanning by means of cost-plus-loss analyses. *Scand. J. For. Res.* **2004**, *19*, 512–523. [CrossRef]
10. Tompalski, P.; Coops, N.C.; White, J.C.; Wulder, M.A. Simulating the impacts of error in species and height upon tree volume derived from airborne laser scanning data. *For. Ecol. Manag.* **2014**, *327*, 167–177. [CrossRef]
11. Gougeon, F.A.; Cormier, R.; Labrecque, P.; Cole, B.; Pitt, D.; Leckie, D. Individual Tree Crown (ITC) delineation on Ikonos and QuickBird imagery: The Cockburn Island Study. In Proceedings of the 25th Canadian Symposium on Remote Sensing, Montreal, QC, Canada, 14–16 October 2003; pp. 14–16.
12. Ahmad Zawawi, A.; Shiba, M.; Jemali, N.J.N. Accuracy of LiDAR-based tree height estimation and crown recognition in a subtropical evergreen broad-leaved forest in Okinawa, Japan. *For. Syst.* **2015**, *24*. [CrossRef]
13. Barnes, C.; Balzter, H.; Barrett, K.; Eddy, J.; Milner, S.; Suárez, J. Individual Tree Crown Delineation from Airborne Laser Scanning for Diseased Larch Forest Stands. *Remote Sens.* **2017**, *9*, 231. [CrossRef]
14. Budei, B.C.; St-Onge, B.; Audet, F.-A.; Hopkinson, C. Identifying the genus or species of individual trees using a three-wavelength airborne lidar system. *Remote Sens. Environ.* **2018**, *204*, 632–647. [CrossRef]
15. Koch, B.; Heyder, U.; Weinacker, H. Detection of Individual Tree Crowns in Airborne Lidar Data. *Photogramm. Eng. Remote Sens.* **2006**, *72*, 357–363. [CrossRef]
16. Jakubowski, M.; Li, W.; Guo, Q.; Kelly, M. Delineating Individual Trees from Lidar Data: A Comparison of Vector- and Raster-based Segmentation Approaches. *Remote Sens.* **2013**, *5*, 4163–4186. [CrossRef]
17. Rana, P.; Prieur, J.-F.; Budei, B.C.; St-Onge, B. Towards a Generalized Method for Tree Species Classification Using Multispectral Airborne Laser Scanning in Ontario, Canada. In Proceedings of the IGARSS 2018-2018 IEEE International Geoscience Remote Sensing Symposium, Valencia, Spain, 22–27 July 2018; pp. 5–8749.
18. Diedershagen, O.; Koch, B.; Weinacker, H. Automatic segmentation and characterisation of forest stand parameters using airborne lidar data, multispectral and fogis data. *Int. Arch. Photogramm. Remote Sens. Spat. Inf. Sci.* **2004**, *36*, 208–212.
19. Machala, M.; Zejdová, L. Forest Mapping Through Object-based Image Analysis of Multispectral and LiDAR Aerial Data. *Eur. J. Remote Sens.* **2014**, *47*, 117–131. [CrossRef]
20. Hyyppa, J.; Kelle, O.; Lehikoinen, M.; Inkinen, M. A segmentation-based method to retrieve stem volume estimates from 3-D tree height models produced by laser scanners. *IEEE Trans. Geosci. Remote Sens.* **2001**, *39*, 969–975. [CrossRef]
21. Gulbe, L. Identification and delineation of individual tree crowns using Lidar and multispectral data fusion. In Proceedings of the 2015 IEEE International Geoscience and Remote Sensing Symposium (IGARSS), Milano, Italy, 26–31 July 2015; pp. 3294–3297.

22. Sačkov, I.; Sedliak, M.; Kulla, L.; Bucha, T. Inventory of Close-to-Nature Forests Based on the Combination of Airborne LiDAR Data and Aerial Multispectral Images Using a Single-Tree Approach. *Forests* **2017**, *8*, 467. [CrossRef]
23. Hamraz, H.; Contreras, M.A.; Zhang, J. A robust approach for tree segmentation in deciduous forests using small-footprint airborne LiDAR data. *Int. J. Appl. Earth Obs. Geoinf.* **2016**, *52*, 532–541. [CrossRef]
24. Bunting, P.; Lucas, R. The delineation of tree crowns in Australian mixed species forests using hyperspectral Compact Airborne Spectrographic Imager (CASI) data. *Remote Sens. Environ.* **2006**, *101*, 230–248. [CrossRef]
25. Dalponte, M.; Orka, H.O.; Gobakken, T.; Gianelle, D.; Naesset, E. Tree Species Classification in Boreal Forests With Hyperspectral Data. *Geosci. Remote Sens. IEEE Trans.* **2013**, *51*, 2632–2645. [CrossRef]
26. Ghosh, A.; Fassnacht, F.E.; Joshi, P.K.; Koch, B. A framework for mapping tree species combining hyperspectral and LiDAR data: Role of selected classifiers and sensor across three spatial scales. *Int. J. Appl. Earth Obs. Geoinf.* **2014**, *26*, 49–63. [CrossRef]
27. Jones, T.G.; Coops, N.C.; Sharma, T. Assessing the utility of airborne hyperspectral and LiDAR data for species distribution mapping in the coastal Pacific Northwest, Canada. *Remote Sens. Environ.* **2010**, *114*, 2841–2852. [CrossRef]
28. Matsuki, T.; Yokoya, N.; Iwasaki, A. Hyperspectral Tree Species Classification of Japanese Complex Mixed Forest With the Aid of Lidar Data. *IEEE J. Sel. Top. Appl. Earth Obs. Remote Sens.* **2015**, *8*, 2177–2187. [CrossRef]
29. Verlic, A.; Duric, N.; Kokalj, Z.; Marsetic, A.; Simoncic, P.; Ostir, K. Tree species classification using worldview-2 satellite images and laser scanning data in a natural urban forest. *Prethod. Priopćenje Prelim. Commun. Šumarski List* **2014**, *138*, 477–488.
30. Fassnacht, F.E.; Latifi, H.; Stereńczak, K.; Modzelewska, A.; Lefsky, M.; Waser, L.T.; Straub, C.; Ghosh, A. Review of studies on tree species classification from remotely sensed data. *Remote Sens. Environ.* **2016**, *186*, 64–87. [CrossRef]
31. Hartling, S.; Sagan, V.; Sidike, P.; Maimaitijiang, M.; Carron, J. Urban tree species classification using a worldview-2/3 and liDAR data fusion approach and deep learning. *Sensors* **2019**, *19*, 1284. [CrossRef] [PubMed]
32. Pham, L.T.H.; Brabyn, L.; Ashraf, S. Combining QuickBird, LiDAR, and GIS topography indices to identify a single native tree species in a complex landscape using an object-based classification approach. *Int. J. Appl. Earth Obs. Geoinf.* **2016**, *50*, 187–197. [CrossRef]
33. Varin, M.; Joanisse, G.; Dupuis, M.; Perrot, Y.; Gadbois-Langevin, R.; Brochu, J.; Painchaud, L.; Chalghaf, B. *Identification Semi-Automatisée D'essences Forestières à Partir D'images Hyperspectrales, Cas du Témiscamingue*; Centre D'enseignement et de Recherche en Foresterie de Sainte-Foy Inc. (CERFO): Québec, QC, Canada, 2019; p. 10.
34. Varin, M.; Gadbois-Langevin, R.; Joanisse, G.; Chalghaf, B.; Perrot, Y.; Marcotte, J.-M.; Painchaud, L.; Cullen, A. *Approche Orientée-Objet pour Cartographier le Frêne et L'épinette en Zone Urbaine*; Centre D'enseignement et de Recherche en Foresterie de Sainte-Foy Inc. (CERFO): Québec, QC, Canada, 2018; p. 8.
35. Dalponte, M.; Bruzzone, L.; Gianelle, D. Tree species classification in the Southern Alps based on the fusion of very high geometrical resolution multispectral/hyperspectral images and LiDAR data. *Remote Sens. Environ.* **2012**, *123*, 258–270. [CrossRef]
36. Arenas-Castro, S.; Fernández-Haeger, J.; Jordano-Barbudo, D. Evaluation and Comparison of QuickBird and ADS40-SH52 Multispectral Imagery for Mapping Iberian Wild Pear Trees (Pyrus bourgaeana, Decne) in a Mediterranean Mixed Forest. *Forests* **2014**, *5*, 1304–1330. [CrossRef]
37. Dube, T.; Mutanga, O.; Elhadi, A.; Ismail, R. Intra-and-Inter Species Biomass Prediction in a Plantation Forest: Testing the Utility of High Spatial Resolution Spaceborne Multispectral RapidEye Sensor and Advanced Machine Learning Algorithms. *Sensors* **2014**, *14*, 15348–15370. [CrossRef]
38. van Ewijk, K.Y.; Randin, C.F.; Treitz, P.M.; Scott, N.A. Predicting fine-scale tree species abundance patterns using biotic variables derived from LiDAR and high spatial resolution imagery. *Remote Sens. Environ.* **2014**, *150*, 120–131. [CrossRef]
39. Alonso-Benito, A.; Arroyo, L.; Arbelo, M.; Hernández-Leal, P. Fusion of WorldView-2 and LiDAR Data to Map Fuel Types in the Canary Islands. *Remote Sens.* **2016**, *8*, 669. [CrossRef]

40. He, Y.; Yang, J.; Caspersen, J.; Jones, T. An Operational Workflow of Deciduous-Dominated Forest Species Classification: Crown Delineation, Gap Elimination, and Object-Based Classification. *Remote Sens.* **2019**, *11*, 2078. [CrossRef]
41. van Deventer, H.; Cho, M.A.; Mutanga, O. Improving the classification of six evergreen subtropical tree species with multi-season data from leaf spectra simulated to WorldView-2 and RapidEye. *Int. J. Remote Sens.* **2017**, *38*, 4804–4830. [CrossRef]
42. Mutlu, M.; Popescu, S.C.; Stripling, C.; Spencer, T. Mapping surface fuel models using lidar and multispectral data fusion for fire behavior. *Remote Sens. Environ.* **2008**, *112*, 274–285. [CrossRef]
43. Ke, Y.; Quackenbush, L.J.; Im, J. Synergistic use of QuickBird multispectral imagery and LIDAR data for object-based forest species classification. *Remote Sens. Environ.* **2010**, *114*, 1141–1154. [CrossRef]
44. Fang, F.; McNeil, B.E.; Warner, T.A.; Maxwell, A.E. Combining high spatial resolution multi-temporal satellite data with leaf-on LiDAR to enhance tree species discrimination at the crown level. *Int. J. Remote Sens.* **2018**, *39*, 1–19. [CrossRef]
45. Li, D.; Ke, Y.; Gong, H.; Li, X. Object-Based Urban Tree Species Classification Using Bi-Temporal WorldView-2 and WorldView-3 Images. *Remote Sens.* **2015**, *7*, 16917–16937. [CrossRef]
46. Kukunda, C.B.; Duque-Lazo, J.; González-Ferreiro, E.; Thaden, H.; Kleinn, C. Ensemble classification of individual Pinus crowns from multispectral satellite imagery and airborne LiDAR. *Int. J. Appl. Earth Obs. Geoinf.* **2018**, *65*, 12–23. [CrossRef]
47. Maxwell, A.E.; Warner, T.A.; Fang, F. Implementation of machine-learning classification in remote sensing: An applied review. *Int. J. Remote Sens.* **2018**, *39*, 2784–2817. [CrossRef]
48. Hawryło, P.; Bednarz, B.; Wężyk, P.; Szostak, M. Estimating defoliation of Scots pine stands using machine learning methods and vegetation indices of Sentinel-2. *Eur. J. Remote Sens.* **2018**, *51*, 194–204. [CrossRef]
49. Nay, J.; Burchfield, E.; Gilligan, J. A machine-learning approach to forecasting remotely sensed vegetation health. *Int. J. Remote Sens.* **2018**, *39*, 1800–1816. [CrossRef]
50. Vaughn, R.N.; Asner, P.G.; Brodrick, G.P.; Martin, E.R.; Heckler, W.J.; Knapp, E.D.; Hughes, F.R. An Approach for High-Resolution Mapping of Hawaiian Metrosideros Forest Mortality Using Laser-Guided Imaging Spectroscopy. *Remote Sens.* **2018**, *10*, 502. [CrossRef]
51. Wu, C.; Chen, W.; Cao, C.; Tian, R.; Liu, D.; Bao, D. Diagnosis of Wetland Ecosystem Health in the Zoige Wetland, Sichuan of China. *Wetlands* **2018**, *38*, 469–484. [CrossRef]
52. Anderson, K.E.; Glenn, N.F.; Spaete, L.P.; Shinneman, D.J.; Pilliod, D.S.; Arkle, R.S.; McIlroy, S.K.; Derryberry, D.R. Estimating vegetation biomass and cover across large plots in shrub and grass dominated drylands using terrestrial lidar and machine learning. *Ecol. Indic.* **2018**, *84*, 793–802. [CrossRef]
53. Matasci, G.; Hermosilla, T.; Wulder, M.A.; White, J.C.; Coops, N.C.; Hobart, G.W.; Zald, H.S.J. Large-area mapping of Canadian boreal forest cover, height, biomass and other structural attributes using Landsat composites and lidar plots. *Remote Sens. Environ.* **2018**, *209*, 90–106. [CrossRef]
54. Zhang, C.; Denka, S.; Cooper, H.; Mishra, D.R. Quantification of sawgrass marsh aboveground biomass in the coastal Everglades using object-based ensemble analysis and Landsat data. *Remote Sens. Environ.* **2018**, *204*, 366–379. [CrossRef]
55. Franklin, S.E.; Skeries, E.M.; Stefanuk, M.A.; Ahmed, O.S. Wetland classification using Radarsat-2 SAR quad-polarization and Landsat-8 OLI spectral response data: A case study in the Hudson Bay Lowlands Ecoregion. *Int. J. Remote Sens.* **2018**, *39*, 1615–1627. [CrossRef]
56. Liu, T.; Abd-Elrahman, A.; Morton, J.; Wilhelm, V.L. Comparing fully convolutional networks, random forest, support vector machine, and patch-based deep convolutional neural networks for object-based wetland mapping using images from small unmanned aircraft system. *GISsci. Remote Sens.* **2018**, *55*, 243–264. [CrossRef]
57. Whyte, A.; Ferentinos, K.P.; Petropoulos, G.P. A new synergistic approach for monitoring wetlands using Sentinels-1 and 2 data with object-based machine learning algorithms. *Environ. Model. Softw.* **2018**, *104*, 40–54. [CrossRef]
58. Ada, M.; San, B.T. Comparison of machine-learning techniques for landslide susceptibility mapping using two-level random sampling (2LRS) in Alakir catchment area, Antalya, Turkey. *Nat. Hazards* **2018**, *90*, 237–263. [CrossRef]

59. Kalantar, B.; Pradhan, B.; Naghibi, S.A.; Motevalli, A.; Mansor, S. Assessment of the effects of training data selection on the landslide susceptibility mapping: A comparison between support vector machine (SVM), logistic regression (LR) and artificial neural networks (ANN). *Geomat. Nat. Hazards Risk.* **2018**, *9*, 49–69. [CrossRef]
60. Wessel, M.; Brandmeier, M.; Tiede, D. Evaluation of different machine learning algorithms for scalable classification of tree types and tree species based on Sentinel-2 data. *Remote Sens.* **2018**, *10*, 1419. [CrossRef]
61. Vapnik, V.N. *The Nature of Statistical Learning Theory*; Springer: Berlin, Germany, 1995; ISBN 0-387-94559-8.
62. Bennett, K.P.; Campbell, C. Support Vector Machines: Hype or Hallelujah? *SIGKDD Explor. Newsl.* **2000**, *2*, 1–13. [CrossRef]
63. Huang, C.-L.; Wang, C.-J. A GA-based feature selection and parameters optimizationfor support vector machines. *Expert Syst. Appl.* **2006**, *31*, 231–240. [CrossRef]
64. Scholkopf, B.; Smola, A.J. *Learning with Kernels: Support Vector Machines, Regularization, Optimization, and Beyond*; MIT Press: Cambridge, MA, USA, 2001; p. 644, ISBN 0262194759.
65. Pedregosa, F.; Varoquaux, G.; Gramfort, A.; Michel, V.; Thirion, B.; Grisel, O.; Blondel, M.; Prettenhofer, P.; Weiss, R.; Dubourg, V.; et al. Scikit-learn: Machine Learning in Python. *J. Mach. Learn. Res.* **2011**, *12*, 2825–2830.
66. Melgani, F.; Bruzzone, L. Classification of hyperspectral remote sensing images with support vector machines. *IEEE Trans. Geosci. Remote Sens.* **2004**, *42*, 1778–1790. [CrossRef]
67. Camps-Valls, G.; Bruzzone, L. Kernel-based methods for hyperspectral image classification. *IEEE Trans. Geosci. Remote Sens.* **2005**, *43*, 1351–1362. [CrossRef]
68. Lawrence, R.L.; Wright, A. Rule-based classification systems using classification and regression tree (CART) analysis. *Photogramm. Eng. Remote. Sens.* **2001**, *67*, 1137–1142.
69. Breiman, L.; Friedman, J.; Stone, C.J.; Olshen, R.A. *Classification and Regression Trees*; Taylor & Francis: Abingdon, UK, 1984; p. 368, ISBN 9780412048418.
70. Breiman, L. Random Forests. *Mach. Learn.* **2001**, *45*, 5–32. [CrossRef]
71. Mao, W.; Wang, F.-Y. (Eds.) Chapter 8-Cultural Modeling for Behavior Analysis and Prediction. In *New Advances in Intelligence and Security Informatics*; Academic Press: Boston, MA, USA, 2012; pp. 91–102, ISBN 978-0-12-397200-2.
72. Cutler, D.R.; Edwards, T.C.; Beard, K.H.; Cutler, A.; Hess, K.T.; Gibson, J.; Lawler, J.J. Random forests for classification in ecology. *Ecology* **2007**, *88*, 2783–2792. [CrossRef]
73. Kavzoglu, T. Chapter 33-Object-Oriented Random Forest for High Resolution Land Cover Mapping Using Quickbird-2 Imagery. In *Handbook of Neural Computation*; Samui, P., Sekhar, S., Balas, V.E., Eds.; Academic Press: Cambridge, MA, USA, 2017; pp. 607–619, ISBN 978-0-12-811318-9.
74. Altman, N.S. An Introduction to Kernel and Nearest-Neighbor Nonparametric Regression. *Am. Stat.* **1992**, *46*, 175–185. [CrossRef]
75. Meerdink, S.K.; Roberts, D.A.; Roth, K.L.; King, J.Y.; Gader, P.D.; Koltunov, A. Classifying California plant species temporally using airborne hyperspectral imagery. *Remote Sens. Environ.* **2019**, *232*, 111308. [CrossRef]
76. Pu, R.; Landry, S.; Yu, Q. Assessing the potential of multi-seasonal high resolution Pléiades satellite imagery for mapping urban tree species. *Int. J. Appl. Earth Obs. Geoinf.* **2018**, *71*, 144–158. [CrossRef]
77. Ferreira, M.P.; Zortea, M.; Zanotta, D.C.; Shimabukuro, Y.E.; de Souza Filho, C.R. Mapping tree species in tropical seasonal semi-deciduous forests with hyperspectral and multispectral data. *Remote Sens. Environ.* **2016**, *179*, 66–78. [CrossRef]
78. Immitzer, M.; Stepper, C.; Böck, S.; Straub, C.; Atzberger, C. Use of WorldView-2 stereo imagery and National Forest Inventory data for wall-to-wall mapping of growing stock. *For. Ecol. Manag.* **2016**, *359*, 232–246. [CrossRef]
79. Tharwat, A.; Gaber, T.; Ibrahim, A.; Hassanien, A.E. Linear discriminant analysis: A detailed tutorial. *AI Commun.* **2017**, *30*, 169–190. [CrossRef]
80. Hidayat, S.; Matsuoka, M.; Baja, S.; Rampisela, D.A. Object-based image analysis for sago palm classification: The most important features from high-resolution satellite imagery. *Remote Sens.* **2018**, *10*, 1319. [CrossRef]
81. Gosselin, J. *Guide de Reconnaissance des Types Écologiques de la Région Écologique 2a–Collines de la Basse-Gatineau*; Ministère des Ressources Naturelles, de la Faune et des Parcs, Forêt Québec, Direction des Inventaires Forestiers, Division de la Classification Écologique et Productivité des Stations: Québec, QC, Canada, 2004; p. 134, ISBN 2-551-22454-3.

82. Lin, C.; Wu, C.-C.; Tsogt, K.; Ouyang, Y.-C.; Chang, C.-I. Effects of atmospheric correction and pansharpening on LULC classification accuracy using WorldView-2 imagery. *Inf. Process. Agric.* **2015**, *2*, 25–36. [CrossRef]
83. Koukoulas, S.; Blackburn, G.A. Mapping individual tree location, height and species in broadleaved deciduous forest using airborne LIDAR and multi-spectral remotely sensed data. *Int. J. Remote Sens.* **2005**, *26*, 431–455. [CrossRef]
84. Zhou, Y.; Qiu, F. Fusion of high spatial resolution WorldView-2 imagery and LiDAR pseudo-waveform for object-based image analysis. *ISPRS J. Photogramm. Remote Sens.* **2015**, *101*, 221–232. [CrossRef]
85. Azevedo, S.C.; Silva, E.A.; Pedrosa, M. Shadow detection improvment using spectral indices and morphological operators in urban areas in high resolution images. *Int. Arch. Photogramm. Remote Sens. Spat. Inf. Sci.* **2015**, *XL-7/W3*, 587–592. [CrossRef]
86. Mora, B.; Wulder, M.A.; White, J.C. Identifying leading species using tree crown metrics derived from very high spatial resolution imagery in a boreal forest environment. *Can. J. Remote Sens.* **2010**, *36*, 332–344. [CrossRef]
87. R Development Core Team. *R: A Language and Environment for Statistical Computing*; R Foundation for Statistical Computing: Vienna, Austria, 2018.
88. SAS Institute Inc. (Ed.) *SAS-STAT User's Guide: Release 9.4 Edition*; SAS Institute Inc.: Cary, NC, USA, 2013; ISBN 1-55544-094-0.
89. Hill, R.A.; Thomson, A.G. Mapping woodland species composition and structure using airborne spectral and LiDAR data. *Int. J. Remote Sens.* **2005**, *26*, 3763–3779. [CrossRef]
90. Leckie, D.; Gougeon, F.; Hill, D.; Quinn, R.; Armstrong, L.; Shreenan, R. Combined high-density lidar and multispectral imagery for individual tree crown analysis. *Can. J. Remote Sens.* **2003**, *29*, 633–649. [CrossRef]
91. McCombs, J.W.; Roberts, S.D.; Evans, D.L. Influence of Fusing Lidar and Multispectral Imagery on Remotely Sensed Estimates of Stand Density and Mean Tree Height in a Managed Loblolly Pine Plantation. *For. Sci.* **2003**, *49*, 457–466. [CrossRef]
92. Holmgren, J.; Persson, Å.; Söderman, U. Species identification of individual trees by combining high resolution LiDAR data with multi-spectral images. *Int. J. Remote Sens.* **2008**, *29*, 1537–1552. [CrossRef]
93. Weinacker, H.; Koch, B.; Heyder, U.; Weinacker, R. Development of filtering, segmentation and modelling modules for lidar and multispectral data as a fundament of an automatic forest inventory system. *Int. Arch. Photogramm. Remote Sens. Spat. Inf. Sci.* **2004**, *36*, 1682–1750.
94. Trimble Germany GmbH. *Trimble Germany GmbH eCognition®Developer 9.3.2 for Windows Operating System: Reference Book*; Trimble Germany GmbH: Munich, Germany, 2018; p. 510.
95. Wang, L.; Sousa, W.P.; Gong, P. Integration of object-based and pixel-based classification for mapping mangroves with IKONOS imagery. *Int. J. Remote Sens.* **2004**, *25*, 5655–5668. [CrossRef]
96. Brillinger, D.R. *International Encyclopedia of Political Science*; SAGE Publications Inc.: Thousand Oaks, CA, USA, 2011; p. 4032.
97. Doherty, M.C. Automating the process of choosing among highly correlated covariates for multivariable logistic regression. In Proceedings of the SAS Conference Proceedings: Western Users of SAS Software 2008, Los Angeles, CA, USA, 5–7 November 2008; p. 7.
98. Kursa, M.B.; Rudnicki, W.R. Feature Selection with the Boruta Package. *J. Stat. Softw.* **2010**, *11*. [CrossRef]
99. Chen, Y.-W.; Lin, C.-J. Combining SVMs with Various Feature Selection Strategies. In *Feature Extraction: Foundations and Applications*; Guyon, I., Nikravesh, M., Gunn, S., Zadeh, L.A., Eds.; Springer: Berlin, Germany, 2006; pp. 315–324, ISBN 978-3-540-35488-8.
100. Prasad, A.M.; Iverson, L.R.; Liaw, A. Newer Classification and Regression Tree Techniques: Bagging and Random Forests for Ecological Prediction. *Ecosystems* **2006**, *9*, 181–199. [CrossRef]
101. Liaw, A.; Wiener, M. Classification and Regression by randomForest. *R News* **2002**, *2*, 18–22.
102. Venables, W.N.; Ripley, B.D. *Modern Applied Statistics with S*; Springer: New York, NY, USA, 2002.
103. Therneau, T.M.; Atkinson, E.J. *An Introduction to Recursive Partitioning Using the RPART Routines*; Springer: New York, NY, USA, 2015.
104. Congalton, R.G. A review of assessing the accuracy of classifications of remotely sensed data. *Remote Sens. Environ.* **1991**, *37*, 35–46. [CrossRef]
105. Monserud, R.A.; Leemans, R. Comparing global vegetation maps with the Kappa statistic. *Ecol. Model.* **1992**, *62*, 275–293. [CrossRef]
106. McHugh, M.L. Interrater reliability: The kappa statistic. *Biochem. Med.* **2012**, *22*, 276–282. [CrossRef]

107. Cohen, J. A Coefficient of Agreement for Nominal Scales. *Educ. Psychol. Meas.* **1960**, *20*, 37–46. [CrossRef]
108. Henrich, V.; Jung, A.; Götze, C.; Sandow, C.; Thürkow, D.; Gläßer, C. Development of an online indices database: Motivation, concept and implementation. In Proceedings of the 6th EARSeL Imaging Spectroscopy SIG Workshop Innovative Tool for Scientific and Commercial Environment Applications, Tel Aviv, Israel, 16–18 March 2009.
109. Hughes, N.M.; Smith, W.K. Seasonal Photosynthesis and Anthocyanin Production in 10 Broadleaf Evergreen Species. *Funct. Plant Biol.* **2007**, *34*, 1072–1079. [CrossRef] [PubMed]
110. Baatz, M.; Schape, A. Multiresolution segmentation—An optimization approach for high quality multi-scale image segmentation. *Angew. Geogr. Inf.* **2000**, *12*, 12–23.
111. Haralick, R.M. Statistical image texture analysis haralick. In *Handbook of Pattern Recognition and Image Processing*; Young, T.Y., Ed.; Elsevier Science: New York, NY, USA, 1986; pp. 247–279.
112. Leboeuf, A.; Vaillancourt, É.; Morissette, A.; Pomerleau, I.; Roy, V.; Leboeuf, A. *Photographic Interpretation Guide for Forest Species in Southern Québec*; Direction des Inventaires Forestiers: Québec, QC, Canada, 2015; ISBN 97825507278732550727878.
113. Gitelson, A.A.; Merzlyak, M.N. Remote estimation of chlorophyll content in higher plant leaves. *Int. J. Remote Sens.* **1997**, *18*, 2691–2697. [CrossRef]
114. Minocha, R.; Martinez, G.; Lyons, B.; Long, S. Development of a standardized methodology for quantifying total chlorophyll and carotenoids from foliage of hardwood and conifer tree species. *Can. J. For. Res.* **2009**, *39*, 849–861. [CrossRef]
115. Gitelson, A.A.; Merzlyak, M.N.; Zur, Y.; Stark, R.H.; Gritz, U. Non-Destructive and Remote Sensing Techniques for Estimation of Vegetation Status. In Proceedings of the 3rd European Conference on Precision Agriculture, Montpelier, France, 8–11 July 2001; pp. 205–210, 273.
116. Peñuelas, J.; Gamon, J.A.; Fredeen, A.L.; Merino, J.; Field, C.B. Reflectance indices associated with physiological changes in nitrogen- and water-limited sunflower leaves. *Remote Sens. Environ.* **1994**, *48*, 135–146. [CrossRef]
117. Gamon, J.A.; Surfus, J.S. Assessing leaf pigment content and activity with a reflectometer. *New Phytol.* **1999**, *143*, 105–117. [CrossRef]
118. Kauth, R.J.; Thomas, G.S. The tasselled cap-A graphic description of the spectral-temporal development of agricultural crops as seen by Landsat. In Proceedings of the Symposium on Machine Processing of Remotely Sensed Data, West Lafayette, IN, USA, 29 June–1 July 1976; Institute of Electrical and Electronics Engineers: Piscataway, NJ, USA, 1976; pp. 4B-41–4B-51.
119. Pham, A.T.; De Grandpre, L.; Gauthier, S.; Bergeron, Y. Gap dynamics and replacement patterns in gaps of the northeastern boreal forest of Quebec. *Can. J. For. Res.* **2004**, *34*, 353–364. [CrossRef]
120. Coops, N.C.; Wulder, M.A.; Culvenor, D.S.; St-Onge, B. Comparison of forest attributes extracted from fine spatial resolution multispectral and lidar data. *Can. J. Remote Sens.* **2004**, *30*, 855–866. [CrossRef]
121. Heinzel, J.; Koch, B. Investigating multiple data sources for tree species classification in temperate forest and use for single tree delineation. *Int. J. Appl. Earth Obs. Geoinf.* **2012**, *18*, 101–110. [CrossRef]
122. Soille, P. *Morphological Image Analysis: Principles and Applications*, 2nd ed.; Springer: Berlin, Germany, 2003; ISBN 3540429883.
123. Lebourgeois, V.; Dupuy, S.; Vintrou, É.; Ameline, M.; Butler, S.; Bégué, A. A combined random forest and OBIA classification scheme for mapping smallholder agriculture at different nomenclature levels using multisource data (simulated Sentinel-2 time series, VHRS and DEM). *Remote Sens.* **2017**, *9*, 259. [CrossRef]
124. Krahwinkler, P.; Rossmann, J. Tree species classification and input data evaluation. *Eur. J. Remote Sens.* **2013**, *46*, 535–549. [CrossRef]
125. Kim, M.; Xu, B.; Madden, M. Object-based Vegetation Type Mapping from an Orthorectified Multispectral IKONOS Image using Ancillary Information. In Proceedings of the GEOBIA 2008—GEOgraphic Object Based Image Analysis for the 21st Century, Calgary, AB, Canada, 6 August 2008; pp. 1682–1777.
126. Ferreira, M.P.; Zortea, M.; Zanotta, D.C.; Feret, J.B.; Souza Filho, C.R. On the use of shortwave infrared for tree species discrimination in tropical semideciduous forest. *Int. Arch. Photogramm. Remote Sens. Spat. Inf. Sci.* **2015**, *XL-3/W3*. [CrossRef]

127. Ferreira, M.P.; Zanotta, D.C.; Zortea, M.; Körting, T.S.; Fonseca, L.M.G.; Shimabukuro, Y.E.; Filho, C.R.S. Automatic tree crown delineation in tropical forest using hyperspectral data. In Proceedings of the 2014 IEEE Geoscience and Remote Sensing Symposium, Quebec City, QC, Canada, 13–18 July 2014; pp. 784–787.
128. Cho, M.A.; Sobhan, I.; Skidmore, A.K.; Leeuw, J. Discriminating species using hyperspectral indices at leaf and canopy scales. *Int. Arch. Photogramm. Remote Sens. Spat. Inf. Sci.* **2008**, *37*, 369–376.
129. Rama Rao, N.; Garg, P.K.; Ghosh, S.K.; Dadhwal, V.K. Estimation of leaf total chlorophyll and nitrogen concentrations using hyperspectral satellite imagery. *J. Agric. Sci.* **2008**, *146*, 65–75. [CrossRef]
130. Apan, A.; Held, A.; Phinn, S.; Markley, J. Detecting sugarcane 'orange rust' disease using EO-1 Hyperion hyperspectral imagery. *Int. J. Remote Sens.* **2004**, *25*, 489–498. [CrossRef]
131. Hovi, A.; Korhonen, L.; Vauhkonen, J.; Korpela, I. LiDAR waveform features for tree species classification and their sensitivity to tree- and acquisition related parameters. *Remote Sens. Environ.* **2016**, *173*, 224–237. [CrossRef]
132. Korpela, I.; Ole Ørka, H.; Maltamo, M.; Tokola, T.; Hyyppä, J. Tree species classification using airborne LiDAR-effects of stand and tree parameters, downsizing of training set, intensity normalization, and sensor type. *Silva Fenn.* **2010**, *44*, 319–339. [CrossRef]
133. Shi, L.; Wan, Y.; Gao, X.; Wang, M. Feature Selection for Object-Based Classification of High-Resolution Remote Sensing Images Based on the Combination of a Genetic Algorithm and Tabu Search. *Comput. Intell. Neurosci.* **2018**. [CrossRef] [PubMed]
134. Tao, Q.; Wu, G.-W.; Wang, F.-Y.; Wang, J. Posterior Probability SVM for Unbalanced Data. *IEEE Trans. Neural Netw.* **2005**, *16*, 1561–1573. [CrossRef] [PubMed]
135. Farquad, M.A.H.; Bose, I. Preprocessing unbalanced data using support vector machine. *Decis. Support Syst.* **2012**, *53*, 226–233. [CrossRef]
136. Chalghaf, B.; Varin, M.; Joanisse, G. *Cartographie Fine des Essences Individuelles par une Approche de Modélisation de type «Random Forest», à partir du lidar et de RapidEye*; Rapport 2019-04; Centre D'enseignement et de Recherche en Foresterie de Sainte-Foy Inc. (CERFO): Québec, QC, Canada, 2019; p. 29.
137. Zhang, Y. Problems in the Fusion of Commercial High-Resolution Satellite Images as Well as LANDSAT 7 Images and Initial Solutions. In Proceedings of the commission IV Symposium on Geospatial Theory, Processing and Applications, Ottawa, ON, Canada, 9–12 July 2002; pp. 587–592.
138. St-Onge, B.; Grandin, S. Estimating the Height and Basal Area at Individual Tree and Plot Levels in Canadian Subarctic Lichen Woodlands Using Stereo WorldView-3 Images. *Remote Sens.* **2019**, *11*, 248. [CrossRef]
139. Paquette, A.; Joly, S.; Messier, C. Explaining forest productivity using tree functional traits and phylogenetic information: Two sides of the same coin over evolutionary scale? *Ecol. Evol.* **2015**, *5*, 1774–1783. [CrossRef]
140. Gitelson, A.A.; Kaufman, Y.J.; Stark, R.; Rundquist, D. Novel algorithms for remote estimation of vegetation fraction. *Remote Sens. Environ.* **2002**, *80*, 76–87. [CrossRef]
141. Datt, B. A New Reflectance Index for Remote Sensing of Chlorophyll Content in Higher Plants: Tests using Eucalyptus Leaves. *J. Plant Physiol.* **1999**, *154*, 30–36. [CrossRef]
142. Hunt, E.R.; Rock, B.N. Detection of changes in leaf water content using Near- and Middle-Infrared reflectances. *Remote Sens. Environ.* **1989**, *30*, 43–54. [CrossRef]
143. Colombo, R.; Meroni, M.; Marchesi, A.; Busetto, L.; Rossini, M.; Giardino, C.; Panigada, C. Estimation of leaf and canopy water content in poplar plantations by means of hyperspectral indices and inverse modeling. *Remote Sens. Environ.* **2008**, *112*, 1820–1834. [CrossRef]
144. Serrano, L.; Ustin, S.L.; Roberts, D.A.; Gamon, J.A.; Peñuelas, J. Deriving Water Content of Chaparral Vegetation from AVIRIS Data. *Remote Sens. Environ.* **2000**, *74*, 570–581. [CrossRef]
145. Serrano, L.; Peñuelas, J.; Ustin, S.L. Remote sensing of nitrogen and lignin in Mediterranean vegetation from AVIRIS data. *Remote Sens. Environ.* **2002**, *81*, 355–364. [CrossRef]
146. Gao, B. NDWI—A normalized difference water index for remote sensing of vegetation liquid water from space. *Remote Sens. Environ.* **1996**, *58*, 257–266. [CrossRef]
147. Chen, D.; Huang, J.; Jackson, T.J. Vegetation water content estimation for corn and soybeans using spectral indices derived from MODIS near-and short-wave infrared bands. *Remote Sens. Environ.* **2005**, *98*, 225–236. [CrossRef]
148. Wang, L.; Qu, J.J. NMDI: A normalized multi-band drought index for monitoring soil and vegetation moisture with satellite remote sensing. *Geophys. Res. Lett.* **2007**, *34*. [CrossRef]

149. Peñuelas, J.; Filella, I.; Gamon, J.A. Assessment of photosynthetic radiation-use efficiency with spectral reflectance. *New Phytol.* **1995**, *131*, 291–296. [CrossRef]
150. Sims, D.A.; Gamon, J.A. Relationships between leaf pigment content and spectral reflectance across a wide range of species, leaf structures and developmental stages. *Remote Sens. Environ.* **2002**, *81*, 337–354. [CrossRef]
151. Elvidge, C.D.; Lyon, R.J.P. Estimation of the vegetation contribution to the 1•65/2•22 μm ratio in airborne thematic-mapper imagery of the Virginia Range, Nevada. *Int. J. Remote Sens.* **1985**, *6*, 75–88. [CrossRef]
152. Zarco-Tejada, P.J.; Ustin, S L. Modeling canopy water content for carbon estimates from MODIS data at land EOS validation sites. In Proceedings of the IEEE 2001 International Geoscience and Remote Sensing Symposium, Sydney, Australia, 9–13 July 2001; pp. 342–344.
153. Penuelas, J.; Pinol, J.; Ogaya, R.; Filella, I. Estimation of plant water concentration by the reflectance Water Index WI (R900/R970). *Int. J. Remote Sens.* **1997**, *18*, 2869–2875. [CrossRef]

remote sensing

MDPI

Article

Tree Species Classification of Forest Stands Using Multisource Remote Sensing Data

Haoming Wan [1,2], Yunwei Tang [1,*], Linhai Jing [1], Hui Li [1], Fang Qiu [3] and Wenjin Wu [1,4]

1 Key Laboratory of Digital Earth Science, Aerospace Information Research Institute, Chinese Academy of Sciences, Beijing 100094, China; wanghm@aircas.ac.cn (H.W.); jinglh@aircas.ac.cn (L.J.); lihui@radi.ac.cn (H.L.); wuwj@radi.ac.cn (W.W.)
2 School of Electronic, Electrical and Communication Engineering, University of Chinese Academy of Sciences, Beijing 100094, China
3 Geospatial Information Sciences, The University of Texas at Dallas, 800 West Campbell Road, Richardson, TX 75080, USA; ffqiu@utdallas.edu
4 Sanya Institute of Remote Sensing, Sanya 572029, China
* Correspondence: tangyw@aircas.ac.cn

Abstract: The spatial distribution of forest stands is one of the fundamental properties of forests. Timely and accurately obtained stand distribution can help people better understand, manage, and utilize forests. The development of remote sensing technology has made it possible to map the distribution of tree species in a timely and accurate manner. At present, a large amount of remote sensing data have been accumulated, including high-spatial-resolution images, time-series images, light detection and ranging (LiDAR) data, etc. However, these data have not been fully utilized. To accurately identify the tree species of forest stands, various and complementary data need to be synthesized for classification. A curve matching based method called the fusion of spectral image and point data (FSP) algorithm was developed to fuse high-spatial-resolution images, time-series images, and LiDAR data for forest stand classification. In this method, the multispectral Sentinel-2 image and high-spatial-resolution aerial images were first fused. Then, the fused images were segmented to derive forest stands, which are the basic unit for classification. To extract features from forest stands, the gray histogram of each band was extracted from the aerial images. The average reflectance in each stand was calculated and stacked for the time-series images. The profile curve of forest structure was generated from the LiDAR data. Finally, the features of forest stands were compared with training samples using curve matching methods to derive the tree species. The developed method was tested in a forest farm to classify 11 tree species. The average accuracy of the FSP method for ten performances was between 0.900 and 0.913, and the maximum accuracy was 0.945. The experiments demonstrate that the FSP method is more accurate and stable than traditional machine learning classification methods.

Keywords: forest stands classification; curve matching; data fusion; multisource remote sensing data; segmentation; tree species mapping

Citation: Wan, H.; Tang, Y.; Jing, L.; Li, H.; Qiu, F.; Wu, W. Tree Species Classification of Forest Stands Using Multisource Remote Sensing Data. *Remote Sens.* **2021**, *13*, 144. https://doi.org/10.3390/rs13010144

Academic Editor: Lin Cao
Received: 5 November 2020
Accepted: 30 December 2020
Published: 4 January 2021

1. Introduction

Forests, an important type of land cover and a key part of ecosystems, have a decisive influence on maintaining carbon dioxide balance, biodiversity, and ecological balance. Forests play a vital role in the survival and development of human civilization. According to the report by the Food and Agriculture Organization (FAO) of the United Nations, forest ecosystems cover approximately one-third of the earth's land surface [1]. The composition and spatial distribution of forest tree species have a great impact on the forest ecological environment, biodiversity, resource utilization efficiency, production and carbon storage capacity, and nutrition cycle [2–8]. The basic unit for the forest inventory is the forest stands, which is a large forested area of homogeneous tree species composition [9].

Classification of tree species, one of the main tasks of forest science, is important to forest management [10]. Tree species information obtained from classification can be served as fundamental dataset. For example, the productivity of forest biomass can be improved based on some tree species-specific models [7]. Timely and accurate identification of forest stand types and tree species can help people better understand, manage, and protect forests. Therefore, effective and efficient techniques for delineating forest stands and classifying tree species are highly demanded [11,12].

In traditional forest surveys, forest stand distributions were obtained through field investigation, which is a time-consuming and laborious process [13,14]. Remote sensing technology can easily obtain forest information over large areas [15], even in dense and inaccessible forests. Spectral images obtained from remote sensing systems offer a practical and economical method to draw the distribution of tree species [13,16], thus reducing the field workload [17,18]. According to the assumption that different tree species have different spectral feature characteristics [19], the distribution of forest species can be extracted from spectral images.

Spectral images include multispectral and hyperspectral images. Multispectral images generally contain mid to low-spatial-resolution images, such as GaoFen-1/4, Sentinel-2, Landsat-7/8, and SPOT 1/2/3/4 and high-spatial-resolution images, such as GaoFen-2, IKONOS-2, QuickBird, RapidEye, and Airborne [20]. In the early stage, a Landsat Multispectral Scanner System (MSS) image was applied to forest cover mapping, but the classification results were limited due to the coarse spatial and spectral resolution of the image [21–23]. With the improvement in multispectral sensors, high-spatial-resolution multispectral images with wider band ranges have become available; thus, more details of forest stands can be obtained from images [3,24–26]. A high-spatial-resolution multispectral image provides rich spectral and textural information [27–29], which can improve the accuracy of tree species classification [24,25]. Object-based image analysis (OBIA) is usually first applied to high-spatial-resolution images, partitioning the image into segments (i.e., objects) according to the textural and spectral information [26]. Each segment can be seen as a forest stand. Classification is applied to each forest stand to obtain the tree species. Qian et al. (2006) showed that the classification accuracy can be improved by 10–15% after introducing textural information compared with using only spectral information [30]. However, traditional OBIA uses only statistical features, such as the mean and standard deviation of the pixels in the objects [27,28,31], so the rich information in the objects is not fully used. Additionally, for the features of objects represented by those statistical features, it assumes that the pixel values in the objects follow a Gaussian distribution [32]. When the spatial resolution is high, the heterogeneity in the objects is large, such as in the forest stand unit, and the traditional OBIA is no longer applicable [33]. In addition to the improvements in spatial and spectral resolutions, the increase in time resolution also benefits forest stand classification. Distinctive spectral-temporal features of tree species can be extracted from time-series images. Karlson et al. (2016) used two seasonal WorldView-2 images for mapping five tree species in West Africa [27]. Madonsela et al. (2017) concluded that two seasonal WorldView-2 images can improve the accuracy of tree species [34]. Pu et al. (2018) evaluated the potential of five seasonal images for classifying tree species in an urban area [35]. These studies have shown that information on the phenology changes in forest stands over the growing season can improve classification accuracy [26].

High-spatial-resolution images usually contain a few bands with a wide bandwidth, thus providing poor spectral information. More importantly, similar spectral information may exist among different tree species in high-spatial-resolution images. Hyperspectral images contain nanometer-level spectral resolutions and the rich spectral information of ground objects. Hyperspectral images have also been used for forest stand classification [36,37]. However, hyperspectral images usually have low spatial resolution. Due to the large number of hyperspectral bands and strong correlation between bands, the increase of feature dimension may cause the performance of the classifier to deteriorate when the feature dimension reaches to a certain critical point. This is the so-called Hughes phenomenon, occurring in traditional machine learning classification methods that rely on spectral features and sample size [38,39]. Additionally, the variance of the spectra within the same class is usually large for hyperspectral images, leading a poor separability of hyperspectral features [39–41]. These problems might greatly affect the accuracy of forest stand classification [42–44], and the classification result is not robust to noise when using hyperspectral images [36]. In addition, multispectral and hyperspectral images do not contain three-dimensional structural information, such as canopy height and vertical structures.

To describe the three-dimensional structure of trees, light detection and ranging (LiDAR) data were introduced to forest stand classification. LiDAR data, a collection of points, are a three-dimensional representation of an object. The LiDAR system became mature in approximately 2000 [45]. Numerous works have pointed out the effectiveness of LiDAR data in tree species classification [42,43,46], and some researchers used LiDAR data to classify forest in a large area [47]. LiDAR data can be used alone for classification, but the accuracy is lower than that using spectral images. More frequently, LiDAR points are regarded as ancillary data to classify forest stands with remotely sensed images. At present, LiDAR has become an important tool in forestry applications. Valuable forest geometric information is obtained from LiDAR data, such as tree height [44], canopy diameter [48], leaf area index [49], and canopy volume profiles [43]. For example, Blomley et al. (2017) analyzed multi-scale geometrical features, revealing that representative features extracted from LiDAR data can improve the accuracy of tree species identification. Rami et al. (2018) and Pu et al. (2018) concluded that the height information extracted from LiDAR data is helpful for mapping urban tree species [50,51]. Shi et al. (2018) evaluated some frequently used LiDAR features for discriminating forest tree species, and these features are useful in a mixed temperate forest [52]. However, the use of LiDAR data has some shortcomings. For example, the features extracted from LiDAR can vary among tree species, which may reduce the classification accuracy [53]. Additionally, it is difficult to fuse LiDAR data with remotely sensed images.

As mentioned above, high-spatial-resolution images include spectral and textural information, making it possible to extract forest stands. Time-series images provide phonological features, and LiDAR data contain information about the geometric structure of tree species. The information provided by the three types of data is complementary [43]. Consequently, combining these three types of data may hold great promise for improving forest inventories, particularly at stand-level discrimination [54–57]. However, different spectral images have different spatial, spectral, and time resolutions, making the fusion of multisource images difficult. Additionally, LiDAR data are in point cloud format, which is different from spectral images. Therefore, the traditional classification methods that fuse spectral images and LiDAR data often sacrifice rich forest information. To fuse with images, LiDAR data are usually transformed to raster formats, such as canopy height modules (CHM) [56,58–61] and canopy volume profiles (CVP) [43]. These characteristics only can describe one aspect of trees. Fassnacht et al. (2016) pointed out that few studies have combined spectral images and LiDAR data in a more complicated way for forest classifications [53].

Consequently, a comprehensive fusion method was expected to utilize the characteristics of various types of data. Curve matching methods have shown promising outlooks for object-based classification [62]. In previous studies, a histogram curve was generated for each object across multispectral bands. Classification was performed based on a comparison of the histogram curves of the object to be classified and the sample objects. The curve matching method includes richer information than traditional classification methods based on statistical measures (e.g., mean value of objects). For LiDAR data, a frequency distribution map that describes the structure of trees can be generated [63]. This profile curve is called the profile curve because it mainly reflects information about the profile of the tree. Compared with feature maps extracted from LiDAR data, such as CHM and CVP, the profile curve can reflect more forest characteristics, and can be applied to estimate the leaf area index and biomass [63–67]. Some researchers have fused the profile curve of LiDAR data with WorldView-2 to classify land cover types [32]. However, these cover types are typical land cover classes, such as buildings, grass, water, trees, and pavement. Although curve matching methods have achieved good results in classifications, there are no related studies focusing on complex forest stand classification. Additionally, using curve matching methods to fuse various types of data has not been explored.

Although a large amount of remote sensing data is available, they have not been fully utilized. Comprehensive utilization of multiple data is expected to more accurately classify forest stands. Currently, there are few studies on fusing time-series images with high-spatial-resolution images to synergize spectral and phenology information for forest stand classification. Additionally, there is a lack of multisource heterogeneous data fusion methods to integrate images and point cloud data (i.e., LiDAR). Therefore, to solve these problems and further improve the classification accuracy of forest stands, a forest stands classification method that fuses high-spatial-resolution images, time-series multispectral images and LiDAR data is developed. We define this method as the Fusion of Spectral image and point data (FSP) method.

This paper is organized as follows: the study areas and experimental data are introduced in Sections 2.1 and 2.2; the method we propose is described in Section 2.3; experimental results and analysis are demonstrated in Section 3; the applicability of this method is discussed in Section 4; the conclusion is provided in Section 5.

2. Materials and Methods

2.1. Study Area

The study area is in the Gaofeng forest farm (22°58′20.54″ N, 108°23′16.26″ E) in Nanning, Guangxi Zhuang Autonomous Region, China (Figure 1). The area, which is in a subtropical monsoon climate zone, is composed of a hilly landform with an elevation varying from 100 to 300 m and a falling gradient of 6° to 35°. The average annual temperature is 21.6 °C, and the annual sunshine time is between 1450 h and 1650 h. Additionally, the annual rainfall, which is mainly concentrated in summer, is 1304.2 mm. The average humidity is above 80%, and the annual evaporation is slightly higher than the rainfall.

This area, with typical characteristics of forests in southern China, is suitable for the growth of a variety of timber trees, especially tropical and subtropical tree species. The forest farm is rich in forest resources, with a forest coverage rate of 87.5%. The number of tree species in the forest mainly includes *Eucalyptus robusta Smith, Illicium verum Hook. f., Mytilaria laosensis Lec, Cunninghamia lanceolata, Pinus massoniana Lamb, Pinus elliottii*, and other broad-leaved tree species.

2.2. Experimental Data

2.2.1. Sentinel-2 Data

Sentinel-2 images are widely available. The multispectral bands of Sentinel-2 images include 13 bands, with bands 2, 3, 4, and 8 having a 10 m spatial resolution; bands 5, 6, 7, 8a, 11, and 12 having a 20 m spatial resolution; and bands 1, 9, and 10 having a 60 m spatial resolution. Due to cloud coverage, only four images which were come from 2015 to 2017 were selected. The four images were acquired on 2 September 2016, 2 June 2016, 1 April 2017, and 30 July 2017, as shown in Figure 2. April and June are the flowering periods of many tree species. In the midsummer in July, the growth of trees is vigorous and leafy. September is the mature period of most trees in the study area. The selected periods are typical time nodes of tree growth, and the spectra of these periods are of equal importance. Time-series multispectral images were stacked in a monthly wise chronological order to provide rich phonological information and spectral–temporal features.

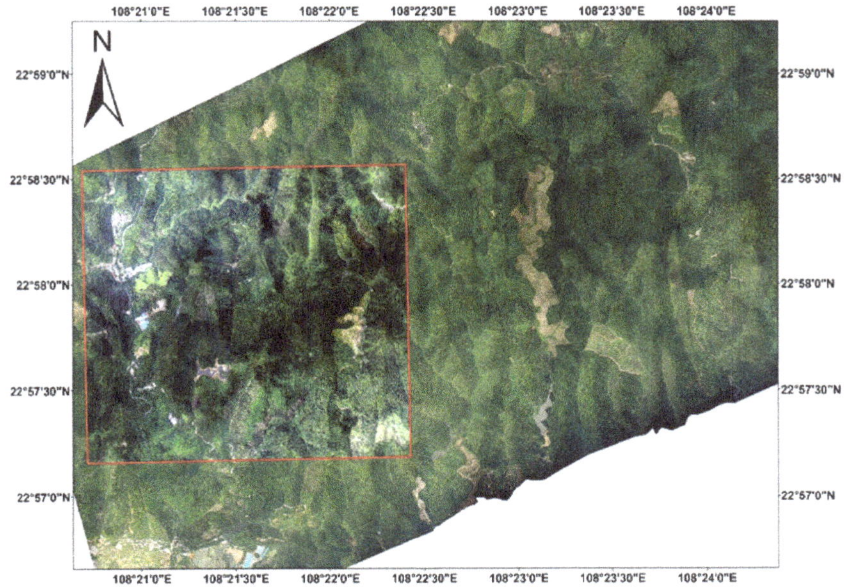

Figure 1. The study site in Gaofeng forest farm, Nanning, Guangxi Zhuang Autonomous Region, China.

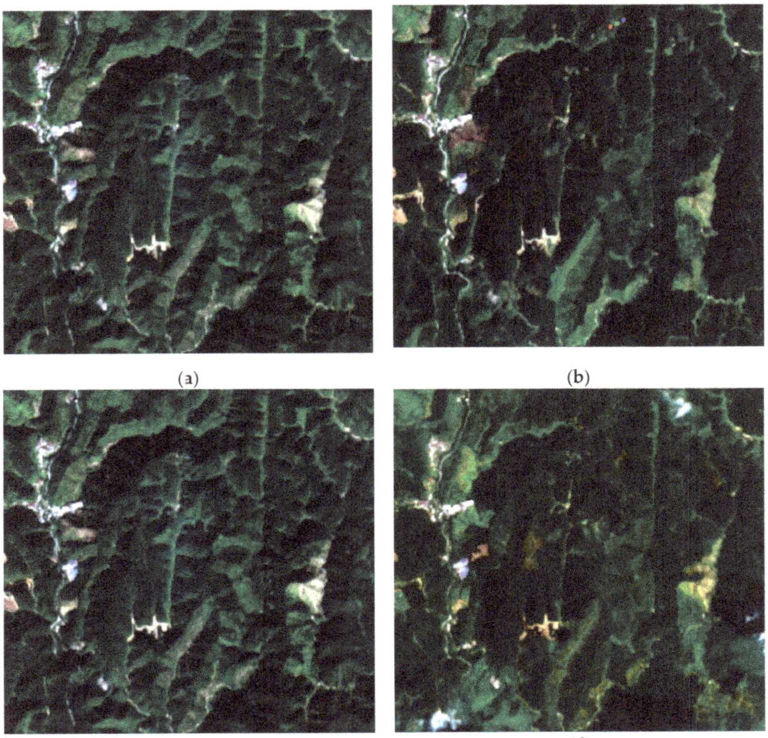

(a) (b) (c) (d)

Figure 2. Time-series Sentinel-2 images: (**a**) 1 April 2017; (**b**) 2 June 2016; (**c**) 30 July 2017; (**d**) 2 September 2016.

2.2.2. Aerial Image and LiDAR Data

High-resolution aerial images and LiDAR data were acquired by the CAF (The Chinese Academy of Forestry)-LiCHy(LiDAR, Charge-Coupled Device (CCD) and Hyperspectral) airborne remote sensing system platform in June 2016. The LiCHy system, developed by the Chinese Academy of Forestry, includes one full-waveform airborne LiDAR (RIEGL LMS-Q680i) and one high-resolution charge-coupled device camera. The CCD sensor is a DigiCAM-60, and the heading and lateral overlap rates are 60% and 30%, respectively. All sensors share the same position and altitude system [68]. The parameters are shown in Table 1.

Table 1. The parameters of LiCHy airborne remote sensing system platform.

CCD: DigiCAM-60		LiDAR: Riegl LMS-Q680i	
Frame size	8956 × 6708	Wavelength	1550 nm
Pixel size	6 µm	Laser beam divergence	0.5 mrad
Imaging sensor size	40.30 mm × 53.78 mm	Laser pulse length	3 ns
Feld of view (FOV)	56.2°	Cross-track FOV	±30°
Ground resolution @1000 m altitude	0.12 m	Vertical resolution	0.15 m
Focal length	50 mm	Point density @1000 m altitude	3.6 pts/m^2
——	——	Waveform Sampling interval	1 ns
——	——	Maximum scanning speed	200 lines/s
——	——	Maximum laser pulse repetition rate	400 kHz

2.2.3. Field Data Collection

In this study, two types of ground reference data are included: (i) field sampling points and (ii) points interpreted from images. The field sampling plot is a square with a side length of 30 m. The three-dimensional coordinates of the four corner points of the sample plot were measured using dual-frequency differential global positioning system (GPS). The surroundings of forest stands were observed when each plot was sampled. If the forest stands within 30 m around the sampling center were the same tree species, the center was sampled. Otherwise, the center was moved to a suitable place where the sampling plot contained only one tree species. If the sample plot could not contain only one tree species by moving to other places, two tree species can be included. Finally, the coordinates of the four corner points of each sample point, dominant tree species, average breast diameter, and average tree height were recorded.

The field samples were collected in August 2016. The samples include 11s tree species (*Illicium verum Hook. f.*, *Eucalyptus urophylla*, *Eucalyptus grandis*, *Cunninghamia lanceolata*, *Linden*, *Pinus elliottii*, *Michelia macclurei*, *Manglietia glauca*, *Mytilaria laosensis*, *Tsoongiodendrom odorum*, *Pinus massoniana*) and a total of 30 sampling areas. Figure 3 shows the spatial distribution of all samples, and the range of the sample plot was determined by the diagonal point and its adjacent point. For convenience, all tree species in the following text are abbreviated as shown in Table 2. The ratio of training samples to test samples is 1:4 in this study.

Table 2. Abbreviations of tree species.

Species	*Illicium verum*	*Tilia tuan*	*Eucalyptus urophylla*	*Michelia odora*
Abbreviation	*I. verum*	*T. tuan*	*E. urophylla*	*M. odora*
Species	*Eucalyptus grandis*	*Pinus massoniana*	*Mytilaria laosensis*	*Cunninghamia lanceolata*
Abbreviation	*E. grandis*	*P. massoniana*	*M. laosensis*	*C. lanceolata*
Species	*Manglietia glauca*	*Michelia macclurei*	*Pinus elliottii*	——
Abbreviation	*M. glauca*	*M. macclurei*	*P. elliottii*	——

Figure 3. Distribution of sample points.

2.3. Methods

The flowchart of the FSP method is shown in Figure 4. First, a high-resolution aerial image was fused with a single-time Sentinel-2 image, and the forest stand was obtained by the fractal net evolution approach (FNEA) segmentation. The features of the three types of data were extracted for each forest stand. The histogram was generated using all pixel values in a forest stand (i.e., one segment) for the aerial image across all multispectral bands. The average reflectance of each band was calculated in a forest stand for the time-series images, and the reflectance curve was generated by stacking all the bands of the time-series images. The profile curve of height was generated from the LiDAR data for each forest stand. Finally, the curve matching classifier was applied to classify the forest stands based on the extracted feature curves. The details of the FSP method are described in the following subsections, including data preprocessing, multisource image fusion, forest stand segmentation, feature extraction, and classification.

2.3.1. Data Preprocessing

Atmospheric correction and resampling were applied to the Sentinel-2 image. The Level-1C products of Sentinel-2 images were used. The Sen2Cor plug-in (v255) was used to manually correct the atmosphere on all bands through the Sentinel application platform (SNAP, v6.0.4, available online: http://step.esa.int/main/third-party-plugins-2/sen2cor/). The water vapor band was removed because it mainly reflects the water vapor in the atmosphere. Since multi-spectral bands of the Sentinel-2 image have different spatial resolutions, a third-party plug-in super-resolution tool Sen2Res was used for resampling. This tool can synthesize all bands with different resolutions to 10 m through super-resolution technology [69].

The LiDAR data were registered with images, and non-signal points were removed. Therefore, the preprocess for the LiDAR data is to classify ground points and forest points. The improved progressive triangulated irregular network (TIN) densification filtering algorithm was applied to classify point clouds [70].

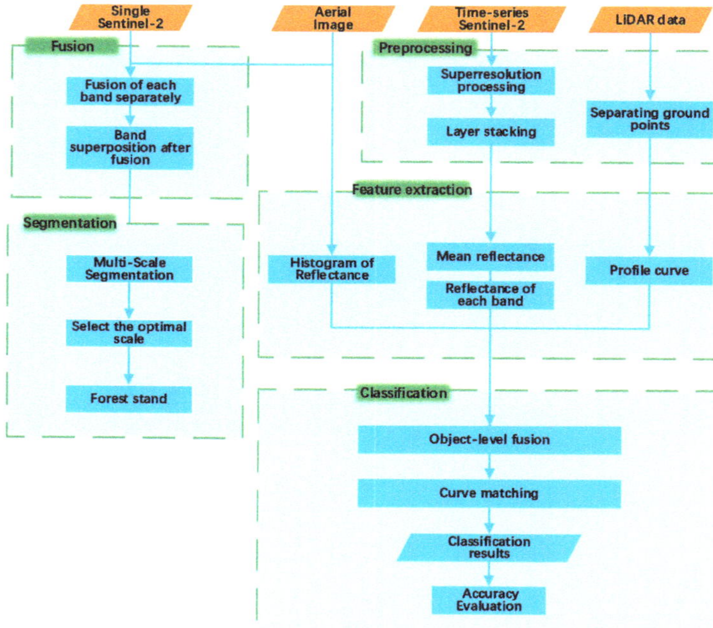

Figure 4. The flowchart of the fusion of spectral image and point data (FSP) method.

In this algorithm, an appropriate grid size was selected to split the LiDAR data, and the initial grid size was 20 m. The lowest point in each grid was selected as the initial seed point. The seed points were used to construct an initial TIN. To iteratively densify the TIN, all points to be classified were traversed, and the triangles into which the horizontal projection of each point fell were queried. The distance k from the point to the triangle was calculated, and the maximum value of the angle was formed by the point and plane of the triangle. The calculated distance and the angle were compared with the iteration distance (the threshold of the distance was 1.5 m) and iteration angle (the threshold of the value was 8°), respectively. If the distance and angle were less than the thresholds, the point was classified as a ground point and added to the TIN. Thus, the ground points and the points returned from the forest were separated. Finally, the values for these forest points were normalized to 0–1. The final height was obtained by subtracting the digital elevation model (DEM) to remove the influence of terrain.

2.3.2. Multisource Image Fusion

As mentioned before, aerial images have rich textural information, and time-series images contain rich spectral information. The color (spectral) is as important because the texture when using segmentation. Therefore, the Sentinel-2 image obtained on 2 June 2016, was fused with the aerial image since both images were acquired in June. The twelve bands of the Sentinel-2 image were used to fuse with the aerial image. The fusion can make the best use of the spectral and textural information for an accurate segmentation. The fusion method adopted in the experiments was a nonlinear transform and multivariate analysis algorithm (NMV) [71]. The NMV method could minimize the spectral distortion in the fusion image. The steps of the NMV algorithm are described as follows.

(1) The spatial details were obtained by the difference between the band and its degraded version:

$$M_{i,h} = M_i + M_{i,L}$$

where M_i is the ith band, and $M_{i,L}$ is an upsampled image using the bicubical method to match the pixel size of the reflective band. The spatial details of the multiple reflective bands can be expressed as follows:

$$M_{i,h,t} = (M_i)^t + (M_{i,L})^t$$

where t is the coefficient.

(2) A multivariate regression of a low-resolution image and multiple reflective bands was established.

$$M_{\text{low}} = \sum_{i=1}^{n} [c_i(M_i{}^t) + a_i M_i + b + e]$$

where c_i, a_i and b are coefficients; e is the residual; and M_{low} is the low-spatial-resolution image. Given value t, the coefficients can be estimated using the least squares approach.

(3) The low-spatial-resolution image was fused to the final image with a high spatial resolution by the following equation:

$$M_{\text{low},f} = M_{\text{low}} + \sum_{i=1}^{n} [c_i M_{i,h,t} + a_i M_i]$$

2.3.3. Forest Stand Segmentation

The FNEA [72] was applied to segment the forest stand. This algorithm grows from bottom to top, following the principle of minimum heterogeneity and adjacent heterogeneity. Pixels with similar spectral information are merged into a homogenous object, during which the textural, spectral, and shape features of the image object are simultaneously considered. The scale parameter was selected using the automated Estimation of Scale Parameter (ESP2) tool. The scale factor was set to 80. The shape factor was 0.3 and the compactness was 0.1.

2.3.4. Feature Extraction

By generating the histograms from the aerial image, the brightness of each multispectral band was projected on the x-axis, and the histogram frequency was projected on the y-axis. One hundred histogram bins were set between 0 and 1, and the number of pixels was counted in each bin. Finally, the total number of pixels was used to normalize the generated histogram, so that the effect of different sizes of objects can be eliminated. Figure 5 shows the histograms of a forest stand for three multispectral bands of the aerial image.

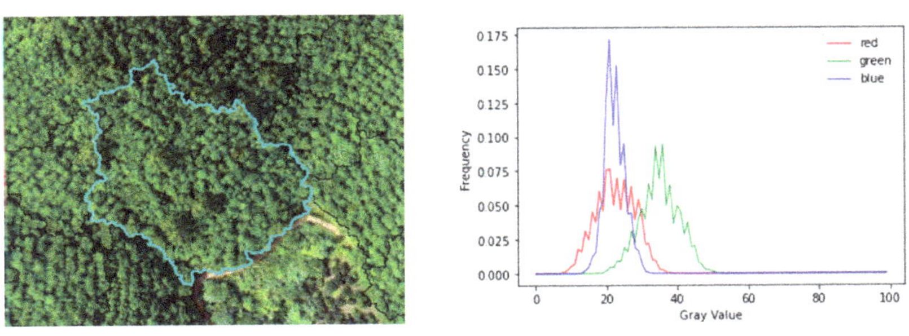

(a) (b)

Figure 5. Grayscale histogram generated from aerial image of a forest stand. (**a**) A forest stand. (**b**) The histogram of the forest stand in bands R, G, and B.

The average reflectance of each band for each forest stand was calculated. Figure 6 shows a forest stand for the time-series Sentinel-2 images and the stacked time-series reflectance curve.

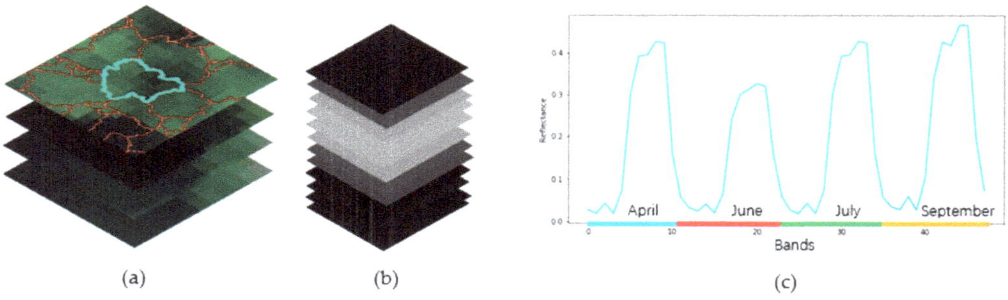

Figure 6. The reflectance curve of time-series image of a forest stand. (**a**) A forest stand. (**b**) The average reflectance of a forest stand for each band. (**c**) The reflectance curve across the stacked time-series bands.

For the LiDAR data, the profile curve for each stand was generated to extract the structural features of forest stands. The profile curve was generated from the vertical frequency distribution of the LiDAR data. The profile curve is essentially a histogram of height. Since this curve can characterize the vertical structure of trees, we use the term profile curve. The same tree species have similar structural features, and different tree species have different structural characteristics, as illustrated in Figure 7. Figure 8 shows the profile curve extracted in a forest stand. The steps for generating the profile curve are described as follows. First, the elevation was uniformly discretized, and the value of each elevation interval was calculated. In this study, N was set to 100. The number of point clouds contained in each discrete height interval of each forest stand was calculated, and the vertical profile curve of the point cloud was generated. Finally, after being divided by the total number of point clouds in this forest stand, the profile curve was obtained. The x-axis in Figure 8 is the height bin, and the y-axis is the frequency distribution of the point cloud. The average point number in a stand is 2687, thus the generated curves are rather smooth.

Figure 7. The profile curve in the vertical direction with different characteristics of a tree's spatial structure.

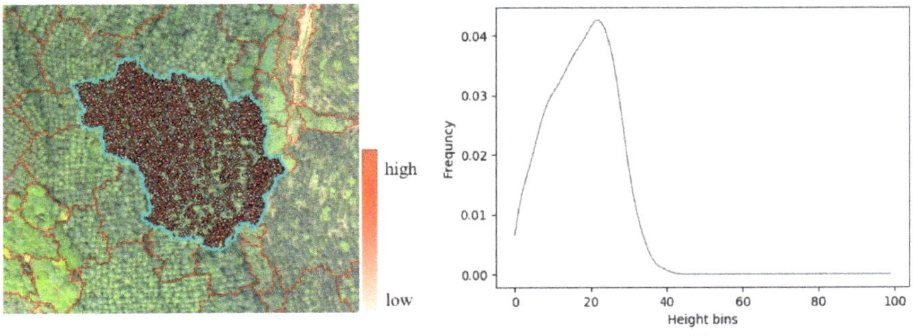

(a) (b)

Figure 8. Frequency distribution map generated from light detection and ranging (LiDAR) data of a forest stand. (**a**) A forest stand and the LiDAR data. (**b**) The profile curve of the forest stand.

2.3.5. Classification

Five feature curves (three for aerial images, one for time-series images, and one for LiDAR) were extracted for each forest stand. To identify the species that the curves belong to, a fusion method was developed to fuse all features using three curve matching classifiers: The Kullback–Leibler divergence (KL), root sum squared differential area (RSSDA), and curve angle mapper (CAM). In the curve matching classifier, the similarity between the known sample and the sample to be classified was measured. In this study, *P*1 represents the feature curves of the reference forest stand, and *P*2 refers to the feature curves of the forest stand to be classified. The three curve matching classifiers are described below.

KL divergence, also known as cross entropy, is a method used to describe the difference between two probability distributions. This method measures the distance between two random distributions. If two random distributions are the same, their relative entropy is zero. As the difference between two random distributions increases, their relative entropy also increases.

$$d_{\mathrm{KL}} = \sum_{i}^{n} \log\left(\frac{P1(i)}{P2(i)}\right) P1(i)$$

CAM calculates the similarity between two discrete curves. The calculation result represents the angle between the curve to be classified and the sample curve in n-dimensional space. The smaller the difference between the two curves, the smaller the angle.

$$d_{\mathrm{CAM}} = \cos^{-1}\left[\frac{\sum_{i=1}^{n} P1_i P2_i}{\sqrt{\left(\sum_{i=1}^{n} P1_i{}^2\right)}\sqrt{\sum_{i=1}^{n} P2_i{}^2}}\right]$$

RSSDA calculates the difference between the area integrals of two curves. This classifier uses discrete intervals as the differential unit to approximate the area. The RSSDA classifier was originally applied to match spectral curves [73], and was improved by Douglas [74].

$$d_{\mathrm{RSSDA}} = \sqrt{\sum_{i=1}^{n} (P1_i - P2_i)^2}$$

In Formulas (5)–(7), i is a discrete interval of the curves, and n is the number of intervals. For the histogram of the aerial image, i refers to the ith gray interval of a spectral histogram, n is the number of intervals of this histogram. For the time-series reflectance curve, i refers to the ith band, n is the total number of the stacked bands. For the profile curve, i refers to the ith height bin, n is number of total bins. Moreover, d_{KL}, d_{CAM}, d_{RSSDA} are the similarities of two curves measured using the KL, CAM, RSSDA curve matching methods.

Five feature curves were obtained for each forest stand. For a forest stand to be classified (i.e., a testing sample), its feature curves were compared with the feature curves of all training samples, using one of the above curve matching classifiers. The FSP method is defined as follows:

$$d_{FSP} = f_1 d^R + f_2 d^G + f_3 d^B + f_4 d^{TS} + f_5 d^{LD}$$

where f_1, f_2, f_3, f_4, and f_5 are proportional weights for different features. These weights were determined by the controlled variable method. In this study, the weights from f_1 to f_5 were set to 0.2, 0.23, 0.23, 0.1, and 0.24, respectively. Moreover, d^R, d^G, and d^B are the similarities of R, G and B bands of the aerial image, respectively; d^{TS} is the similarity of time-series image; and d^{LD} is the similarity of the LiDAR data.

Finally, the maximum value of fusion for each stand to be classified was found, and the category of the training sample corresponding to the maximum value was assigned to the stand to be classified.

3. Results

3.1. The Results of Fusion and Segmentation

Figure 9 shows the fusion results and detailed parts of the aerial image and Sentinel-2 image. These detailed images (Figure 9d) show that textures in the fused image are very clear. The fused image has a high resolution (0.2 m) and richer spectra than the aerial images. The segmentation results and some representative details are shown in Figure 10. As seen in the detailed images, the segmentation results divide the forest stands with different textures, and each forest stand can largely maintain its internal consistency.

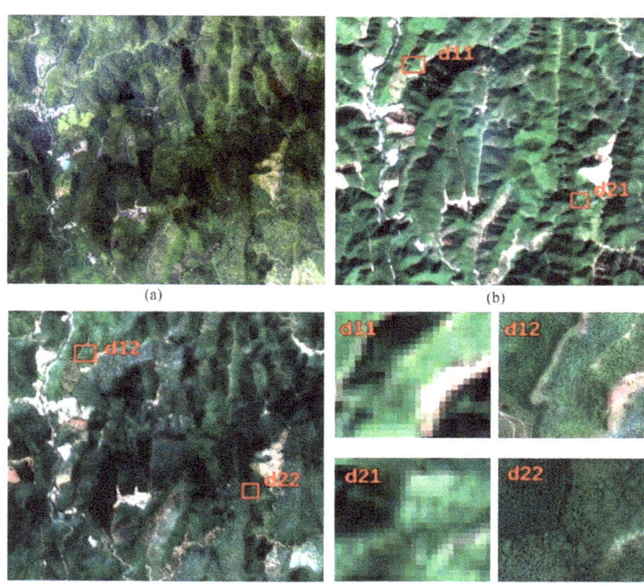

Figure 9. The relative image of fusion. (**a**) Aerial image. (**b**) Sentinel-2 image. (**c**) The results of Sentinel-2 image fused with aerial image. (**d**) The details of fusion results. The left part is the Sentinel-2 image; the right part shows the image after fusion.

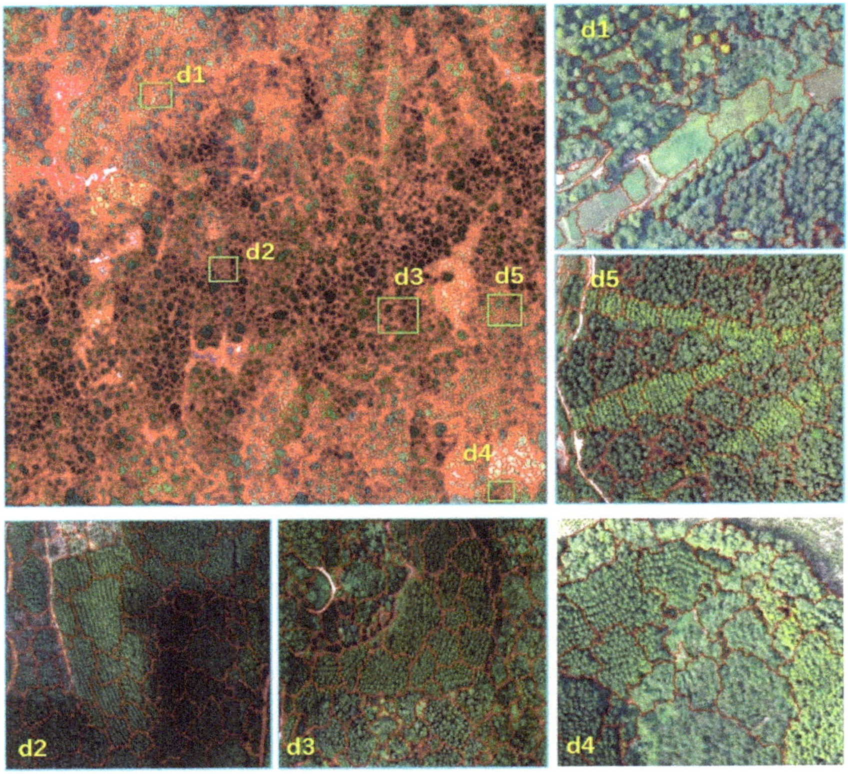

Figure 10. The results of segmentation and the enlarged details.

3.2. Feature Extraction Results

Figure 11 shows the histograms of 11 tree species in the R, G, and B bands. Figure 12 shows the histograms generated by a single sample for each class in the R, G, and B bands. The histograms in the R and G bands are similar. However, the peak of the R band appears more quickly than the peak of the G band, and the gray value of the B band is more concentrated than the values of the R and G bands. Therefore, the maximum value of the histogram in the B band is larger than that in the R and G bands, and the wave crest appears more quickly in the B band. This similarity can be regarded as the vegetation commonality. In addition to the commonalities, the histogram shapes for different tree species show diversity for different bands. In general, the histograms generated by each category show similarity in the overall distribution. Different categories of tree species have certain differences in each waveband, and some tree species in some wavebands have high degrees of similarity.

Figure 13a–k shows the curves of the time series of eleven tree species. Figure 13l shows the time-series curve generated by a single sample of 11 tree species. In the time-series curves, distinguishability is greatest in the red-edged band (bands 4, 5, 6, and 7), and the spectra show distinctive differences in different seasons.

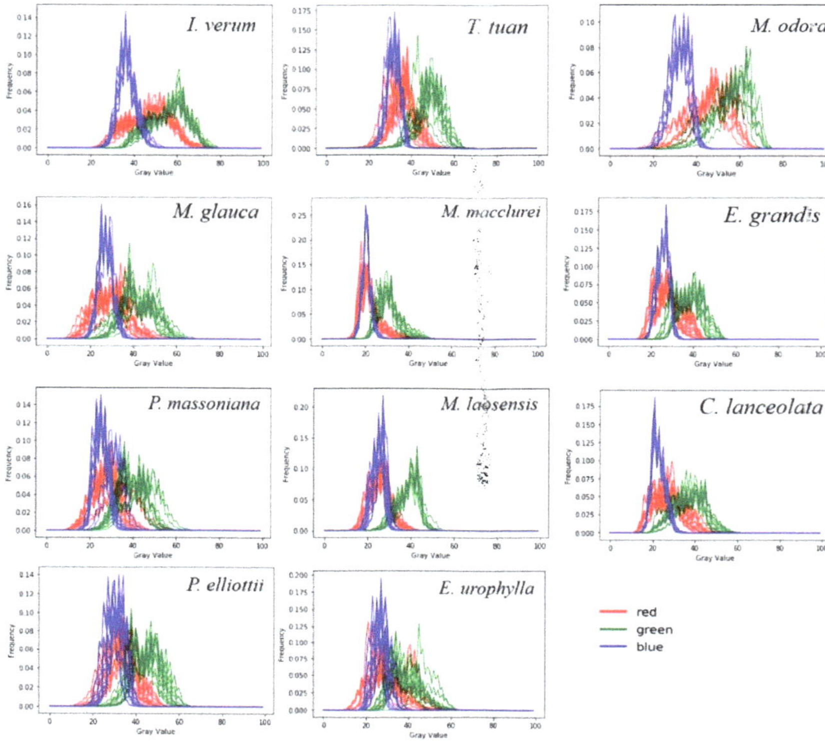

Figure 11. The histograms of 11 tree species in the R, G, B bands.

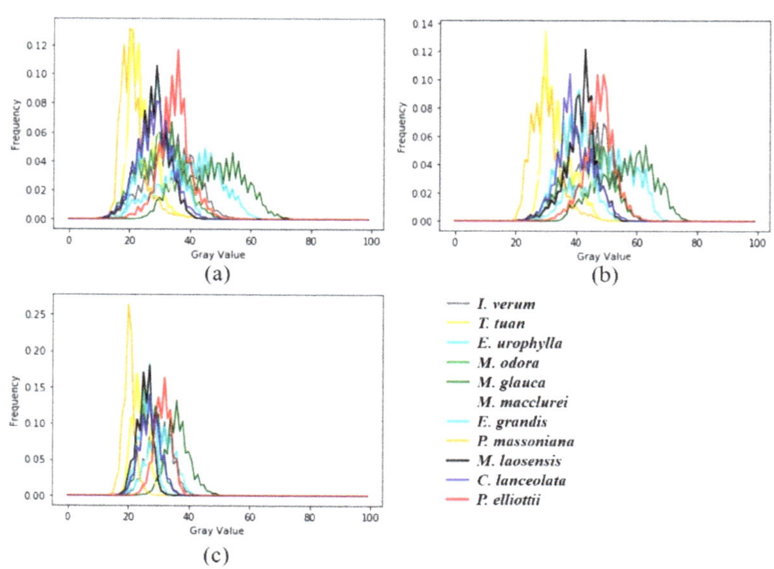

Figure 12. The histograms generated in the R, G, B, bands of 11 types of tree species. (**a**) The histograms in the R band. (**b**) The histograms in the G band. (**c**) The histograms in the B band.

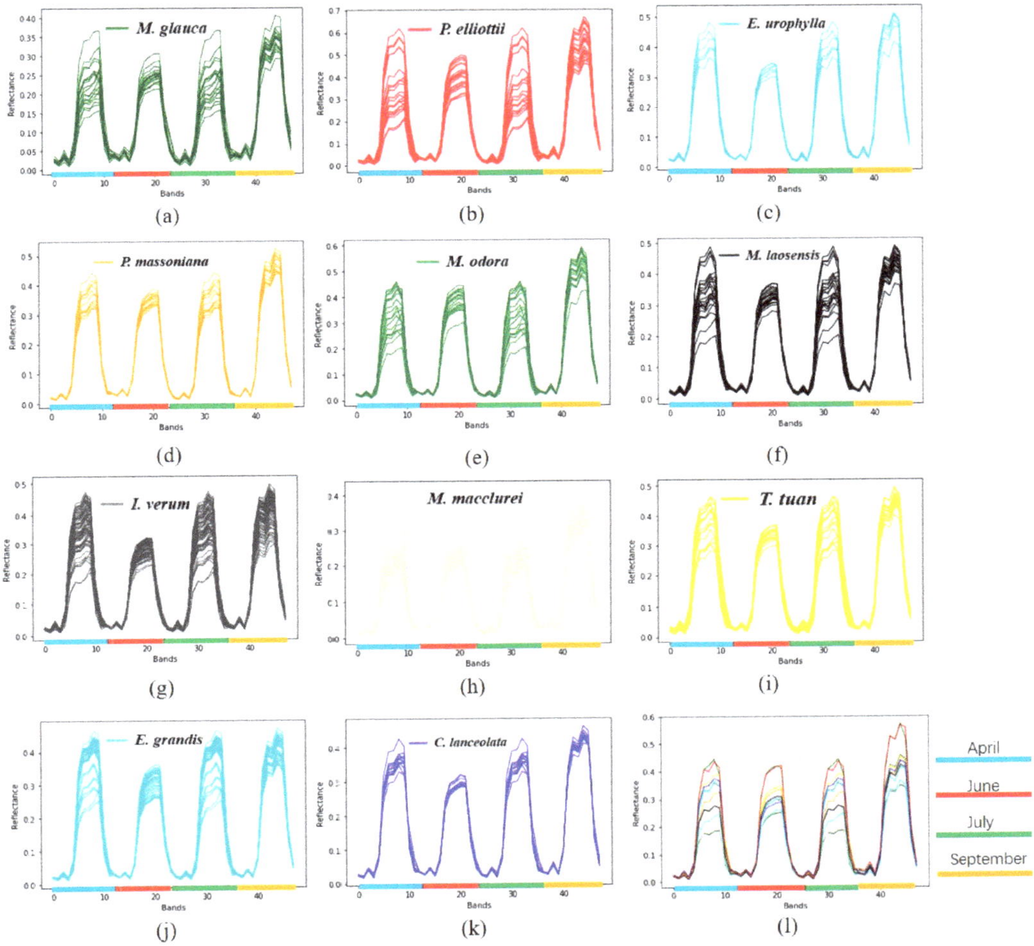

Figure 13. (**a–k**) The spectral curve of time-series image of 11 tree species, and (**l**) a comparison chart of the curves for 11 tree species, shows the curves of the time series of eleven tree species.

Figure 14a–k shows the profile curve generated by the LiDAR data for different tree species. The profile curves of the same tree species are similar. This similarity demonstrates that the features extracted from the LiDAR data are effective for distinguishing different tree species. Among the profile curves of 11 tree species, nine have one peak, and two have double peaks. In the curve with two peaks, the first peak is caused by the vegetation under the trees, such as small shrubs. The second peak is caused by the characteristics of trees. When the forest stands are of the same species, the structure below the canopy might be different, which may cause the deviation of the profile curve, such as for *M. glauca* and *I. verum*. In addition, the tree species in a forest stand may not be all the same, causing the waveform to deviate. Generally, the amount of deviation in the profile curve is only a small part of the total forest stand.

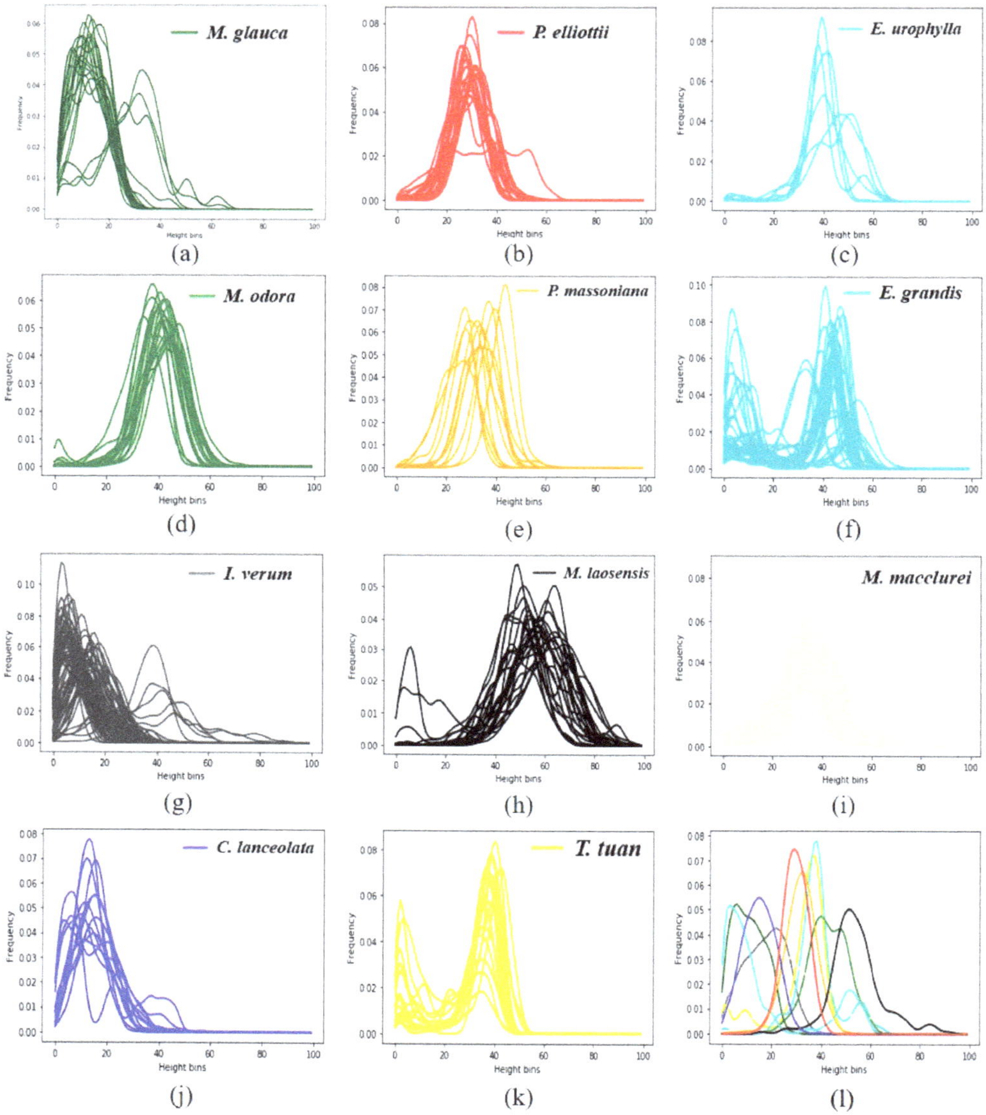

Figure 14. (**a**–**k**) The profile curve generated by the LiDAR data of 11 tree species and (**l**) a comparison chart of the curves for 11 tree species.

To clearly show the difference in profile curves among the different tree species, Figure 14l presents the results that integrate the profile curve of different tree species. Generally, different tree species have different waveforms, including the locations of wave peaks and the shape of the waveform. Sometimes, the profile curves of certain tree species are nearly identical, such as *P. elliottii* and *P. massoniana* (Figure 14b,e). Therefore, the utilization of the profile curve alone cannot distinguish between *P. elliottii* and *P. massoniana* Fortunately, the histograms of *P. elliottii* and *P. massoniana* in the R, G, and B bands are distinctive. *M. laosensis* and *C. lanceolata* (Figure 14h,j) can be easily distinguished using profile curves even though they have similar histogram and time-series curves.

3.3. Classification Results of FSP

Figure 15a shows the classification result obtained using the FSP method KL classifier. The white areas are non-forest areas. *C. lanceolata*, *P. elliottii*, *I. verum* and *E. grandis* have the widest distributions. Table 3 shows the classification results using the FSP method based on three curve matching classifiers. The overall accuracy of the KL matching result is 0.937, and the kappa coefficient is 0.926. The overall accuracy of the CAM matching result is 0.902, with a kappa coefficient of 0.884, and the overall accuracy of the RSSDA matching result is 0.925, with a kappa coefficient 0.911. The overall classification results of the FSP methods based on the three curve matching classifiers reach 0.900, and the RSSDA and KL classifiers are better than the CAM. Among all species, *E. urophylla* and *P. elliottii* have the worst classification results. The KL and CAM classifiers classify a large part of these two tree species into *E. grandis* because the two species have greater similarities in the histograms for the R, G and B bands. The product accuracy of *P. elliottii* is fairly high (0.900), but the user accuracy is poor (0.562) because the number of samples for *P. elliottii* is small, and the number of samples for *P. massoniana* is five times more than that of *P. elliottii* This imbalance of samples caused a part of the *P. massoniana* to be incorrectly classified as *P. elliottii* The indices of the FSP method based on three curve matching classifiers indicate that all classifications are well classified except for *P. elliottii* and *E. urophylla* However, the F1-score for these two tree species achieves 0.75 and 0.83 in the RSSDA classifier. This test indicates that the FSP method is suited to classify the tree species of forest stands and that the classification accuracy is rather high.

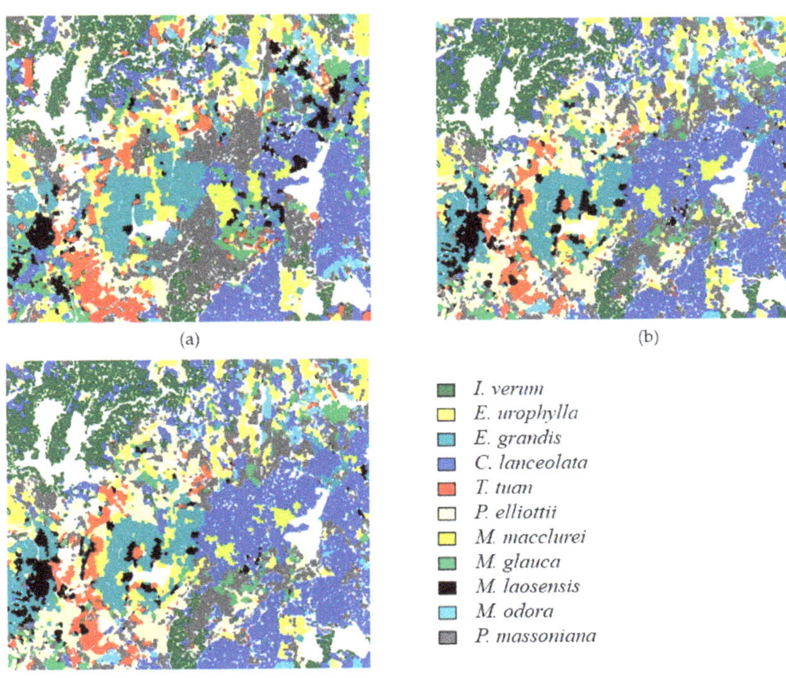

Figure 15. The classification results of three methods. (**a**) The FSP result based on KL. (**b**) The results using aerial and time-series image based on KL. (**c**) The results of using aerial image alone based on KL.

Table 3. Classification accuracies of the FSP method based on the Kullback–Leibler divergence (KL), curve angle mapper (CAM), and root sum squared differential area (RSSDA) curve matching classifiers.

FSP Method	KL-Based			CAM-Based			RSSDA-Based		
Class	UA	PA	F1-Score	UA	PA	F1-Score	UA	PA	F1-Score
I. verum	1	0.933	0.966	1	0.867	0.929	0.938	1	0.968
T. tuan	1	1	1	1	0.941	0.970	1	0.824	0.903
E. urophylla	0.818	0.692	0.750	0.800	0.615	0.696	0.909	0.769	0.833
M. odora	1	1	1	1	0.800	0.889	1.000	1	1
M. glauca	1	0.933	0.966	1	0.933	0.966	0.875	0.933	0.903
M. macclurei	1	1	1	0.889	0.889	0.889	1	1	1
E. grandis	0.885	0.920	0.902	0.800	0.960	0.873	0.958	0.920	0.939
P. massoniana	0.980	0.877	0.926	0.942	0.860	0.899	0.943	0.877	0.909
M. laosensis	0.905	1	0.950	0.900	0.947	0.923	1	0.947	0.973
C. lanceolata	0.86	1	0.993	0.932	0.971	0.951	0.920	0.986	0.952
P. elliottii	0.562	0.900	0.692	0.571	0.800	0.667	0.643	0.900	0.750
	Overall accuracy: 0.937 Kappa coefficient: 0.926			Overall accuracy: 0.902 Kappa coefficient: 0.884			Overall accuracy: 0.925 Kappa coefficient: 0.911		
UA: user accuracy; PA: product accuracy									

3.4. Comparison between Fusion Results of Different Types of Data

To determine whether the fusion can effectively improve the classification accuracy, we further compared the proposed FSP method with the method based on aerial images alone and the method based on the fusion of aerial images and time-series images. The same curve matching classifiers (KL, CAM, and RSSDA) were used. To reduce the effects of sampling, ten performances were applied with different random samples. The ratio of training samples to test samples is 1:4.

Table 4 shows the accuracy of the assessment results. When only the aerial image was used, the average classification accuracies (AVG) of ten performances were 0.795, 0.788, and 0.794 based on KL, CAM, and RSSDA, respectively, and the highest accuracy (MAX) reached 0.835. The fusion of time-series images and aerial images slightly improved the classification accuracy. The classification accuracies of FSP were 0.911, 0.900, and 0.913 based on the KL, CAM, and RSSDA classifiers, respectively. The accuracy of FSP was higher than that of the method that uses only aerial images. The SD column shows the standard deviation of the accuracies for the ten performances. The standard deviation decreased as more data were fused. The standard deviation of FSP was significantly lower than the other two methods, suggesting that the FSP method is more robust and less affected by sampling.

Table 4. The average of overall accuracy for ten performances using the KL, CAM, and RSSDA classifiers.

	Aerial Alone			Fusion of Aerial Image and Time-Series Images			Fusion of Aerial Image, Time-Series Images, and LiDAR Data		
	AVG	SD	MAX	AVG	SD	MAX	AVG	SD	MAX
KL	0.795	0.023	0.835	0.805	0.021	0.839	0.911	0.017	0.937
CAM	0.788	0.016	0.808	0.788	0.017	0.812	0.900	0.014	0.925
RSSDA	0.794	0.019	0.820	0.797	0.017	0.824	0.913	0.017	0.945

From the results of fusing different types of data in Table 4, it can be seen that the most helpful information for classification mainly comes from the aerial image and LiDAR data. The improvement contributed by the time-series images is limited. Nevertheless, the standard deviation of ten classification results is reduced when introducing time-series images, indicating a more robust result can be obtained.

Figure 15b,c shows the classification results using the KL classifier based on the fusion of the aerial image and time-series images and the aerial image alone. After fusing the time-series images, the details are improved. By comparing these two results, the classification result of FSP provides a more accurate description of the distribution of tree species.

3.5. Comparison with Traditional Methods

The FSP method was also compared with other traditional object-level classifiers, including the random forest (RF) [27], support vector machine (SVM) [75], and eXtreme Gradient Boosting (XGBoost) algorithms [51]. These three classifiers are commonly used. For traditional OBIA methods, the spectral and textural information for each forest stand were used for all bands of the aerial image and time-series images, whereas only the mean and standard deviation of the heights were used for the LiDAR data [7]. The spectral feature includes the mean and standard deviation, and the textural information includes contrast, entropy, homogeneity, angular second moment, dissimilarity, and correlation based on the gray-level co-occurrence matrix (GLCM) [28]. Therefore, the multi-dimensional summarized characteristics of forest stands are obtained. For a fair comparison, all the classifications were performed on the same ten samplings as previously mentioned.

The FSP and the benchmark classifications were coded in Python 3.7. The main package includes scikit-learn and gdal. The results of traditional classification methods are shown in Table 5. The worst classification result was SVM (0.814), and the best result was RF (0.824). RF also had the highest classification accuracy in ten performances (0.875); regardless RF has the largest standard deviation of 0.034. The overall accuracy of the FSP was 0.900, which was 0.09 higher than that of the traditional method. The highest classification accuracy of the FSP was 0.06 higher than that of the traditional method. Additionally, the standard deviation of the ten performances shows that the FSP was more stable than traditional methods. In general, the FSP method we proposed has a higher accuracy than the RF, SVM, XGBoost classifiers based on traditional summarized features (i.e. mean, standard deviation, etc.).

Table 5. The accuracy assessment of ten performances of support vector machine (SVM), random forest (RF), eXtreme Gradient Boosting (XGBoost).

	Algorithm	AVG	SD	MAX
Traditional methods	SVM	0.814	0.025	0.855
	RF	0.824	0.034	0.875
	XGBoost	0.817	0.025	0.855
FSP	KL	0.911	0.017	0.937
	CAM	0.900	0.014	0.925
	RSSDA	0.913	0.017	0.945

To compare the separability of the summarized features used in the traditional classifications and comprehensive features (i.e., feature curves) used in the FSP method, a projection of these two types of features was performed to visualize their separability. The summarized features (120 dimension) include spectral and textual features of images and two height features of LiDAR data. The comprehensive features (448 dimension) consist of all the points on three types of feature curves in the FSP method. The t-SNE tool [76] was used to downscale the extracted features to two dimensions at the best visual aspect. The final visualization results are shown in Figure 16. The red circles mark some points with poor separability. In general, the points characterized by the comprehensive features in the FSP method are more concentrated, even those in the red circle remain aggregated (Figure 16b). In contrast, the summarized features show a confusion of many tree species (Figure 16a). Therefore, the comprehensive features using the FSP method have better separability than the summarized features based on the traditional classification methods.

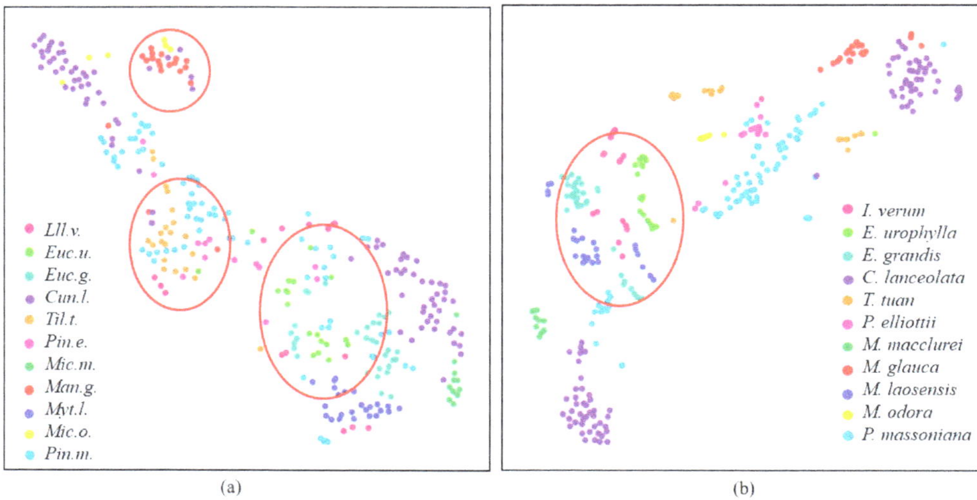

Figure 16. Separability of features: (**a**) the separability of summarized features used in traditional methods. (**b**) The separability of comprehensive features used in the FSP method.

4. Discussion

The study area is a managed forest, and there are not many mixed forests in this area. Therefore, the classification accuracy can reach over 90% for 11 tree species. If in a mixed forest, the accuracy may be compromised. Some improvements can be made from the following aspects.

First, the histogram curves from high-spatial-resolution image reflect spectral variability of forest stand. However, the histogram is essentially a disorder expression of the pixels, the spatial relationship lies in the forest stand is ignored. Sometimes forest stands belonging to different tree species may have similar spectral histograms but different textural information. Therefore, the rich textural information contained in the high-spatial-resolution image is not fully utilized. Therefore, in the follow-up research, a more sophisticated feature extraction method is expected to extract and incorporate textural information.

Second, the time-series Sentinel-2 images reflect the phenology features. When introducing phenology features, the classification accuracy is improved 1% and the standard deviation of accuracy is reduced comparing with that used the high-spatial-resolution image alone. It is known that the wavelength beyond 2000 nm is distinctive for many tree species [53]. Sentinel-2 images, however, do not cover such a wide wavelength range. If hyperspectral images are available, the FSP method can be applied similarly and probably derive a better result.

Third, the profile curve from LiDAR data is generated by counting the number of point clouds in forest stands, which means that the profile curves greatly rely on the density of point clouds. The shape of the profile curves is also affected by the shape of the tree and the size of the stand area. If in a mixed forest, the density of the point clouds is required to be high to characterize different structures of different tree species in the profile curve.

Finally, if the tree species are mixed seriously, it would be difficult to classify tree species at the forest stand level. Instead, classification can be performed at the individual tree level. In such case, the current FNEA segmentation algorithm is not suited, and the individual tree delineation algorithm is required. The FSP method can be extended to the delineation result, but higher requirements for the spatial resolution of images and the density of the LiDAR data are required to extract distinct features of individual trees.

5. Conclusions

This paper proposed an FSP method to synthesize high-spatial-resolution multispectral images, time-series images, and LiDAR data. The developed FSP method first extracts rich information in the form of curves from three types of data. The histogram for the multispectral band is generated in a stand for the high-spatial-resolution image, the average reflectance is calculated in each stand for a single band of time-series images, and a reflectance curve is generated by stacking time-series bands, and the profile curve from the point cloud LiDAR data is generated for each stand. Then, the fusion method is used based on curve matching classifiers for forest mapping. The performance of the three curve matching classifiers is evaluated, including KL, CAM, and RSSDA.

The features provided by different types of data contain a large amount of key information. The histograms extracted from the aerial image have richer spectral information than those of traditional OBIA methods based on some statistical measures, such as the mean and standard deviation. The phenology information is contained in time-series images and, thus, distinctive features can be reflected for some tree species from the reflectance curves. The profile curve generated from LiDAR data includes rich forest structure information and is effective in distinguishing tree species. Additionally, the features in the form of the curves facilitate the fusion of disparate data on the stand unit by introducing curve matching classifiers. The results show that the FSP method fused with three types of data can achieve higher accuracy and is more stable than the methods fused with less data or using only aerial images. The FSP method also shows a great advantage over traditional OBIA classification methods.

Author Contributions: H.W. designed and completed the experiment and drafted the manuscript; Y.T., L.J., and F.Q. designed the methodology; Y.T., L.J., and H.L. provided comments on the method; Y.T., L.J., and W.W. modified the manuscript and provided feedback on the manuscript. All authors have read and agreed to the published version of the manuscript.

Funding: This research was funded by the Key Research and Development Program of Hainan Province (ZDYF2019005); the Aerospace Information Research Institute, Chinese Academy of Sciences (Y951150Z2F); the Science and Technology Major Project of Xinjiang Uygur Autonomous Region (2018A03004); and the National Natural Science Foundation of China (41972308, 42071312).

Institutional Review Board Statement: Not applicable.

Informed Consent Statement: Not applicable.

Data Availability Statement: Data sharing is not applicable to this article.

Acknowledgments: The authors thank the Key Laboratory of Digital Earth Science for supporting this research with the hardware device, and the two anonymous reviewers for providing helpful comments and suggestions to improve the manuscript.

Conflicts of Interest: The authors declare no conflict of interest.

References

1. Schiefer, F.; Kattenborn, T.; Frick, A.; Frey, J.; Schall, P.; Koch, B.; Schmidtlein, S. Mapping forest tree species in high resolution UAV-based RGB-imagery by means of convolutional neural networks. *ISPRS J. Photogramm. Remote Sens.* **2020**, *170*, 205–215. [CrossRef]
2. Puumalainen, J.; Kennedy, P.; Folving, S. Monitoring forest biodiversity: A European perspective with reference to temperate and boreal forest zone. *J. Env. Manag.* **2003**, *67*, 5–14. [CrossRef]
3. Ørka, H.O.; Dalponte, M.; Gobakken, T.; Næsset, E.; Ene, L.T. Characterizing forest species composition using multiple remote sensing data sources and inventory approaches. *Scand. J. For. Res.* **2013**, *28*, 677–688. [CrossRef]
4. Dalponte, M.; Bruzzone, L.; Gianelle, D. Tree species classification in the Southern Alps based on the fusion of very high geometrical resolution multispectral/hyperspectral images and LiDAR data. *Remote Sens. Environ.* **2012**, *123*, 258–270. [CrossRef]
5. Li, W.K.; Guo, Q.H.; Jakubowski, M.K.; Kelly, M. A New Method for Segmenting Individual Trees from the Lidar Point Cloud. *Photogramm. Eng. Remote Sens.* **2012**, *78*, 75–84. [CrossRef]
6. Yu, X.W.; Hyyppa, J.; Kaartinen, H.; Maltamo, M. Automatic detection of harvested trees and determination of forest growth using airborne laser scanning. *Remote Sens. Environ.* **2004**, *90*, 451–462. [CrossRef]

7. Torabzadeh, H.; Leiterer, R.; Hueni, A.; Schaepman, M.E.; Morsdorf, F. Tree species classification in a temperate mixed forest using a combination of imaging spectroscopy and airborne laser scanning. *Agric. For. Meteorol.* **2019**, *279*, 107744. [CrossRef]
8. Crabbe, R.A.; Lamb, D.; Edwards, C. Discrimination of species composition types of a grazed pasture landscape using Sentinel-1 and Sentinel-2 data. *Int. J. Appl. Earth Obs. Geoinf.* **2020**, *84*, 101978. [CrossRef]
9. Dechesne, C.; Mallet, C.; Le Bris, A.; Gouet, V.; Hervieu, A. Forest Stand Segmentation Using Airborne Lidar Data and Very High Resolution Multispectral Imagery. *ISPRS—Int. Arch. Photogramm. Remote Sens. Spat. Inf. Sci.* **2016**, *XLI-B3*, 207–214. [CrossRef]
10. Liu, J.; Wang, X.; Wang, T. Classification of tree species and stock volume estimation in ground forest images using Deep Learning. *Comput. Electron. Agric.* **2019**, *166*, 105012. [CrossRef]
11. Shi, Y.F.; Skidmore, A.K.; Wang, T.J.; Holzwarth, S.; Heiden, U.; Pinnel, N.; Zhu, X.; Heurich, M. Tree species classification using plant functional traits from LiDAR and hyperspectral data. *Int. J. Appl. Earth Obs. Geoinf.* **2018**, *73*, 207–219. [CrossRef]
12. Lin, Y.; Hyyppä, J. A comprehensive but efficient framework of proposing and validating feature parameters from airborne LiDAR data for tree species classification. *Int. J. Appl. Earth Obs. Geoinf.* **2016**, *46*, 45–55. [CrossRef]
13. Immitzer, M.; Atzberger, C.; Koukal, T. Tree Species Classification with Random Forest Using Very High Spatial Resolution 8-Band WorldView-2 Satellite Data. *Remote Sens.* **2012**, *4*, 2661–2693. [CrossRef]
14. Pu, R.; Landry, S. A comparative analysis of high spatial resolution IKONOS and WorldView-2 imagery for mapping urban tree species. *Remote Sens. Environ.* **2012**, *124*, 516–533. [CrossRef]
15. Wang, K.; Akar, G. Gender gap generators for bike share ridership: Evidence from Citi Bike system in New York City. *J. Transp. Geogr.* **2019**, *76*, 1–9. [CrossRef]
16. Heinzel, J.; Koch, B. Investigating multiple data sources for tree species classification in temperate forest and use for single tree delineation. *Int. J. Appl. Earth Obs. Geoinf.* **2012**, *18*, 101–110. [CrossRef]
17. Clark, M.L.; Roberts, D.A. Species-Level Differences in Hyperspectral Metrics among Tropical Rainforest Trees as Determined by a Tree-Based Classifier. *Remote Sens.* **2012**, *4*, 1820–1855. [CrossRef]
18. Pu, R.; Landry, S. Mapping urban tree species by integrating multi-seasonal high resolution pléiades satellite imagery with airborne LiDAR data. *Urban For. Urban Green.* **2020**, *53*, 126675. [CrossRef]
19. Asner, G.P.; Martin, R.E. Airborne spectranomics: Mapping canopy chemical and taxonomic diversity in tropical forests. *Front. Ecol. Environ.* **2009**, *7*, 269–276. [CrossRef]
20. Nagendra, H. Using remote sensing to assess biodiversity. *Int. J. Remote Sens.* **2010**, *22*, 2377–2400. [CrossRef]
21. Wolter, P.T.; Mladenoff, D.J.; Host, G.E.; Crow, T.R. Improved Forest stamped classification in the northern lake states using multi-temporal Landsat imagery. *Photogramm. Eng. Remote Sens.* **1995**, *61*, 1129–1143.
22. Mead, R.A. *LANDSAT Digital Data Application to Forest Vegetation and Land Use Classification in Minnesota*; Purdue University: West Lafayette, IN, USA, 1977.
23. Roller, N.E.G.; Visser, L. Accuracy of landsat forest cover type mapping. *Cell Biol. Int.* **1994**, *18*, 289–290.
24. Johansen, K.; Phinn, S. Mapping structural parameters and species composition of riparian vegetation using IKONOS and landsat ETM plus data in Australian tropical savannahs. *Photogramm. Eng. Remote Sens.* **2006**, *72*, 71–80. [CrossRef]
25. Mallinis, G.; Koutsias, N.; Tsakiri-Strati, M.; Karteris, M. Object-based classification using Quickbird imagery for delineating forest vegetation polygons in a Mediterranean test site. *ISPRS J. Photogramm. Remote Sens.* **2008**, *63*, 237–250. [CrossRef]
26. Huang, W.; Li, H.; Lin, G. Classifying Forest Stands Based on Multi-Scale Structure Features Using Quickbird Image. In Proceedings of the 2015 2nd IEEE International Conference on Spatial Data Mining and Geographical Knowledge Services, Fuzhou, China, 8–10 July 2015; pp. 202–208.
27. Karlson, M.; Ostwald, M.; Reese, H.; Bazié, H.R.; Tankoano, B. Assessing the potential of multi-seasonal WorldView-2 imagery for mapping West African agroforestry tree species. *Int. J. Appl. Earth Obs. Geoinf.* **2016**, *50*, 80–88. [CrossRef]
28. Michez, A.; Piegay, H.; Lisein, J.; Claessens, H.; Lejeune, P. Classification of riparian forest species and health condition using multi-temporal and hyperspatial imagery from unmanned aerial system. *Env. Monit. Assess.* **2016**, *188*, 146. [CrossRef] [PubMed]
29. Ferreira, M.P.; Wagner, F.H.; Aragão, L.E.O.C.; Shimabukuro, Y.E.; de Souza Filho, C.R. Tree species classification in tropical forests using visible to shortwave infrared WorldView-3 images and texture analysis. *ISPRS J. Photogramm. Remote Sens.* **2019**, *149*, 119–131. [CrossRef]
30. Yu, Q.; Gong, P.; Clinton, N.; Biging, G.; Kelly, M.; Schirokauer, D. Object-based Detailed Vegetation Classification with Airborne High Spatial Resolution Remote Sensing Imagery. *Photogramm. Eng. Remote Sens.* **2006**, *72*, 799–811. [CrossRef]
31. Franklin, S.E.; Ahmed, O.S. Deciduous tree species classification using object-based analysis and machine learning with unmanned aerial vehicle multispectral data. *Int. J. Remote Sens.* **2017**, *39*, 5236–5245. [CrossRef]
32. Zhou, Y.H.; Qiu, F. Fusion of high spatial resolution WorldView-2 imagery and LiDAR pseudo-waveform for object-based image analysis. *ISPRS J. Photogramm. Remote Sens.* **2015**, *101*, 221–232. [CrossRef]
33. Tang, Y.; Qiu, F.; Jing, L.; Shi, F.; Li, X. Integrating spectral variability and spatial distribution for object-based image analysis using curve matching approaches. *ISPRS J. Photogramm. Remote Sens.* **2020**, *169*, 320–336. [CrossRef]
34. Madonsela, S.; Cho, M.A.; Mathieu, R.; Mutanga, O.; Ramoelo, A.; Kaszta, Ż.; Kerchove, R.V.D.; Wolff, E. Multi-phenology WorldView-2 imagery improves remote sensing of savannah tree species. *Int. J. Appl. Earth Obs. Geoinf.* **2017**, *58*, 65–73. [CrossRef]
35. Pu, R.; Landry, S.; Yu, Q. Assessing the potential of multi-seasonal high resolution Pléiades satellite imagery for mapping urban tree species. *Int. J. Appl. Earth Obs. Geoinf.* **2018**, *71*, 144–158. [CrossRef]

36. Cochrane, M.A. Using vegetation reflectance variability for species level classification of hyperspectral data. *Int. J. Remote Sens.* **2010**, *21*, 2075–2087. [CrossRef]
37. Ustin, S.L.; Roberts, D.A.; Gamon, J.A.; Asner, G.P.; Green, R.O. Using imaging spectroscopy to study ecosystem processes and properties. *Bioscience* **2004**, *54*, 523–534. [CrossRef]
38. Fasnacht, L.; Renard, P.; Brunner, P. Robust input layer for neural networks for hyperspectral classification of data with missing bands. *Appl. Comput. Geosci.* **2020**, *8*, 100034. [CrossRef]
39. Zhao, Q.; Jia, S.; Li, Y. Hyperspectral remote sensing image classification based on tighter random projection with minimal intra-class variance algorithm. *Pattern Recognit.* **2021**, *111*, 107635. [CrossRef]
40. Li, W.; Du, Q.; Xiong, M. Kernel Collaborative Representation With Tikhonov Regularization for Hyperspectral Image Classification. *IEEE Geosci. Remote Sens. Lett.* **2014**, *12*, 48–52.
41. Lixin, G.; Weixin, X.; Jihong, P. Segmented minimum noise fraction transformation for efficient feature extraction of hyperspectral images. *Pattern Recognit.* **2015**, *48*, 3216–3226. [CrossRef]
42. Dalponte, M.; Bruzzone, L.; Gianelle, D. Fusion of hyperspectral and LIDAR remote sensing data for classification of complex forest areas. *IEEE Trans. Geosci. Remote Sens.* **2008**, *46*, 1416–1427. [CrossRef]
43. Jones, T.G.; Coops, N.C.; Sharma, T. Assessing the utility of airborne hyperspectral and LiDAR data for species distribution mapping in the coastal Pacific Northwest, Canada. *Remote Sens. Environ.* **2010**, *114*, 2841–2852. [CrossRef]
44. Morsdorf, F.; Meier, E.; Kotz, B.; Itten, K.I.; Dobbertin, M.; Allgower, B. LIDAR-based geometric reconstruction of boreal type forest stands at single tree level for forest and wildland fire management. *Remote Sens. Environ.* **2004**, *92*, 353–362. [CrossRef]
45. Vauhkonen, J.; Ørka, H.O.; Holmgren, J.; Dalponte, M.; Heinzel, J.; Koch, B. Tree Species Recognition Based on Airborne Laser Scanning and Complementary Data Sources. In *Forestry Applications of Airborne Laser Scanning*; Springer: Berlin/Heidelberg, Germany, 2014; pp. 135–156.
46. Ke, Y.H.; Quackenbush, L.J.; Im, J. Synergistic use of QuickBird multispectral imagery and LIDAR data for object-based forest species classification. *Remote Sens. Environ.* **2010**, *114*, 1141–1154. [CrossRef]
47. Nilsson, M.; Nordkvist, K.; Jonzén, J.; Lindgren, N.; Axensten, P.; Wallerman, J.; Egberth, M.; Larsson, S.; Nilsson, L.; Eriksson, J.; et al. A nationwide forest attribute map of Sweden predicted using airborne laser scanning data and field data from the National Forest Inventory. *Remote Sens. Environ.* **2017**, *194*, 447–454. [CrossRef]
48. Popescu, S.C.; Wynne, R.H.; Nelson, R.F. Measuring individual tree crown diameter with lidar and assessing its influence on estimating forest volume and biomass. *Can. J. Remote Sens.* **2003**, *29*, 564–577. [CrossRef]
49. Morsdorf, F.; Kotz, B.; Meier, E.; Itten, K.I.; Allgower, B. Estimation of LAI and fractional cover from small footprint airborne laser scanning data based on gap fraction. *Remote Sens. Environ.* **2006**, *104*, 50–61. [CrossRef]
50. Piiroinen, R.; Fassnacht, F.E.; Heiskanen, J.; Maeda, E.; Mack, B.; Pellikka, P. Invasive tree species detection in the Eastern Arc Mountains biodiversity hotspot using one class classification. *Remote Sens. Environ.* **2018**, *218*, 119–131. [CrossRef]
51. Karasiak, N.; Sheeren, D.; Fauvel, M.; Willm, J.; Monteil, C. Mapping tree species of forests in southwest France using Sentinel-2 image time series. In Proceedings of the 2017 9th International Workshop on the Analysis of Multitemporal Remote Sensing Images, Brugge, Belgium, 27–29 June 2017.
52. Shi, Y.; Wang, T.; Skidmore, A.K.; Heurich, M. Important LiDAR metrics for discriminating forest tree species in Central Europe. *ISPRS J. Photogramm. Remote Sens.* **2018**, *137*, 163–174. [CrossRef]
53. Fassnacht, F.E.; Latifi, H.; Sterenczak, K.; Modzelewska, A.; Lefsky, M.; Waser, L.T.; Straub, C.; Ghosh, A. Review of studies on tree species classification from remotely sensed data. *Remote Sens. Environ.* **2016**, *186*, 64–87. [CrossRef]
54. Plourde, L.C.; Ollinger, S.V.; Smith, M.L.; Martin, M.E. Martin. Estimating Species Abundance in a Northern Temperate Forest Using Spectral Mixture Analysis. *Photogramm. Eng. Remote Sens.* **2007**, *73*, 829–840. [CrossRef]
55. Andrew, M.E.; Ustin, S.L. Habitat suitability modelling of an invasive plant with advanced remote sensing data. *Divers. Distrib.* **2009**, *15*, 627–640. [CrossRef]
56. Lucas, R.M.; Lee, A.C.; Bunting, P.J. Retrieving forest biomass through integration of CASI and LiDAR data. *Int. J. Remote Sens.* **2008**, *29*, 1553–1577. [CrossRef]
57. Asner, G.P.; Knapp, D.E.; Kennedy-Bowdoin, T.; Jones, M.O.; Martin, R.E.; Boardman, J.; Hughes, R.F. Invasive species detection in Hawaiian rainforests using airborne imaging spectroscopy and LiDAR. *Remote Sens. Environ.* **2008**, *112*, 1942–1955. [CrossRef]
58. Hill, R.A.; Thomson, A.G. Mapping woodland species composition and structure using airborne spectral and LiDAR data. *Int. J. Remote Sens.* **2011**, *26*, 3763–3779. [CrossRef]
59. Naidoo, L.; Cho, M.A.; Mathieu, R.; Asner, G. Classification of savanna tree species, in the Greater Kruger National Park region, by integrating hyperspectral and LiDAR data in a Random Forest data mining environment. *ISPRS J. Photogramm. Remote Sens.* **2012**, *69*, 167–179. [CrossRef]
60. Hudak, A.T.; Evans, J.S.; Smith, A.M.S. LiDAR Utility for Natural Resource Managers. *Remote Sens.* **2009**, *1*, 934–951. [CrossRef]
61. Machala, M.; Zejdová, L. Forest Mapping Through Object-based Image Analysis of Multispectral and LiDAR Aerial Data. *Eur. J. Remote Sens.* **2017**, *47*, 117–131. [CrossRef]
62. Sridharan, H.; Qiu, F. Developing an Object-based Hyperspatial Image Classifier with a Case Study Using WorldView-2 Data. *Photogramm. Eng. Remote Sens.* **2013**, *79*, 1027–1036. [CrossRef]
63. Blair, J.B.; Hofton, M.A. Modeling laser altimeter return waveforms over complex vegetation using high-resolution elevation data. *Geophys. Res. Lett.* **1999**, *26*, 2509–2512. [CrossRef]

64. Lovell, J.L.; Jupp, D.L.B.; Culvenor, D.S.; Coops, N.C. Using airborne and ground-based ranging lidar to measure canopy structure in Australian forests. *Can. J. Remote Sens.* **2003**, *29*, 607–622. [CrossRef]
65. Farid, A.; Goodrich, D.C.; Bryant, R.; Sorooshian, S. Using airborne lidar to predict Leaf Area Index in cottonwood trees and refine riparian water-use estimates. *J. Arid Environ.* **2008**, *72*, 1–15. [CrossRef]
66. Muss, J.D.; Mladenoff, D.J.; Townsend, P.A. A pseudo-waveform technique to assess forest structure using discrete lidar data. *Remote Sens. Environ.* **2011**, *115*, 824–835. [CrossRef]
67. Popescu, S.C.; Zhao, K.G.; Neuenschwander, A.; Lin, C.S. Satellite lidar vs. small footprint airborne lidar: Comparing the accuracy of aboveground biomass estimates and forest structure metrics at footprint level. *Remote Sens. Environ.* **2011**, *115*, 2786–2797. [CrossRef]
68. Pang, Y.; Li, Z.Y.; Ju, H.B.; Lu, H.; Jia, W.; Si, L.; Guo, Y.; Liu, Q.W.; Li, S.M.; Liu, L.X.; et al. LiCHy: The CAF's LiDAR, CCD and Hyperspectral Integrated Airborne Observation System. *Remote Sens.* **2016**, *8*, 398. [CrossRef]
69. Brodu, N. Super-Resolving Multiresolution Images With Band-Independent Geometry of Multispectral Pixels. *IEEE Trans. Geosci. Remote Sens.* **2017**, *55*, 4610–4617. [CrossRef]
70. Zhao, X.Q.; Guo, Q.H.; Su, Y.J.; Xue, B.L. Improved progressive TIN densification filtering algorithm for airborne LiDAR data in forested areas. *ISPRS J. Photogramm. Remote Sens.* **2016**, *117*, 79–91. [CrossRef]
71. Jing, L.H.; Cheng, Q.M. A technique based on non-linear transform and multivariate analysis to merge thermal infrared data and higher-resolution multispectral data. *Int. J. Remote Sens.* **2010**, *31*, 6459–6471. [CrossRef]
72. Schäpe, M.B.A. Multiresolution Segmentation: An optimization approach for high quality multi-scale image segmentation. In *Beutrage zum AGIT-Symposium. Salzburg, Heidelberg*; Wichmann: Lotte, Germany, 2000; pp. 12–23.
73. Hamada, Y.; Stow, D.A.; Coulter, L.L.; Jafolla, J.C.; Hendricks, L.W. Detecting Tamarisk species (*Tamarix* spp.) in riparian habitats of Southern California using high spatial resolution hyperspectral imagery. *Remote Sens. Environ.* **2007**, *109*, 237–248. [CrossRef]
74. Stow, D.A.; Toure, S.I.; Lippitt, C.D.; Lippitt, C.L.; Lee, C.R. Frequency distribution signatures and classification of within-object pixels. *Int. J. Appl. Earth Obs. Geoinf.* **2012**, *15*, 49–56. [CrossRef]
75. Wessel, M.; Brandmeier, M.; Tiede, D. Evaluation of Different Machine Learning Algorithms for Scalable Classification of Tree Types and Tree Species Based on Sentinel-2 Data. *Remote Sens.* **2018**, *10*, 1419. [CrossRef]
76. Diamand, M. The Solution to Overpopulation, The Depletion of Resources and Global Warming. *J. Neurol. Neurosci.* **2016**, *7*, 140. [CrossRef]

remote sensing

MDPI

Communication

Combination of Sentinel-1 and Sentinel-2 Data for Tree Species Classification in a Central European Biosphere Reserve

Michael Lechner [1], Alena Dostálová [2], Markus Hollaus [2], Clement Atzberger [1] and Markus Immitzer [1,*]

1 Institute of Geomatics, University of Natural Resources and Life Sciences, Vienna (BOKU), Peter-Jordan-Straße 82, 1190 Vienna, Austria; michael.lechner@boku.ac.at (M.L.); clement.atzberger@boku.ac.at (C.A.)
2 Department of Geodesy and Geoinformation, Vienna University of Technology, 1040 Vienna, Austria; alena.dostalova@geo.tuwien.ac.at (A.D.); markus.hollaus@geo.tuwien.ac.at (M.H.)
* Correspondence: markus.immitzer@boku.ac.at

Abstract: Microwave and optical imaging methods react differently to different land surface parameters and, thus, provide highly complementary information. However, the contribution of individual features from these two domains of the electromagnetic spectrum for tree species classification is still unclear. For large-scale forest assessments, it is moreover important to better understand the domain-specific limitations of the two sensor families, such as the impact of cloudiness and low signal-to-noise-ratio, respectively. In this study, seven deciduous and five coniferous tree species of the Austrian Biosphere Reserve *Wienerwald* (105,000 ha) were classified using Breiman's random forest classifier, labeled with help of forest enterprise data. In nine test cases, variations of Sentinel-1 and Sentinel-2 imagery were passed to the classifier to evaluate their respective contributions. By solely using a high number of Sentinel-2 scenes well spread over the growing season, an overall accuracy of 83.2% was achieved. With ample Sentinel-2 scenes available, the additional use of Sentinel-1 data improved the results by 0.5 percentage points. This changed when only a single Sentinel-2 scene was supposedly available. In this case, the full set of Sentinel-1-derived features increased the overall accuracy on average by 4.7 percentage points. The same level of accuracy could be obtained using three Sentinel-2 scenes spread over the vegetation period. On the other hand, the sole use of Sentinel-1 including phenological indicators and additional features derived from the time series did not yield satisfactory overall classification accuracies (55.7%), as only coniferous species were well separated.

Keywords: tree species classification; Sentinel-1; Sentinel-2; multitemporal; random forest; *Wienerwald* biosphere reserve; BPWW

Citation: Lechner, M.; Dostálová, A.; Hollaus, M.; Atzberger, C.; Immitzer, M. Combination of Sentinel-1 and Sentinel-2 Data for Tree Species Classification in a Central European Biosphere Reserve. *Remote Sens.* **2022**, *14*, 2687. https://doi.org/10.3390/rs14112687

Academic Editor: Carlos Alberto Silva

Received: 9 May 2022
Accepted: 1 June 2022
Published: 3 June 2022

1. Introduction

The ongoing species loss and the continued degradation of many terrestrial ecosystems make it increasingly important to monitor changes on the Earth's surface on a large scale, with high accuracy and low latency [1]. The multispectral image data generated by the Sentinel-2 (S2) twin satellites are provided free of charge through the European Copernicus program. These data provide a great opportunity to monitor the entire Earth's surface with high spatial and spectral as well as temporal resolution [2,3]. Several studies have already shown that the use of multispectral imagery generates highly informative data for land cover and tree species classification [4,5]. Further improvements in classification accuracy can be achieved by using multispectral time series [6–8].

While the high 5-day-temporal resolution of the S2 satellites leads to dense time series, it is not guaranteed that areas larger than 10^3–10^4 km^2 are fully covered by the very same cloud-free acquisitions. The selection of suitable image material is, therefore, often easy and straightforward for smaller areas, but for large areas, additional pre-processing steps are needed to ensure a homogenous and gap-free set of features. Suitable techniques are, for example, compositing techniques [9], gap-filling procedures [10,11], or the use of descriptive temporal metrics [12].

Radar sensors, on the other hand, provide a more continuous data stream with however a lower signal-to-noise-ratio (SNR), as well as terrain- and observation-geometry-related artifacts [13,14]. The two Sentinel-1 satellites (S1), which possess high spatial resolution and high revisit frequency, are also provided free of charge by the Copernicus program and generate microwave images under almost all weather conditions. A number of studies have demonstrated a good potential for the differentiation of deciduous trees and conifers [14,15]. Rüetschi et al. [16] obtained an overall accuracy of 72% for a test site in Switzerland. Udali et al. [17] presented a forest type and tree species classification in a test area in southern Sweden using multitemporal S1 data with overall accuracies of 94% and 66%, respectively.

The combination of S1 and S2 data was used in several studies [7,15] for forest type and tree species classifications. Bjerreskov et al. [7] used a combination of multitemporal S1 and S2 data to classify nemoral forests in Denmark into broadleaf and coniferous forest types as well as into predefined tree species groups with overall accuracies of 95% and 63%, respectively. A combination of S1 and optical Landsat imagery was used for the classification of dominant tree species in broadleaf deciduous forests in Vietnam with an overall accuracy of 79% [18]. Systematic ablation studies are lacking investigating the potential of the two electromagnetic domains for the classification of a higher number ($\geq$10) of tree species in forests with high diversity.

The objective of this paper is to study the benefits of combining S1 and S2 data for tree species classification in the mid-altitude forests of Austria. Various S1 and S2 data combinations are used to classify 12 different tree species with the help of Breiman's random forest classifier [19] to evaluate the discriminative power of the two data streams:

(1) a set of S1-derived parameters and 14 cloud-free S2 scenes were classified individually and in combination;
(2) the monotemporal S2 scenes were classified separately as well as paired with S1 data;
(3) the accuracies obtained from the monotemporal S2 scenes were used to determine the most- and least-accurate S2 scenes from spring, summer, and autumn seasons. Combinations of the least- and most-accurate seasonal S2 scenes were classified with and without S1 data.

2. Materials and Methods

2.1. Study Site, Reference Data

The Biosphere Reserve *Wienerwald* (BPWW) is located southwest of Vienna (Austria) and covers an area of approximately 105,000 hectares with a geographical extension of ca. 42 km × 47 km and an elevation between 162 and 893 m above sea level. The broadleaf-dominated forest is characterized by more than 20 forest communities [20]. Due to its diversity of tree species and its location within the overlapping area of two S2 orbits, it is particularly well suited for investigating remote sensing methods.

The study site and the reference data are shown in Figure 1. An existing dataset, created by using information from several forest enterprises such as forest inventories and stand-wise description of the forest management plans, was used as reference data [6]. Additional samples were added to better balance the different classes. While it was not always possible for sparsely represented tree species, a maximum of one pixel per forest stand was selected. The final dataset consisted of 1283 individual pixels, representing a total of 12 tree species—seven deciduous and five coniferous. Although not all tree species occurring in the BPWW were represented in the samples, they nevertheless provided a good overview of the main tree species prevailing in the park.

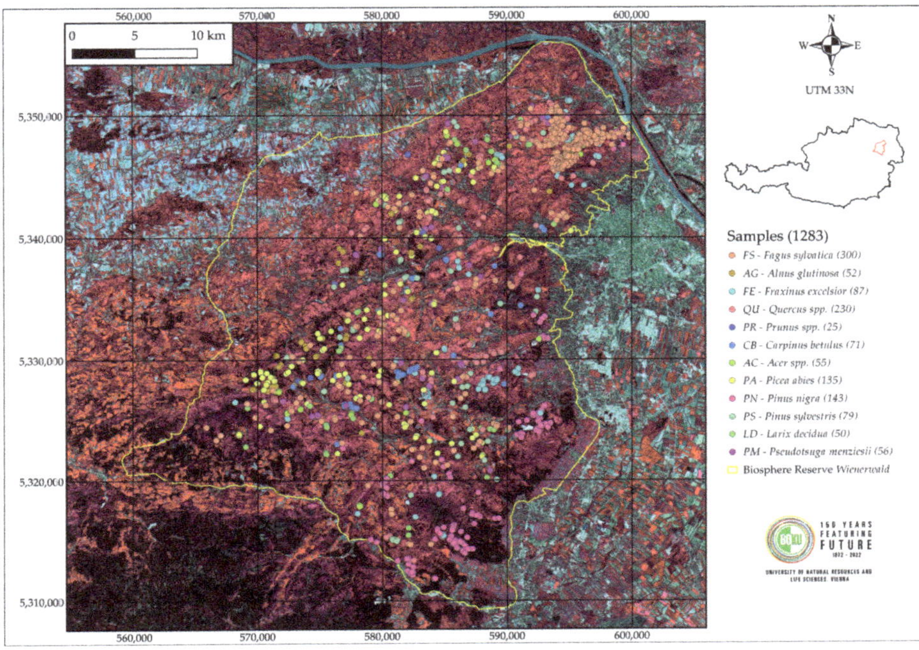

Figure 1. Overview of the Biosphere Reserve *Wienerwald* in the southwestern area of Vienna as well as the 12 classes of tree species reference samples. Background: Color Infrared composite of Sentinel-2.

2.2. Sentinel-2 Data

All 14 completely cloud-free S2 scenes from the 2018 growing period (April to October) were selected (Table 1). Since Nkosi et al. [21] found band 9 to have a high capability in discriminating tree species, band 9 was left in the dataset, while bands 1 and 10 were excluded. The remaining eleven S2 spectral bands were resampled to a unique 10 m spatial resolution and corrected using the Sen2Cor atmospheric correction [22]. The resulting dataset was expanded by two biophysical vegetation variables calculated from visible and NIR spectral channels: FAPAR and the LAI [23,24]. In addition, 30 vegetation indices were calculated and added to the dataset (Table A1).

Table 1. Summary of the Sentinel-2 (S2) scenes of the 2018 vegetation period used for classification.

S2 Satellite	Date	Orbit	Sun Zenith Angle	Sun Azimuth Angle
B	8 April 2018	79	43.02	157.29
B	21 April 2018	122	37.72	160.39
A	6 May 2018	122	33.06	159.37
A	2 July 2018	79	28.23	147.73
B	9 August 2018	122	34.38	155.97
A	21 August 2018	79	38.49	154.81
B	29 August 2018	122	40.40	160.44
A	13 September 2018	122	45.59	163.93
B	18 September 2018	122	47.41	164.98
B	28 September 2018	122	51.10	166.96
A	30 September 2018	79	52.25	164.21
B	5 October 2018	79	54.08	165.12
A	10 October 2018	79	55.90	165.94
A	30 October 2018	79	62.82	168.14

2.3. *Sentinel-1 Data*

In this study, 250 S1 ground-range-detected (GRD) interferometric wide (IW) swath mode acquisitions from the year 2018 were used. The pre-processed data were available via the Austrian Data Cube [25]. The pre-processing steps included precise orbit correction, border noise removal, radiometric correction to β^0 values, radiometric terrain flattening, range-Doppler terrain correction, and conversion to the decibel scale. A terrain model based on airborne laser scanning resampled onto a 10 m grid was used for the radiometric terrain flattening and the range-Doppler terrain correction steps. From the multitemporal S1 data, several parameters were computed. These included temporally averaged backscatter values for given time periods, phenological parameters, and harmonic regression model parameters.

2.3.1. Backscatter Averages and Ratios

For each repeat cycle of the S1 satellites (12 days), the temporal average of S1 backscatter was computed (Table 2). Two values per polarization (VV and VH)—one representing snow free, leaf-off conditions (14 to 26 March 2018) and one representing leaf-on conditions (18 to 30 June 2018) were selected. Furthermore, in case of the leaf-on period, the cross ratio (CPR) was computed as the backscatter ratio between VH and VV polarization [26,27], and in the case of VH polarization, the backscatter ratio between the leaf-on (18 to 30 June 2018) and leaf-off (14 to 26 March 2018) conditions was included.

Table 2. Summary of the Sentinel-1 temporal average backscatter parameters of the 2018 vegetation period used for classification.

Temporal Average Backscatter	Leaf-Off	Leaf-On
VH	20180314_20180326_VH	20180618_20180630_VH
VV	20180314_20180326_VV	20180618_20180630_VV
VH/VV		20180618_20180630_CPR

2.3.2. Phenological Parameters

Deciduous forest classes show distinct backscatter behavior where the VH backscatter drops during the summer period by 1–2 dB when compared to the leaf-off period [16,28]. Several studies assumed that the drop in backscatter is connected to the leaf emergence, while the backscatter increase is caused by the leaf fall [16,28–30]. We applied the breakpoints algorithm described in Zeileis et al. [31] and successfully tested on the annual time series of the 12-day VH backscatter averages [16]. The computation was limited to pixels, where the average backscatter in the leaf-on period was lower than that in the leaf-off period (hence the ratio between the temporally averaged backscatter for leaf-on and leaf-off conditions was positive). The first breakpoint in the time series was assumed to represent the start of the season, while the second breakpoint represented the end of the season. The length of the season was computed as the difference between the two values. Start of season, end of season, and length of season were used in this study (Table 3).

Table 3. Summary of the Sentinel-1 phenological parameters of the 2018 vegetation period used for classification.

Backscatter ratio	VH or VV
Leaf-on/Leaf-off	Rat_Leaf_on_off
Phenology	VH or VV
Start of season—day of the year	sos_doy
End of season—day of the year	eos_doy
Length of season—days	sos_doy
	correlation_winter
	slope_winter

2.3.3. Harmonic Parameters

Changes in the vegetation structure and environmental conditions cause temporal changes in radar backscatter. Especially in the case of vegetation, these changes typically have a strong seasonal character. Harmonic models (Equation (1)) can be used to describe this seasonal backscatter variation [32].

$$\hat{\gamma}^0_{t_{day}} = \overline{\gamma^0} + \sum_{i=1}^{k} \left(C_i \cos \frac{2\pi i t_{day}}{n} + S_i \sin \frac{2\pi i t_{day}}{n} \right) \tag{1}$$

The model estimates the most probable radar backscatter, $\hat{\gamma}^0_{t_{day}}$, for a given day of the year, t_{day}, from the average backscatter, $\overline{\gamma^0}$, for the given time period (year 2018 in case of this study) and the harmonic coefficients of the cosine and sine components, C_i and S_i. Harmonic coefficients and average backscatter are referred to as harmonic parameters (HPAR). As suggested, k is set to 3, representing the processes of a time scale of four months [32].

The HPARs (Table 4) are derived from a least-squares estimation based on the backscatter values and corresponding observation times of the input S1 time series. As opposed to Schlaffer et al. [32], we used the backscatter observations directly instead of using 10-day composites. Due to the strong dependency of the backscatter on the acquisition geometry, the parameter estimation was performed separately for each unique acquisition geometry (e.g., relative S1 orbit). As an additional parameter, the standard deviation of the residual error of the harmonic model was calculated as the square root of the sum of squared errors (SSE), divided by the number of data points (N_{points}) adjusted for the degrees of freedom of the model (Equation (2)). The SSE was derived from the pixel's backscatter time-series $\gamma^0_{t,\,r}$ and its harmonic model $\hat{\gamma}^0_{t,\,r}$.

For this study, we computed HPARs for both VV and VH polarization and for one ascending (relative orbit number 73) and one descending (relative orbit number 22) orbit.

$$s = \sqrt{\frac{SSE\left(\sigma^0_{t,\,r}, \hat{\sigma}^0_{t,\,r}\right)}{N_{points} - 2}} \tag{2}$$

Table 4. Summary of the Sentinel-1 harmonic parameters of the 2018 vegetation period used for classification.

		Ascending (Orbit 73)	Descending (Orbit 22)
HPAR			
Cosine 1	VH	HPAR-C1_2018_VH_A073	HPAR-C1_2018_VH_D022
	VV	HPAR-C1_2018_VV_A073	HPAR-C1_2018_VV_D022
Cosine 2	VH	HPAR-C2_2018_VH_A073	HPAR-C2_2018_VH_D022
	VV	HPAR-C2_2018_VV_A073	HPAR-C2_2018_VV_D022
Cosine 3	VH	HPAR-C3_2018_VH_A073	HPAR-C3_2018_VH_D022
	VV	HPAR-C3_2018_VV_A073	HPAR-C3_2018_VV_D022
Sine 1	VH	HPAR-S1_2018_VH_A073	HPAR-S1_2018_VH_D022
	VV	HPAR-S1_2018_VV_A073	HPAR-S1_2018_VV_D022
Sine 2	VH	HPAR-S2_2018_VH_A073	HPAR-S2_2018_VH_D022
	VV	HPAR-S2_2018_VV_A073	HPAR-S2_2018_VV_D022
Sine 3	VH	HPAR-S3_2018_VH_A073	HPAR-S3_2018_VH_D022
	VV	HPAR-S3_2018_VV_A073	HPAR-S3_2018_VV_D022
HPAR temporal average			
	VH	HPAR-M0_2018_VH_A073	HPAR-M0_2018_VH_D022
	VV	HPAR-M0_2018_VV_A073	HPAR-M0_2018_VV_D022
HPAR model error			
	VH	HPAR-STD_2018_VH_A073	HPAR-STD_2018_VH_D022
	VV	HPAR-STD_2018_VV_A073	HPAR-STD_2018_VV_D022

2.4. Classification Approach

Different test cases were defined (Table 5), which included various band and parameter combinations of the two satellite systems.

Table 5. Scenario of the nine test cases evaluated using different combinations of Sentinel-1 (S1) and Sentinel-2 (S2) data.

Test Case	Acronym	Features	Comment
1	S1	43	multitemporal S1 parameters
2	S2 (MULTI)	574	multitemporal image data of S2
3	S1 + S2 (MULTI)	617	multitemporal S1 parameters + multitemporal image data of S2
4	S2 (MONO)	41	monotemporal image data of S2
5	S1 + S2 (MONO)	84	multitemporal S1 parameters + monotemporal image data of S2
6	S2 (MAS)	123	image data of Most Accurate S2 Scene of each growing season
7	S1 + S2 (MAS)	166	multitemporal S1 parameters + image data of Most Accurate S2 Scene of each growing season
8	S2 (LAS)	123	image data of Least Accurate S2 Scene of each growing season
9	S1 + S2 (LAS)	166	multitemporal S1 parameters + image data of Least Accurate S2 Scene of each growing season

Test case 1 to Test case 3 represent various combinations of the respective total data of the two satellite systems. While Test case 4 and Test case 5 were defined to evaluate the influence of additional S1 data on a monotemporal S2 scene, they were also necessary to identify the S2 scenes to be used in Test case 6 to Test case 9. Test cases 6–9 used a selection of the three scenes from the spring, summer, and autumn seasons, which had the highest/lowest total accuracy. Of these, classification models were created as the "most accurate scene of season" (MAS) and the "least accurate scene of season" (LAS) with and without the S1 features, respectively.

These 9 test cases (Table 5) were passed to Breiman's [19] random forest (RF) algorithm, a widely used ensemble learning approach. To avoid overfitting, a recursive "mean decrease in accuracy" feature selection (MDA) was performed similarly to other studies [6,33–35]. The number of trees in the random forests was chosen with ntree = 1000, and for mtry (number of predictors randomly sampled for each node), the default value was used, that is the square root of available input variables. The accuracy assessment was done based on the out-of-bag-results (OOB) of the random forest models calculating common metrics based on confusion matrices.

3. Results

3.1. Full S1/S2 Dataset Comparison

In Table 6, the results of the model created exclusively with S1 data are shown (Test case 1). An overall accuracy (OA) of 55.7%, with a Cohen's kappa of 0.469, was achieved. While good class-specific accuracies were achieved, especially for the conifers, the deciduous trees, apart from European beech (*Fagus sylvatica* FS), could only be separated with significantly lower accuracy. Furthermore, not a single sample of the two tree species maple (*Acer* spp. AC) and alder (*Alnus glutinosa* AG) could be assigned to the correct class.

If the sample classes were stratified in broadleaved (BL) and coniferous (CO) groups, the random forest model was able to separate them very well (BL = 96.5%, CO = 92.7%).

In the model built using only the S2 data (Test case 2), the overall accuracy was 83.2% with a Cohen's kappa of 0.806. We also observed an increase in overall accuracy between the broadleaved and coniferous groups (BL = 99.4%, CO = 97.2%) compared to Test case 1, while a significant increase occurred in both the user's and producer's accuracies of all tree species. The two species that could not be classified using the S1 data were also relatively well separated (Table 7).

Table 6. Confusion matrix based on the out-of-bag-result of Test case 1, using only Sentinel-1 data. UA: user's accuracy, PA: producer's accuracy, OA: overall accuracy; for the abbreviations of tree species names see Figure 1.

Sentinel-1 – Test case 1 / Classification	Reference													
	FS	**AG**	**FE**	**QU**	**PR**	**CP**	**AC**	**PA**	**PN**	**PS**	**LD**	**PM**	**UA**	**F$_1$-Score**
FS	236	33	36	86	16	39	34	0	10	4	14	0	**46.5%**	0.584
AG	0	0	0	1	0	1	0	0	0	0	0	1	**NA**	NA
FE	3	2	6	2	1	1	4	0	0	0	0	0	**31.6%**	0.113
QU	47	14	40	140	2	18	13	0	4	1	0	0	**50.2%**	0.550
PR	1	0	0	0	1	0	0	0	0	0	0	0	**50.0%**	0.074
CP	0	0	2	0	0	4	0	1	0	0	0	1	**50.0%**	0.101
AC	0	0	0	0	0	0	0	0	0	0	0	0	**NA**	NA
PA	1	0	0	0	0	2	0	119	2	7	0	24	**76.8%**	0.821
PN	7	2	2	1	0	5	2	1	114	19	1	5	**71.7%**	0.755
PS	0	0	0	0	1	0	0	5	11	44	1	5	**65.7%**	0.603
LD	5	1	1	0	4	1	2	1	0	2	31	1	**63.3%**	0.626
PM	0	0	0	0	0	0	0	8	2	2	3	19	**55.9%**	0.422
∑ Reference	300	52	87	230	25	71	55	135	143	79	50	56		
PA	**78.7%**	**0.0%**	**6.9%**	**60.9%**	**4.0%**	**5.6%**	**0.0%**	**88.1%**	**79.7%**	**55.7%**	**62.0%**	**33.9%**		
								OA 55.7%			Kappa 0.469			

Table 7. Confusion matrix based on the out-of-bag-result of Test case 2, using only Sentinel-2 data. UA: user's accuracy, PA: producer's accuracy, OA: overall accuracy; for the abbreviations of tree species names see Figure 1.

Sentinel-2 – Test case 2 / Classification	Reference													
	FS	**AG**	**FE**	**QU**	**PR**	**CP**	**AC**	**PA**	**PN**	**PS**	**LD**	**PM**	**UA**	**F$_1$-Score**
FS	272	4	10	26	5	22	15	0	2	0	6	0	**75.1%**	0.822
AG	1	42	3	1	1	1	0	0	0	0	0	0	**85.7%**	0.832
FE	6	2	64	5	2	4	3	0	2	0	2	0	**71.1%**	0.723
QU	13	1	7	195	5	4	4	0	0	0	1	0	**84.8%**	0.848
PR	0	0	0	0	12	0	0	0	0	0	0	0	**100.0%**	0.649
CP	3	2	1	1	0	40	1	0	0	0	0	0	**83.3%**	0.672
AC	5	1	0	1	0	0	30	0	0	0	0	0	**81.1%**	0.652
PA	0	0	0	0	0	0	0	127	1	4	1	3	**93.4%**	0.937
PN	0	0	0	0	0	0	2	3	133	1	1	1	**94.3%**	0.937
PS	0	0	0	0	0	0	0	3	3	72	3	5	**83.7%**	0.873
LD	0	0	2	1	0	0	0	0	2	2	36	2	**80.0%**	0.758
PM	0	0	0	0	0	0	0	2	0	0	0	45	**95.7%**	0.874
∑ Reference	300	52	87	230	25	71	55	135	143	79	50	56		
PA	**90.7%**	**80.8%**	**73.6%**	**84.8%**	**48.0%**	**56.3%**	**54.5%**	**94.1%**	**93.0%**	**91.1%**	**72.0%**	**80.4%**		
								OA 83.2%			Kappa 0.806			

The results of Test case 3, combining all data, show a marginal 0.5-percentage-point improvement in overall accuracy with OA = 83.7 % and kappa = 0.811, compared to Test case 2 (Table 8). The within-coniferous-group accuracy slightly gained by 1.1 percentage points compared to that for the sole use of optical data (Test case 2). The results of the individual classes remained more or less constant.

To better understand the inputs, Figure 2 lists the 15 most important out of the 50 remaining variables after the MDA-Feature selection of Test case 3. The analysis reveals that, with "20180618_20180630_VH" and "20180618_20180630_CPR" (red), only two S1 parameters were among the 50 remaining variables of the model. Interestingly, they were in the first and third place of importance. Band 9 was not included in the remaining variables.

Table 8. Confusion matrix based on the out-of-bag-result of Test case 3, using Sentinel-1 data and Sentinel-2 data. UA: user's accuracy, PA: producer's accuracy, OA: overall accuracy; for the abbreviations of tree species names see Figure 1.

		Reference													
		FS	**AG**	**FE**	**QU**	**PR**	**CP**	**AC**	**PA**	**PN**	**PS**	**LD**	**PM**	**UA**	**F$_1$-Score**
Sentinel-3 – Test case 3 Classification	**FS**	271	2	9	28	8	19	16	0	1	0	3	0	**75.9%**	0.825
	AG	0	43	1	0	0	1	0	0	0	0	0	0	**95.6%**	0.887
	FE	6	2	64	6	0	3	4	0	1	0	2	0	**72.7%**	0.731
	QU	16	1	9	193	6	4	3	0	0	0	0	0	**83.2%**	0.835
	PR	0	0	0	0	10	0	0	0	0	0	0	0	**100%**	0.571
	CP	2	3	3	2	0	44	1	0	1	0	0	0	**78.6%**	0.693
	AC	4	1	0	0	1	0	29	0	0	0	0	0	**82.9%**	0.644
	PA	0	0	0	0	0	0	0	129	1	4	1	5	**92.1%**	0.938
	PN	1	0	0	0	0	0	2	3	135	2	2	0	**93.1%**	0.938
	PS	0	0	0	0	0	0	0	2	3	70	2	3	**87.5%**	0.881
	LD	0	0	1	1	0	0	0	0	1	3	40	2	**83.3%**	0.816
	PM	0	0	0	0	0	0	0	1	0	0	0	46	**97.9%**	0.893
∑ Reference		300	52	87	230	25	71	55	135	143	79	50	56		
PA		**90.3%**	**82.7%**	**73.6%**	**83.9%**	**40.0%**	**62.0%**	**52.7%**	**95.6%**	**94.4%**	**88.6%**	**80.0%**	**82.1%**		
									OA **83.7%**			Kappa 0.811			

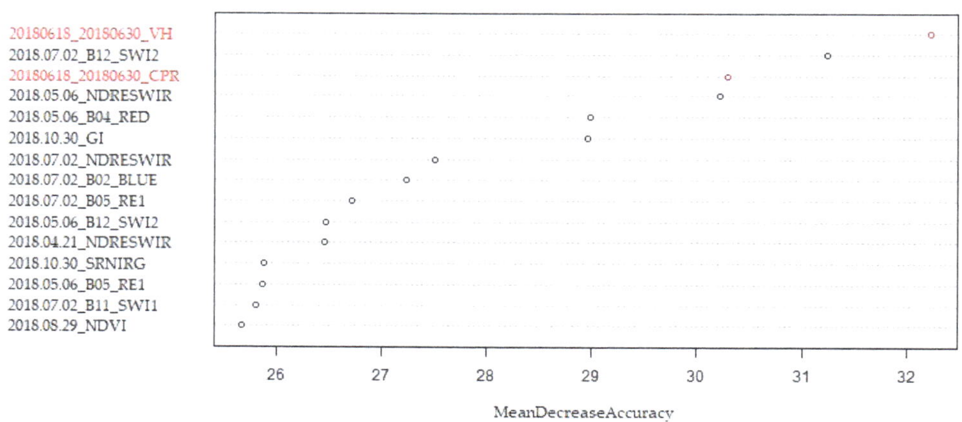

Figure 2. Importance plot of the remaining variables of Test case 3. The two Sentinel-1 parameters 20180618_20180630_VH and 20180618_20180630_CPR (in red letters) remain in the classification model.

3.2. Added Value of S1 on Monotemporal S2 Datasets

To evaluate the contributions of S1 data in the (extreme) case of only one available S2 scene, the individual S2 scenes were each classified separately with (Test case 5) and without S1 data (Test case 4). Figure 3 compares the two variants for each of the 14 S2 acquisitions of Test case 4 and Test case 5, each with (blue) and without S1 data (green). Furthermore, as surrogates for compositing approaches, the three most-accurate (MAS) and the three least-accurate scenes (LAS) of each growing season are shown, which serve as S2 data material for Test case 6 to Test case 9. The last bar shows the result of using all available S2 data, with (blue) and without S1 data (green).

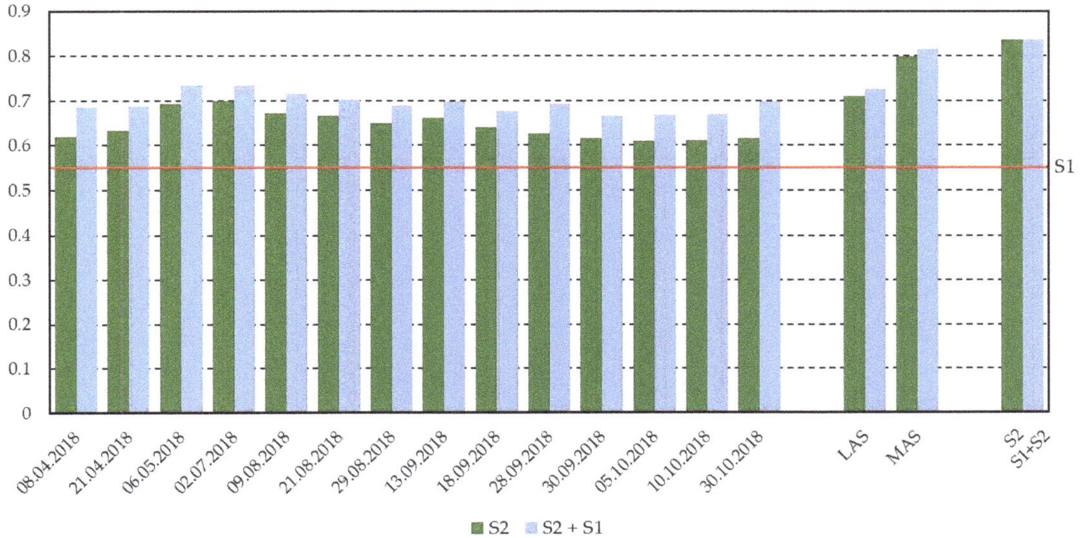

Figure 3. Overall accuracy of models using monotemporal Sentinel-2 (S2) scenes as well as two seasonal selections using the three least- (LAS) and the three most-performing S2 scenes (MAS), with and without Sentinel-1 (S1) combination. Also shown are the results when using all available S2 scenes with and without S1 combination. The red horizontal line highlights the results of the model based on S1 data only.

The monotemporal S2 results show an increase of the OA from spring to summer and again slightly lower values in the fall. Adding S1 data, the overall accuracy increased by around 5 percentage points but still fell far short of the accuracies achieved in Test case 2 (all S2 scenes without any S1; Figure 3 right, green). Each of the 14 monotemporal S2 scenes largely outperformed (by 5.2 to 14.3 percentage points) the full S1 dataset (horizontal red line in Figure 3).

3.3. Added Value of S1 on Multitemporal S2 Dataset

The two variants of seasonal S2 data (LAS vs. MAS) were separated by roughly 10 percentage points from each other (71.1% vs. 80.0%) with very minor improvements when the suite of S1 features were included (Figure 3). The LAS variant without S1 data, in which for each of the three seasons only the worst performing S2 image was retained (Test case 8), still performed better than the single best performing S2 scene (July image) and largely better (15.4%) compared to the full S1 dataset. The differences between S1 and S2 data were further accentuated (by almost 9%) when the three seasonal images were composed of the best performing individual S2 images (MAS; Test case 6). For this test case, classification results were almost on par with Test case 2 (all S2 scenes) and Test case 3 (all S1 and S2 data).

When comparing the sample-class F_1-scores of Test cases 5 to Test case 9, shown in Figure 4a, it becomes apparent that already three S2 MASs were sufficient to eliminate the added value of S1 data on the classification result. If only three S2 LASs were available, there was a marginal improvement in the F_1-scores of individual sample classes, but the results of already well-classified sample-classes were not further improved.

Nevertheless, it was not possible to reach the highest overall accuracy of Test case 3 by using only three individual S2 scenes, with and without additional S1 data.

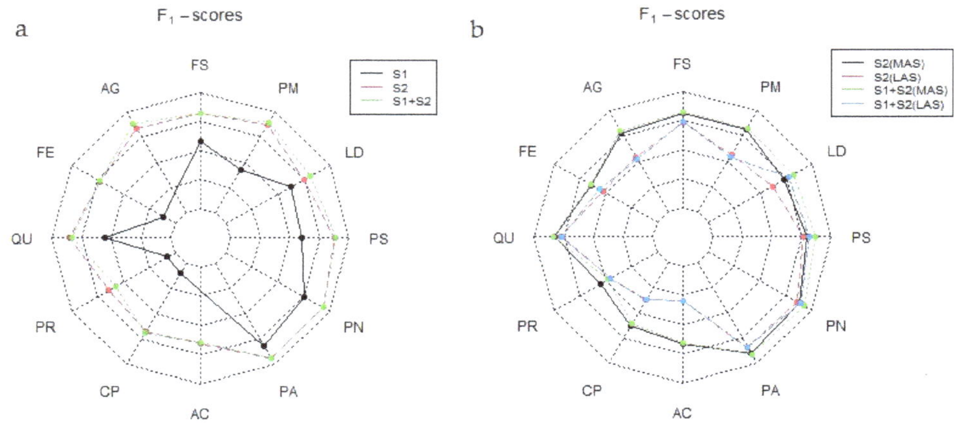

Figure 4. F_1-scores of Test cases 1 to Test case 3 (**a**), F_1-scores of all expressions of Test cases 6 to Test case 9 (**b**); for the abbreviations of tree species names see Figure 1, and definitions of test cases are provided in Table 5.

4. Discussion

The classification based only on S1 data (Test case 1) did not achieve the same high accuracy values as the classification with either monotemporal and/or combined S2 image data. Nevertheless, it was already possible to separate the five coniferous species with a moderate degree of accuracy. This underlines the potential of S1 for separating different conifers. Furthermore, the separation of the two groups, deciduous and coniferous forests, was already at a very high level and exceeded the results from previous S1-based studies [14,15]. The increased accuracy could be related to the higher number of samples and/or the significantly smaller study area compared to the aforementioned studies. The larger number of derived S1 features could also play a role here, albeit comparisons across datasets are generally to be taken with caution.

The fact that the species within the conifer group were separated with satisfactory accuracies using microwave data—but not the deciduous species—is possibly related to the more distinct canopy surface roughness between conifers (as compared to the smoother deciduous species canopy surface). These results are in line with previous studies on tree species classification from S1 with slightly higher overall accuracies but fewer species (Table 9).

Test case 2, which was only based on S2 data, delivered significantly better results than Test case 1. This was expected as optical data reflect both structural and biochemical forest traits [46] and their temporal evolution, and S2 data are known to have a very high SNR and a good temporal coverage [47]. The differentiation of the two strata was again very good, but more importantly, now all species (deciduous and conifers) were separable. Separating the individual tree species yielded a satisfactory result with an OA of 83.2%, kappa of 0.806 and an improvement of 27.5 percentage points compared to the sole use of S1 (Test case 1). The result of Test case 2 could not reach the high accuracy of 88.7%, which was reached by using the original sample dataset and S2 scenes from several years [6]. However, the samples were increased in size and more balanced. Compared to other studies presented in Table 9, the OA of Test case 2 is within the range usually achieved with multitemporal S2 data.

The result of Test case 2 could be improved only marginally, 0.5 percentage points, by using additional S1 data (Test case 3). Including S1 data, however, we observed a shift in contributing features. Indeed, the MDA analysis revealed that Test case 3 included two very high ranking S1 parameters as input variables, which, therefore, had to replace at least two optical features from Test case 2. The high number of highly correlated features makes it very difficult to fully understand the MDA findings. After the MDA-Feature selection, band 9 never remained in the dataset. Therefore, the added value of band 9 in the discrimination of tree species depicted by Nkosi et al. [21] could not be confirmed.

Table 9. Summary of previous studies using Sentinel-1 (S1), Sentinel-2 (S2), and both combined for tree species classification and their achieved overall accuracies (OAs).

Satellite	No. of Species	Species Names	OA	Reference
S1	3	*Quercus* spp., *Fagus sylvatica*, *Picea abies*	72%	[16]
S1	4	*Quercus robur*, *Betula* spp., *Picea abies*, *Pinus sylvestris*	66%	[17]
S1	12	*Fagus sylvatica*, *Alnus glutinosa*, *Fraxinus excelsior*, *Quercus* spp., *Prunus* spp., *Carpinus betulus*, *Acer* spp., *Picea abies*, *Pinus nigra*, *Pinus sylvestris*, *Larix decidua*, *Pseudotsuga menziesii*	58%	Table 6
S2	4	*Fagus sylvatica*, *Quercus* spp., other broadleaf trees, coniferous trees	88%	[36]
S2	4	*Sabina przewalskii*, *Picea crassifolia*, *Betula* spp., *Populus* spp.	90%	[37]
S2	5	*Larix* spp., *Pinus* spp., *Pinus mugo*, *Abies alba/Picea abies*, broadleaf trees	84%	[38]
S2	5	*Picea abies*, *Pinus silvestris*, *Larix* × *marschlinsii*, *Betula* sp., *Quercus robur*	88%	[39]
S2	7	*Picea* sp., *Pinus* sp., *Larix* sp., *Abies* sp., *Fagus* sp., *Quercus* sp., other broadleaf trees	66%	[3]
S2	7	*Acacia mearnsii*, *Eucalyptus dunnii*, *Eucalyptus grandiis*, *Eucalyptus* mix, *Pinus tecunumanii*, *Pinus elliotii*, *Pinus taedea*	84%	[40]
S2	8	*Fagus sylvatica*, *Quercus* spp., *Alnus* spp., *Betula pendula*, *Picea abies*, *Pinus sylvestris*, *Abies alba*, *Larix decidua*	82%	[8]
S2	9	*Fagus sylvatica*, *Betula pendula*, *Carpinus betulus*, *Abies alba*, *Acer pseudoplatanus*, *Larix decidua*, *European larch*, *Alnus incana*, *Pinus sylvestris*, *Picea abies*	92%	[41]
S2	11	*Alnus* spp., *Acer pseudoplatanus*, *Fagus sylvatica*, *Betula pendula*, *Carpinus betulus*, *Quercus* spp., *Picea abies*, *Pinus sylvestris*, *Larix decidua*, *Pseudotsuga menziesii*	87%	[42]
S2	12	*Fagus sylvatica*, *Alnus glutinosa*, *Fraxinus excelsior*, *Quercus* spp., *Prunus* spp., *Carpinus betulus*, *Acer* spp., *Picea abies*, *Pinus nigra*, *Pinus sylvestris*, *Larix decidua*, *Pseudotsuga menziesii*	83%	Table 7
S2	12	*Fagus sylvatica*, *Alnus glutinosa*, *Fraxinus excelsior*, *Quercus* spp., *Prunus* spp., *Carpinus betulus*, *Acer* spp., *Picea abies*, *Pinus nigra*, *Pinus sylvestris*, *Larix decidua*, *Pseudotsuga menziesii*	90%	[6]
S2	12	*Betula pendula*, *Quercus robur/pubescens/petraea*, *Quercus rubra*, *Populus* spp., *Fraxinus excelsior*, *Robinia pseudoacacia*, *Salix* spp., *Eucalyptus* spp., *Pinus nigra* subsp. *laricio*, *Pinus pinaster*, *Pinus nigra*, *Abies alba*, *Pseudotsuga menziesii*, *Cupressus* spp.	0.90 *	[43]
S2	17	*Fagus sylvatica*, *Alnus* spp., *Quercus petraea/robur*, *Quercus rubra*, *Betula pendula*, *Robinia pseudoacacia*, *Tilia cordata*, *Acer pseudoplatanus*, *Fraxinus excelsior*, *Populus* spp., *Carpinus betulus*, *Picea abies*, *Larix* spp., *Pseudotsuga menziesii*, *Pinus sylvestris*, *Pinus strobus*, *Pinus nigra*	96% **	[44]
S1 + LS	4	*Shorea siamensis*, *Shorea obtuse*, *Dipterocarpus tuberculatus*, semi-evergreen/evergreen	79%	[18]
S1 + S2	6	*Fagus sylvatica*, *Quercus* spp., other broadleaves, *Picea* sp., *Pinus* sp., other conifers	63%	[7]
S1 + S2	6	*Quercus mongolia*, *Betula* spp., *Populus* spp., *Armeniaca sibirica Larix* spp., *Pinus tabulaeformis*	78%	[45]
S1 + S2	7	*Acacia mearnsii*, *Eucalyptus dunnii*, *Eucalyptus grandiis*, *Eucalyptus* mix, *Pinus tecunumanii*, *Pinus elliotii*, *Pinus taedea*	88%	[40]
S1 + S2	12	*Fagus sylvatica*, *Alnus glutinosa*, *Fraxinus excelsior*, *Quercus* spp., *Prunus* spp., *Carpinus betulus*, *Acer* spp., *Picea abies*, *Pinus nigra*, *Pinus sylvestris*, *Larix decidua*, *Pseudotsuga menziesii*	84%	Table 8

* mean F1-score instead of OA; ** OA influenced by high dominance of one class.

For the study area, *Wienerwald*, and the employed RF classifier, it can be stated that if there are a sufficient number of S2 scenes available, no more added value is generated by using additional S1 data. We did, however, not evaluate the impact of the S2 orbit overlap on the classification accuracy, which certainly would negatively impact the obtained accuracies. On the other hand, temporal metrics (both parametric and using harmonics) in this study were only employed for S1 data.

The best monotemporal S2 result was achieved by the scenes from May and July, often a period with only few cloud-free data in Central Europe. Whenever data from several dates are available, the classification performance can be improved. Both variants of seasonal data (i.e., combining the three best and the three worst S2 scenes per season) demonstrated the added-value of multitemporal data.

Interestingly, in (optical) data-poor regions, the low S1 performance can be boosted significantly, if at least one S2 scene can be added as this adds valuable biochemical trait information. In this case, the OA of the classification can be increased by ca. 5.0 percentage points on average. This finding is significant in that S1 scenes, unlike S2 scenes, are always available due to the nature of their sensor's active microwaves, but are only marginally performant when used alone. If, however, several S2 scenes are available, the added value of S1 is increasingly equalized and decreases to 1.7 percentage points with a composite of MAS-S2 scenes and to 1.5 percentage points with a composite of LAS-S2 scenes. This is in line with the findings of Mngadi et al. [40] and Waser et al. [15]. Even using seasonal S2-composites as in other studies [48], overall accuracies of the full multitemporal datasets as in Test case 2 and Test case 3 were far from achieved.

5. Conclusions

In this study, the performance of the Sentinel-1 and Sentinel-2 satellite pairs, both individually and in various combinations, is presented. Twelve tree species, seven deciduous and five coniferous, of the Austrian Biosphere Reserve Wienerwald were classified using Breiman's random forest. While the results using only Sentinel-1 data were not satisfactory, the ability of the Sentinel-2 satellites to classify tree species was demonstrated once again. The greatest increase in accuracy can be achieved by using multitemporal Sentinel-2 data. In areas with insufficient coverage of optical satellites, Sentinel-1 can add value to classification accuracy. Seasonal Sentinel-2 composites have advantages over monotemporal classifications, but preference should be given to a full time series whenever possible. If sufficient Sentinel-2 data are available, the added value of Sentinel-1 data is only marginal, so that the effort to acquire data is offset by the added value of increased accuracy. However, for large-scale applications, the possibilities of acquiring cloud-free Sentinel-2 time series are often the limiting factor. In such cases, the advantage of the Sentinel-1 time series is obvious. The next steps would be further combinations with other datasets such as LiDAR or hyperspectral data (e.g., EnMAP). For a better understanding of the results and the relationship between vegetation structure and reflectance properties, radiative transfer models (RTMs) should be consulted.

Author Contributions: Conceptualization, M.I.; methodology, M.I. and M.L.; software, M.L. and M.I.; validation, A.D. and M.H.; formal analysis, M.L. and A.D.; investigation, M.L. and M.I.; resources, M.I.; data curation, M.L. and A.D.; writing—original draft preparation, M.L.; writing—review and editing, M.I., A.D., M.H. and C.A.; visualization, M.L. and M.I.; supervision, M.I.; project administration, M.I.; funding acquisition, M.I. All authors have read and agreed to the published version of the manuscript.

Funding: The study was partly funded by the Austrian Academy of Sciences (ÖAW), under the Earth System Science—Man and the Biosphere Programme (Project name: BRmon: Land cover classification and -monitoring of the Austrian biosphere reserves based on satellite data).

Data Availability Statement: Not applicable.

Acknowledgments: We thank our project partners: *Biosphärenpark Wienerwald Management*, Austrian federal forests (ÖBf AG), Forestry Office and Urban Agriculture of Vienna (MA 49), the forest enterprise of *Klosterneuburg* Abbey, and the forest enterprise of *Heiligenkreuz* Abbey for providing reference information. We are grateful to Alexandra Wieshaider (ÖBf AG) and Harald Brenner (*Biosphärenpark Wienerwald Management*) for supporting the project.

Conflicts of Interest: The authors declare no conflict of interest.

Appendix A

Table A1. Summary of the additional vegetation indices used for the classification, together with the corresponding formula and references (band 8 was used for the NIR = Near-Infrared; RE = Red-Edge).

Index-Name	Formula	Reference
Built-up Area Index (BAI)	$\frac{BLUE-NIR}{BLUE+NIR}$	[49]
Chlorophyll Green Index (CGI)	$\frac{NIR}{GREEN+RE1}$	[50]
Greenness Index (GI)	$\frac{GREEN}{RED}$	[51]
Green Normalized-Difference Vegetation Index (gNDVI)	$\frac{NIR-GREEN}{NIR+GREEN}$	[52]
Leaf Chlorophyll Content Index (LCCI)	$\frac{RE3}{RE1}$	[53]
Moisture Stress Index (MSI)	$\frac{SWIR1}{NIR}$	[54]
Normalized-Difference Red-Edge and SWIR2(NDRESWIR)	$\frac{RE2-SWIR2}{RE2+SWIR2}$	[55]
Normalized-Difference Tillage Index (NDTI)	$\frac{SWIR1-SWIR2}{SWIR1+SWIR2}$	[56]
Normalized-Difference Vegetation Index (NDVI)	$\frac{NIR-RED}{NIR+RED}$	[57]
Red-Edge Normalized-Difference Vegetation Index(reNDVI)	$\frac{NIR-RE1}{NIR+RE1}$	[52]
Normalized-Difference Water Index 1 (NDWI1)	$\frac{NIR-SWIR1}{NIR+SWIR1}$	[58]
Normalized-Difference Water Index 2 (NDWI2)	$\frac{NIR-SWIR2}{NIR+SWIR2}$	[52]
Normalized Humidity Index (NHI)	$\frac{SWIR2-GREEN}{SWIR2+GREEN}$	[59]
Red-Edge Peak Area (REPA)	$RED + RE1 + RE2 + RE3 + NIR$	[55,60]
Red SWIR1 Difference (DIRESWIR)	$RED + SWIR1$	[61]
Red-Edge Triangular Vegetation Index (RETVI)	$100(NIR - RE1) - 10(NIR - GREEN)$	[62]
Soil Adjusted Vegetation Index (SAVI)	$\frac{NIR-RED}{NIR+GREEN+0.5}1.5$	[63]
Blue and RE1 Ratio (SRBRE1)	$\frac{BLUE}{RE1}$	[51]
Blue and RE2 Ratio (SRBRE2)	$\frac{BLUE}{RE2}$	[64]
Blue and RE3 Ratio (SRBRE3)	$\frac{BLUE}{RE3}$	[55]
NIR and Blue Ratio (SRNIRB)	$\frac{NIR}{BLUE}$	[65]
NIR and Green Ratio (SRNIRG)	$\frac{NIR}{GREEN}$	[51]
NIR and Red Ratio (SRNIRR)	$\frac{NIR}{RED}$	[65]
NIR and RE1 Ratio (SRNIRRE1)	$\frac{NIR}{RE1}$	[50]
NIR and RE2 Ratio (SRNIRRE2)	$\frac{NIR}{RE2}$	[55]
NIR and RE3 Ratio (SRNIRRE3)	$\frac{NIR}{RE3}$	[55]
Soil Tillage Index (STI)	$\frac{SWIR1}{SWIR2}$	[56]
Water Body Index (WBI)	$\frac{BLUE-RED}{BLUE+RED}$	[66]

References

1. Secretariat of the Convention on Biological Diversity. *Global Biodiversity Outlook 5*; Secretariat of the Convention on Biological Diversity: Montreal, Canada, 2020.
2. Breidenbach, J.; Waser, L.T.; Debella-Gilo, M.; Schumacher, J.; Rahlf, J.; Hauglin, M.; Puliti, S.; Astrup, R. National Mapping and Estimation of Forest Area by Dominant Tree Species Using Sentinel-2 Data. *Can. J. For. Res.* **2021**, *51*, 365–379. [CrossRef]
3. Immitzer, M.; Vuolo, F.; Atzberger, C. First Experience with Sentinel-2 Data for Crop and Tree Species Classifications in Central Europe. *Remote Sens.* **2016**, *8*, 166. [CrossRef]
4. Maschler, J.; Atzberger, C.; Immitzer, M. Individual Tree Crown Segmentation and Classification of 13 Tree Species Using Airborne Hyperspectral Data. *Remote Sens.* **2018**, *10*, 1218. [CrossRef]
5. Grybas, H.; Congalton, R.G. A Comparison of Multi-Temporal RGB and Multispectral UAS Imagery for Tree Species Classification in Heterogeneous New Hampshire Forests. *Remote Sens.* **2021**, *13*, 2631. [CrossRef]
6. Immitzer, M.; Neuwirth, M.; Böck, S.; Brenner, H.; Vuolo, F.; Atzberger, C. Optimal Input Features for Tree Species Classification in Central Europe Based on Multi-Temporal Sentinel-2 Data. *Remote Sens.* **2019**, *11*, 2599. [CrossRef]

7. Bjerreskov, K.S.; Nord-Larsen, T.; Fensholt, R. Classification of Nemoral Forests with Fusion of Multi-Temporal Sentinel-1 and 2 Data. *Remote Sens.* **2021**, *13*, 950. [CrossRef]
8. Hościło, A.; Lewandowska, A. Mapping Forest Type and Tree Species on a Regional Scale Using Multi-Temporal Sentinel-2 Data. *Remote Sens.* **2019**, *11*, 929. [CrossRef]
9. Kohrs, R.A.; Lazzara, M.A.; Robaidek, J.O.; Santek, D.A.; Knuth, S.L. Global Satellite Composites—20 Years of Evolution. *Atmospheric Res.* **2014**, *135–136*, 8–34. [CrossRef]
10. Vuolo, F.; Ng, W.-T.; Atzberger, C. Smoothing and Gap-Filling of High Resolution Multi-Spectral Time Series: Example of Landsat Data. *Int. J. Appl. Earth Obs. Geoinf.* **2017**, *57*, 202–213. [CrossRef]
11. Moreno-Martínez, Á.; Izquierdo-Verdiguier, E.; Maneta, M.P.; Camps-Valls, G.; Robinson, N.; Muñoz-Marí, J.; Sedano, F.; Clinton, N.; Running, S.W. Multispectral High Resolution Sensor Fusion for Smoothing and Gap-Filling in the Cloud. *Remote Sens. Environ.* **2020**, *247*, 111901. [CrossRef]
12. Griffiths, P.; Nendel, C.; Hostert, P. Intra-Annual Reflectance Composites from Sentinel-2 and Landsat for National-Scale Crop and Land Cover Mapping. *Remote Sens. Environ.* **2019**, *220*, 135–151. [CrossRef]
13. Bauer-Marschallinger, B.; Cao, S.; Navacchi, C.; Freeman, V.; Reuß, F.; Geudtner, D.; Rommen, B.; Vega, F.C.; Snoeij, P.; Attema, E.; et al. The Normalised Sentinel-1 Global Backscatter Model, Mapping Earth's Land Surface with C-Band Microwaves. *Sci. Data* **2021**, *8*, 277. [CrossRef] [PubMed]
14. Dostálová, A.; Lang, M.; Ivanovs, J.; Waser, L.T.; Wagner, W. European Wide Forest Classification Based on Sentinel-1 Data. *Remote Sens.* **2021**, *13*, 337. [CrossRef]
15. Waser, L.T.; Rüetschi, M.; Psomas, A.; Small, D.; Rehush, N. Mapping Dominant Leaf Type Based on Combined Sentinel-1/-2 Data—Challenges for Mountainous Countries. *ISPRS J. Photogramm. Remote Sens.* **2021**, *180*, 209–226. [CrossRef]
16. Rüetschi, M.; Schaepman, M.; Small, D. Using Multitemporal Sentinel-1 C-Band Backscatter to Monitor Phenology and Classify Deciduous and Coniferous Forests in Northern Switzerland. *Remote Sens.* **2017**, *10*, 55. [CrossRef]
17. Udali, A.; Lingua, E.; Persson, H.J. Assessing Forest Type and Tree Species Classification Using Sentinel-1 C-Band SAR Data in Southern Sweden. *Remote Sens.* **2021**, *13*, 3237. [CrossRef]
18. Tran, A.T.; Nguyen, K.A.; Liou, Y.A.; Le, M.H.; Vu, V.T.; Nguyen, D.D. Classification and Observed Seasonal Phenology of Broadleaf Deciduous Forests in a Tropical Region by Using Multitemporal Sentinel-1A and Landsat 8 Data. *Forests* **2021**, *12*, 235. [CrossRef]
19. Breiman, L. Random Forests. *Mach. Learn.* **2001**, *45*, 5–32. [CrossRef]
20. Mrkvicka, A.; Drozdowski, I.; Brenner, H. Kernzonen im Biosphärenpark Wienerwald—Urwälder von morgen. *Wiss. Mitt. Aus Niederösterreichischen Landesmus.* **2014**, *25*, 41–88.
21. Nkosi, S.E.; Adam, E.; Barrett, A.S.; Brown, L.R. Mapping the Spatial Distribution of Tree Species Selected by Elephants (Loxodonta Africana) in Venetia-Limpopo Nature Reserve Using Sentinel-2 Imagery. *Appl. Geomat.* **2021**, *13*, 701–713. [CrossRef]
22. Louis, J.; Debaecker, V.; Pflug, B.; Main-Knorn, M.; Bieniarz, J.; Mueller-Wilm, U.; Cadau, E.; Gascon, F. Sentinel-2 Sen2Cor: L2A Processor for Users. In Proceedings of the Living Planet Symposium 2016, Prague, Czech Republic, 9–13 May 2016; Spacebooks Online. 2016; pp. 1–8.
23. Myneni, R.B.; Nemani, R.R.; Shabanov, N.V.; Knyazikhin, Y.; Morisette, J.T.; Privette, J.L.; Running, S.W. LAI and FPAR. Available online: https://cce.nasa.gov/mtg2008_ab_presentations/LAI-FPAR_Myneni_whitepaper.pdf (accessed on 4 May 2022).
24. Watson, D.J. Comparative Physiological Studies on the Growth of Field Crops: I. Variation in Net Assimilation Rate and Leaf Area between Species and Varieties, and within and between Years. *Ann. Bot.* **1947**, *11*, 41–76. [CrossRef]
25. EODC GmbH Austrian Data Cube. Available online: https://acube.eodc.eu/ (accessed on 1 April 2022).
26. Le Toan, T.; Beaudoin, A.; Riom, J.; Guyon, D. Relating Forest Biomass to SAR Data. *IEEE Trans. Geosci. Remote Sens.* **1992**, *30*, 403–411. [CrossRef]
27. Vreugdenhil, M.; Wagner, W.; Bauer-Marschallinger, B.; Pfeil, I.; Teubner, I.; Rüdiger, C.; Strauss, P. Sensitivity of Sentinel-1 Backscatter to Vegetation Dynamics: An Austrian Case Study. *Remote Sens.* **2018**, *10*, 1396. [CrossRef]
28. Dostálová, A.; Wagner, W.; Milenković, M.; Hollaus, M. Annual Seasonality in Sentinel-1 Signal for Forest Mapping and Forest Type Classification. *Int. J. Remote Sens.* **2018**, *39*, 7738–7760. [CrossRef]
29. Frison, P.-L.; Fruneau, B.; Kmiha, S.; Soudani, K.; Dufrêne, E.; Toan, T.L.; Koleck, T.; Villard, L.; Mougin, E.; Rudant, J.-P. Potential of Sentinel-1 Data for Monitoring Temperate Mixed Forest Phenology. *Remote Sens.* **2018**, *10*, 2049. [CrossRef]
30. Ahern, F.J.; Leckie, D.J.; Drieman, J.A. Seasonal Changes in Relative C-Band Backscatter of Northern Forest Cover Types. *IEEE Trans. Geosci. Remote Sens.* **1993**, *31*, 668–680. [CrossRef]
31. Zeileis, A.; Leisch, F.; Hornik, K.; Kleiber, C. Strucchange: An *R* Package for Testing for Structural Change in Linear Regression Models. *J. Stat. Softw.* **2002**, *7*, 1–38. [CrossRef]
32. Schlaffer, S.; Matgen, P.; Hollaus, M.; Wagner, W. Flood Detection from Multi-Temporal SAR Data Using Harmonic Analysis and Change Detection. *Int. J. Appl. Earth Obs. Geoinf.* **2015**, *38*, 15–24. [CrossRef]
33. Einzmann, K.; Immitzer, M.; Böck, S.; Bauer, O.; Schmitt, A.; Atzberger, C. Windthrow Detection in European Forests with Very High-Resolution Optical Data. *Forests* **2017**, *8*, 21. [CrossRef]
34. Immitzer, M.; Böck, S.; Einzmann, K.; Vuolo, F.; Pinnel, N.; Wallner, A.; Atzberger, C. Fractional Cover Mapping of Spruce and Pine at 1 Ha Resolution Combining Very High and Medium Spatial Resolution Satellite Imagery. *Remote Sens. Environ.* **2018**, *204*, 690–703. [CrossRef]

35. Guyon, I.; Weston, J.; Barnhill, S. Gene Selection for Cancer Classification Using Support Vector Machines. *Mach. Learn.* **2002**, *46*, 389–422. [CrossRef]
36. Wessel, M.; Brandmeier, M.; Tiede, D. Evaluation of Different Machine Learning Algorithms for Scalable Classification of Tree Types and Tree Species Based on Sentinel-2 Data. *Remote Sens.* **2018**, *10*, 1419. [CrossRef]
37. Ma, M.; Liu, J.; Liu, M.; Zeng, J.; Li, Y. Tree Species Classification Based on Sentinel-2 Imagery and Random Forest Classifier in the Eastern Regions of the Qilian Mountains. *Forests* **2021**, *12*, 1736. [CrossRef]
38. Kollert, A.; Bremer, M.; Löw, M.; Rutzinger, M. Exploring the Potential of Land Surface Phenology and Seasonal Cloud Free Composites of One Year of Sentinel-2 Imagery for Tree Species Mapping in a Mountainous Region. *Int. J. Appl. Earth Obs. Geoinf.* **2021**, *94*, 102208. [CrossRef]
39. Persson, M.; Lindberg, E.; Reese, H. Tree Species Classification with Multi-Temporal Sentinel-2 Data. *Remote Sens.* **2018**, *10*, 1794. [CrossRef]
40. Mngadi, M.; Odindi, J.; Peerbhay, K.; Mutanga, O. Examining the Effectiveness of Sentinel-1 and 2 Imagery for Commercial Forest Species Mapping. *Geocarto Int.* **2021**, *36*, 1–12. [CrossRef]
41. Grabska, E.; Hostert, P.; Pflugmacher, D.; Ostapowicz, K. Forest Stand Species Mapping Using the Sentinel-2 Time Series. *Remote Sens.* **2019**, *11*, 1197. [CrossRef]
42. Grabska, E.; Frantz, D.; Ostapowicz, K. Evaluation of Machine Learning Algorithms for Forest Stand Species Mapping Using Sentinel-2 Imagery and Environmental Data in the Polish Carpathians. *Remote Sens. Environ.* **2020**, *251*, 112103. [CrossRef]
43. Karasiak, N.; Fauvel, M.; Dejoux, J.-F.; Monteil, C.; Sheeren, D. Optimal Dates for Deciduous Tree Species Mapping Using Full Years Sentinel-2 Time Series in South West France. *ISPRS Ann. Photogramm. Remote Sens. Spat. Inf. Sci.* **2020**, *V-3-2020*, 469–476. [CrossRef]
44. Hemmerling, J.; Pflugmacher, D.; Hostert, P. Mapping Temperate Forest Tree Species Using Dense Sentinel-2 Time Series. *Remote Sens. Environ.* **2021**, *267*, 112743. [CrossRef]
45. Xie, B.; Cao, C.; Xu, M.; Duerler, R.S.; Yang, X.; Bashir, B.; Chen, Y.; Wang, K. Analysis of Regional Distribution of Tree Species Using Multi-Seasonal Sentinel-1&2 Imagery within Google Earth Engine. *Forests* **2021**, *12*, 565. [CrossRef]
46. Schlerf, M.; Atzberger, C. Inversion of a Forest Reflectance Model to Estimate Structural Canopy Variables from Hyperspectral Remote Sensing Data. *Remote Sens. Environ.* **2006**, *100*, 281–294. [CrossRef]
47. ESA Sentinel-2 L1C Data Quality Report. Available online: https://sentinel.esa.int/documents/247904/685211/Sentinel-2_L1C_Data_Quality_Report (accessed on 2 May 2022).
48. Ghassemi, B.; Dujakovic, A.; Żółtak, M.; Immitzer, M.; Atzberger, C.; Vuolo, F. Designing a European-Wide Crop Type Mapping Approach Based on Machine Learning Algorithms Using LUCAS Field Survey and Sentinel-2 Data. *Remote Sens.* **2022**, *14*, 541. [CrossRef]
49. Shahi, K.; Shafri, H.Z.M.; Taherzadeh, E.; Mansor, S.; Muniandy, R. A Novel Spectral Index to Automatically Extract Road Networks from WorldView-2 Satellite Imagery. *Egypt. J. Remote Sens. Space Sci.* **2015**, *18*, 27–33. [CrossRef]
50. Datt, B. A New Reflectance Index for Remote Sensing of Chlorophyll Content in Higher Plants: Tests Using Eucalyptus Leaves. *J. Plant Physiol.* **1999**, *154*, 30–36. [CrossRef]
51. Le Maire, G.; François, C.; Dufrêne, E. Towards Universal Broad Leaf Chlorophyll Indices Using PROSPECT Simulated Database and Hyperspectral Reflectance Measurements. *Remote Sens. Environ.* **2004**, *89*, 1–28. [CrossRef]
52. Gitelson, A.A.; Kaufman, Y.J.; Merzlyak, M.N. Use of a Green Channel in Remote Sensing of Global Vegetation from EOS-MODIS. *Remote Sens. Environ.* **1996**, *58*, 289–298. [CrossRef]
53. Wulf, H.; Stuhler, S. Sentinel-2: Land Cover, Preliminary User Feedback on Sentinel-2a Data. In Proceedings of the Sentinel-2A Expert Users Technical Meeting, Frascati, Italy, 27 February–1 March 2015; pp. 29–30.
54. Vogelmann, J.E.; Rock, B.N. Spectral Characterization of Suspected Acid Deposition Damage in Red Spruce (Picea Rubens) Stands from Vermont. In Proceedings of the Airborne Imaging Spectrometer Data Anal. Workshop, Pasadena, CA, USA, 8–10 April 1985.
55. Radoux, J.; Chomé, G.; Jacques, D.; Waldner, F.; Bellemans, N.; Matton, N.; Lamarche, C.; d'Andrimont, R.; Defourny, P. Sentinel-2's Potential for Sub-Pixel Landscape Feature Detection. *Remote Sens.* **2016**, *8*, 488. [CrossRef]
56. Van Deventer, A.P.; Ward, A.D.; Gowda, P.H.; Lyon, J.G. Using Thematic Mapper Data to Identify Contrasting Soil Plains and Tillage Practices. *Photogramm. Eng. Remote Sens.* **1997**, *63*, 87–93.
57. Tucker, C.J. Red and Photographic Infrared Linear Combinations for Monitoring Vegetation. *Remote Sens. Environ.* **1979**, *8*, 127–150. [CrossRef]
58. Gao, B. NDWI—A Normalized Difference Water Index for Remote Sensing of Vegetation Liquid Water from Space. *Remote Sens. Environ.* **1996**, *58*, 257–266. [CrossRef]
59. Lacaux, J.P.; Tourre, Y.M.; Vignolles, C.; Ndione, J.A.; Lafaye, M. Classification of Ponds from High-Spatial Resolution Remote Sensing: Application to Rift Valley Fever Epidemics in Senegal. *Remote Sens. Environ.* **2007**, *106*, 66–74. [CrossRef]
60. Filella, I.; Penuelas, J. The Red Edge Position and Shape as Indicators of Plant Chlorophyll Content, Biomass and Hydric Status. *Int. J. Remote Sens.* **1994**, *15*, 1459–1470. [CrossRef]
61. Jacques, D.C.; Kergoat, L.; Hiernaux, P.; Mougin, E.; Defourny, P. Monitoring Dry Vegetation Masses in Semi-Arid Areas with MODIS SWIR Bands. *Remote Sens. Environ.* **2014**, *153*, 40–49. [CrossRef]
62. Chen, P.-F.; Nicolas, T.; Wang, J.-H.; Philippe, V.; Huang, W.-J.; Li, B.-G. New Index for Crop Canopy Fresh Biomass Estimation. *Spectrosc. Spectr. Anal.* **2010**, *30*, 512–517.

63. Huete, A.R. A Soil-Adjusted Vegetation Index (SAVI). *Remote Sens. Environ.* **1988**, *25*, 295–309. [CrossRef]
64. Lichtenthaler, H.K.; Lang, M.; Sowinska, M.; Heisel, F.; Miehé, J.A. Detection of Vegetation Stress Via a New High Resolution Fluorescence Imaging System. *J. Plant Physiol.* **1996**, *148*, 599–612. [CrossRef]
65. Blackburn, G.A. Quantifying Chlorophylls and Caroteniods at Leaf and Canopy Scales. *Remote Sens. Environ.* **1998**, *66*, 273–285. [CrossRef]
66. Domenech, E.; Mallet, C. *Change Detection in High Resolution Land Use/Land Cover Geodatabases (at Object Level)*; Official Publication No. 64; EuroSDR: Leuven, Belgium, 2014.

remote sensing

MDPI

Article

Optimal Input Features for Tree Species Classification in Central Europe Based on Multi-Temporal Sentinel-2 Data

Markus Immitzer [1,*], Martin Neuwirth [1], Sebastian Böck [1], Harald Brenner [2], Francesco Vuolo [1] and Clement Atzberger [1]

1 University of Natural Resources and Life Sciences, Vienna (BOKU), Institute of Geomatics, Peter-Jordan-Straße 82, 1190 Vienna, Austria; martin.neuwirth@boku.ac.at (M.N.); sebastian.boeck@boku.ac.at (S.B.); francesco.vuolo@boku.ac.at (F.V.); clement.atzberger@boku.ac.at (C.A.)

2 Biosphärenpark Wienerwald Management GmbH, Norbertinumstraße 9, 3013 Tullnerbach, Austria; hb@bpww.at

* Correspondence: markus.immitzer@boku.ac.at

Received: 19 September 2019; Accepted: 4 November 2019; Published: 6 November 2019

Abstract: Detailed knowledge about tree species composition is of great importance for forest management. The two identical European Space Agency (ESA) Sentinel-2 (S2) satellites provide data with unprecedented spectral, spatial and temporal resolution. Here, we investigated the potential benefits of using high temporal resolution data for classification of five coniferous and seven broadleaved tree species in a diverse Central European Forest. To run the classification, 18 cloud-free S2 acquisitions were analyzed in a two-step approach. The available scenes were first used to stratify the study area into six broad land-cover classes. Subsequently, additional classification models were created separately for the coniferous and the broadleaved forest strata. To permit a deeper analytical insight in the benefits of multi-temporal datasets for species identification, classification models were developed taking into account all 262,143 possible permutations of the 18 S2 scenes. Each model was fine-tuned using a stepwise recursive feature reduction. The additional use of vegetation indices improved the model performances by around 5 percentage points. Individual mono-temporal tree species accuracies range from 48.1% (January 2017) to 78.6% (June 2017). Compared to the best mono-temporal results, the multi-temporal analysis approach improves the out-of-bag overall accuracy from 72.9% to 85.7% for the broadleaved and from 83.8% to 95.3% for the coniferous tree species, respectively. Remarkably, a combination of six–seven scenes achieves a model quality equally high as the model based on all data; images from April until August proved most important. The classes European Beech and European Larch attain the highest user's accuracies of 96.3% and 95.9%, respectively. The most important spectral variables to distinguish between tree species are located in the Red (coniferous) and short wave infrared (SWIR) bands (broadleaved), respectively. Overall, the study highlights the high potential of multi-temporal S2 data for species-level classifications in Central European forests.

Keywords: tree species classification; Sentinel-2; multi-temporal; Wienerwald biosphere reserve

1. Introduction

The current Global Assessment Report on Biodiversity and Ecosystem Services again depicts an alarming picture of the Earth with accelerating rates of biodiversity loss [1]. Earth observation (EO) has a high potential for biodiversity assessments, mainly for the description of vegetation habitats [2]. The synoptic view, and the delivery of detailed, objective and cost-efficient information over large areas, makes EO data one of the most useful tools for biodiversity assessments [3–5]. Depending on the spectral, spatial and temporal resolution of the EO data, various categorical and biophysical traits can be mapped [6,7]. In forest ecosystems, tree species diversity is a key parameter for ecologists, conservationists and also for forest managers [8,9]. In addition to the occurrences of tree species, information about the distribution and the spatial pattern of tree species within larger geographic extents is also essential.

In the last few years, the number and variety of commercially and publicly funded EO sensors has increased dramatically. As a result, data with higher spatial, spectral and temporal resolutions are available. Analysis of hyperspectral data demonstrated the added value of the dense spectral sampling for the separation of tree species [10–13]. Multi-spectral, very high resolution (VHR) satellite data were successfully used for mapping tree species distribution for up to ten different species [14–17]. The small pixel size of VHR data enables the classification of individual tree crowns. However, the use of VHR satellite data or airborne hyperspectral data is often limited by high data costs and limited area coverage.

Studies covering larger regions by combining data with different spatial resolution have thus far only focused on a small number of tree species [18], respectively, on tree species groups [19]. Likewise, existing continental-scale forest maps such as the Copernicus high resolution forest layer [20], only distinguish broadleaf and deciduous forests. Studies analyzing several tree species and covering large geographic extents are still missing [21].

With the launch of the twin Sentinel-2A and 2B satellites since 2015, high quality data with high spatial, spectral and temporal resolution are now freely available. Despite the fact that individual tree crowns cannot be separated with the 10–20 m data, the rich spectral information with bands in the visible, Red-Edge, Near-Infrared (NIR) and Shortwave-Infrared (SWIR) wavelengths has a high potential for tree species separation [22–29]. An additional advantage is the very high revisit interval of the two satellites. The twins cover the entire Earth surface every 5 days, with even higher number of observations in the overlap areas of adjacent orbits.

In many (partly) cloudy areas of the world, the availability of dense time series is paramount to obtaining reliable and cloud-free observations during key phenological periods [30]. An adequate number of cloud-free observations also enables a better description of the actual situation and historical evolution and moreover helps to detect changes [31]. Consequently, the use of multi-temporal Sentinel-2 data also improves tree species identification. Nelson [32], for example, analyzed six tree species classes in Sweden testing all possible combinations of three Sentinel-2 scenes from May, July and August. They achieved overall accuracies of up to 86%. Bolyn et al. [22] classified eleven forest classes in Belgium with an overall accuracy of 92% using Sentinel-2 scenes from May and October. In a German test site, Wessel et al. [33] achieved up to 88% overall accuracy for four tree species classes using Sentinel-2 scenes from May, August, and September. Persson et al. [27] used four scenes from April, May, July, and October for the separation of five tree species in Sweden and obtained an overall accuracy of 88%. In a Mediterranean forest, four forest types were separated with accuracies of over 83% by Puletti et al. [28] using the Sentinel-2 bands together with vegetation indices. Hościło and Lewandowska [34] used four scenes to classify eight tree species in southern Poland with an overall accuracy of 76%. Using additional topographic features and a stratification in broadleaf and coniferous species, the accuracy increased to 85%. Grabska et al. [23] achieved, with five (from 18) Sentinel-2 images, an overall accuracy of 92% for the classification of nine tree species in a Carpathian test site. The most important band was the Red-Edge 2 and most important scene were acquisitions from October. All studies clearly demonstrated the benefit of multi-temporal data and gave some hints about the importance of individual bands and optimum acquisition times. However, the number of identified tree species was still relatively small (2–11), and generally only a few (3–5) Sentinel-2 scenes were analyzed.

The aim of this study is to assess the suitability of dense multi-temporal Sentinel-2 data for a detailed description of tree species and other vegetation/land cover classes in the Wienerwald biosphere reserve in Austria. In protected areas, detailed information on the actual land cover—and possible changes—are of high importance. Up-to-now, the forest description of the biosphere reserve was mainly based on management plans from different forest enterprises. These data do not cover the entire biosphere reserve and are sometimes outdated. The biosphere management would tremendously benefit from consistent and reliable information about the spatial distribution of the major coniferous and broadleaf tree species.

The main objectives of our research were:

- To evaluate the potential of multi-temporal Sentinel-2 data for mapping 12 tree species at 10 m spatial resolution for the entire Wienerwald biosphere reserve.
- To identify the best acquisition dates and scene combinations for tree species separation.
- To identify the most important Sentinel-2 bands for tree species classification and the added value of several vegetation indices.
- To evaluate the benefits of stratified classifications.
- To apply an additional short-term change detection analysis to monitor forest management activities and to ensure that the final tree species maps are up-to-date.

2. Materials and Methods

For the land cover classification and the tree species mapping in the Wienerwald biosphere reserve, 18 cloud-free Sentinel-2 scenes acquired between 2015 and 2017 were used. The mapping was done in three steps using different reference data sets (Figure 1). In the first step, six broad land cover classes were mapped. Subsequently, the individual tree species were identified within the resulting forest strata. In the final step, change detection was applied to identify areas where forest activities took place.

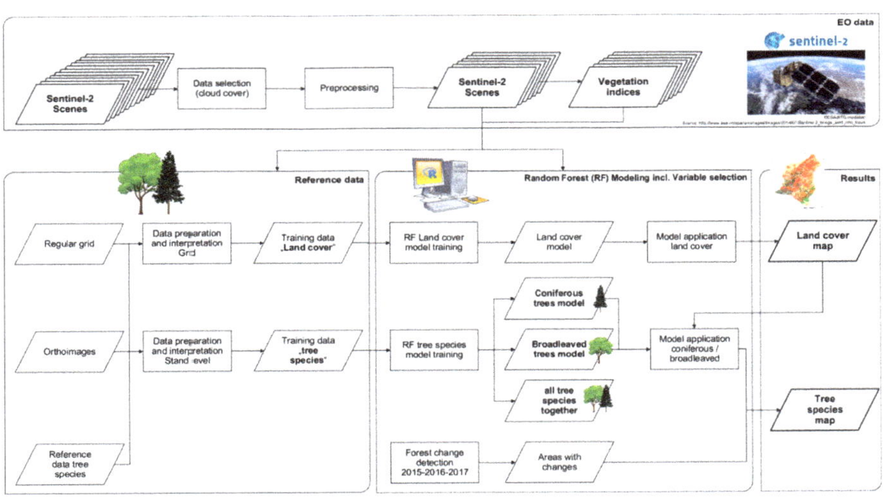

Figure 1. Workflow diagram of the classification approach with three main steps: (1) broad land cover classification, (2) tree species identification within the forest strata and (3) change detection to mask out areas where forest activities took place.

For the broad land cover classification, reference data were visually interpreted in a regular grid using four-band orthoimages with a spatial resolution of 20 cm acquired in the course of the national aerial image campaign and provided by Austria's Federal Office of Metrology and Surveying. The reference data for the tree species were derived from stand maps and other forest management databases. To enhance the data quality, the reference points were cross-checked by visual interpretation of color-infrared (CIR) orthoimages. With these reference data, the coniferous and broadleaved tree species were classified both separately and together, while testing all possible combinations of the Sentinel-2 data. The best classification results were merged together and areas where changes could be detected were masked out.

2.1. Study Site Wienerwald Biosphere Reserve

The Wienerwald biosphere reserve is one of the largest contiguous deciduous beech woodlands in Central Europe. It is located in the south-west of Vienna (Austria) and covers an area of 105,645 ha. The location of such a large forest on the edge of a metropolitan area is unique. The range of (micro) climatic and geological conditions in the Wienerwald is the main reason for the large diversity of vegetation types [35]. The Biosphere Reserve has more than 20 types of woodland—with beech, oak and hornbeam being dominant—and more than 23 types of meadow [36]. Concerning the forest, particularly rare woods can be found, such as Austrian's largest downy oak forests (*Quercus pubescens*) and unique stands of Austrian Black Pine (*Pinus nigra subsp. nigra*) occurring in the eastern part of the Wienerwald [37]. The inlet in Figure 2 shows the location of the study area within a region characterized by its diversity of nature and culture, and sustainable ecosystem management.

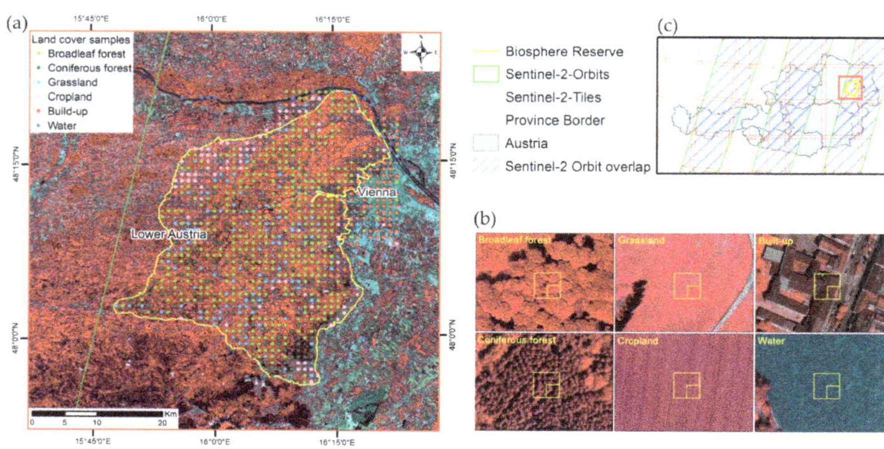

Figure 2. Overview of the study area and the 6-class land cover reference data. (**a**) Regular grid for reference data collection for the land cover classification covering the biosphere reserve and some surrounding areas (background: Sentinel-2 bands 8-4-3). (**b**) Examples (10 and 20 m grid cells) for each class (background: CIR orthoimage). (**c**) Location of the biosphere reserve Wienerwald within Austria and Sentinel-2 orbit cover.

2.2. Reference Data Sets

For the reference data creation, a regular grid (1 km × 1 km) was laid over the entire biosphere reserve as well as some surrounding areas (Figure 2a). At each point, the grid cell was visually interpreted using CIR orthoimages (Figure 2b). Table 1 presents the number of samples and class definitions of the six land cover classes.

Table 1. Summary of the reference data for the land cover classification.

Class Name	Definition	Samples	Amount [%]
Broadleaf forest	Broadleaf-dominated forests	388	48.68
Coniferous forest	Conifer-dominated forests	97	12.17
Grassland	Grassland, meadows, lawns, pastures, parks, etc.	104	13.05
Cropland	Agricultural crops, wine yards	77	9.66
Built-up	Sealed surfaces - buildings, roads, infrastructure, etc.	116	14.56
Water	Lakes, rivers, ponds, etc.	15	1.88
	$\sum$	797	100.00

Six target classes were distinguished for the land cover classification: deciduous forest, broadleaf forest, grassland, cropland, build-up areas and water bodies. To receive adequate numbers of trainings samples for the classes cropland, build-up areas and water, the grid was extended to surrounding areas in the north and east of the study area. Only clearly interpretable samples which contain only one class were retained for the training of the classification model. In the end, 797 out of 1360 pixels were useable.

For the tree species classification, additional reference samples were necessary and were derived from forest management data. First, pure stands of the 12 tree species were identified in the forest management maps. Next, one or two Sentinel-2 pixels were chosen in the center of each stand and the correctness of the information was checked using CIR orthoimages (Figure 3).

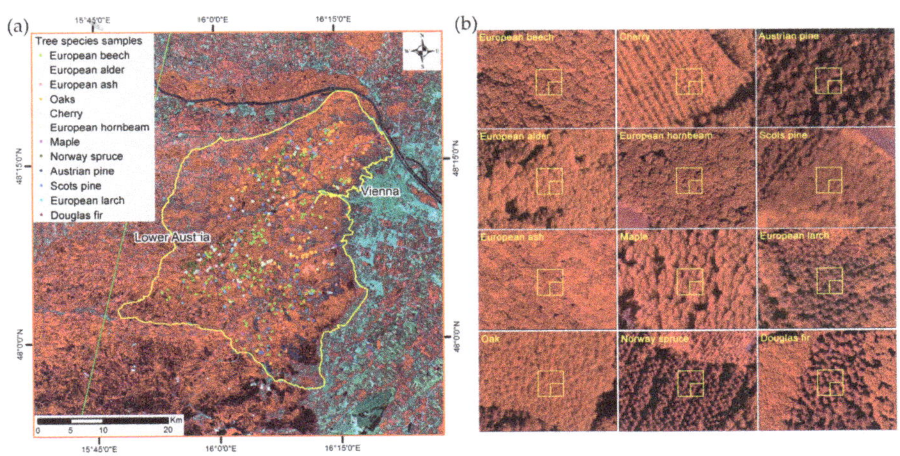

Figure 3. (**a**) Distribution of the reference data set for the tree species classification and (**b**) examples (10 and 20 m grid cells) for each tree species. Background images: Color Infrared composites of Sentinel-2 (**a**) and orthoimages (**b**).

In this way, on average 85 reference samples per tree species were distinguished, well distributed over the entire biosphere reserve (Table 2). The variation in the number of available reference data reflects the difficulties to identify sufficiently large and pure stands for some of the species.

Table 2. Summary of the reference data for the tree species classification.

Tree Species	Scientific Name	Acronym	Samples	Amount [%]
European beech	*Fagus sylvatica*	FS	215	21.37
European alder	*Alnus glutinosa*	AG	52	5.17
European ash	*Fraxinus excelsior*	FE	60	5.96
Oaks	*Quercus sp.*	QU	130	12.92
Cherry	*Prunus sp.*	PR	25	2.49
European hornbeam	*Carpinus betulus*	CP	65	6.46
Maple	*Acer sp.*	AC	33	3.28
Norway spruce	*Picea abies*	PA	135	13.42
Austrian pine	*Pinus nigra*	PN	107	10.64
Scots pine	*Pinus sylvestris*	PS	79	7.85
European larch	*Larix decidua*	LD	49	4.87
Douglas fir	*Pseudotsuga menziesii*	PM	56	5.57
		Σ	1006	100.00

2.3. Sentinel-2 Data Sets

The study area is located in the overlapping area of two Sentinel-2 orbits (122 and 79—Figure 2), and therefore, the number of acquisitions twice as high as normal in this latitude. For the analysis, all available Sentinel-2 scenes were visually checked. Only cloud-free data were selected. From the 188 scenes acquired between June 2015 to end of 2017, 18 scenes were perfectly useable. Summary information about selected scenes can be found in Table 3. All scenes were atmospherically corrected using Sen2Cor [38] Version 2.4 using the data service platform operated by BOKU [39] on the Earth Observation Data Centre (EODC) [40]. The 20 m bands B5, B6, B7, B8a, B11 and B12 were resampled to 10 m and the 60 m-bands B1, B9 and B10 were excluded from the analyses.

Table 3. Summary of the selected Sentinel-2 data sets (granule T33UWP). Over the region of interest, the images were free of clouds. The percentage cloud cover of the entire scenes was in the range 0-15%.

Sentinel-2 Satellite	Date	Orbit	Sun Zenith Angle	Sun Azimuth Angle
A	30.08.2015	122	40.64	160.67
A	25.12.2015	79	72.89	165.72
A	27.03.2016	122	46.92	161.03
A	13.04.2016	79	40.99	157.03
A	06.05.2016	122	32.93	159.34
A	31.08.2016	79	41.81	157.47
A	13.09.2016	122	45.77	164.04
A	30.09.2016	79	52.43	164.31
A	11.01.2017	122	71.18	165.83
A	01.04.2017	122	45.04	160.96
A	28.05.2017	79	29.06	151.92
A	20.06.2017	122	26.83	153.18
A	01.08.2017	79	33.19	150.41
A	29.08.2017	122	40.48	160.55
A	08.09.2017	122	43.90	162.88
A	28.09.2017	122	51.20	167.02
B	30.09.2017	79	52.35	164.20
A	15.10.2017	79	57.79	166.70

2.4. Random Forest Classification Approach

For all classifications, the ensemble learning random forest (RF) approach developed by Breiman [41] was used. The two hyper-parameters *mtry* (number of predictors randomly sampled for each node) and *ntree* (the number of trees) were set to the square root of available input variables (default) and, to 1000, respectively.

One advantage of the bootstrapping is that it yields relatively unbiased 'out-of-bag' (OOB) results, as long as representative reference data are provided [42]. Another benefit is the computation of importance measures which can be used for the evaluation of the input data and subsequent feature reduction. In this study, a recursive feature selection process using the 'Mean decrease in Accuracy' (MDA) was applied similarly to other studies [18,43,44]. More information about the algorithm and its advantages, such as the importance measure for the input variables and the integrated bootstrapping, can be found in the literature [16,41,45,46].

To classify the six land cover classes, first a model based on all Sentinel-2 datasets was developed using the land cover reference data from the visually interpreted regular grid. The tree species models were developed separately for the broadleaf and coniferous species—for testing purposes we also pooled all tree species together. The tree species classification models were based on the tree species reference data set and only applied to areas previously mapped as *broadleaf* or *coniferous forest*.

To find the best combinations for the tree species classification, we tested all possible combinations of the 18 Sentinel-2 scenes. We tested for example 18 individual scenes, 153 combinations of two scenes, 816 of three and so on. In total, 262,143 different models for each of the two forest strata were developed. The training of each model, including the feature selection, took on average about 5 min on a standard PC (CPU i7-2600 3.40 GHz, 16 GB RAM), and therefore, a high-performance computing (HPC) environment was used.

The modeling was done with two data sets: one contains only the 10 spectral bands, the second combines the 10 spectral bands with 28 widely used vegetation indices (Table A1 in the Appendix A).

2.5. Input Data Evaluation

The classification models were assessed using the OOB results. Previous studies had demonstrated that the OOB results of RF classifiers compare well against an assessment based on a separate validation data set [42,47,48].

To assess the importance of individual Sentinel-2 bands and acquisition times, the 'Mean decrease in Accuracy' (MDA) importance values of the final RF models (after feature selection) were normalized for each model to 1 by dividing all values by the maximum value of the specific model. Variables which were eliminated by the feature selection procedure were assigned an importance value of 0. All normalized values were summed up for all tested combinations and divided by the total number of tested combinations (Equation (1)):

$$IMP_i = \frac{\sum_{j=0}^{n} \frac{MDA_{ji}}{MDA_{jmax}}}{n} \quad (1)$$

where IMP_i is the normalized and aggregated importance value for variable i (= one band of one specific Sentinel-2 scene); MDA_{ji} the MDA importance value of variable i in the model j; MDA_{jmax} the maximum MDA importance value in the model j; and n the number of models (combinations of Sentinel-2 scenes) considering variable i (= 131,072).

When evaluating the importance of the spectral bands, models involving the vegetation indices were discarded to avoid skewing the results by the chosen indices. This was deemed particularly important as most indices include the NIR band. However, we applied the evaluation also to the models with vegetation indices to investigate the most important vegetation indices for tree species classification.

2.6. Change Detection

As outlined above, for the tree species classification, data from a three-year period (2015 to 2017) were used. To avoid interference of possibly occurring changes during the three years, a simple change detection was applied. Changes in the forest cover were detected by comparing the NDVI values from the respective August scenes of the years 2015, 2016 and 2017. Based on the difference between the

NDVI of the actual and the previous year, pixels with absolute differences of ≤0.05 were flagged as 'change'. Negative values indicate a decrease in leaf biomass and were interpreted as an indicator of forest management activities such as thinning, harvesting or calamities. This interpretation was cross-checked by visual interpretation of the data sets and consultations with the forest management. All areas where forest management activities were detected were masked out from the tree species map.

3. Results

3.1. Land Cover Classification

The land cover classification based on all input data using the random forest modeling approach including the feature selection achieved an overall accuracy of 96%, and nearly all class-specific accuracies were higher than 90% (Table 4). The highest misclassifications can be found between the two agricultural classes grassland and cropland. The two forest classes (*broadleaf* and *conifer forest*) achieved very high producer and user accuracies (>93%).

Table 4. Confusion matrix based on OOB results of the land cover classification model based on all 18 Sentinel-2 scenes. (UA: user's accuracy, PA: producer's accuracy, OA: overall accuracy).

		Reference						
		BF	**CF**	**GL**	**CL**	**BU**	**WB**	**UA**
Classification	Broadleaf forest (BF)	**387**	7	2	0	2	0	**97.3%**
	Conifer forest (CF)	0	**90**	0	0	0	0	**100.0%**
	Grassland (GL)	1	0	**94**	5	0	0	**94.0%**
	Cropland (CL)	0	0	6	**70**	3	0	**88.6%**
	Build-up (BU)	0	0	2	2	**111**	0	**96.5%**
	Waterbody (WB)	0	0	0	0	0	**15**	**100.0%**
	∑ reference data	388	97	104	77	116	15	**797**
	PA	**99.7%**	**92.8%**	**90.4%**	**90.9%**	**95.7%**	**100%**	
				OA		**96.2%**	Kappa 0.946	

3.2. Tree Species Classification

The boxplots in Figure 4 illustrate the overall accuracies based on the out-of-bag results after feature selection. Each bar summarizes the different image combinations (1–18 images). For the coniferous species (middle row), overall accuracies of around 90% were achieved. Occasionally, a combination of five to six scenes was sufficient for such high classification accuracies. For the broadleaf trees (top row), the overall accuracies leveled out at around 80%. Here, a slightly higher number of scenes was necessary to reach optimum performance (7–8 scenes). The OA of the model trained on the pooled set of tree species was somewhere between the two class-specific results (bottom row). In all three cases, the use of vegetation indices (right column) improved the classification compared to the sole use of the reflectance data (left column). The average improvement of the OA was around 5 percentage points: The highest OA of the best models improved from 82.1% to 85.9% for the broadleaf strata, from 90.4% to 95.3% for the coniferous strata and from 83.5% to 88.7% for all tree species pooled together. Compared to the best mono-temporal results the use of multi-temporal data (both including vegetation indices) improved the out-of-bag overall accuracy from 72.9% to 85.7% for the broadleaved, from 83.8% to 95.3% for the coniferous and from 74.4% to 88.7% for all tree species together.

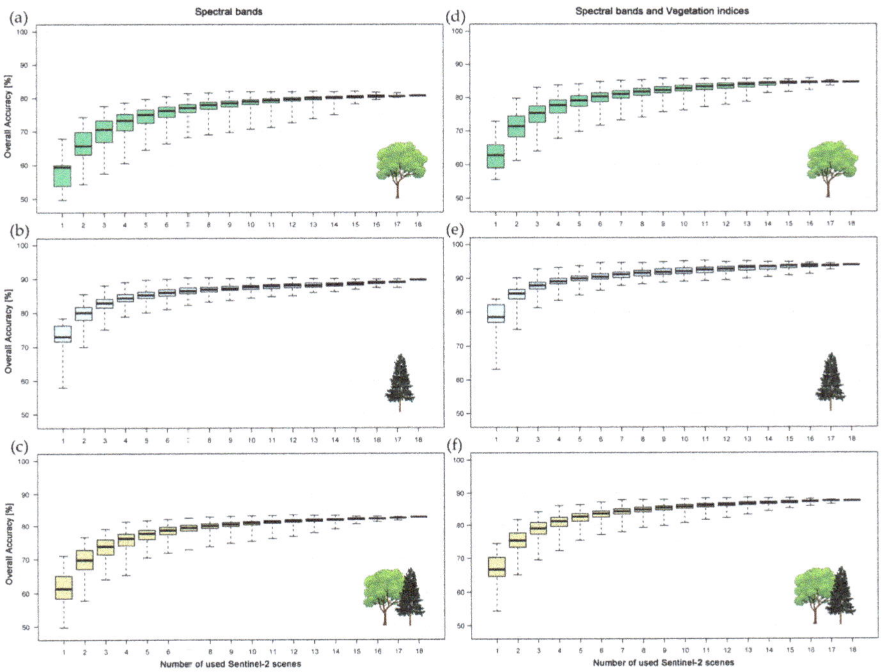

Figure 4. Overall accuracies of all possible Sentinel-2 combinations based on models using only spectral bands (**a–c**) and using spectral bands and vegetation indices (**d–f**). Strata-specific results are displayed in the rows: results for broadleaf species (**a**,**d**), coniferous species (**b**,**e**) and all tree species together (**c**,**f**).

The best results for the broadleaf trees obtained an overall accuracy of 86%. For all species, the achieved Producer's and User's accuracies are good (>70%) to very good (>90%) except for maple (Acer sp.). For the coniferous the best model reached an overall accuracy of 95.3%. All class specific accuracies were above 90%. For comparison reason the two results of the best broadleaf and the best coniferous models were combined into a single confusion matrix in Table 5. The differences to the result of the best model including all tree species (Table 6) are small. The aggregated overall accuracy of the stratified approach is slightly higher (89.9% versus 88.7%) and some classes with small numbers of reference data also benefit from the separated modeling. The best models for the three groups are based on the following input data after feature selection: (1) broadleaf trees: 159 variables from nine dates including all spectral bands and indices, (2) coniferous trees: 24 variables from seven dates including three bands and 13 indices, (3) all tree species together: 126 variables from 13 dates including eight bands and 26 indices.

Table 5. Confusion matrix based on the combined OOB results of the best broadleaf and the best coniferous model using spectral bands and vegetation indices. (UA: user's accuracy, PA: producer's accuracy, OA: overall accuracy).

		Reference												
		FS	**AG**	**FE**	**QU**	**PR**	**CP**	**AC**	**PA**	**PN**	**PS**	**LD**	**PM**	**UA**
Classification	*Fagus sylvatica (FS)*	**211**	4	7	7	4	12	8						**83.4%**
	Alnus glutinosa (AG)	0	**44**	0	1	0	0	0						**97.8%**
	Fraxinus excelsior (FE)	0	0	**44**	3	0	0	7						**81.5%**
	Quercus sp. (QU)	0	1	5	**117**	1	4	2						**90.0%**
	Prunus sp. (PR)	0	0	0	0	**18**	0	0						**100.0%**
	Carpinus betulus (CB)	4	2	3	1	1	**48**	0						**81.4%**
	Acer sp. (AC)	0	1	1	1	1	1	**16**						**76.2%**
	Picea abies (PA)								**132**	0	3	1	1	**96.4%**
	Pinus nigra (PN)								1	**101**	1	1	2	**95.3%**
	Pinus sylvestris (PS)								1	4	**75**	0	0	**93.8%**
	Larix decidua (LD)								1	1	0	**46**	1	**93.9%**
	Pseudotsuga menziesii (PM)								0	1	0	1	**52**	**96.3%**
	∑ Reference data	215	52	60	130	25	65	33	135	107	79	49	56	
	PA	**98.1%**	**84.6%**	**73.3%**	**90.0%**	**72.0%**	**73.8%**	**48.5%**	**97.8%**	**94.4%**	**94.9%**	**93.9%**	**92.9%**	
								OA	**89.9%**		Kappa	**0.885**		

Table 6. Confusion matrix based on the OOB result of the best model for all tree species together using spectral bands and vegetation indices. (UA: user's accuracy, PA: producer's accuracy, OA: overall accuracy).

		Reference												
		FS	**AG**	**FE**	**QU**	**PR**	**CP**	**AC**	**PA**	**PN**	**PS**	**LD**	**PM**	**UA**
Classification	*Fagus sylvatica (FS)*	210	3	4	7	4	11	8	0	0	0	0	0	**85.0%**
	Alnus glutinosa (AG)	0	43	0	1	0	0	1	0	0	0	0	0	**95.6%**
	Fraxinus excelsior (FE)	0	0	46	2	0	0	7	0	0	0	0	0	**83.6%**
	Quercus sp. (QU)	2	2	7	115	1	4	2	1	0	0	0	0	**85.8%**
	Prunus sp. (PR)	0	0	1	0	17	0	0	0	0	0	1	0	**89.5%**
	Carpinus betulus (CB)	2	3	0	3	2	49	0	0	0	0	0	0	**83.1%**
	Acer sp. (AC)	0	1	2	1	1	1	15	0	0	0	0	0	**71.4%**
	Picea abies (PA)	0	0	0	0	0	0	0	132	0	3	1	5	**93.6%**
	Pinus nigra (PN)	0	0	0	0	0	0	0	1	98	1	1	1	**96.1%**
	Pinus sylvestris (PS)	0	0	0	0	0	0	0	1	7	74	1	0	**89.2%**
	Larix decidua (LD)	1	0	0	1	0	0	0	0	2	1	45	2	**86.5%**
	Pseudotsuga menziesii (PM)	0	0	0	0	0	0	0	0	0	0	0	48	**100.0%**
	∑ Reference data	215	52	60	130	25	65	33	135	107	79	49	56	
	PA	**97.7%**	**82.7%**	**76.7%**	**88.5%**	**68.0%**	**75.4%**	**45.5%**	**97.8%**	**91.6%**	**93.7%**	**91.8%**	**85.7%**	
								OA	**88.7%**		Kappa	**0.871**		

Huge differences in feature importance were found within and between the model tree species groups (Figure 5). The dot size indicates the importance of a specific band (y-axis) and date (x-axis) in the classification models. Larger dots indicate a higher importance. To generate this information, the results of all combinations were aggregated, excluding models with vegetation indices. For the broadleaf species, the most important Sentinel-2 bands are the SWIR bands B12 and B11 followed by the Red-Edge band 1 (B5) and acquisitions from May, April and June are more useful compared to those from late summer and autumn. For the coniferous species, a higher number of scenes showed high importance compared to the broadleaf species. Again, the acquisition End of May, as well as the August acquisitions, showed the highest contribution. For the separation of the coniferous species the Red band is the most relevant band. The results for the aggregated modeling with all tree species together is a mixture of the results of the modeling for the two strata. Only the high importance of the NIR band B8 is striking.

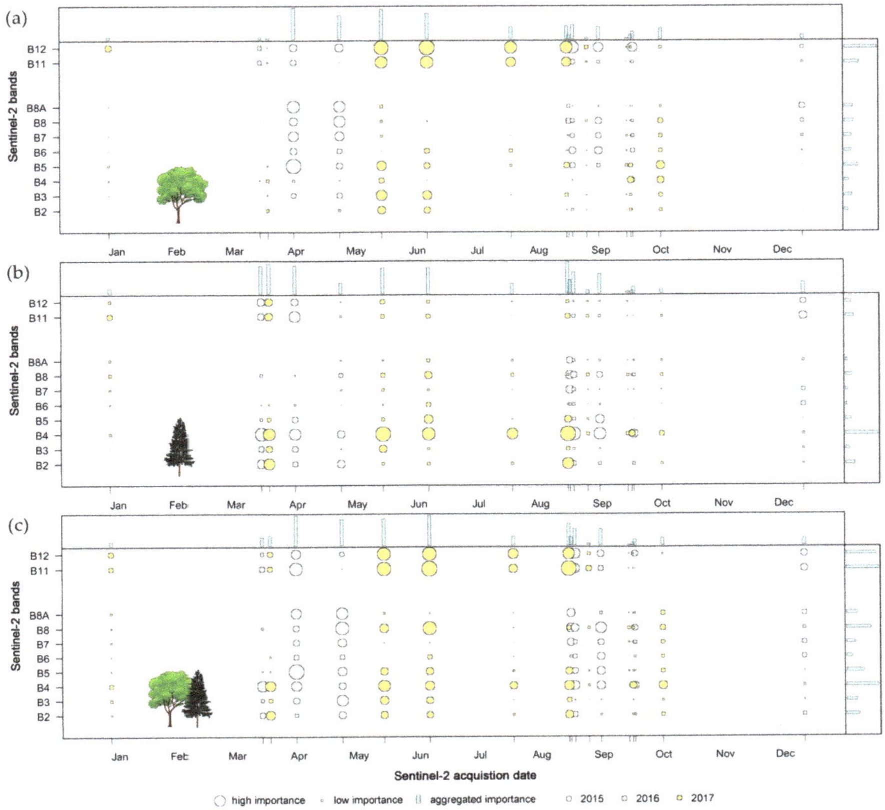

Figure 5. Aggregated feature importance derived from the combination of all classification models, excluding models involving spectral vegetation indices. A larger dot size indicates a higher importance of the specific band and date combination. The bars on the top and right side of the graphs summarize the importance of the individual months and spectral bands, respectively. (**a**) Results for the broadleaf stratum, (**b**) coniferous species and (**c**) all tree species pooled together. Different colors indicate the year of the Sentinel-2 acquisition.

Similar results can be found for the models using spectral data and vegetation indices together. For all three groups, the same Sentinel-2 bands show the highest importance (Figures A1–A3). The vegetation indices with the highest contributions to the classification performance for the broadleaf species were again indices which consider the SWIR, NIR and Red-Edge bands in different variations such as simple ratios, differences, and normalized differences (Figure A1). For the coniferous species mainly indices based on the NIR and Red bands (simple ratios and normalized differences) and the ratio between Green and Red showed the highest importance values (Figure A2). The result for all tree species together is again a mixture of the two strata. The highest ranked variable in the aggregated modeling was the Difference between Red and SWIR (Figure A3).

Figure 6 shows the final result of the tree species mapping for the entire Wienerwald biosphere reserve. The map combines the results of the land cover classification, the stratified tree species models (separated for broadleaf and coniferous species) and the change detection (forest management activities).

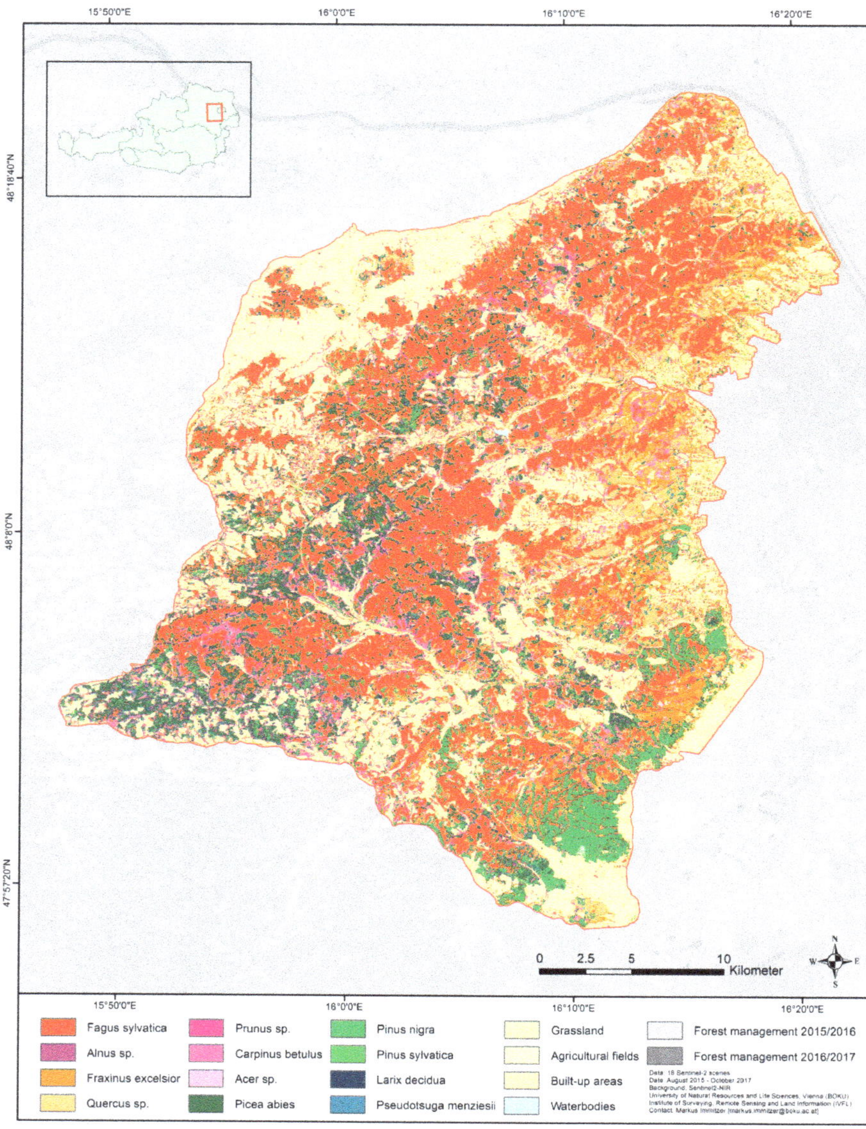

Figure 6. Final map based on aggregation of the best models of the land cover, broadleaf trees and coniferous trees classifications and the results of the change detection.

A qualitative check was done by the biosphere reserve management and the validity was confirmed. The different forest types mainly result from geological differences (e.g., Black pine forests in the southeast grow on limestone) and different management strategies (e.g., higher amount of coniferous trees in other regions).

For the detection of forest changes, the simple comparison of NDVI values from August scenes of different years was useful, as for each year, changes were captured on around 1% of the forest area. Areas where changes were detected in first period (2015–2016) often showed a clear regrowth in the second period. First trials aggregating the values to stand maps and interpreting the absolute values showed promising results to distinguish between different forest management activities (not shown). Both the affected area and the grade of the thinning can be determined based on the change in the NDVI values and the number of pixels with changes.

4. Discussion

4.1. Classification Accuracy

The classification of 12 tree species revealed very good results. We obtained an overall accuracy of 89% in line with comparable studies. However, most of the previous studies used only three to five S2 scenes and considered only fewer classes: Nelson [32] achieved 86% for six tree species classes, Bolyn et al. [22] 92% for 11 forest classes, Wessel et al. [33] 88% for four tree species classes, Persson et al. [27] 88% for five tree species, Soleimannejad et al. [49] 77% for three tree species and Hościło and Lewandowska [34] 76% (using only S2 bands) and 85% (using a stratified approach and including topographic features), for eight tree species. Grabska et al. [23] achieved with five (out of 18) S2 scenes an overall accuracy of 92% for nine tree species. We attribute our favorable results to the high quality of the reference data and the acquisition density, which allowed covering the temporal changes in the spectral signatures well, which in turn contributed to the successful classifications.

Several studies showed that higher accuracies can be achieved using data with higher spatial resolution such as WorldView-2 or Pleiades [25,26,49]. The accuracies of the present study are even higher compared to studies using WorldView-2 data for tree species classification under similar ecological conditions. For example, Immitzer et al. [16] obtained for 10 tree species overall accuracies of 82%, Waser et al. [17] for seven species 83% and Fassnacht et al. [15] for 10 species 80%. In the present study, we found clear indications that the use of multi-temporal data contributed to the successful classifications, further enhanced by the availability of spectral bands in the SWIR.

Only maple, cherry and European hornbeam were classified with low accuracies (46–75%). All other classes reached high (≥77%) to very high (≥85%) producer's and user's accuracies. Coniferous species were generally very well identified, in line with results published by Grabska et al. [23] and in contrast to Hościło and Lewandowska [34]. Further research is warranted to determine why not all species show distinct spectral-temporal features.

Similar to other studies [15,16,23–25,30], we found that class imbalances negatively affect the class-specific results. For 'hard-to-separate classes', the class with more samples is obviously preferred by the RF classifier and consequently obtains higher accuracies. This underlines again the importance of an adequate number of high quality reference samples. Although desirable, in practice, such a request is not always feasible, as some tree species hardly occur in pure stands, respectively, do not cover large enough areas for being detectable with 10 m resolution data.

4.2. Acquisition Date

In particular for the multi-date classifications, it is difficult to depict the importance of the acquisition date, as counter-balancing effects occur. For example, for mono-temporal classifications, the dataset from May achieved the best result for all tree species together. Images acquired in May were also used in models achieving the highest accuracies based on combinations of two and more scenes, together with October images. The sole use of October images, however, gave only moderate results. Hence, whereas individual acquisition dates can be well interpreted in mono-temporal classification, this is not always possible when multiple scenes are involved. To sustain interpretation, we aggregated the feature importance information in a new way (Equation (1) and Figure 5). As a general trend we confirm findings of previous studies, that a well-balanced data set, involving scenes from spring to autumn, are preferable [23,34,50]. When enough processing power and storage is available, a simple straight forward solution is to use all available images. Obviously, for very large areas such an approach will fail, as clouds will occur at least in parts of the area, making it necessary to use either compositing techniques [51] or spatio-temporal gap-filling procedures [52].

Species-specific temporal changes of the spectral signatures can be visualized nicely using dense time-series [23]. The date of leaf flush, for example, varies from species to species. This permits to distinguish tree species that otherwise show very similar spectral signatures during the summer months. The same holds for the timing of the leaf coloration. Therefore, both phenomena can be very helpful for the separation of tree species [23,50,53–57]. However, regional differences related to altitude, aspect or soil conditions co-influence the phenological development and timing of leaf flush and coloration, making additional information necessary [58]. Therefore, the quality of the available reference data is very important, as well as means to define strata in which individual models perform well. Alternatively, one has to use (auxiliary) proxy variables describing the species-independent phenological variations that are typically encountered when working across larger regions [59].

In general terms, it is beneficial to acquire and use large representative samples for each species covering as good as possible the site-specific variations. To leverage species-dependent differences in phenological development, the highest possible revisit frequencies are optimum. Very dense time series not only mitigate cloud effects but also permit to extract key phenological indicators such as start and end of season [50,60]. In our study, the available cloud-free scenes did not cover the entire year in regular intervals: mainly in July, only very few cloud-free data sets were acquired due to convective cloud formation; additionally, in spring and autumn, spells of bad weather prevented the acquisition of useful scenes.

4.3. Sentinel-2 Bands and Vegetation Indices

Our study demonstrates the high suitability of the Sentinel-2 bands for the separation of broad land cover classes as well as for the identification of various tree species. The most important bands are the SWIR, Red and NIR. The importance of the NIR and SWIR bands for species classification was also highlighted by other studies [22,27,28,49]. Our results reveal that the SWIR bands are mainly necessary for the separation of the individual broadleaf species; the NIR band is useful for the separation of the two strata coniferous and broadleaf species. In our work, the broader NIR band (band 8 at 10 m) achieves higher importance values than the narrow NIR band (band 8a at 20 m). This is in contrast to the work of Bolyn et al. [22] and Persson et al. [27], who found that band 8a was more important. Additionally, these two studies and Grabska et al. [23] highlighted the significance of the Red-Edge bands, which cannot be fully confirmed by our study. As the mentioned bands are recorded at different spatial resolutions, the contradictions may stem from site-specific differences in the patchiness of the forests. On the other hand, the high importance of the Red band for the coniferous species is in agreement with Hościło and Lewandowska [34].

We found that the use of vegetation indices improved the classification performance compared to the sole use of spectral signatures. Similar results were reported by Puletti et al. [28] and Maschler et al. [12]. The most relevant vegetation indices were based on the same bands, which show high importance in the models based only on spectral bands. For the broadleaf species, band combinations involving the SWIR, NIR and Red Edge bands were most useful, for the coniferous indices based on Red and NIR bands. The highest ranked indices considering all types of indices (simple ratios, differences, normalize differences).

5. Conclusions

The introduced Sentinel-2 workflow for tree species classification is robust, cost-efficient and scalable. The workflow was already successfully applied in several test areas across Germany. In the highly diverse Wienerwald biosphere reserve in Austria, the classification method achieved high classification accuracies for most of the 12 investigated tree species. However, a noticeable dependence on tree species, but also on the number and quality of the reference data, was found.

The NIR band is useful to separate the two tree species groups, coniferous and broadleaf trees, but much less so for identifying individual species. For the identification of the seven broadleaf species, the two SWIR bands are the most important. To separate the five individual coniferous species, the highest importance was found for the Red band. The use of additional vegetation indices further improved the performance of the classification models, and is therefore highly recommended.

The sensors on-board of Sentinel-2A and 2B provide rich spectral information in 10 spectral bands. The data are extremely useful for tree species mapping over large areas. The data is freely available, and due to the regular and dense acquisition pattern (five-day), images covering all phenological stages can be acquired and used for the classification. Although a few well-placed acquisitions can possibly yield very good classification results, the easiest way to ensure high classification accuracies is to combine and use all available cloud-free images simultaneously. This avoids the definition of the 'perfect' acquisition date(s), which far too often cannot be obtained due to local weather conditions.

With two Sentinel-2 satellites and their high revisit rate, the acquisition of several scenes per year should be possible for most Central European regions. However, especially over mountainous regions such as the Alps, the revisit frequency is probably still not high enough. The high cloudiness also prevents the application of methods for the extraction of land surface phenology (LSP) parameters, and their subsequent use in classification models. Unfortunately, in mountainous regions such as the Alps, microwave sensors are also only of limited use, due to strong terrain effects and radar shadows. Hence, it is requested to further increase the temporal revisit frequency of optical sensors and to make better use of virtual constellations, if possible involving the fleet of commercial VHR satellites.

From a methodological point of view, research should focus on further increasing the number of tree species involved in such classifications, while finding ways to handle strong class imbalances and missing values. The handling of mixed classes should also have high priority, as long as sensor resolution does not permit to resolve individual tree crowns. Ultimately, the full benefits of Earth Observation data will only become visible if such maps are produced and regularly updated at continental scale. This requires progress in terms of data standardization and feature extraction, and implementation of suitable code within HPC environments having direct access to the data storage. More research is warranted to identify the bio-physical factors leading to the observed large variations in species-specific classification accuracies.

Author Contributions: Conceptualization: M.I., M.N., H.B. and C.A.; methodology: M.I., M.N. and S.B.; formal analysis: M.I., M.N. and S.B., validation: M.I. and H.B.; investigation: M.I., S.B. and M.N.; resources: H.B.; writing—original draft preparation: M.I. and C.A.; writing—review and editing: F.V. and H.B.; visualization: M.I. and S.B.; project administration: M.I.; funding acquisition: M.I. and C.A.

Funding: This research was partly funded by the grant 854027 EO4Forest from the Austrian Research Promotion Agency (FFG).

Acknowledgments: We thank our project partners: Austrian federal forests (ÖBf AG), Forestry Office and Urban Agriculture of Vienna (MA 49) and forest enterprise of Heiligenkreuz Abbey for providing reference information. We are grateful to Dr. Alexandra Wieshaider (ÖBf AG) for supporting the project.

Conflicts of Interest: The authors declare no conflict of interest.

Appendix A

Table A1. Overview of the used vegetation indices and band combinations together with the corresponding formula and references (band 8 was used for the NIR = Near − Infrared; RE = Red − Edge).

Name	Formula	Reference
Built-up Area Index (BAI)	$\frac{BLUE-NIR}{BLUE+NIR}$	[61]
Chlorophyll Green index (CGI)	$\frac{NIR}{GREEN+RE1}$	[62]
Global Environmental Monitoring Index (GEMI)	$\eta - 0.25\eta^2 - \frac{RED-0.125}{1-RED}$ $\eta = \frac{2(NIR^2-RED^2)+1.5NIR+0.5RED}{NIR+RED+0.5}$	[63]
Greenness Index (GI)	$\frac{GREEN}{RED}$	[64]
Green Normalized Difference Vegetation Index (gNDVI)	$\frac{NIR-GREEN}{NIR+GREEN}$	[65]
Leaf Chlorophyll Content Index (LCCI)	$\frac{RE3}{RE1}$	[29]
Moisture Stress Index (MSI)	$\frac{SWIR1}{NIR}$	[66]
Normalized Difference Red-Edge and SWIR2 (NDRESWIR)	$\frac{RE2-SWIR2}{RE2+SWIR2}$	[67]
Normalized Difference Tillage Index (NDTI)	$\frac{SWIR1-SWIR2}{SWIR1+SWIR2}$	[68]
Normalized Difference Vegetation Index (NDVI)	$\frac{NIR-RED}{NIR+RED}$	[69]
Red-Edge Normalized Difference Vegetation Index (reNDVI)	$\frac{NIR-RE1}{NIR+RE1}$	[65]
Normalized Difference Water Index 1 (NDWI1)	$\frac{NIR-SWIR1}{NIR+SWIR1}$	[70]
Normalized Difference Water Index 2 (NDWI2)	$\frac{NIR-SWIR2}{NIR+SWIR2}$	[65]
Normalized Humidity Index (NHI)	$\frac{SWIR1-GREEN}{SWIR1+GREEN}$	[71]
Red-Edge Peak Area (REPA)	$RED + RE1 + RE2 + RE3 + NIR$	[67,72]
Red SWIR1 Difference (DIRESWIR)	$RED - SWIR1$	[73]
Red-Edge Triangular Vegetation Index (RETVI)	$100\ (NIR - RE1) - 10\ (NIR - GREEN)$	[74]
Soil Adjusted Vegetation Index (SAVI)	$\frac{NIR-RED}{NIR+RED+0.5}\ 1.5$	[75]
Blue and RE1 ratio (SRBRE1)	$\frac{BLUE}{RE1}$	[64]
Blue and RE2 ratio (SRBRE2)	$\frac{BLUE}{RE2}$	[76]
Blue and RE3 ratio (SRBRE3)	$\frac{BLUE}{RE3}$	[67]
NIR and Blue ratio (SRNIRB)	$\frac{NIR}{BLUE}$	[77]
NIR and Green ratio (SRNIRG)	$\frac{NIR}{GREEN}$	[64]
NIR and Red ratio (SRNIRR)	$\frac{NIR}{RED}$	[77]
NIR and RE1 ratio (SRNIRRE1)	$\frac{NIR}{RE1}$	[62]
NIR and RE2 ratio (SRNIRRE2)	$\frac{NIR}{RE2}$	[67]
NIR and RE3 ratio (SRNIRR3)	$\frac{NIR}{RE3}$	[67]
Soil Tillage Index (STI)	$\frac{SWIR1}{SWIR2}$	[68]
Water Body Index (WBI)	$\frac{BLUE-RED}{BLUE+RED}$	[78]

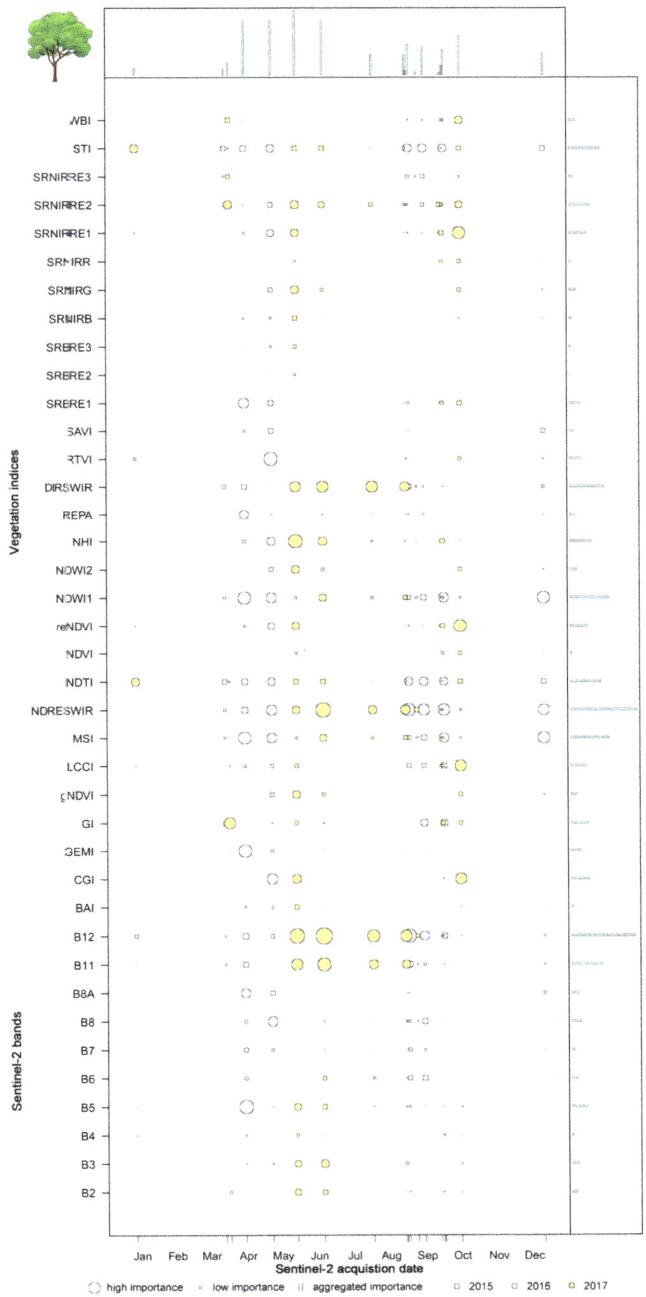

Figure A1. Aggregated feature importance for the broadleaf stratum derived from the combination of all classification models, based on spectral bands and vegetation indices (please see Figure 4 for more details about the graph and Table A1 for the Vegetation indices description).

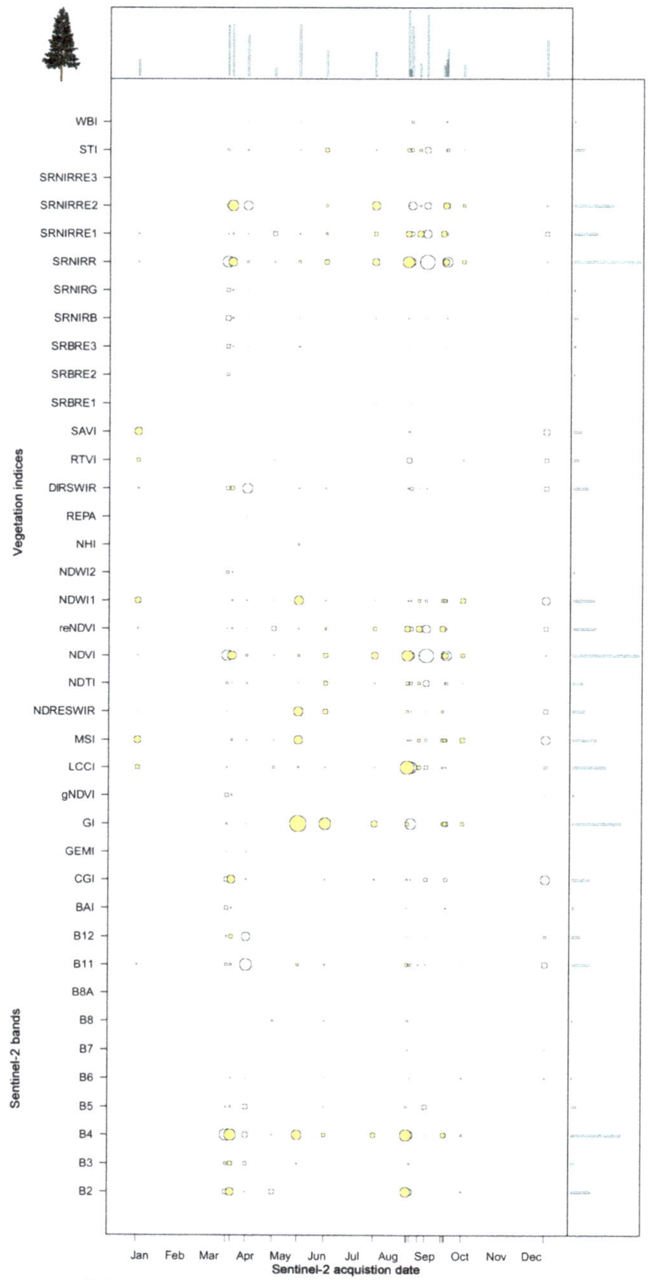

Figure A2. Aggregated feature importance for the coniferous stratum derived from the combination of all classification models, based on spectral bands and vegetation indices (please see Figure 4 for more details about the graph and Table A1 for the Vegetation indices description).

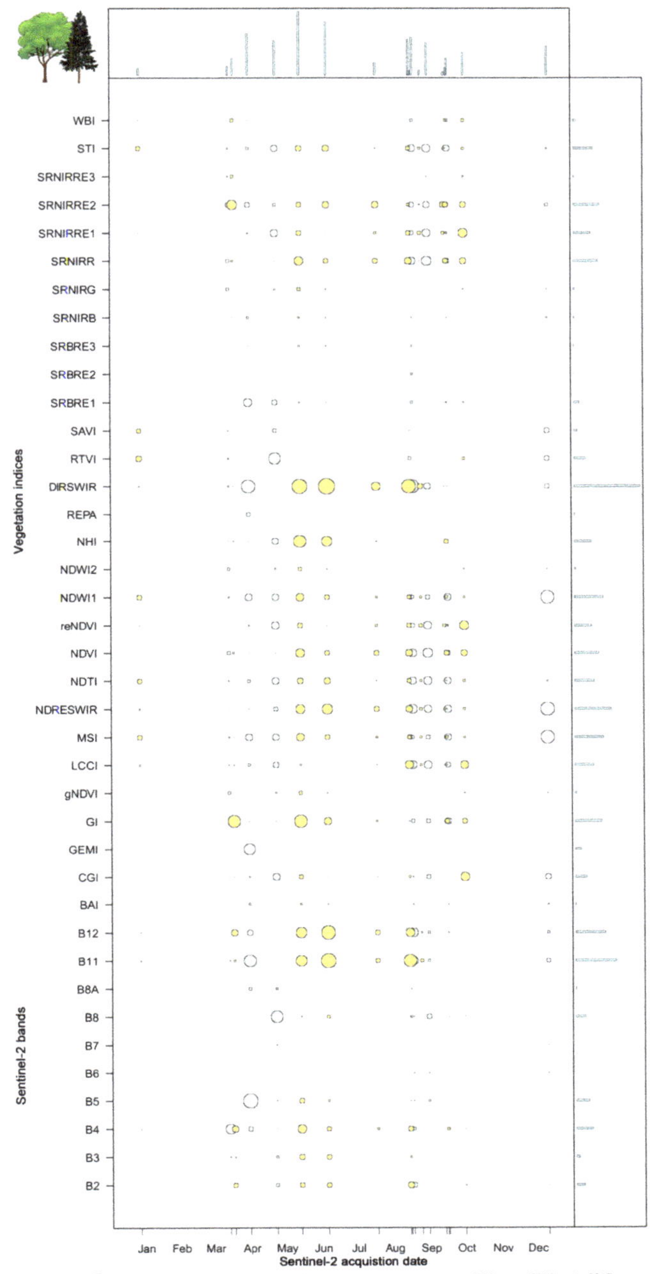

Figure A3. Aggregated feature importance for all tree species together derived from the combination of all classification models based on spectral bands and vegetation indices (please see Figure 4 for more details about the graph and Table A1 for the Vegetation indices description).

References

1. IPBES. *Summary for policymakers of the global assessment report on biodiversity and ecosystem services of the Intergovernmental Science-Policy Platform on Biodiversity and Ecosystem Services*; Díaz, S., Settele, J., Brondizio, E.S., Ngo, H.T., Guèze, M., Agard, J., Arneth, A., Balvanera, P., Brauman, K.A., Butchart, S.H.M., et al., Eds.; IPBES secretariat: Bonn, Germany, 2019; p. 45.
2. Kuenzer, C.; Ottinger, M.; Wegmann, M.; Guo, H.; Wang, C.; Zhang, J.; Dech, S.; Wikelski, M. Earth observation satellite sensors for biodiversity monitoring: Potentials and bottlenecks. *Int. J. Remote Sens.* **2014**, *35*, 6599–6647. [CrossRef]
3. Nagendra, H. Using remote sensing to assess biodiversity. *Int. J. Remote Sens.* **2001**, *22*, 2377–2400. [CrossRef]
4. Pettorelli, N.; Wegmann, M.; Skidmore, A.; Mücher, S.; Dawson, T.P.; Fernandez, M.; Lucas, R.; Schaepman, M.E.; Wang, T.; O'Connor, B.; et al. Framing the concept of satellite remote sensing essential biodiversity variables: Challenges and future directions. *Remote Sens. Ecol. Conserv.* **2016**, *2*, 122–131. [CrossRef]
5. Pettorelli, N.; Safi, K.; Turner, W. Satellite remote sensing, biodiversity research and conservation of the future. *Phil. Trans. R. Soc. B* **2014**, *369*, 20130190. [CrossRef] [PubMed]
6. Kissling, W.D.; Walls, R.; Bowser, A.; Jones, M.O.; Kattge, J.; Agosti, D.; Amengual, J.; Basset, A.; van Bodegom, P.M.; Cornelissen, J.H.C.; et al. Towards global data products of Essential Biodiversity Variables on species traits. *Nat. Ecol. Evol.* **2018**, *2*, 1531–1540. [CrossRef] [PubMed]
7. Schlerf, M.; Atzberger, C. Inversion of a forest reflectance model to estimate structural canopy variables from hyperspectral remote sensing data. *Remote Sens. Environ.* **2006**, *100*, 281–294. [CrossRef]
8. Lindenmayer, D.B.; Margules, C.R.; Botkin, D.B. Indicators of biodiversity for ecologically sustainable forest management. *Conserv. Biol.* **2000**, *14*, 941–950. [CrossRef]
9. Wulder, M.A.; Hall, R.J.; Coops, N.C.; Franklin, S.E. High spatial resolution remotely sensed data for ecosystem characterization. *BioScience* **2004**, *54*, 511–521. [CrossRef]
10. Dalponte, M.; Ørka, H.O.; Ene, L.T.; Gobakken, T.; Næsset, E. Tree crown delineation and tree species classification in boreal forests using hyperspectral and ALS data. *Remote Sens. Environ.* **2014**, *140*, 306–317. [CrossRef]
11. Fassnacht, F.; Neumann, C.; Forster, M.; Buddenbaum, H.; Ghosh, A.; Clasen, A.; Joshi, P.; Koch, B. Comparison of Feature Reduction Algorithms for Classifying Tree Species With Hyperspectral Data on Three Central European Test Sites. *IEEE J. Sel. Top. Appl. Earth Obs. Remote Sens.* **2014**, *7*, 2547–2561. [CrossRef]
12. Maschler, J.; Atzberger, C.; Immitzer, M. Individual Tree Crown Segmentation and Classification of 13 Tree Species Using Airborne Hyperspectral Data. *Remote Sens.* **2018**, *10*, 1218. [CrossRef]
13. Peerbhay, K.Y.; Mutanga, O.; Ismail, R. Commercial tree species discrimination using airborne AISA Eagle hyperspectral imagery and partial least squares discriminant analysis (PLS-DA) in KwaZulu–Natal, South Africa. *ISPRS J. Photogramm. Remote Sens.* **2013**, *79*, 19–28. [CrossRef]
14. Carleer, A.; Wolff, E. Exploitation of very high resolution satellite data for tree species identification. *Photogramm. Eng. Remote Sens.* **2004**, *70*, 135–140. [CrossRef]
15. Fassnacht, F.E.; Mangold, D.; Schäfer, J.; Immitzer, M.; Kattenborn, T.; Koch, B.; Latifi, H. Estimating stand density, biomass and tree species from very high resolution stereo-imagery—Towards an all-in-one sensor for forestry applications? *For. Int. J. For. Res.* **2017**, *90*, 613–631. [CrossRef]
16. Immitzer, M.; Atzberger, C.; Koukal, T. Tree species classification with Random Forest using very high spatial resolution 8-band WorldView-2 satellite data. *Remote Sens.* **2012**, *4*, 2661–2693. [CrossRef]
17. Waser, L.T.; Küchler, M.; Jütte, K.; Stampfer, T. Evaluating the Potential of WorldView-2 Data to Classify Tree Species and Different Levels of Ash Mortality. *Remote Sens.* **2014**, *6*, 4515–4545. [CrossRef]
18. Immitzer, M.; Böck, S.; Einzmann, K.; Vuolo, F.; Pinnel, N.; Wallner, A.; Atzberger, C. Fractional cover mapping of spruce and pine at 1ha resolution combining very high and medium spatial resolution satellite imagery. *Remote Sens. Environ.* **2018**, *204*, 690–703. [CrossRef]
19. Metzler, J.W.; Sader, S.A. Model development and comparison to predict softwood and hardwood per cent cover using high and medium spatial resolution imagery. *Int. J. Remote Sens.* **2005**, *26*, 3749–3761. [CrossRef]
20. EEA Forests—Copernicus Land Monitoring Service. Available online: http://land.copernicus.eu/pan-european/high-resolution-layers/forests (accessed on 8 February 2017).

21. Fassnacht, F.E.; Latifi, H.; Stereńczak, K.; Modzelewska, A.; Lefsky, M.; Waser, L.T.; Straub, C.; Ghosh, A. Review of studies on tree species classification from remotely sensed data. *Remote Sens. Environ.* **2016**, *186*, 64–87. [CrossRef]
22. Bolyn, C.; Michez, A.; Gaucher, P.; Lejeune, P.; Bonnet, S. Forest mapping and species composition using supervised per pixel classification of Sentinel-2 imagery. *Biotechnol. Agron. Soc. Environ.* **2018**, *22*, 172–187.
23. Grabska, E.; Hostert, P.; Pflugmacher, D.; Ostapowicz, K. Forest Stand Species Mapping Using the Sentinel-2 Time Series. *Remote Sens.* **2019**, *11*, 1197. [CrossRef]
24. Immitzer, M.; Vuolo, F.; Atzberger, C. First Experience with Sentinel-2 Data for Crop and Tree Species Classifications in Central Europe. *Remote Sens.* **2016**, *8*, 166. [CrossRef]
25. Immitzer, M.; Vuolo, F.; Einzmann, K.; Ng, W.-T.; Böck, S.; Atzberger, C. Verwendung von multispektralen Sentinel-2 Daten für die Baumartenklassifikation und Vergleich mit anderen Satellitensensoren. In Proceedings of the Beiträge zur 36. Wissenschaftlich-Technischen Jahrestagung der DGPF, Bern, Switzerland, 7–9 June 2016; Volume 25.
26. Ng, W.-T.; Rima, P.; Einzmann, K.; Immitzer, M.; Atzberger, C.; Eckert, S. Assessing the Potential of Sentinel-2 and Pléiades Data for the Detection of Prosopis and Vachellia spp. in Kenya. *Remote Sens.* **2017**, *9*, 74. [CrossRef]
27. Persson, M.; Lindberg, E.; Reese, H. Tree Species Classification with Multi-Temporal Sentinel-2 Data. *Remote Sens.* **2018**, *10*, 1794. [CrossRef]
28. Puletti, N.; Chianucci, F.; Castaldi, C. Use of Sentinel-2 for forest classification in Mediterranean environments. *Ann. Silvic. Res.* **2018**, *42*, 32–38.
29. Wulf, H.; Stuhler, S. Sentinel-2: Land Cover, Preliminary User Feedback on Sentinel-2A Data. In Proceedings of the Sentinel-2A Expert Users Technical Meeting, Frascati, Italy, 29–30 September 2015; pp. 29–30.
30. Sheeren, D.; Fauvel, M.; Josipović, V.; Lopes, M.; Planque, C.; Willm, J.; Dejoux, J.-F. Tree Species Classification in Temperate Forests Using Formosat-2 Satellite Image Time Series. *Remote Sens.* **2016**, *8*, 734. [CrossRef]
31. Gómez, C.; White, J.C.; Wulder, M.A. Optical remotely sensed time series data for land cover classification: A review. *ISPRS J. Photogramm. Remote Sens.* **2016**, *116*, 55–72. [CrossRef]
32. Nelson, M. Evaluating Multitemporal Sentinel-2 Data for Forest Mapping Using Random Forest. Master's Thesis, Stockholm University, Stockholm, Sweden, 2017.
33. Wessel, M.; Brandmeier, M.; Tiede, D. Evaluation of Different Machine Learning Algorithms for Scalable Classification of Tree Types and Tree Species Based on Sentinel-2 Data. *Remote Sens.* **2018**, *10*, 1419. [CrossRef]
34. Hościło, A.; Lewandowska, A. Mapping Forest Type and Tree Species on a Regional Scale Using Multi-Temporal Sentinel-2 Data. *Remote Sens.* **2019**, *11*, 929. [CrossRef]
35. Staudinger, M.; Scheiblhofer, J. Artenreichtum, Artenverteilung und räumliche Aspekte der Biodiversität der Gefäßpflanzen in Wäldern des Biosphärenpark Wienerwald. *Wiss. Mitteilungen Niederösterreichischen Landesmus.* **2014**, *25*, 249–268.
36. Mrkvicka, A.; Drozdowski, I.; Brenner, H. Kernzonen im Biosphärenpark Wienerwald—Urwälder von morgen. *Wiss. Mitteilungen Niederösterreichischen Landesmus.* **2014**, *25*, 41–88.
37. Drozdowski, I.; Mrkvicka, A. Der Wienerwald ist UNESCO-Biosphärenpark—Eine Modellregion für Nachhaltigkeit. *Wiss. Mitteilungen Niederösterreichischen Landesmus.* **2014**, *25*, 9–40.
38. Pflug, B.; Bieniarz, J.; Debaecker, V.; Louis, J.; Müller-Wilms, U. Some experience using sen2cor. In Proceedings of the EGU General Assembly Conference Abstracts, Vienna, Austria, 17–22 April 2016; Volume 18.
39. Vuolo, F.; Żółtak, M.; Pipitone, C.; Zappa, L.; Wenng, H.; Immitzer, M.; Weiss, M.; Baret, F.; Atzberger, C. Data Service Platform for Sentinel-2 Surface Reflectance and Value-Added Products: System Use and Examples. *Remote Sens.* **2016**, *8*, 938. [CrossRef]
40. Bucur, A.; Wagner, W.; Elefante, S.; Naeimi, V.; Briese, C. Development of an Earth Observation Cloud Platform in Support to Water Resources Monitoring. In *Earth Observation Open Science and Innovation*; Mathieu, P.-P., Aubrecht, C., Eds.; ISSI Scientific Report, Series; Springer International Publishing: Cham, Switzerland, 2018; pp. 275–283. ISBN 978-3-319-65633-5.
41. Breiman, L. Random forests. *Mach. Learn.* **2001**, *45*, 5–32. [CrossRef]
42. Toscani, P.; Immitzer, M.; Atzberger, C. Texturanalyse mittels diskreter Wavelet Transformation für die objektbasierte Klassifikation von Orthophotos. *Photogramm. Fernerkund. Geoinf.* **2013**, *2*, 105–121. [CrossRef]
43. Guyon, I.; Weston, J.; Barnhill, S.; Vapnik, V. Gene Selection for Cancer Classification using Support Vector Machines. *Mach. Learn.* **2002**, *46*, 389–422. [CrossRef]

44. Einzmann, K.; Immitzer, M.; Böck, S.; Bauer, O.; Schmitt, A.; Atzberger, C. Windthrow Detection in European Forests with Very High-Resolution Optical Data. *Forests* **2017**, *8*, 21. [CrossRef]
45. Hastie, T.; Tibshirani, R.; Friedman, J. *The Elements of Statistical Learning: Data Mining, Inference, and Prediction*, 2nd ed.; Springer: New York, NY, USA, 2009; ISBN 978-0-387-84858-7.
46. Pal, M. Random forest classifier for remote sensing classification. *Int. J. Remote Sens.* **2005**, *26*, 217–222. [CrossRef]
47. Vuolo, F.; Neuwirth, M.; Immitzer, M.; Atzberger, C.; Ng, W.-T. How much does multi-temporal Sentinel-2 data improve crop type classification? *Int. J. Appl. Earth Obs. Geoinf.* **2018**, *72*, 122–130. [CrossRef]
48. Belgiu, M.; Drăguţ, L. Random forest in remote sensing: A review of applications and future directions. *ISPRS J. Photogramm. Remote Sens.* **2016**, *114*, 24–31. [CrossRef]
49. Soleimannejad, L.; Ullah, S.; Abedi, R.; Dees, M.; Koch, B. Evaluating the potential of sentinel-2, landsat-8, and irs satellite images in tree species classification of hyrcanian forest of iran using random forest. *J. Sustain. For.* **2019**, *38*, 615–628. [CrossRef]
50. Pasquarella, V.J.; Holden, C.E.; Woodcock, C.E. Improved mapping of forest type using spectral-temporal Landsat features. *Remote Sens. Environ.* **2018**, *210*, 193–207. [CrossRef]
51. Pflugmacher, D.; Rabe, A.; Peters, M.; Hostert, P. Mapping pan-European land cover using Landsat spectral-temporal metrics and the European LUCAS survey. *Remote Sens. Environ.* **2019**, *221*, 583–595. [CrossRef]
52. Vuolo, F.; Ng, W.-T.; Atzberger, C. Smoothing and gap-filling of high resolution multi-spectral time series: Example of Landsat data. *Int. J. Appl. Earth Obs. Geoinf.* **2017**, *57*, 202–213. [CrossRef]
53. Elatawneh, A.; Rappl, A.; Rehush, N.; Schneider, T.; Knoke, T. Forest tree species identification using phenological stages and RapidEye data: A case study in the forest of Freising. In Proceedings of the 5th RESA Workshop, From the Basics to the Service, DLR e.V., Neustrelitz, Germany, 20–21 March 2013; pp. 21–38.
54. Hill, R.A.; Wilson, A.K.; George, M.; Hinsley, S.A. Mapping tree species in temperate deciduous woodland using time-series multi-spectral data. *Appl. Veg. Sci.* **2010**, *13*, 86–99. [CrossRef]
55. Lisein, J.; Pierrot-Deseilligny, M.; Bonnet, S.; Lejeune, P. A Photogrammetric Workflow for the Creation of a Forest Canopy Height Model from Small Unmanned Aerial System Imagery. *Forests* **2013**, *4*, 922–944. [CrossRef]
56. Schriever, J.R.; Congalton, R.G. Evaluating seasonal variability as an aid to cover-type mapping from Landsat Thematic Mapper data in the Northeast. *Photogramm. Eng. Remote Sens.* **1995**, *61*, 321–327.
57. Stoffels, J.; Hill, J.; Sachtleber, T.; Mader, S.; Buddenbaum, H.; Stern, O.; Langshausen, J.; Dietz, J.; Ontrup, G. Satellite-Based Derivation of High-Resolution Forest Information Layers for Operational Forest Management. *Forests* **2015**, *6*, 1982–2013. [CrossRef]
58. Li, D.; Ke, Y.; Gong, H.; Li, X. Object-Based Urban Tree Species Classification Using Bi-Temporal WorldView-2 and WorldView-3 Images. *Remote Sens.* **2015**, *7*, 16917–16937. [CrossRef]
59. Mascaro, J.; Asner, G.P.; Knapp, D.E.; Kennedy-Bowdoin, T.; Martin, R.E.; Anderson, C.; Higgins, M.; Chadwick, K.D. A Tale of Two "Forests": Random Forest Machine Learning Aids Tropical Forest Carbon Mapping. *PLoS ONE* **2014**, *9*, e85993. [CrossRef]
60. Guerif, M.; Gu, X.F.; Inra, J.P.G. Crop-system characterization by multitemporal SPOT data in the South-East of France. *Int. J. Remote Sens.* **1992**, *13*, 1843–1851. [CrossRef]
61. Shahi, K.; Shafri, H.Z.M.; Taherzadeh, E.; Mansor, S.; Muniandy, R. A novel spectral index to automatically extract road networks from WorldView-2 satellite imagery. *Egypt. J. Remote Sens. Space Sci.* **2015**, *18*, 27–33. [CrossRef]
62. Datt, B. A New Reflectance Index for Remote Sensing of Chlorophyll Content in Higher Plants: Tests using Eucalyptus Leaves. *J. Plant Physiol.* **1999**, *154*, 30–36. [CrossRef]
63. Pinty, B.; Verstraete, M.M. GEMI: A non-linear index to monitor global vegetation from satellites. *Vegetatio* **1992**, *101*, 15–20. [CrossRef]
64. Le Maire, G.; François, C.; Dufrêne, E. Towards universal broad leaf chlorophyll indices using PROSPECT simulated database and hyperspectral reflectance measurements. *Remote Sens. Environ.* **2004**, *89*, 1–28. [CrossRef]
65. Gitelson, A.A.; Kaufman, Y.J.; Merzlyak, M.N. Use of a green channel in remote sensing of global vegetation from EOS-MODIS. *Remote Sens. Environ.* **1996**, *58*, 289–298. [CrossRef]

66. Vogelmann, J.E.; Rock, B.N. Spectral Characterization of Suspected Acid Deposition Damage in Red Spruce (picea Rubens) Stands from Vermont. In Proceedings of the Airborne Imaging Spectrometer Data Anal. Workshop, Pasadena, CA, USA, 8–10 April 1985; pp. 51–55.
67. Radoux, J.; Chomé, G.; Jacques, D.C.; Waldner, F.; Bellemans, N.; Matton, N.; Lamarche, C.; d'Andrimont, R.; Defourny, P. Sentinel-2's Potential for Sub-Pixel Landscape Feature Detection. *Remote Sens.* **2016**, *8*, 488. [CrossRef]
68. Van Deventer, A.P.; Ward, A.D.; Gowda, P.M.; Lyon, J.G. Using thematic mapper data to identify contrasting soil plains and tillage practices. *Photogramm. Eng. Remote Sens.* **1997**, *63*, 87–93.
69. Tucker, C.J. Red and photographic infrared linear combinations for monitoring vegetation. *Remote Sens. Environ.* **1979**, *8*, 127–150. [CrossRef]
70. Gao, B. NDWI—A normalized difference water index for remote sensing of vegetation liquid water from space. *Remote Sens. Environ.* **1996**, *58*, 257–266. [CrossRef]
71. Lacaux, J.P.; Tourre, Y.M.; Vignolles, C.; Ndione, J.A.; Lafaye, M. Classification of ponds from high-spatial resolution remote sensing: Application to Rift Valley Fever epidemics in Senegal. *Remote Sens. Environ.* **2007**, *106*, 66–74. [CrossRef]
72. Filella, I.; Penuelas, J. The red edge position and shape as indicators of plant chlorophyll content, biomass and hydric status. *Int. J. Remote Sens.* **1994**, *15*, 1459–1470. [CrossRef]
73. Jacques, D.C.; Kergoat, L.; Hiernaux, P.; Mougin, E.; Defourny, P. Monitoring dry vegetation masses in semi-arid areas with MODIS SWIR bands. *Remote Sens. Environ.* **2014**, *153*, 40–49. [CrossRef]
74. Chen, P.-F.; Tremblay, N.; Wang, J.-H.; Vigneault, P.; Huang, W.-J.; Li, B.-G. New index for crop canopy fresh biomass estimation. *Spectrosc. Spectr. Anal.* **2010**, *30*, 512–517.
75. Huete, A.R. A soil-adjusted vegetation index (SAVI). *Remote Sens. Environ.* **1988**, *25*, 295–309. [CrossRef]
76. Lichtenthaler, H.; Lang, M.; Sowinska, M.; Heisel, F.; Miehé, J. Detection of Vegetation Stress Via a New High Resolution Fluorescence Imaging System. *J. Plant Physiol.* **1996**, *148*, 599–612. [CrossRef]
77. Blackburn, G.A. Quantifying Chlorophylls and Caroteniods at Leaf and Canopy Scales: An Evaluation of Some Hyperspectral Approaches. *Remote Sens. Environ.* **1998**, *66*, 273–285. [CrossRef]
78. Domenech, E.; Mallet, C. Change Detection in High resolution land use/land cover geodatabases (at object level). *EuroSDR Off. Publ.* **2014**, *64*.

remote sensing

MDPI

Article

Statistical Stability and Spatial Instability in Mapping Forest Tree Species by Comparing 9 Years of Satellite Image Time Series

Nicolas Karasiak [1,*], Jean-François Dejoux [2], Mathieu Fauvel [2], Jérôme Willm [1], Claude Monteil [1] and David Sheeren [1]

[1] DYNAFOR, Université de Toulouse, INRA, 31326 Castanet-Tolosan, France; jerome.willm@inra.fr (J.W.); Claude.Monteil@ensat.fr (C.M.); david.sheeren@ensat.fr (D.S.)

[2] CESBIO, Université de Toulouse, CNES/CNRS/INRA/IRD/UPS, 31401 Toulouse, France; jean-francois.dejoux@cesbio.cnes.fr (J.-F.D.); mathieu.fauvel@cesbio.cnes.fr (M.F.)

* Correspondence: nicolas.karasiak@inra.fr

Received: 30 September 2019; Accepted: 24 October 2019; Published: 26 October 2019

Abstract: Mapping forest composition using multiseasonal optical time series remains a challenge. Highly contrasted results are reported from one study to another suggesting that drivers of classification errors are still under-explored. We evaluated the performances of single-year Formosat-2 time series to discriminate tree species in temperate forests in France and investigated how predictions vary statistically and spatially across multiple years. Our objective was to better estimate the impact of spatial autocorrelation in the validation data on measurement accuracy and to understand which drivers in the time series are responsible for classification errors. The experiments were based on 10 Formosat-2 image time series irregularly acquired during the seasonal vegetation cycle from 2006 to 2014. Due to lot of clouds in the year 2006, an alternative 2006 time series using only cloud-free images has been added. Thirteen tree species were classified in each single-year dataset based on the Support Vector Machine (SVM) algorithm. The performances were assessed using a spatial leave-one-out cross validation (SLOO-CV) strategy, thereby guaranteeing full independence of the validation samples, and compared with standard non-spatial leave-one-out cross-validation (LOO-CV). The results show relatively close statistical performances from one year to the next despite the differences between the annual time series. Good agreements between years were observed in monospecific tree plantations of broadleaf species versus high disparity in other forests composed of different species. A strong positive bias in the accuracy assessment (up to 0.4 of Overall Accuracy (OA)) was also found when spatial dependence in the validation data was not removed. Using the SLOO-CV approach, the average OA values per year ranged from 0.48 for 2006 to 0.60 for 2013, which satisfactorily represents the spatial instability of species prediction between years.

Keywords: tree species; forest; biodiversity; time series; spatial autocorrelation; cross-validation; accuracy

1. Introduction

Forest ecosystems play a major role in global biodiversity [1]. They provide several services to humanity including carbon sequestration (which regulates climate [2]), timber production [3], soil protection [4], and recreation. They also have an impact on human health and well-being. However, the provision of such ecosystem services depends on several factors including the diversity of tree species [5]. Therefore, knowing the distribution of tree species in forests is crucial to assess ecosystem functions and services. More broadly, information on tree species is required for forest management and also for long-term forest monitoring, especially in the current context of climate change and related disturbances (forest fires, windstorms, drought, pests and diseases) [6].

Remote sensing has long been used to collect information on forest resources including stand composition [7,8]. Nevertheless, accurately distinguishing tree species is still challenging [9]. In the past, maps of tree species were based on field surveys completed by computer-aided analysis of aerial photographs [10,11]. While this approach provides accurate operational results for forest managers, it is limited to small spatial extent because it is costly and time consuming, which also affects its updating. In the last few decades, various types of remotely sensed images have been used to automate the identification of forest tree species. Some authors focused on the spatial resolution using very high-resolution satellite or airborne imagery [12–15]. They assumed the classification would benefit from the spatially detailed information and would therefore be accurate. Despite some successful results, this approach revealed itself to be of limited interest when only a single date was used due to the low spectral resolution of the data because of the reduced spectral and temporal information. Alternatively, as tree morphology and biochemical traits have a subtle influence on spectral reflectance [16], several authors explored airborne hyperspectral imagery [17–19]. Depending on the number of classes of species, on the methodology used for classification, and on the characteristics of the images (pixel size, number of spectral bands), the accuracy of the classification varied. Nevertheless, studies based on hyperspectral imagery were typically more accurate than those based on single-date multispectral data [9].

Taking advantage of the temporal dimension of the satellite data was another way to separate tree species [9]. Time series can capture the phenological behavior of the vegetation and this functional trait can be useful to discriminate the forest types. Changes in pigment contents, water and leaf morphology across seasons can vary from one species to another. Time series with images covering all phenological events from green-up to senescence (leaf-on, spring flush, autumn senescence, leaf-off) can produce detailed classification results. The use of multitemporal data for this purpose is not new. This approach has been explored from various image datasets of different spatial and temporal resolutions based on spaceborne sensors such as MODIS [20,21], Landsat [22–26], RapidEye [27], WorldView [28], as airborne sensor [29,30] or unmanned aerial systems [31]. More recently, the potential of the new freely available high spatial resolution Sentinel-2 (S2) data has been investigated [32–36]. In general, the authors found it advantageous to combine images acquired in spring and autumn, at the key phenological stages of temperate forests, since it had a positive influence on the accuracy of the classification. Images acquired in summer are also frequently selected in features ranking procedures, particularly for conifer species [36], but also for deciduous species [30]. From a spectral point of view, red-edge bands and SWIR bands are reported to be important variables when S2 time series are used [32–34].

Despite the increasing number of studies that use time series to identify forest types, the true predictive power of these kinds of data remain to be demonstrated. Even though it is difficult to compare studies because of the use of different methods, sensors, and classes of tree species, we observed very contrasted results from one study to another. For instance, using four dates for S2 data in 2017, Persson et al. [34] obtained a kappa value of 0.83 to classify five species (Norway spruce—*Picea abies*; Scots pine—*Pinus sylvestris*; Hybrid larch—*Larix x marschlinsii*; Birch—*Betula sp.* and Pedunculate Oak—*Quercus robur*). This differs substantially from the Immitzer's results [32] (kappa = 0.59) as they used only two S2 images to map seven species including Norway spruce, Scots pine, European larch (*Larix decidua*) and Oak (*Oak sp*). There was also a difference of almost 0.2 points of overall accuracy (OA) between a study by Persson et al. [34] and one by Liu et al. [35] who classified eight types of forest in China with the same number of S2 images. In another study, using only two S2 images to separate 11 forest classes of broadleaves and conifers, Bolyn et al. [33] obtained very accurate results (OA of 0.93) in contrast with previous works but in line with others based on dense time series acquired using different sensors [27,37,38].

The notable difference in accuracy among past studies suggests a better understanding is required of the factors that affect the classification of species, as recommended by [9]. Several drivers of classification errors remain insufficiently explored, among which, spatial autocorrelation of reference data has long been identified but rarely quantified [39,40]. Spatial dependence in the reference data due to an inadequate sampling strategy to split training and validation sets can wrongly increase classification accuracy [39,41,42]. Despite different approaches addressing this issue by imposing a spatial stratification to select samples for training and testing [41,42], the spatial autocorrelation is not always estimated explicitely. This can lead to keep a residual spatial dependence between training and testing samples and not to fully eliminate the spatial overfitting. Contamination by clouds and cloud shadows in dense image high resolution time series may also have a major impact on classification performances. Because the distribution of such contamination may vary over time and in space across years, a multiyear analysis is required to reliably evaluate their effect.

The first objective of this study is to explore the stability of classification results in tree species discrimination from optical remotely sensed time series data of multiple years. The second objective is to understand the main drivers that affect the classification accuracy in this context. To our knowledge, this is the first study of variability between one-year classifications of tree species based on multiple years using dense image high spatial resolution time series. We evaluated the classification performance of single-year Formosat-2 time series in distinguishing forest types with spatially independent validation data. We also investigated how the predictions vary statistically and spatially across multiple years (from 2006 to 2014). The main contribution of this work is a better estimation of the classification accuracy of the forest maps by reducing optimistic bias due to spatial autocorrelation. The second contribution, resulting from the first, is a finer understanding of the drivers responsible for classification errors. We hypothesize that time series data improve species discrimination compared to single-date image due to seasonal variability in spectral reflectance between species.

2. Material

2.1. Study Area

The study site is located in south-western France, next to Toulouse, and covers an area of 24 km × 24 km (Figure 1). This delimited area was determined by a satellite acquisitions scheme by the *Centre National d'Etudes Spatiales* (CNES) who acquired a Formosat-2 Satellite Image Time Series (SITS) of the site. The Garonne river crosses the eastern part of the study area, influencing soil composition and the nearly flat topography of the area. The climate is sub-Atlantic characterized by sunny autumns, hot dry summers, and mild rainy winters (the average annual temperature is >13 °C; annual precipitation = 656 mm). The landscape is dominated by arable lands (including wheat, sunflower, maize) and grasslands. Forests cover up to 10% of the landscape (53 km^2).

2.2. Satellite Image Time Series

We used a dense optical image dataset composed of Formosat-2 time series acquired in nine consecutive years from 2006 to 2014. This dataset was obtained during preparation for the Sentinel-2 and VENµS mission with cooperation between the Israeli Space Agency (ISA) and the French CNES [43]. A total of 156 dates was acquired with an average of 14 images per year and a maximum of 43 images in 2006. The distribution of the dates over time varied from one year to another and the number of images available during the growth season differed from the number available at the end of vegetation season (Figure 2).

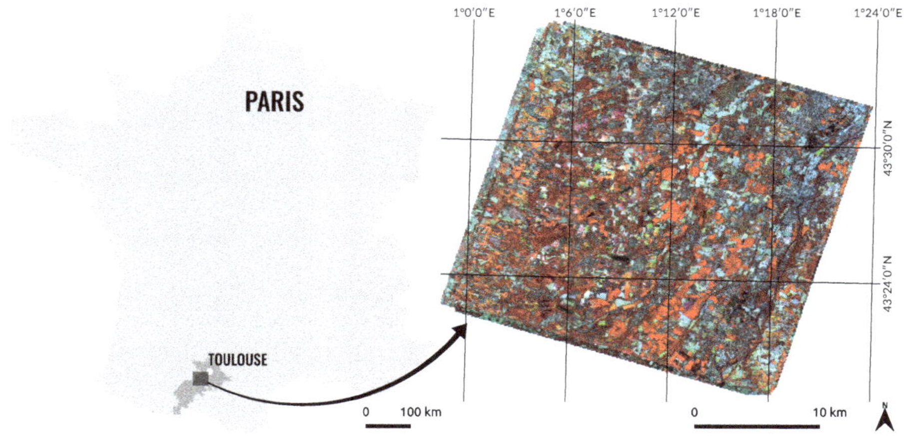

Figure 1. The map on the left shows the location of the study area in the Haute-Garonne district (in dark grey) near Toulouse, France. On the right, a false-color image acquired on 13 August 2013 which represents the entire Formosat-2 scene covering an extent of 24 × 24 km.

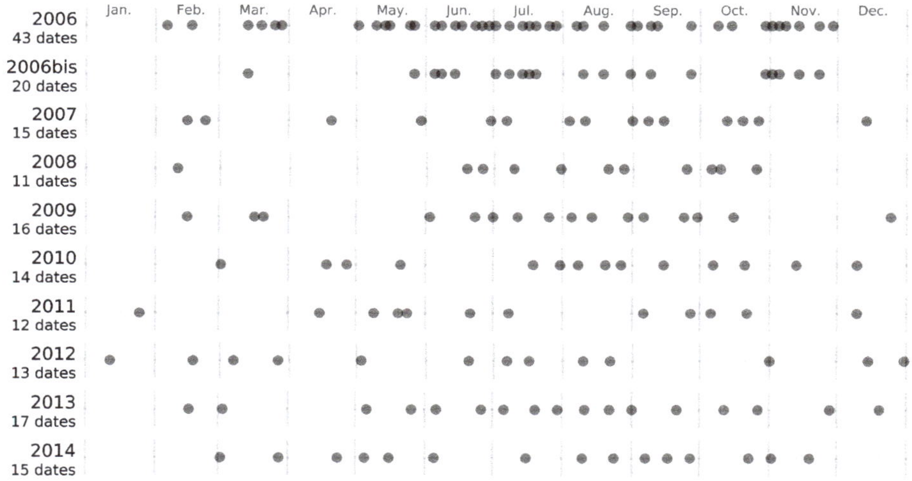

Figure 2. Number and acquisition dates of each image in the Formosat-2 time series from 2006 to 2014.

Cloud coverage also varied considerably from one date to another, ranging from a minimum of 8 cloud-free images in 2011 to a maximum of 20 in 2006. For 2006, by visual inspection, we created an additional dataset (named 2006 bis) by selecting only the cloud-free images, resulting in a time series of 20 dates (compared to the original 46).

The Formosat-2 multispectral images are delivered in an 8-bit radiometric resolution. Each image provides 4 spectral bands ranging from the visible (Blue: 0.45–0.52 μm, Green: 0.53–0.60 μm, Red: 0.63–0.69 μm) to the near-infrared (NIR: 0.76–0.90 μm) with a nominal pixel size of 8 m. All the images were acquired under a constant viewing angle and a field of view of 24 km like Landsat, VENμS and Sentinel-2.

2.3. Ancillary Data

A forest mask produced in 1996 by the French National Forest Inventory database (IGN BDForêt®, v.1) was used to select forest pixels in the SITS (i.e., forest stands with a minimum area of 2.25 hectares) and to exclude non-forested areas. Based on aerial photographs taken in 2006, 2010 and 2013 (IGN BDOrtho®), the mask was manually updated to retain only SITS forest stands that remained stable over the nine year period

2.4. Field Data

Four field campaigns were conducted between November 2013 and January 2017 to identify and locate reference samples of tree species in the study site. All the main forests were visited. Only the dominant broadleaf and conifer tree species were recorded. To insure tree species purity in the training samples, plots were delimited at the center of homogeneous areas covering an area of approximatly 576 m^2 (i.e., nine contiguous 8 m $\times$ 8 m Formosat-2 pixels). Only the pixel at the center of each area was used for the classification protocol. Plots were located using a Garmin GPSMap 62st receiver (3–5 m accuracy) and distributed over 72 distinct forest stands.

Thirteen tree species of which eight were broadleaf species and five conifer species were studied (Table 1). In some species, identification was limited to the genus level because of the existence of cultivars (case of Aspen) and the difficulty involved in determining the exact species of Oak, Willow and Eucalyptus. We acquired a total of 1262 sample plots (named *reference samples* in the rest of the paper). Class distribution was moderately imbalanced reflecting the uneven distribution of species abundances in the forests. The number of samples varied from 50 (the minimum for Willow) to 211 (the maximum for Aspen). Conifers were less well represented with an average of 73 samples per class compared with 112 for broadleaf species.

Table 1. List of tree species with their sample size, in pixels, collected during field surveys ($n = 1262$). The number of forest stands in which the samples were collected is also provided. Stand delimitation is based on the French National Forest Inventory database (IGN BDForêt® v.1).

Species	Sample Size	Forest Stands
Broadleaf		
Silver birch (*Betula pendula*)	85	3
Oak (*Quercus robur/pubescens/petraea*)	115	12
Red Oak (*Quercus rubra*)	147	7
Aspen (*Populus spp.*)	211	6
European Ash (*Fraxinus excelsior*)	80	3
Black locust (*Robinia pseudoacacia*)	63	7
Willow (*Salix spp.*)	50	3
Eucalyptus (*Eucalyptus spp.*)	148	4
Conifer		
Corsican Pine (*Pinus nigra subsp. laricio*)	70	6
Maritime Pine (*Pinus pinaster*)	103	7
Black Pine (*Pinus nigra*)	55	2
Silver Fir (*Abies alba*)	75	5
Douglas Fir (*Pseudotsuga menziesii*)	60	7

3. Classification Protocol

A global overview of the classification protocol applied on each Formosat-2 single-year time series is shown in Figure 3.

3.1. Pre-Processing

In this step, surface reflectance time series were produced from the Formosat-2 level 1A images using the MACCS (Multisensor Atmospheric Correction and Cloud Screening) processing chain developed by the CNES [44,45]. MACCS involved orthorectification, atmospheric correction, detection of clouds and cloud shadows, and reduction of topographic effects on illumination, based on multitemporal and multispectral criteria. Atmospheric correction relies on the estimation of aerosol optical thickness based on a spectro-temporal technique that minimizes (i) variations in surface reflectances between pixels acquired consecutive cloud-free images after correction and (ii) differences between the blue surface reflectance predicted from the red band (empirical relationship) and the blue surface reflectance obtained after correction [46]. Clouds are detected using a multitemporal approach that analyzes the increase in reflectance in the blue spectral band [45]. If high variation is observed, cloud is likely to be present. Based on this method, masks of clouds and related shadows are produced by MACCS for each image in the time series.

In the second step, SITS of each year were filtered using a linear gap-filling algorithm applied to each spectral band to remove noisy data (i.e., cloudy and shady pixels) and to retrieve their surface reflectance [47]. Invalid pixels were replaced by the interpolated values from the closest available valid pixels in the time series. Gap-filling was chosen for its simplicity and its previously demonstrated efficiency already demonstrated when time gaps between consecutive images are limited [48].

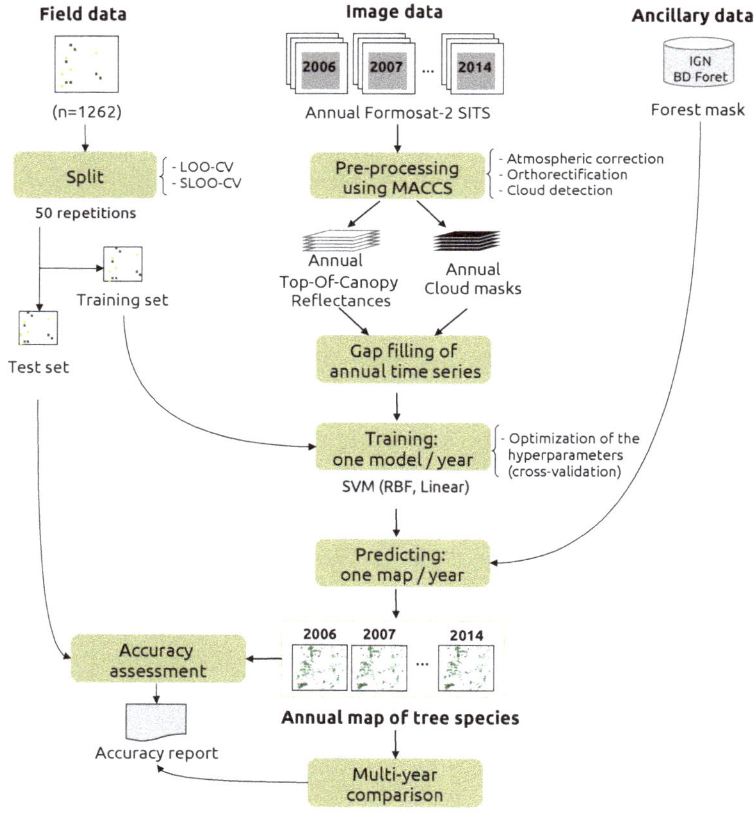

Figure 3. Classification protocol for a single year time series, repeated for the 9 years available from 2006 to 2014. The splitting procedure to create independent training and test sets is based on a spatial and non-spatial leave-one-out cross-validation (SLOO-CV and LOO-CV, respectively). The LOO-CV were trained with exactly the same number of training samples as the SLOO-CV, after random undersampling.

3.2. Training

Classification models were built using all spectral bands of each annual time series as predictors with exactly the same pixels for training and testing. We used the supervised SVM (Support Vector Machine) classifier [49] known to be the best approach in the case of small training data sets with respect to data dimensionality [50]. In this study, we selected the Radial Basis Function (RBF) kernel which is the most frequently used and has already been proven to be effective in the case of similar classification problems [51]. Hyperparameters including the regularization parameter (C) and the kernel bandwidth (γ) were tuned by cross-validation in a search space with the following settings: $C = \{0.01, 0.1, ..., 1^{10}\}$ and $\gamma = \{1^{-9}, 1^{-8}, ..., 1^{3}\}$. A linear kernel was also tested for comparison with RBF. However, since the linear kernel performed worse, the results are not presented here. To account for imbalanced data and to prevent potential bias due to the dominant classes [52], the class weights in the SVM parameters were also modified. Weights were set inversely proportional to class frequencies. SVM was computed using the scikit-learn python library [53]. Vector of features were standardized (i.e., centering and scaling to unit variance) prior to training.

3.3. Estimating Prediction Errors by Spatial Cross-Validation

Because spatial autocorrelation between training and test sets may produce optimistic bias in assessments of classification performance [39,41,42], we used a spatial leave-one-out cross-validation (SLOO-CV) sampling strategy [54,55] to separate the training and test sets to guarantee full independence between them. In this approach, one reference sample is used as the test set and the remaining samples, non-spatially correlated with the test set, are used as the training set (Figure 4). This is repeated n times where n equals the number of samples. The n prediction results are then averaged to obtain an estimation of the prediction error. In our case, the test set was composed of one pixel of each class (i.e., a total of 13 pixels at each iteration) and the procedure was repeated 50 times, this being the number of reference samples of the lowest class size. We compared this splitting procedure with the classical non-spatial leave-one-out cross-validation strategy (LOO-CV) using the same training size per class as in SLOO-CV, by random undersampling. For year-to-year comparison, we also used the same training and test sets related to each sampling approach by setting the same random seed.

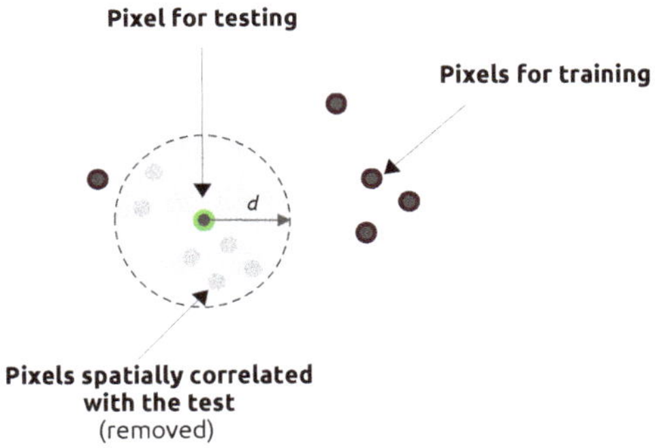

Figure 4. Spatial leave-one-out cross-validation (SLOO-CV) schema for one class. One pixel is used for testing. The other pixels are used for training, except pixels geographically too close to the pixel selected for testing. This procedure is repeated n times where n is the number of reference samples. Spatial autocorrelation between nearby pixels is assumed up to a distance d which can be estimated using Moran's I.

The spatial autocorrelation distance was estimated by computing the Moran's Index from the pixels of forests in the SITS [56]. Moran's I estimates the correlation between the value of a variable at one location and nearby observations. The index ranges from -1 (negative spatial autocorrelation) to

+1 (positive spatial autocorrelation) with a value close to 0 in the absence of spatial autocorrelation (random spatial distribution). More formally, the Moran's I is defined as the ratio of the covariance between neighborhood pixels and the variance of the entire image:

$$I(d) = \frac{n}{S_0} \frac{\sum_{i=1}^{n} \sum_{j=1}^{n} w_{i,j}(x_i - \overline{x})(x_j - \overline{x})}{\sum_{i=1}^{n} (x_i - \overline{x})^2} \tag{1}$$

where, in our case, x_i is the pixel value of x (a spectral band of the SITS for pixels of forests) at location i, x_j is the pixel value of x at location j (a nearby pixel of forest of i), $\overline{x}$ is the average value of x, n is the number of pixels of forests in the image, $w_{i,j}$ is the weight equals to 1 if pixel j is within distance d of pixel i, otherwise $w_{i,j} = 0$, and S_0 the sum of all $w_{i,j}$'s:

$$S_0 = \sum_{i=1}^{n} \sum_{j=1}^{n} w_{i,j} \tag{2}$$

In this study, Moran's I was computed for each spectral band of each year, for neighborhoods (lags) varying from 1 to 100 pixels (i.e., from 8 m to 800 m). Based on correlograms, we evaluated the distance between nearby pixels for which Moran's I equals 0.2, considering the potential effect of spatial autocorrelation as not significantly different from the thresold value of Moran's I [57]. Then, the median distance was calculated for each spectral band, taking all the dates of one year into account (Figure 5). This was done for each year. Finally, the average value of the median distance of each year was kept in the spatial cross-validation procedure to split the training and test sets. This average value was estimated to be 340 m (i.e., 42 pixels).

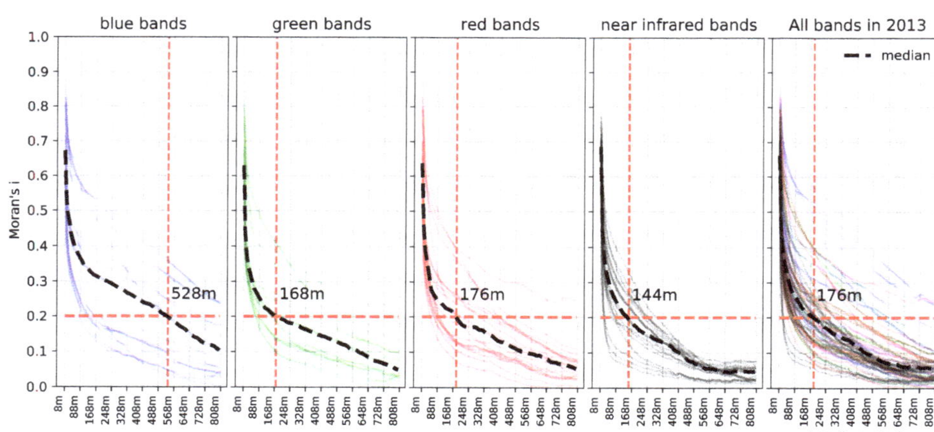

Figure 5. Moran's I correlograms of each Formosat-2 spectral band of the Satellite Image Time Series (SITS) 2013, for pixels representing forests. Each curve represents one date of the SITS. the red dashed line represents the median distance value (in x) where Moran's I = 0.2 (in y). For a Moran's I thresold value of 0.2, spatial independence between nearby pixels was assumed. This is the case beyond to 528 m in the blue band, 168 m in the green, 176 m in the red and 144 m in the near-infrared.

3.4. Accuracy Assessment of One-Year Classifications and Comparison

The results of the classifications were assessed according to the confusion matrix based on Overall Accuracy (OA) and the F1 score (i.e., the harmonic mean of precision and recall varying from 0 for the worst case to 1 for perfect classification), errors of omission and errors of commission. A Wilcoxon signed-rank test was used to determine if the difference in accuracy between annual classifications and sampling strategies (LOO-CV vs SLOO-CV) was statistically significant.

Classifications were also compared spatially to highlight instability between years. A map of uncertainty was produced by computing the number of agreements between the one-year classifications (i.e., the modal value related to the class with the highest frequency) for each pixel. Additionally, the distribution of this uncertainty was examined per class using either all the predicted pixels or only the reference samples. Finally, the maps were visually inspected to identify problem areas and to better understand the errors with the help of field knowledge. The maps shown in results section were produced using the SLOO-CV.

4. Results

4.1. Overall Statistical Performances

The classification performances for each year are presented in Table 2. Generally speaking, the performances were similar between the years but very different between sampling strategies (SLOO-CV vs LOO-CV) in a given year.

When prediction errors were estimated by spatial cross-validation (SLOO-CV), the average OA varied from 0.48 in 2007 to 0.60 in 2013 with high variability in the results (average standard deviation of 0.12). No significant differences were observed between the years 2008–2012, 2012–2014 and between 2006 and 2007 which were the cloudiest SITS ($p < 0.05$; Wilcoxon signed-rank test statistic; see Appendix B for statistical details). For the year 2006, when cloudy images were removed from the SITS (i.e., using the 2006bis dataset), the classification was improved, the performance was similar to that in the other years (average OA = 0.57). The best classification was obtained using the 2013 time series (average OA = 0.60).

When accuracy was computed using the standard leave-one-out cross-validation (LOO-CV), prediction errors were very low compared to when SLOO-CV was used, suggesting a high optimistic bias in the evaluation. The average OA varied from 0.97 in 2011 to 1.00 in 2006 and 2014 with a standard deviation close to zero. The cloudiest years (2006 and 2007) did not differ significantly in performance from the other years in most cases (Appendix B). These results contradict the previous ones: while the year 2007 was the worst with the SLOO-CV, with LOO-CV it had the second best score.

In the following sections, we only detail the results based on the SLOO-CV strategy since it best reflects the true performance of the classifications.

4.2. Accuracy per Species

In most cases, whatever the year, broadleaf tree species were better discriminated than conifers (Figure 6). The highest performances were obtained for monospecific plantations of Red oak (average F1 score = 87%) and Willow (average F1 score = 86%). Aspen was also detected with good accuracy (average F1 score = 68%). Conversely, some species were difficult to identify, including European ash (average F1 score = 26%) and Silver birch (average F1 score = 36%) except in the years 2010 and 2013.

High confusion rates were obtained for conifer tree species. Black pine was the worst class with a F1 score close to zero, except in 2014 (F1 score = 62%). Maritime pines were generally better discriminated but the performances remained low (average F1 score = 40%). The best agreement was obtained for Silver fir (average F1 score = 50%) which reached its best score (average F1 score = 81%) in the year 2010.

Table 2. Accuracy report of spatial leave-one-out cross-validation (SLOO-CV) sampling strategy and leave-one-out cross-validation (LOO-CV) for each single-year classification based on OA statistics. The 2006bis time series only includes cloud-free images of 2006. The average percentage of cloud coverage was estimated by computing for each species the number of time each reference sample was affected by clouds (detected from the MACCS processing chain).

	2006	**2006bis**	**2007**	**2008**	**2009**	**2010**	**2011**	**2012**	**2013**	**2014**
Classification accuracy (average Overall Accuracy ± standard deviation)										
SLOO-CV	0.52 ± 0.13	0.57 ± 0.15	0.48 ± 0.12	0.57 ± 0.10	0.55 ± 0.11	0.56 ± 0.12	0.55 ± 0.11	0.58 ± 0.14	0.60 ± 0.11	0.58 ± 0.11
LOO-CV	1.00 ± 0.02	0.99 ± 0.03	0.99 ± 0.02	0.98 ± 0.04	0.99 ± 0.03	0.98 ± 0.03	0.97 ± 0.04	0.98 ± 0.04	0.99 ± 0.02	1.00 ± 0.02
Characteristics of each SITS										
Number of images	43	20	15	11	16	14	12	13	17	15
Images in spring	13	4	2	1	2	3	4	3	3	5
Images in autumn	10	6	4	4	3	4	4	2	4	4
Cloud coverage	25%	0%	12%	5%	4%	3%	2%	0%	1%	0%

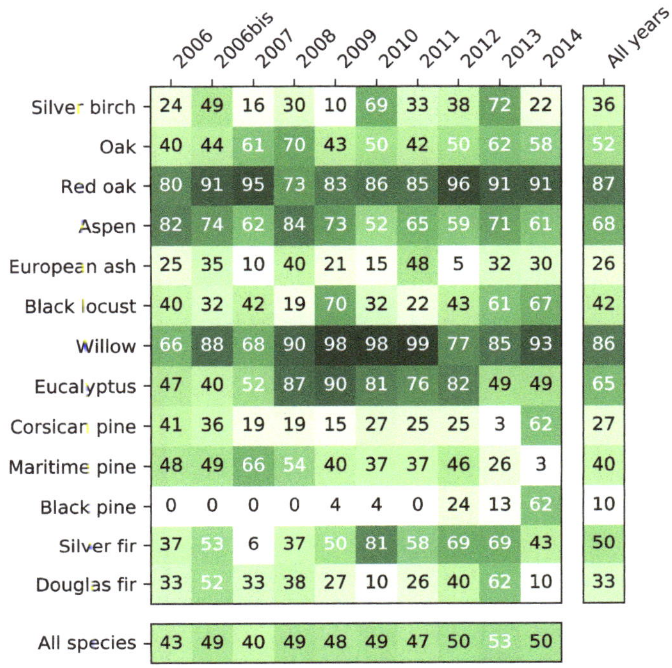

Figure 6. Average F1 score (in %) per species and per year based on the SLOO-CV sampling strategy using the Support Vector Machine (SVM) (RBF kernel) classifier. Values range from white (F1 score = 0%) to dark green (F1 score = 100%). Average values of F1 score per year and per species are also provided (in the bottom row and last column on the right, respectively).

On average, the year 2013 was the best, mainly because of a high score for Silver birch compared to the other years. Year 2007 was the least accurate. Higher performance disparity was observed from one year to another for most species, except Red oak and Willow.

4.3. Confusion between Species

Generally, when errors occurred, the broadleaf tree species were confused with each other as well as with conifers. The main source of omissions for Silver birch was mispredictions as Oak which, in turn, was confused with European ash but also with Black locust and with some pines (see the confusion matrix for the year 2013 in Figure 7, for example). Red oak was the subject of very little confusion. High rates of omissions were observed for European ash with misclassifications as Oak, Aspen and Black locust. Under-detection was also observed for the evergreen Eucalyptus plantations due to confusion with Willow. In conifer species, the errors mainly appeared between species of Pine but also between Pine and Douglas fir.

Confusions between species were similar from one year to another but the commission and omission errors rates varied and accuracy was very low for some species (Figure 8).

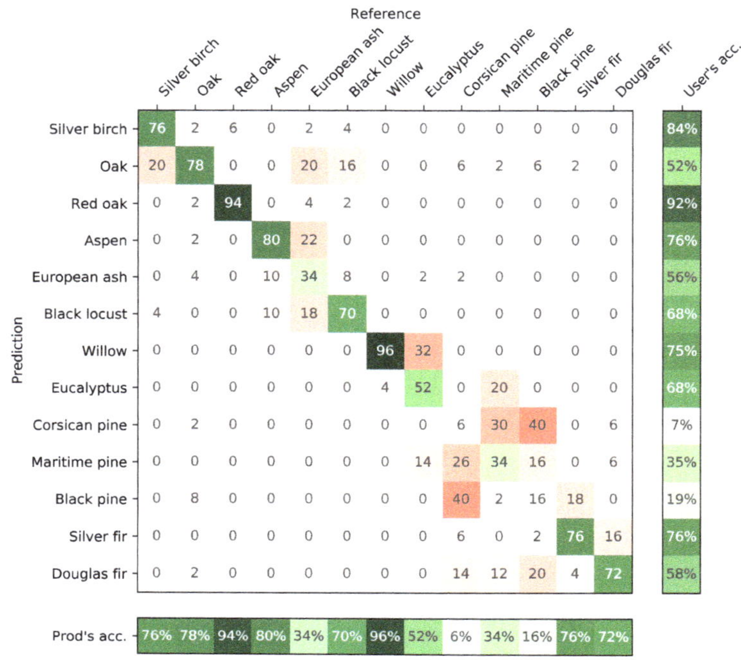

Figure 7. Confusion matrix for 13 tree species for year 2013. Each cell provides an average value of agreement or confusion (in %) based on 50 classifications (number of iterations) using the SLOO-CV sampling strategy.

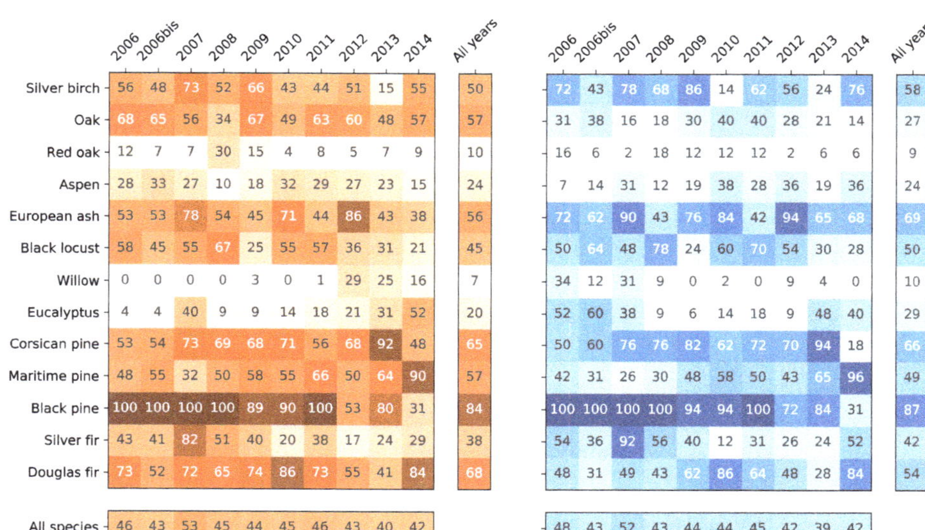

(**a**) Commission (**b**) Omission

Figure 8. Average rate of commission and omission errors (in %) per species and per year based on the SLOO-CV sampling strategy.

4.4. Spatial Agreement between Years

As revealed by the map of the modal class values, Oak was the most representative species in the study region, especially in the small forests, which is consistent with our field observations (Appendix A). Conifers and plantations of broadleaf species were less frequent but pixels of the same class appear to be grouped in homogeneous stands, as expected.

When spatial uncertainty was analyzed using the map of agreements between the one-year classifications, good stability was observed in the monospecific tree plantations of broadleaf species (Figure 9). The stands composed of Aspen, Red oak, and Eucalyptus were clearly differentiated. In contrast, in complex forests including a mix of different species, disagreements between annual classifications were higher, as suggested by the previous statistical assessment. An example is given in Figure 10 showing a mix forest composed of conifers (mainly Black pine but also Douglas fir and Silver fir) and deciduous species (mainly Oak and Silver birch). There was considerable confusions between conifer species from one year to another (low agreement). The extent of Silver birch areas was also highly variable. In this forest, the dominant species were rather well-identified but their exact location was inaccurate at the pixel level.

Significant disagreements between the classifications were also observed in other contexts, especially in thin riparian forests and forest edges where species composition and diversity is high, with lots of species unsampled (Figure 9). This was also true in low density forest stands, for which confusions appeared with the understory vegetation. Finally, disagreements were also observed in areas very affected by clouds and cloud shadows.

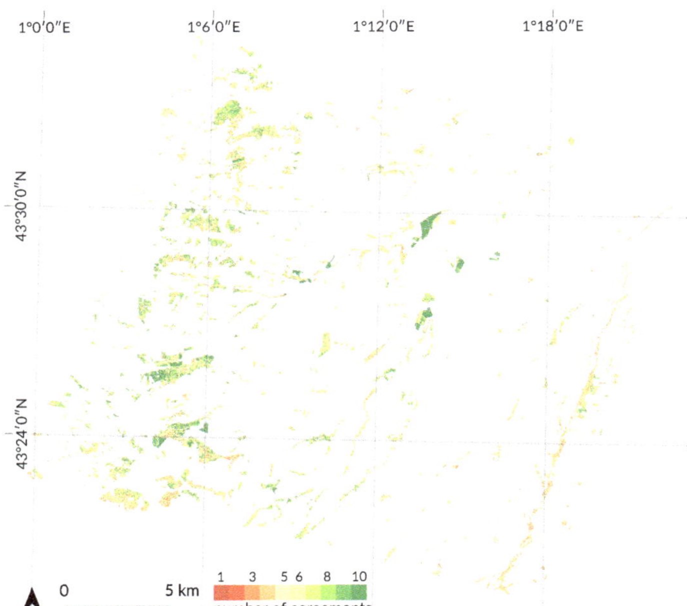

Figure 9. Spatial comparison between the annual classifications of forest tree species from 2006 to 2014 (including 2006bis). The number of agreement is the modal value related to the class with the highest frequency. This map illustrates the stability of predictions from one year to another with a high number of disagreements in red (high uncertainty) and a low number of disagreements in green (high accuracy). In many cases, homogeneous green areas are tree plantations of broadleaf species such as Red Oak and Aspen or Eucalyptus.

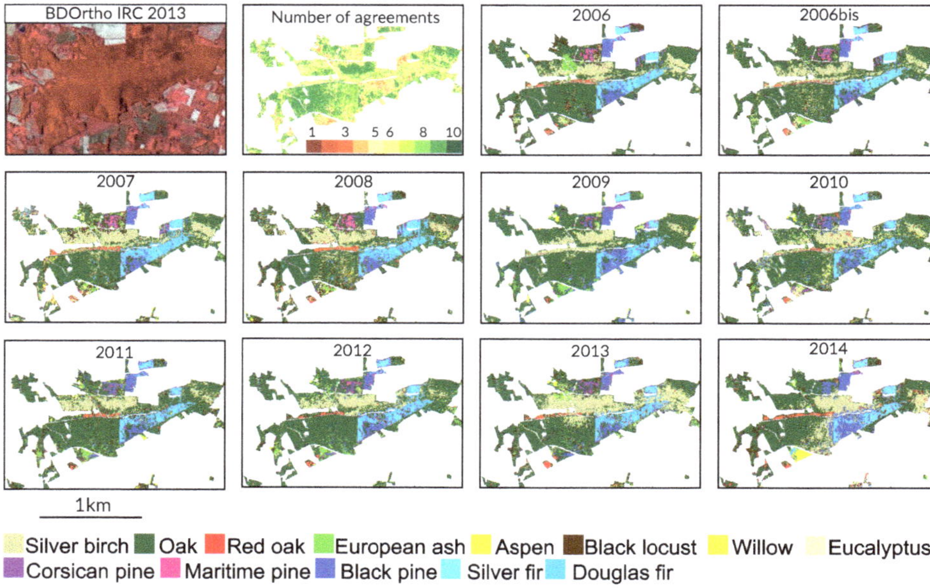

Figure 10. Annual classifications of tree species in a mixed forest composed of different conifers and broadleaf species. Instability was observed in the conifer plantations composed of Black pine, Silver fir and Douglas fir. This was also the case for Silver birch. Part of the forest (in the north) has was excluded from the analysis because of changes during the study period (clear-cuts and reforestation).

5. Discussion

In this study, an archive of Formosat-2 time series was used to classify tree species in temperate forests in nine consecutive years. Each classification was validated using the same spatial leave-one-out cross-validation approach to remove the test samples that were spatially correlated with the training samples. To our knowledge, this is the first study to examine the stability of predictions from one year to another using dense SITS of high spatial resolution with spatially independent validation data. The present study is a first attempt to assess the robustness of tree species discrimination in multiple years and to better understand the drivers that affect classification performances.

5.1. Effect of Spatial Autocorrelation: The SLOO-CV Strategy as a Standard

Our results revealed a strong positive bias in validation based on the usual LOO-CV strategy for splitting reference data. This bias was already suspected in our previous studies when we used stratified-k-fold but was not quantified [38,58]. Regarding the importance of the overestimation in the classification accuracy (ΔOA $>$ 0.4 between LOO and SLOO-CV), the use of spatially independent data for validation should no longer be an option but wherever possible, a requirement, in agreement with the recommendation of [9].

Spatial autocorrelation in the reference data has long been known to affect the classification and accuracy assessment [39,40,59,60]. Different sampling strategies for data splitting have already been studied in the literature including spatial [42,55,61,62] and aspatial approaches [61,63–65]. Although the spatial sampling approach is recommended to reduce the spatial autocorrelation effect, an aspatial (i.e., random, systematic or stratified) sampling strategy assuming independence between training and test sets is usually used in remote sensing analyses for the sake of simplicity [65].

In this study, the SLOO-CV strategy was used to estimate an unbiased prediction performance, similar to used in [54]. We measured the spatial dependence between nearby pixels of forests explicitly, using the Moran's I, as in [66], and we separated training and validation samples that were located geographically too close to one another. In other studies, spatial partitioning was achieved differently, based on *k-means* clustering [42,67] or on the definition of patches [68], or blocks related to the spatial structure [62]. In these approaches, the spatial dependence between training and test sets is supposed to be removed but this is not verified. Whatever the spatial sampling method used, all the studies demonstrated larger errors in predictions with lower spatial autocorrelation between training and test sets, as we observed here. The absence of independence between training and testing samples thus provides an inflated estimate of classification performances as confirmed by our results.

An important point to note is that we guaranteed complete independence between the training and test sets but not among the training samples. Thus, spatial autocorrelation still persisted in the training set. Compared to LOO-CV, the SLOO-CV strategy provides a statistical estimate of accuracy that fits the quality of the map product better (and hence predictive performance) but in terms of predictions, the results of classification results are similar. This is illustrated by the number of agreements between the annual classifications for both LOO-CV and SLOO-CV (Figure 11). With the exception of some classes (e.g., Maritime and Black pine), the distribution of agreements per class of species is rather similar.

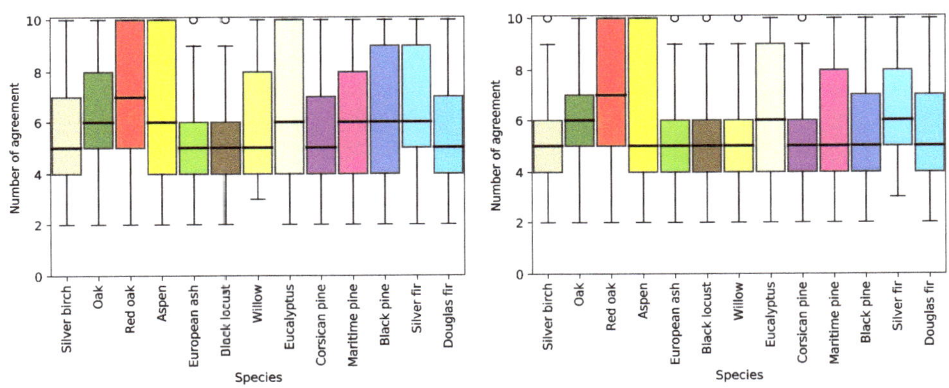

(**a**) LOO-CV sampling (**b**) SLOO-CV sampling

Figure 11. Distribution per species of the number of agreements between the annual classifications using all the predicted pixels for both the LOO-CV and SLOO-CV sampling strategies.

5.2. *Effect of the Size of the Reference Sample*

The size of the training sample dataset is known to be a key factor affecting both classification accuracy and predictive performance [59,69]. In practice, it is hard to adequately judge the optimal training set size which depend on several factors such as the number of features, the degree of imbalance in class distribution, and the machine learning algorithm. In this study, we used the SVM classifier, which is known to be less sensitive to sample size since the decision boundaries rely on only a few support vectors. We also adjusted class weights to avoid bias due to the uneven distribution of tree species. Nevertheless, we obtained a slightly significant positive correlation (r = 0.52; *p-value* = 0.06) between the average number of pixels used for training (see Appendix D) and the average F1 score obtained for each species (including all the years), suggesting a potential effect of training set size on accuracy. We also observed that the least well-identified species were those with a limited number of forest stands (only three for Silver birch, European ash and Black pine). For these small classes, the presence of noise on the data (under detected clouds or cloud shadows under-detected, see below) may have a greater negative impact on their discrimination. However, Willow is an exception, as it was the least populated species with only 21 samples for training and a total of three forest stands but obtained the second highest F1 score (average = 0.86) behind Red oak (average = 0.87) with 118 samples (Appendix D).

More generally, due to the complexity of the learning problem (i.e., partial overlapping between spectro-temporal signatures of species in the feature space), increasing the size of the training sample and reducing the degree of imbalance in class distribution should improve the predictive performance [70]. In a recent study, Bolyn et al. [33] obtained a high degree of accuracy (OA = 88.5%) when they classified 11 forest classes (including seven tree species) in the entire Belgian Ardenne ecoregion with only two Sentinel-2 dates. Although this statistical performance may be partially inflated by spatial autocorrelation, this greater accuracy could also be explained by the large sample size (from 2589 to 7068 pixels for each class with a minimum of 31 forest stands and a maximum of 64). An equivalent level of performance (OA = 88.2%) was obtained by [34] when they identified five tree species in a forest in central Sweden using four S2 images acquired from spring to autumn. However, in their case study, the sample size was very close to ours (from 27 to 98 pixels per species). Spatial overfitting is suspected, as in our previous works [38,58].

5.3. Effect of Clouds and Cloud Shadows

When we compared the stability of predictions from one year to another (i.e., the map of agreements between the annual classifications) with the number of times the pixels were affected by clouds or cloud shadows, we found no significant correlation. This indicates that disagreements between the classifications can not be attributed to this factor. We observed that the years most affected by clouds (2006 and 2007) had the lowest average OA values (52% and 48% respectively, see Table 2). However, for the other years, the F1 score per species was not always consistent with the extent of cloud coverage. For instance, in 2008, 11% of the reference pixels for Oak were affected by clouds (Appendix C) but the F1 score was the highest of all the years. This suggests that when clouds and cloud shadows are detected by the MACCS processing chain, the gap-filling approach is appropriate to correct noisy pixels. This procedure is currently used in the French production center THEIA (https://theia.cnes.fr) for Landsat, VENµS and Sentinel-2 level2A products.

An in-depth visual analysis of the map products in fact revealed that misdetections of clouds and cloud shadows had a major effect on classification performances. When the forests were partially affected by clouds and cloud shadows or when these were under detected (which is what happened in the case of slight fog), spectral signatures were skewed and confusion between species was likely. This issue is illustrated in Figure 12 which shows changes in the reflectance values of an Oak pixel in 2006. On most of the dates, the pixel was free of clouds and shadows (green dots). In some cases, clouds or cloud shadows were found and the pixel values were gap-filled by linear interpolation (orange dots). But on certain dates, clouds or cloud shadows were not detected (red dots) and this influences the spectro-temporal signature. These dates had erroneous values but also had a negative impact on nearby gap-filled values since the dates are considered to be valid (e.g., see the 10th image in Figure 12). Another example showing an erroneous spatial pattern in a forest stand due to undetected clouds is provided in Appendix A3. This noise may influence the training step through the addition of inadequate support vectors, as well as the validation step, if the reference pixel to be tested is impacted by noise but the training pixels are not.

An alternative to the gap-filling approach to reduce noise could be the use of smoothing methods applied to the whole time series and not only to a limited temporal window (i.e., the cloudy and shady pixels). Non-parametric methods such as Whittaker smoother [71] or splines [72] may be appropriated. Another way to limit the effect of noise could be reducing the number of features [73]. Limiting the number of features in the classification protocol could help remove noise but also prevent the Hughes phenomenon [74] (i.e., a decrease in accuracy with the addition of new features after an initial increase). In theory, the Hughes effect should not be observed with the SVM classifier which is robust to the dimensionality of data [75]. However, a previous study demonstrated a positive role of feature selection on classification results with SVM, particularly when the training set used is small [76].

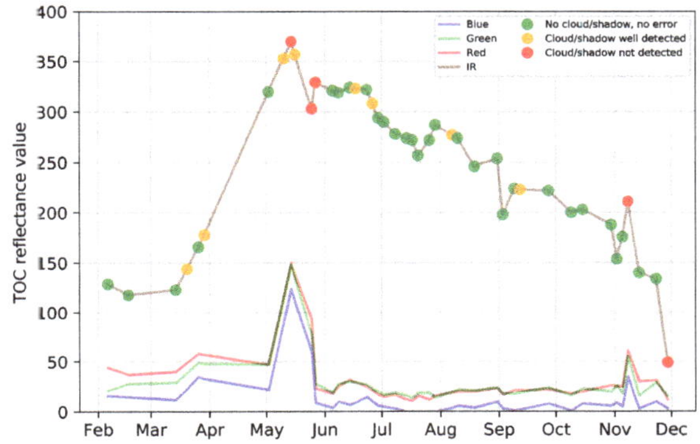

Figure 12. Influence of undetected clouds in the time series, an example with the reflectance evolution of a Oak tree pixel in 2006. If the dot is red in the infrared time series, it means the pixel is in a cloud or in a shadow but this was not detected by the algorithm. So this pixel is taken as a valid pixel to be used to gap-fill nearest missing data.

5.4. Effect of the Available Dates in the SITS

We hypothesized that time series data improve the identification of tree species compared to the use of only one date or a couple of images due to phenological differences in spring or autumn. Using several dates should be better than using one date, whatever the year, as demonstrated in [30,31,34]. However, because of the variation in the dates available from one year to another, the differences in cloud and shadow contamination, and the potential Hughes effect, it is difficult to give a clear answer concerning the most appropriate time windows to separate species. When we examine the relationships between the classification performances and the number of dates acquired during the key seasons, we find no evidence of a positive effect. For example, classification accuracy in the year 2008 was statistically equivalent to that in the year 2014 when there was comparable cloud coverage (<5%) whereas we only had one available image in spring 2008 versus five in 2014. Similarly, classification performance in the year 2009 was statistically identical to that in other years including in 2006bis. However, in autumn 2009, no images were available from end of October to the middle of December and only a very limited number of images were available in spring compared to 2006bis which had four dates in spring and six dates in autumn (Figure 2).

In order to better evaluate the most relevant dates for classification and to assess the sensitivity of our results to the Hughes effect, an additional analysis was carried out using a feature selection approach following the SLOO-CV strategy. Feature selection was based on the simple Sequential Forward Selection (SFS) algorithm which adds at each iteration, the most important feature (*sensu* OA) in the pool of selected ones, starting with an empty set [77]. Each feature is permanently conserved after selection and the process is repeated until all the features are included. Here, the feature selection approach was applied for each year, using all the spectral bands of the available dates. However, to reduce the computation time, we considered as a feature a single-date acquisition composed of four spectral bands.

Globally, this analysis confirmed the difficulty to draw robust conclusions about the tangible contribution of seasonal variations in species discrimination. The most important dates were highly variable from one year to another because of the irregular acquisition dates. For instance, compared to 2008, the maximum classification accuracy of 2011 (OA = 0.62) was statistically equivalent, for the same number of dates (5 dates) after feature selection. However, the image ranking was quite different with more dates selected in autumn in 2008 compared to 2011 (Figure 13a,b). In 2013, the first two most important dates enabling to reach a similar level of accuracy (OA $\approx$ 0.62) were acquired in winter and summer which is still different in 2014 (Figure 13c,d; see Appendix E for full results). What appears to be constant in this additional analysis is the limited number of dates (around five) to reach maximum accuracy and the decrease in accuracy using all the dates. The identification of tree species is improved with the use of multitemporal images compared to single-date image but in a most effective way when feature selection is applied before training. The use of fewer features that contain the maximum discrimination information about the tree species classes is better than the use of all the features of which many of them could be correlated or irrelevant because of noise. These results confirms that the Hughes effect can occur using SVM, as already observed by [76]. This is also in line with the Zhu and Liu's recommendations of selecting the most discriminative features before classifying forest types using dense time series [24].

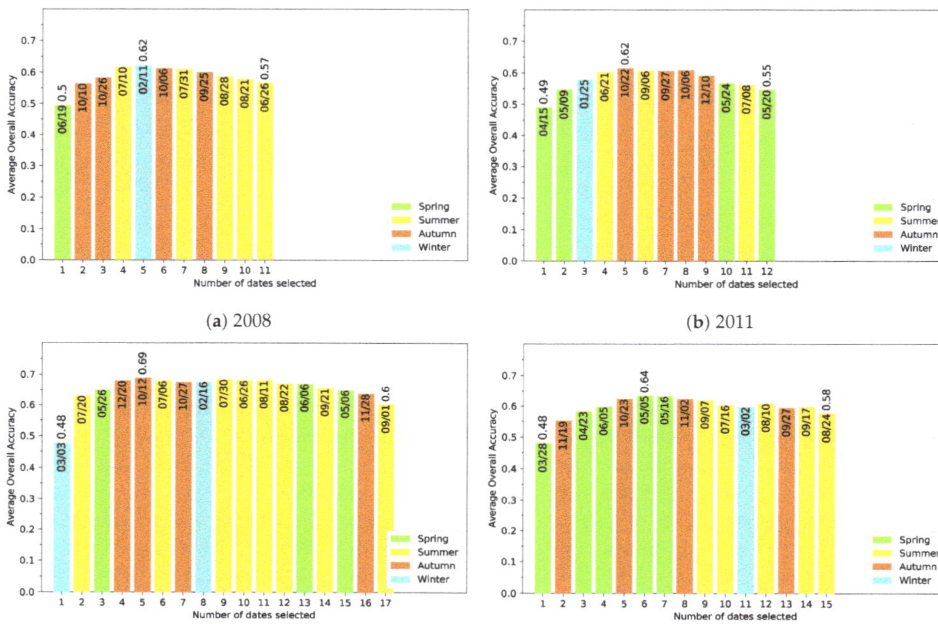

(a) 2008 (b) 2011

(c) 2013 (d) 2014

Figure 13. Ranking-based feature selection of image dates for single-year classification of 2008 (**a**), 2011 (**b**), 2013 (**c**), 2014 (**d**).

In a study, a combination of three aerial images acquired in spring (March 17th), summer (July 16th) and autumn (October 27th) provided the highest classification accuracy compared to all possible combinations based on five dates [30]. When only one image was selected, autumn appeared to be the best period to distinguish between common Oak, English Oak, Field Maple, Silver birch, Aspen and small-leaved Elm. In other recent studies based on Sentinel-2 data, the optimal single date was found in May [33,34]. The same observation was made when discrimating deciduous tree species in a multitemporal image dataset acquired with an unmanned aerial system [31]. When more images were combined, the best datasets included data acquired in different seasons (spring, early summer and autumn) in accordance with [30].

5.5. *Differences between Species*

We found more difficulty to separate conifers than broadleaf species as previously highlighted in other studies when multitemporal data are used [13,26,34,38,78]. Seasonal changes are less pronounced which induces higher overlapping between spectro-temporal profiles. Pine species were the harder to identified (in particular Corsican and Black pines). Among broadleaf species, European Ash and

Silver birch were the most confused (contrary to [13]). The best agreements were obtained for Red Oak, Eucalyptus, Willow and Aspen, in line with [31] for the latter species.

The evergreen phenology of Eucalyptus explains its high rate of agreements among 9 years despite a medium F1-score (average of all years OA = 65%). Because of differences in morphological and anatomical traits, optical properties also differ from those of conifers. The Red oak phenology is also specific, in particular in autumn when leaves turn red due to the production of anthocyanins. This gives a spectral characteristic which help to recognize them among the other species. For Willow, stands are located in well-suited humid areas sometimes waterlogged. The variable moisture conditions associated with a partially recovering canopy provide them a weaker reflectance in the near-infrared band compared to the other broadleaf species which may explain the good classification accuracy. For other species that cannot easily be separated, various factors may be involved, in addition to close spectral signatures. Forest managing practice is one of them. Stand age, density and the existence of understory vegetation are others. Spectral disparity for a given species (intra-species variability) may also influence the classification [31].

6. Conclusions

This study based on temperate forests in France is the first to explore the stability of tree species classification over nine consecutive years using dense high spatial resolution SITS with spatially uncorrelated validation data. The study was based on surface reflectance products derived from Formosat-2 optical time series acquired at irregular intervals from 2006 to 2014. Despite close statistical results in terms of classification accuracy, we observed high spatial disparities from one year to the next reflecting the moderate ability to predict tree species at the pixel level because of various disturbing factors.

Based on our findings, several conclusions can be drawn:

1. Spatial autocorrelation within validation data drastically overestimates the classification accuracy. In our context, an average optimistic bias of 0.4 of OA is observed when spatial dependence remained (LOO-CV strategy vs SLOO-CV). In further studies, we recommend adapting the data-splitting procedure to systematically reduce or eliminate spatial autocorrelation in the validation set in order to provide more robust conclusions about the true predictive performance.
2. Noise in the time series (i.e., undetected clouds and shadows) affects the SVM based classification performances. Despite accurate masks of clouds and shadows and a gap-filling approach to correct invalid pixels, residual noise impacts the learning and prediction processes. Feature selection is a good option to ignore noisy data, reduce data dimension, and to find the optimal subset of images for classification. There is a clear benefit (+0.08 of OA in average) of using fewer images containing the maximum discrimination information about the tree species classes.
3. The use of multitemporal images improves the tree species discrimination compared to single-date image. However, there is no clear evidence that the positive effect is really due to phenological differences between species. The most important dates varied from one year to another with no strong preference for images acquired at the key seasons.
4. The monospecific broadleaf plantations of Aspen, Red Oak and Eucalyptus are the easiest to classify. Conifers are the most difficult. The lowest accuracy was obtained for Silver birch, European ash and Black pines for which only a few forest stands were available.

Perspectives of this study are twofold. The first one is the use of S2 time series to confirm the results and assess the contribution of additional spectral bands such as the red-edge to separate tree species for the same in situ dataset and area. With its 5-day revisit time, S2 provides many more data in one year. These new time series should help better identify the best combination of multitemporal images and check that the combination is consistent with phenological events of the tree species concerned. Work is in progress to collect ground phenological observations on the study site. S2 also offers the possibility to work at a larger scale and will thus give us more reference pixels to reduce the bias due to the spatial autocorrelation. The second is related to the Formosat-2 time series. Annual datasets could be combined to reconstruct a synthesized multiyear time series based on all cloud-free images to combine all the phenological events of the species into one representative year.

In order to reproduce this study or to have tree species ground references suitable for remote sensing, our reference samples are publicly available at Zenodo, the Open Science platform at the CERN Data (https://doi.org/10.5281/zenodo.2581400). An interactive web version of the predicted species map using the 10 SITS with the Spatial Leave-One-Out cross-validation method (SLOO-CV) is available online (https://dynafor1201.github.io/publications/maps/treespeciesformosat2/).

Author Contributions: Conceptualization, N.K. and D.S.; Data curation, N.K., D.S. and J.W.; Methodology, N.K., D.S. and M.F.; Investigation, N.K.; Validation, N.K. and D.S.; Visualization, N.K.; Software, N.K.; Funding acquisition, D.S.; Supervision, D.S., C.M. and J.-F.D.; Writing—original draft, N.K. and D.S.; Writing—review & editing, J.-F.D., M.F. and C.M.

Funding: N.K. received a PhD scholarship from the French Ministry of Higher Education and Research (University of Toulouse). Field work was carried out as part of TOSCA OSO and TOSCA HyperBIO projects funded by the French Space Agency CNES.

Acknowledgments: Special thanks to O. Hagolle and M. Huc from CESBIO Lab. for providing the pre-processed Formosat-2 time series with masks of clouds and cloud shadows.

Conflicts of Interest: The authors declare no conflict of interest. The funders had no role in the design of the study; in the collection, analyses, or interpretation of data; in the writing of the manuscript, or in the decision to publish the results.

Appendix A. Tree Species Map

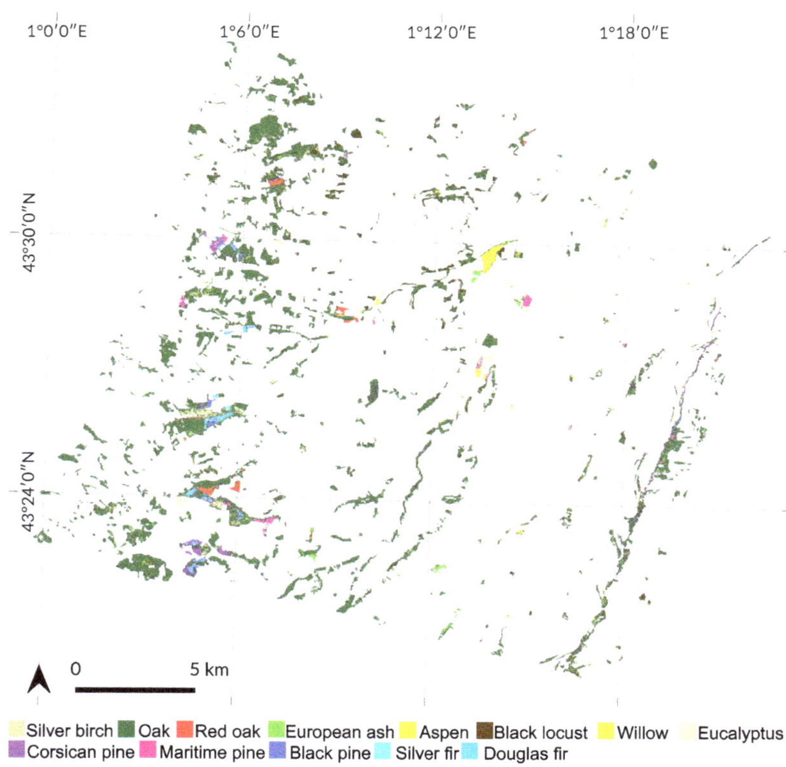

Figure A1. Map of the most predicted species using the 10 SITS with the Spatial Leave-One-Out cross-validation method (SLOO-CV).

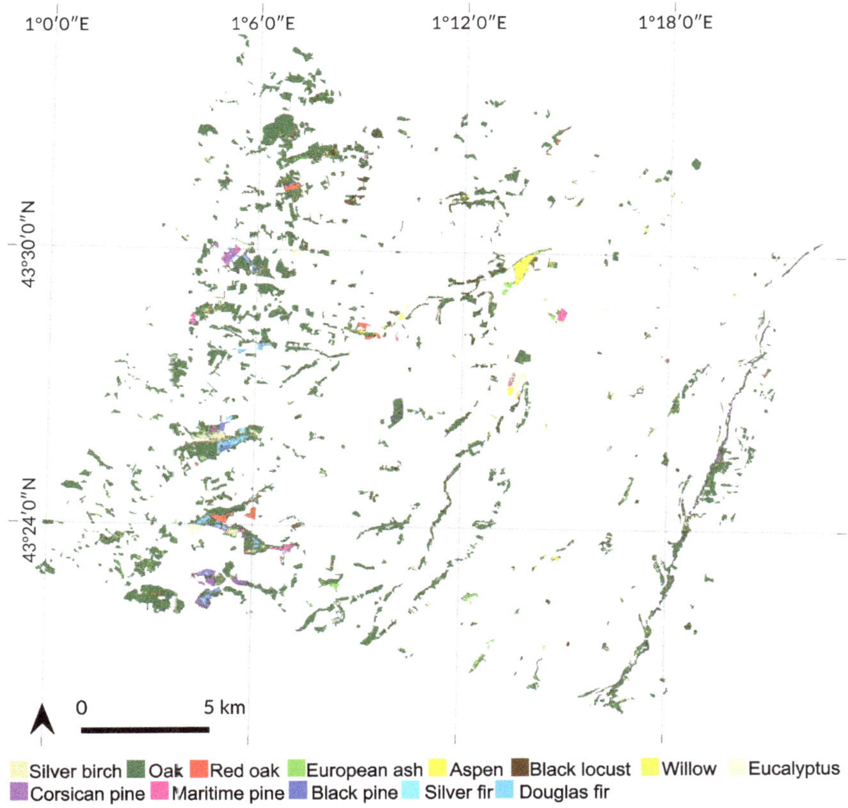

Figure A2. Map of the most predicted species using the 10 SITS with the Leave-One-Out cross-validation method (LOO-CV).

Appendix B. Significance Tables for Prediction between Years

Table A1. Wilcoxon signed-rank test significance table for SLOO-CV computed from the overall accuracy of the 50 predictions for each single-year SITS. In **bold**, where $p < 0.05$.

	2006	2006bis	2007	2008	2009	2010	2011	2012	2013	2014
2006	*nan*	**244**	242	255	302	354	313	**194**	**193**	**309**
2006bis	**244**	*nan*	**198**	551	335	407	307	556	384	459
2007	242	**198**	*nan*	**143**	**112**	**90**	**208**	**120**	**57**	**144**
2008	**255**	551	**143**	*nan*	349	345	365	433	275	257
2009	302	335	**112**	349	*nan*	288	299	348	**173**	229
2010	354	407	**90**	345	288	*nan*	382	284	**215**	384
2011	313	307	**208**	365	299	382	*nan*	315	**166**	**250**
2012	**194**	556	**120**	433	348	284	315	*nan*	356	451
2013	**193**	384	**57**	275	**173**	**215**	**166**	356	*nan*	360
2014	**309**	459	**144**	257	229	384	**250**	451	360	*nan*

Table A2. Wilcoxon signed-rank test significance table for LOO-CV computed from the overall accuracy of the 50 predictions for each single-year SITS. In **bold**, where $p < 0.05$.

	2006	**2006bis**	**2007**	**2008**	**2009**	**2010**	**2011**	**2012**	**2013**	**2014**
2006	*nan*	0	0	**0**	**0**	**0**	**0**	**0**	0	6
2006bis	0	*nan*	6	20	18	30	**29**	30	10	7
2007	0	6	*nan*	**4**	10	**0**	**6**	**5**	0	6
2008	**0**	20	**4**	*nan*	22	27	25	42	**4**	**5**
2009	**0**	18	10	22	*nan*	30	11	32	16	11
2010	**0**	30	**0**	27	30	*nan*	16	30	**0**	**12**
2011	**0**	**29**	**6**	25	**11**	16	*nan*	16	**6**	**7**
2012	**0**	30	**5**	42	32	30	16	*nan*	**4**	**13**
2013	0	10	0	**4**	16	**0**	**6**	**4**	*nan*	7
2014	6	7	6	**5**	11	**12**	**7**	**13**	7	*nan*

Appendix C. Effect of Clouds and Cloud Shadows

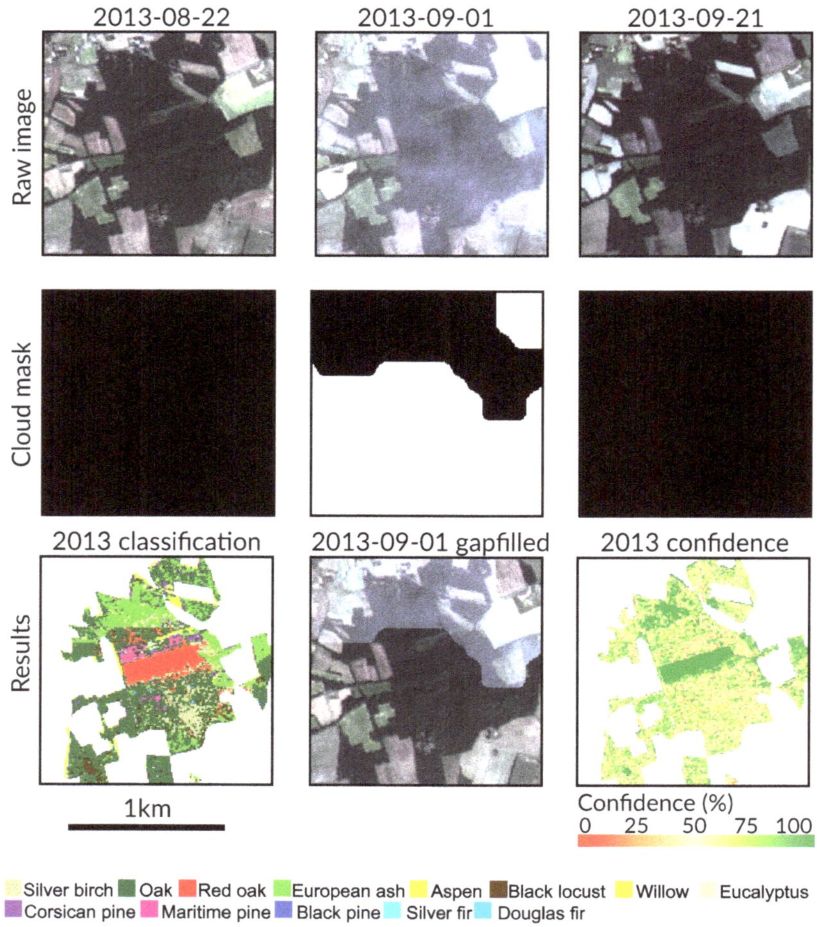

Figure A3. Example of misclassified forest stand (for European ash, in the north) due to under-detected clouds in 1 September 2019. In the cloud mask, pixels in black are cloud and shadow free. Pixels in white are cloudy or shady.

Table A3. Percent of reference pixels affected by clouds or shadows detected by the MACCS algorithm for each species and each year.

Species	2006	2006bis	2007	2008	2009	2010	2011	2012	2013	2014
Silver birch	26	0	17	1	9	0	0	0	4	0
Oak	24	0	20	11	6	5	1	0	2	0
Red Oak	21	0	11	0	1	6	3	0	0	0
Aspen	29	0	8	0	0	1	4	3	0	0
European Ash	22	0	11	6	5	1	0	0	3	1
Black locust	25	0	11	1	0	0	5	0	2	0
Willow	27	0	6	2	2	0	6	2	0	0
Eucalyptus	23	0	7	0	0	5	4	0	2	0
Corsican Pine	16	0	16	11	2	5	5	0	1	4
Maritime Pine	25	0	9	9	3	3	2	0	0	0
Black Pine	30	0	17	5	10	1	0	0	2	0
Silver Fir	28	0	17	5	8	5	0	0	3	0
Douglas	26	0	10	9	5	4	4	0	0	0

Appendix D. Training Size per Species

Table A4. Number of training pixels for each cross-validation method. For both methods, value per species is the mean from the 50-folds. The split between test and train set is exactly the same for each year.

Species	SLOO-CV	LOO-CV
	Broadleaf	
Silver birch	35	35
Oak	97	97
Red oak	118	118
Aspen	142	142
European ash	50	50
Black locust	50	50
Willow	21	21
Eucalyptus	85	85
	Conifer	
Corsican pine	33	33
Maritime pine	79	79
Black pine	26	26
Silver fir	54	54
Douglas fir	46	46
Total	836	836

Appendix E. Ranking-based feature selection of Image Dates for Each Single-Year Classification

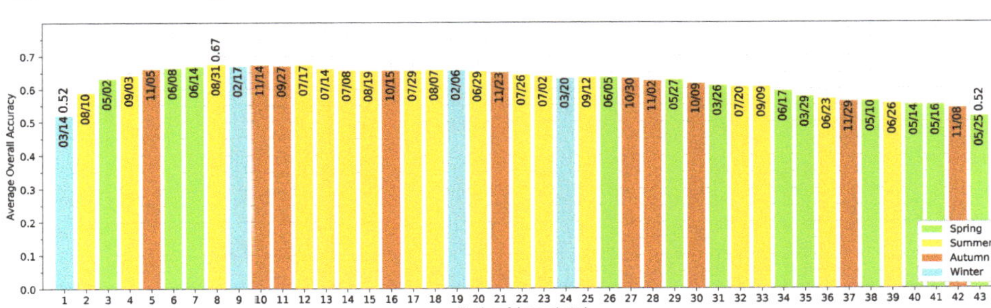

(a) 2006

Figure A4. *Cont.*

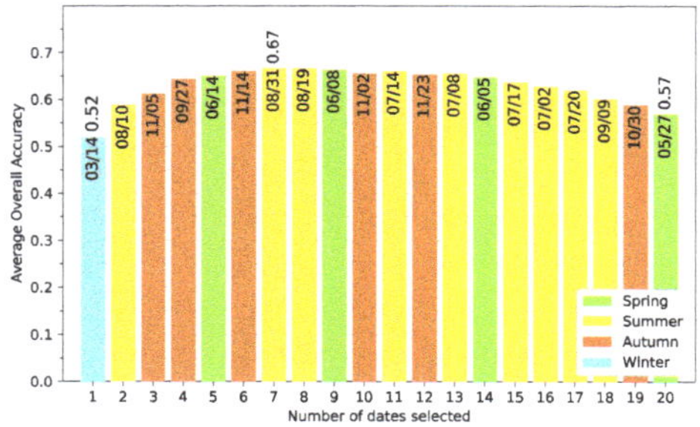

(**b**) 2006bis

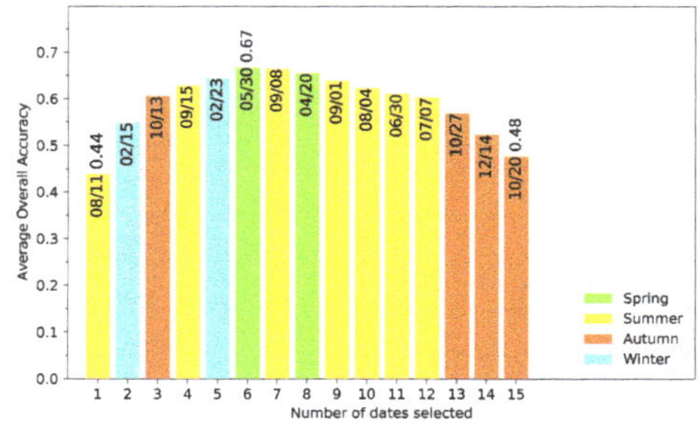

(**c**) 2007

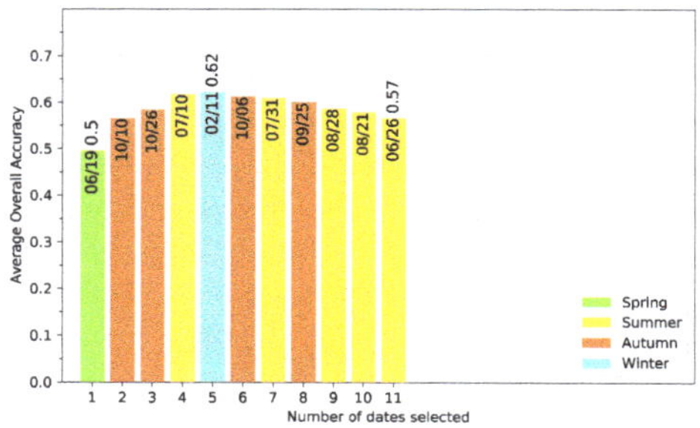

(**d**) 2008

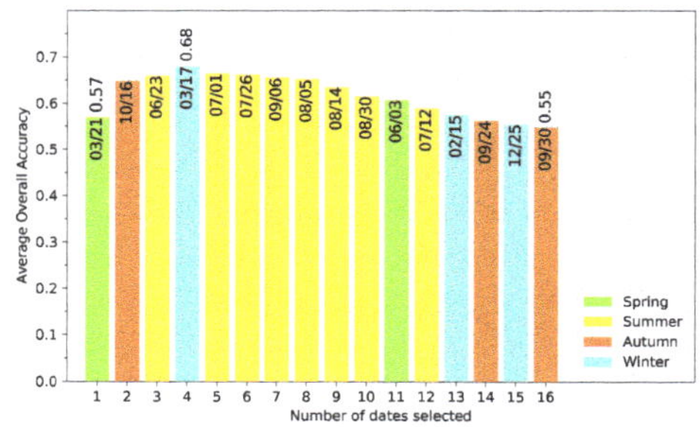

(**e**) 2009

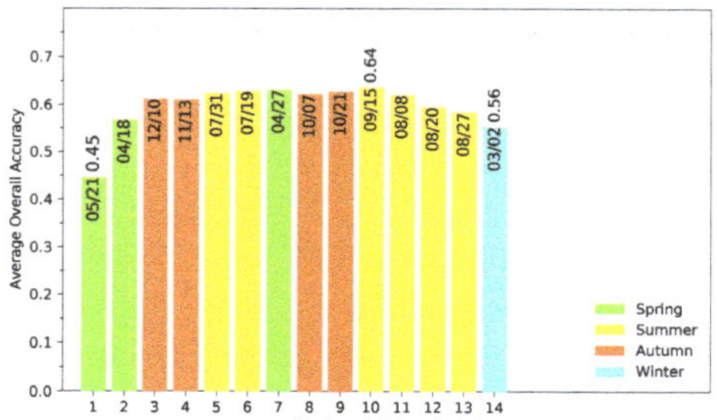

(**f**) 2010

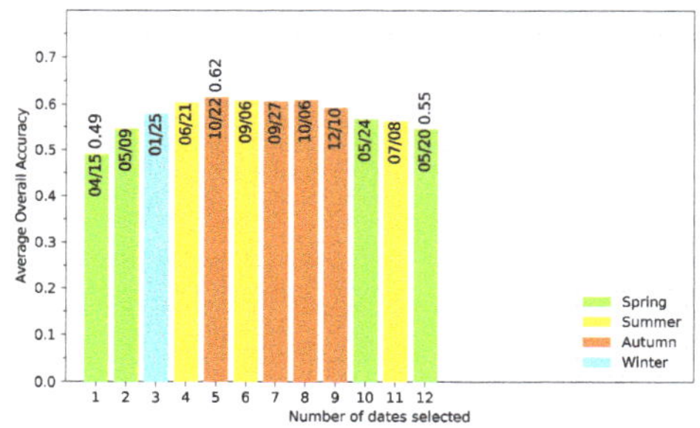

(**g**) 2011

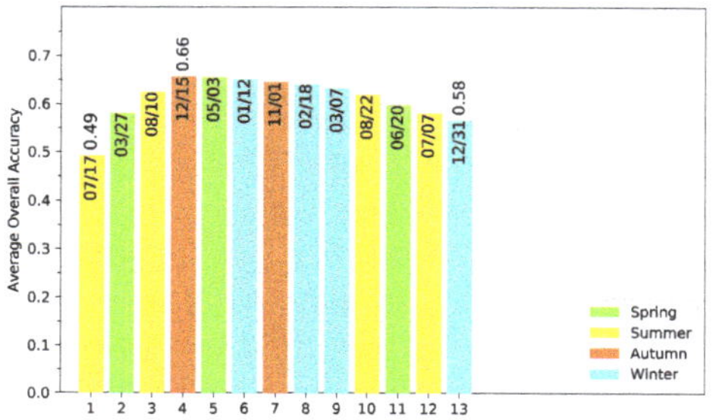

(**h**) 2012

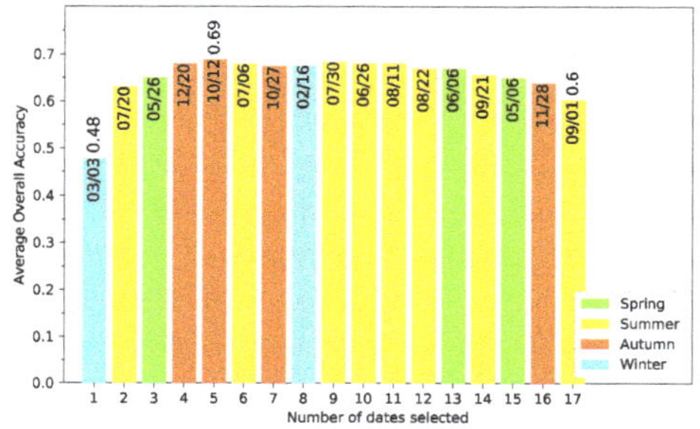

(**i**) 2013

Figure A4. *Cont.*

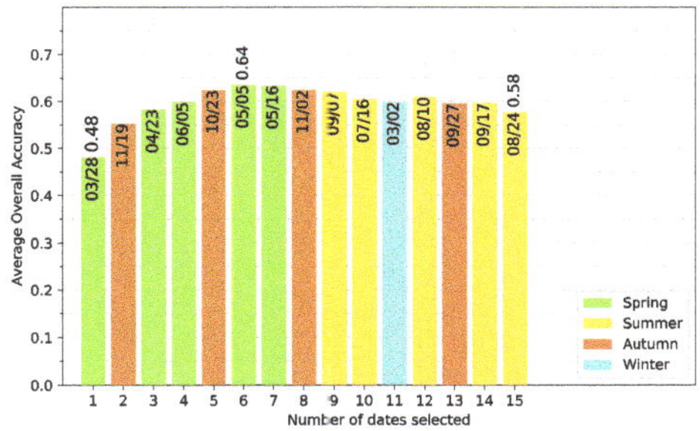

(j) 2014

Figure A4. Ranking-based feature selection of image dates for single-year classification.

References

1. Thompson, I.D.; Okabe, K.; Tylianakis, J.M.; Kumar, P.; Brockerhoff, E.G.; Schellhorn, N.A.; Parrotta, J.A.; Nasi, R. Forest Biodiversity and the Delivery of Ecosystem Goods and Services: Translating Science into Policy. *BioScience* **2011**, *61*, 972–981. [CrossRef]
2. Bunker, D.E.; Declerck, F.; Bradford, J.C.; Colwell, R.K.; Perfecto, I.; Phillips, O.L.; Sankaran, M.; Naeem, S. Species loss and aboveground carbon storage in a tropical forest. *Science* **2005**, *310*, 1029–1031. [CrossRef] [PubMed]
3. Thompson, I.D.; Mackey, B.; McNulty, S.; Mosseler, A. Forest resilience, biodiversity, and climate change. In *A synthesis of the Biodiversity/Resilience/Stability Relationship in Forest Ecosystems*; Technical Series No. 43; Secretariat of the Convention on Biological Diversity: Montreal, QC, Canada, 2009.
4. Harris, J. Soil microbial communities and restoration ecology: Facilitators or followers? *Science* **2009**, *325*, 573–574. [CrossRef] [PubMed]
5. Gamfeldt, L.; Snäll, T.; Bagchi, R.; Jonsson, M.; Gustafsson, L.; Kjellander, P.; Ruiz-Jaen, M.C.; Fröberg, M.; Stendahl, J.; Philipson, C.D.; et al. Higher levels of multiple ecosystem services are found in forests with more tree species. *Nat. Commun.* **2013**, *4*, 1340. [CrossRef]
6. Seidl, R.; Thom, D.; Kautz, M.; Martin-Benito, D.; Peltoniemi, M.; Vacchiano, G.; Wild, J.; Ascoli, D.; Petr, M.; Honkaniemi, J.; et al. Forest disturbances under climate change. *Nat. Clim. Chang.* **2017**, *7*, 395–402. [CrossRef]
7. Boyd, D.S.; Danson, F. Satellite remote sensing of forest resources: Three decades of research development. *Prog. Phys. Geogr.* **2005**, *29*, 1–26. [CrossRef]
8. Walsh, S.J. Coniferous tree species mapping using LANDSAT data. *Remote Sens. Environ.* **1980**, *9*, 11–26. [CrossRef]
9. Fassnacht, F.E.; Latifi, H.; Stereńczak, K.; Modzelewska, A.; Lefsky, M.; Waser, L.T.; Straub, C.; Ghosh, A. Review of studies on tree species classification from remotely sensed data. *Remote Sens. Environ.* **2016**, *186*, 64–87. [CrossRef]
10. Meyera, P.; Staenzb, K.; Ittena, K.I. Semi-automated procedures for tree species identification in high spatial resolution data from digitized colour infrared-aerial photography. *ISPRS J. Photogramm. Remote Sens.* **1996**, *51*, 5–16. [CrossRef]
11. Trichon, V.; Julien, M.P. Tree species identification on large-scale aerial photographs in a tropical rain forest, French Guiana—Application for management and conservation. *For. Ecol. Manag.* **2006**, *225*, 51–61.
12. Waser, L.; Ginzler, C.; Kuechler, M.; Baltsavias, E.; Hurni, L. Semi-automatic classification of tree species in different forest ecosystems by spectral and geometric variables derived from Airborne Digital Sensor (ADS40) and RC30 data. *Remote Sens. Environ.* **2011**, *115*, 76–85. [CrossRef]
13. Immitzer, M.; Atzberger, C.; Koukal, T. Tree Species Classification with Random Forest Using Very High Spatial Resolution 8-Band WorldView-2 Satellite Data. *Remote Sens.* **2012**, *4*, 2661–2693. [CrossRef]
14. Carleer, A.; Wolff, E. Exploitation of very high resolution satellite data for tree species identification. *Photogramm. Eng. Remote Sens.* **2004**, *70*, 135–140. [CrossRef]

15. Lin, C.; Popescu, S.; Thomson, G.; Tsogt, K.; Chang, C. Classification of tree species in overstorey canopy of subtropical forest using QuickBird images. *PLoS ONE* **2015**, *10*, e0125554. [CrossRef]
16. Ustin, S.; Gitelson, A.; Jacquemoud, S.; Schaepman, M.; Asner, G.; Gamon, J.; Zarco-Tejada, P. Retrieval of Foliar Information about Plant Pigment Systems from High Resolution Spectroscopy. *Remote Sens. Environ.* **2009**, *113*, S67–S77. [CrossRef]
17. Ghiyamat, A.; Shafri, H. A review on hyperspectral remote sensing for homogeneous and heterogeneous forest biodiversity assessment. *Int. J. Remote Sens.* **2010**, *31*, 1837–1856. [CrossRef]
18. Féret, J.B.; Asner, G.P. Tree Species Discrimination in Tropical Forests Using Airborne Imaging Spectroscopy. *IEEE Trans. Geosci. Remote Sens.* **2012**, *51*, 73–84. [CrossRef]
19. Aval, J.; Fabre, S.; Zenou, E.; Sheeren, D.; Fauvel, M.; Briottet, X. Object-based fusion for urban tree species classification from hyperspectral, panchromatic and nDSM data. *Int. J. Remote Sens.* **2019**, *40*, 5339–5365. [CrossRef]
20. Cano, E.; Denux, J.P.; Bisquert, M.; Hubert-Moy, L.; Chéret, V. Improved forest-cover mapping based on MODIS time series and landscape stratification. *Int. J. Remote Sens.* **2017**, *38*, 1865–1888. [CrossRef]
21. Aragones, D.; Rodriguez-Galiano, V.; Caparros-Santiago, J.; Navarro-Cerrillo, R. Could land surface phenology be used to discriminate Mediterranean pine species? *Int. J. Appl. Earth Obs. Geoinf.* **2019**, *78*, 281–294. [CrossRef]
22. Wolter, P.; Mladenoff, D.; Host, G.; Crow, T. Improved forest classification in the northern Lake States using multi-temporal Landsat imagery. *Photogramm. Eng. Remote Sens.* **1995**, *61*, 1129–1143.
23. Foody, G.; Hill, R. Classification of tropical forest classes from Landsat TM data. *Int. J. Remote Sens.* **1996**, *17*, 2353–2367. [CrossRef]
24. Zhu, X.; Liu, D. Accurate mapping of forest types using dense seasonal Landsat time-series. *ISPRS J. Photogramm. Remote Sens.* **2014**, *96*, 1–11. [CrossRef]
25. Diao, C.; Wang, L. Incorporating plant phenological trajectory in exotic saltcedar detection with monthly time series of Landsat imagery. *Remote Sens. Environ.* **2016**, *182*, 60–71. [CrossRef]
26. Pasquarella, V.J.; Holden, C.; Woodcock, C. Improved mapping of forest type using spectral-temporal Landsat features. *Remote Sens. Environ.* **2018**, *210*, 193–207. [CrossRef]
27. Tigges, J.; Lakes, T.; Hostert, P. Urban vegetation classification: Benefits of multitemporal RapidEye satellite data. *Remote Sens. Environ.* **2013**, *136*, 66–75. [CrossRef]
28. He, Y.; Yang, J.; Caspersen, J.; Jones, T. An Operational Workflow of Deciduous-Dominated Forest Species Classification: Crown Delineation, Gap Elimination, and Object-Based Classification. *Remote Sens.* **2019**, *11*, 2078. [CrossRef]
29. Key, T.; Warner, T.; McGraw, J.; Fajvan, M. A Comparison of Multispectral and Multitemporal Information in High Spatial Resolution Imagery for Classification of Individual Tree Species in a Temperate Hardwood Forest. *Remote Sens. Environ.* **2001**, *75*, 100–112. [CrossRef]
30. Hill, R.A.; Wilson, A.; George, M.; Hinsley, S. Mapping tree species in temperate deciduous woodland using time-series multi-spectral data. *Appl. Veg. Sci.* **2010**, *13*, 86–99. [CrossRef]
31. Lisein, J.; Michez, A.; Claessens, H.; Lejeune, P. Discrimination of Deciduous Tree Species from Time Series of Unmanned Aerial System Imagery. *PLoS ONE* **2015**, *10*, e0141006. [CrossRef]
32. Immitzer, M.; Vuolo, F.; Atzberger, C. First experience with sentinel-2 data for crop and tree species classifications in central Europe. *Remote Sens.* **2016**, *8*, 166. [CrossRef]
33. Bolyn, C.; Michez, A.; Gaucher, P.; Lejeune, P.; Bonnet, S. Forest mapping and species composition using supervised per pixel classification of Sentinel-2 imagery. *Biotechnol. Agron. Soc. Environ.* **2018**, *22*, 16.
34. Persson, M.; Lindberg, E.; Reese, H. Tree Species Classification with Multi-Temporal Sentinel-2 Data. *Remote Sens.* **2018**, *10*, 1794. [CrossRef]
35. Liu, Y.; Gong, W.; Hu, X.; Gong, J. Forest Type Identification with Random Forest Using Sentinel-1A, Sentinel-2A, Multi-Temporal Landsat-8 and DEM Data. *Remote Sens.* **2018**, *10*, 946. [CrossRef]
36. Spracklen, B.D.; Spracklen, D.V. Identifying European Old-Growth Forests using Remote Sensing: A Study in the Ukrainian Carpathians. *Forests* **2019**, *10*, 127. [CrossRef]
37. Zhen, Z.; Quackenbush, L.J.; Stehman, S.V.; Zhang, L. Impact of training and validation sample selection on classification accuracy and accuracy assessment when using reference polygons in object-based classification. *Int. J. Remote Sens.* **2013**, *34*, 6914–6930. [CrossRef]

38. Sheeren, D.; Fauvel, M.; Josipović, V.; Lopes, M.; Planque, C.; Willm, J.; Dejoux, J.F. Tree Species Classification in Temperate Forests Using Formosat-2 Satellite Image Time Series. *Remote Sens.* **2016**, *8*, 734. [CrossRef]
39. Hammond, T.O.; Verbyla, D.L. Optimistic bias in classification accuracy assessment. *Int. J. Remote Sens.* **1996**, *17*, 1261–1266. [CrossRef]
40. Chen, D.; Wei, H. The effect of spatial autocorrelation and class proportion on the accuracy measures from different sampling designs. *ISPRS J. Photogramm. Remote Sens.* **2009**, *64*, 140–150. [CrossRef]
41. Meyer, H.; Reudenbach, C.; Hengl, T.; Katurji, M.; Nauss, T. Improving performance of spatio-temporal machine learning models using forward feature selection and target-oriented validation. *Environ. Model. Softw.* **2018**, *101*, 1–9. [CrossRef]
42. Schratz, P.; Muenchow, J.; Iturritxa, E.; Richter, J.; Brenning, A. Hyperparameter tuning and performance assessment of statistical and machine-learning algorithms using spatial data. *Ecol. Model.* **2019**, *406*, 109–120. [CrossRef]
43. Dedieu, G.; Karnieli, A.; Hagolle, O.; Jeanjean, H.; Cabot, F.; Ferrier, P.; Yaniv, Y. Venμs: A joint Israel-French Earth Observation scientific mission with High spatial and temporal resolution capabilities. In *Second Recent Advances in Quantitative Remote Sensing*; Sobrino, J.A., Ed.; Publicacions de la Universitat de València: Valencia, Spain, 2006; pp. 517–521.
44. Hagolle, O.; Dedieu, G.; Mougenot, B.; Debaecker, V.; Duchemin, B.; MEYGRET, A. Correction of aerosol effects on multi-temporal images acquired with constant viewing angles: Application to Formosat-2 images. *Remote Sens. Environ.* **2008**, *112*, 1689–1701. [CrossRef]
45. Hagolle, O.; Huc, M.; Pascual, D.; Dedieu, G. A multi-temporal method for cloud detection, applied to FORMOSAT-2, VENμS, LANDSAT and SENTINEL-2 images. *Remote Sens. Environ.* **2010**, *114*, 1747–1755. [CrossRef]
46. Hagolle, O.; Huc, M.; Pascual, D.; Dedieu, G. A multi-temporal and multi-spectral method to estimate aerosol optical thickness over land, for the atmospheric correction of FormoSat-2, LandSat, VENS and sentinel-2 images. *Remote Sens.* **2015**, *7*, 2668–2691. [CrossRef]
47. Grizonnet, M.; Michel, J.; Poughon, V.; Inglada, J.; Savinaud, M.; Cresson, R. Orfeo ToolBox: Open source processing of remote sensing images. *Open Geospat. Data Softw. Stand.* **2017**, *2*, 15. [CrossRef]
48. Kandasamy, S.; Baret, F.; Verger, A.; Neveux, P.; Weiss, M. A comparison of methods for smoothing and gap filling time series of remote sensing observations—Application to MODIS LAI products. *Biogeosciences* **2013**, *10*, 4055–4071. [CrossRef]
49. Vapnik, V.N. *Adaptive and Learning Systems for Signal Processing Communications, and Control*; Statistical Learning Theory; Wiley-Interscience: Hoboken, NJ, USA, 1998.
50. Mountrakis, G.; Im, J.; Ogole, C. Support vector machines in remote sensing: A review. *ISPRS J. Photogramm. Remote Sens.* **2011**, *66*, 247–259. [CrossRef]
51. Kavzoglu, T.; Colkesen, I. A kernel functions analysis for support vector machines for land cover classification. *Int. J. Appl. Earth Obs. Geoinf.* **2009**, *11*, 352–359. [CrossRef]
52. Graves, S.; Asner, G.; Martin, R.; Anderson, C.; Colgan, M.; Kalantari, L.; Bohlman, S. Tree Species Abundance Predictions in a Tropical Agricultural Landscape with a Supervised Classification Model and Imbalanced Data. *Remote Sens.* **2016**, *8*, 161. [CrossRef]
53. Pedregosa, F.; Varoquaux, G.; Gramfort, A.; Michel, V.; Thirion, B.; Grisel, O.; Blondel, M.; Prettenhofer, P.; Weiss, R.; Dubourg, V.; et al. Scikit-learn: Machine learning in Python. *J. Mach. Learn. Res.* **2011**, *12*, 2825–2830.
54. Le Rest, K.; Pinaud, D.; Monestiez, P.; Chadoeuf, J.; Bretagnolle, V. Spatial leave-one-out cross-validation for variable selection in the presence of spatial autocorrelation. *Glob. Ecol. Biogeogr.* **2014**, *23*, 811–820. [CrossRef]
55. Pohjankukka, J.; Pahikkala, T.; Nevalainen, P.; Heikkonen, J. Estimating the prediction performance of spatial models via spatial k-fold cross validation. *Int. J. Geogr. Inf. Sci.* **2017**, *31*, 2001–2019. [CrossRef]
56. Moran, P.A.P. Notes on Continuous Stochastic Phenomena. *Biometrika* **1950**, *37*, 17–23. [CrossRef] [PubMed]
57. Dale, M.R.; Fortin, M.J. *Spatial Analysis: A Guide For Ecologists*, 2nd ed.; Cambridge University Press: Cambridge, UK, 2014.
58. Karasiak, N.; Sheeren, D.; Fauvel, M.; Willm, J.; Dejoux, J.F.; Monteil, C. Mapping tree species of forests in southwest France using Sentinel-2 image time series. In Proceedings of the 2017 9th International Workshop on the Analysis of Multitemporal Remote Sensing Images (MultiTemp), Brugge, Belgium, 27–29 June 2017; pp. 1–4.

59. Congalton, R.G. A review of assessing the accuracy of classifications of remotely sensed data. *Remote Sens. Environ.* **1991**, *37*, 35–46. [CrossRef]
60. Griffith, D.A.; Chun, Y. Spatial Autocorrelation and Uncertainty Associated with Remotely-Sensed Data. *Remote Sens.* **2016**, *8*, 535. [CrossRef]
61. Mu, X.; Hu, M.; Song, W.; Ruan, G.; Ge, Y.; Wang, J.; Huang, S.; Yan, G. Evaluation of Sampling Methods for Validation of Remotely Sensed Fractional Vegetation Cover. *Remote Sens.* **2015**, *7*, 16164–16182. [CrossRef]
62. Roberts, D.R.; Bahn, V.; Ciuti, S.; Boyce, M.S.; Elith, J.; Guillera-Arroita, G.; Hauenstein, S.; Lahoz-Monfort, J.J.; Schröder, B.; Thuiller, W.; et al. Cross-validation strategies for data with temporal, spatial, hierarchical, or phylogenetic structure. *Ecography* **2017**, *40*, 913–929. [CrossRef]
63. Stehman, S. Sampling designs for accuracy assessment of land cover. *Int. J. Remote Sens.* **2009**, *30*, 5243–5272. [CrossRef]
64. Lyons, M.B.; Keith, D.A.; Phinn, S.R.; Mason, T.J.; Elith, J. A comparison of resampling methods for remote sensing classification and accuracy assessment. *Remote Sens. Environ.* **2018**, *208*, 145–153. [CrossRef]
65. Ramezan, C.A.; Warner, T.A.; Maxwell, A.E. Evaluation of Sampling and Cross-Validation Tuning Strategies for Regional-Scale Machine Learning Classification. *Remote Sens.* **2019**, *11*, 185. [CrossRef]
66. Millard, K.; Richardson, M. On the Importance of Training Data Sample Selection in Random Forest Image Classification: A Case Study in Peatland Ecosystem Mapping. *Remote Sens.* **2015**, *7*, 8489–8515. [CrossRef]
67. Brenning, A. Spatial cross-validation and bootstrap for the assessment of prediction rules in remote sensing: The R package sperrorest. In Proceedings of the 2012 IEEE International Geoscience and Remote Sensing Symposium, Munich, Germany, 22–27 July 2012; pp. 5372–5375.
68. Cánovas-García, F.; Alonso-Sarría, F.; Gomariz-Castillo, F.; Oñate-Valdivieso, F. Modification of the random forest algorithm to avoid statistical dependence problems when classifying remote sensing imagery. *Comput. Geosci.* **2017**, *103*, 1–11. [CrossRef]
69. Foody, G.M. Sample size determination for image classification accuracy assessment and comparison. *Int. J. Remote Sens.* **2009**, *30*, 5273–5291. [CrossRef]
70. Sun, Y.; Wong, A.K.; Kamel, M.S. Classification of imbalanced data: A review. *Int. J. Pattern Recognit. Artif. Intell.* **2009**, *23*, 687–719. [CrossRef]
71. Eilers, P. A perfect smoother. *Anal. Chem.* **2011**, *75*, 3299–3304. [CrossRef]
72. Hermance, J.F.; Jacob, R.W.; Bradley, B.A.; Mustard, J.F. Extracting Phenological Signals from Multiyear AVHRR NDVI Time Series: Framework for Applying High-Order Annual Splines with Roughness Damping. *IEEE Trans. Geosci. Remote Sens.* **2007**, *45*, 3264–3276. [CrossRef]
73. Green, A.A.; Berman, M.; Switzer, P.; Craig, M.D. A transformation for ordering multispectral data in terms of image quality with implications for noise removal. *IEEE Trans. Geosci. Remote Sens.* **1988**, *26*, 65–74. [CrossRef]
74. Hughes, G. On the mean accuracy of statistical pattern recognizers. *IEEE Trans. Inf. Theory* **1968**, *14*, 55–63. [CrossRef]
75. Melgani, F.; Bruzzone, L. Classification of hyperspectral remote sensing images with support vector machines. *IEEE Trans. Geosci. Remote Sens.* **2004**, *42*, 1778–1790. [CrossRef]
76. Pal, M.; Foody, G.M. Feature Selection for Classification of Hyperspectral Data by SVM. *IEEE Trans. Geosci. Remote Sens.* **2010**, *48*, 2297–2307. [CrossRef]
77. Whitney, A. A Direct Method of Nonparametric Measurement Selection. *IEEE Trans. Comput.* **1971**, *C-20*, 1100–1103. [CrossRef]
78. Hościło, A.; Lewandowska, A. Mapping Forest Type and Tree Species on a Regional Scale Using Multi-Temporal Sentinel-2 Data. *Remote Sens.* **2019**, *11*, 929. [CrossRef]

remote sensing

MDPI

Article

Inferring Species Diversity and Variability over Climatic Gradient with Spectral Diversity Metrics

Amrita N. Chaurasia [1], Maulik G. Dave [1], Reshma M. Parmar [1], Bimal Bhattacharya [2], Prashanth R. Marpu [3], Aditya Singh [4] and N. S. R. Krishnayya [1,*]

1 Ecology Laboratory, Department of Botany, The Maharaja Sayajirao University of Baroda, Vadodara 390002, Gujarat, India; amrita.n.c-botanyphd@msubaroda.ac.in (A.N.C.); maulikdave-botanyphd@msubaroda.ac.in (M.G.D.); reshma.p-botanyphd@msubaroda.ac.in (R.M.P.)

2 Space Applications Centre, Indian Space Research Organisation, Ahmedabad 380015, Gujarat, India; bkbhattacharya@sac.isro.gov.in

3 Department of Electrical Engineering and Computer Science, Khalifa University of Science and Technology, Abu Dhabi 127788, UAE; prashanth.marpu@ku.ac.ae

4 Department of Agricultural and Biological Engineering, University of Florida, P.O. Box 110570, Gainesville, FL 32611-0570, USA; aditya01@ufl.edu

* Correspondence: nsr.krishnayya-botany@msubaroda.ac.in

Received: 24 May 2020; Accepted: 25 June 2020; Published: 2 July 2020

Abstract: Filling in the void between forest ecology and remote sensing through monitoring biodiversity variables is of great interest. In this study, we utilized imaging spectroscopy data from the ISRO–NASA Airborne Visible InfraRed Imaging Spectrometer—Next Generation (AVIRIS-NG) India campaign to investigate how the measurements of biodiversity attributes of forests over wide areas can be augmented by synchronous field- and spectral-metrics. Three sites, Shoolpaneshwar Wildlife Sanctuary (SWS), Vansda National Park (VNP), and Mudumalai Tiger Reserve (MTR), spread over a climatic gradient (rainfall and temperature), were selected for this study. Abundant species maps of three sites were produced using a support vector machine (SVM) classifier with a 76–80% overall accuracy. These maps are a valuable input for forest resource management. Convex hull volume (CHV) is computed from the first three principal components of AVIRIS-NG spectra and used as a spectral diversity metric. It was observed that CHV increased with species numbers showing a positive correlation between species and spectral diversity. Additionally, it was observed that the abundant species show higher spectral diversity over species with lesser spread, provisionally revealing their functional diversity. This could be one of the many reasons for their expansive reach through adaptation to local conditions. Higher rainfall at MTR was shown to have a positive impact on species and spectral diversity as compared to SWS and VNP. Redundancy analysis explained 13–24% of the variance in abundant species distribution because of climatic gradient. Trends in spectral CHVs observed across the three sites of this study indicate that species assemblages may have strong local controls, and the patterns of co-occurrence are largely aligned along climatic gradient. Observed changes in species distribution and diversity metrics over climatic gradient can help in assessing these forests' responses to the projected dynamics of rainfall and temperature in the future.

Keywords: species diversity; spectral diversity; convex hull volume; AVIRIS-NG; tropical forests; ISRO–NASA campaign; climatic gradient

1. Introduction

Forest communities across the globe are constantly affected by a combination of biotic and abiotic factors that impact their composition, structure, and function [1]. Understanding how regional and global environmental forcing act as drivers towards impacting species abundances [2], and how different facets of biodiversity affect ecosystem functioning at different scales [3], are critical aspects of research. Owing to a longer lifespan, trees have the ability to adapt to spatio-temporal variations in environmental conditions [4], and also play a vital role in determining ecosystem structure and function by large contributions to biomass [5]. These are more important for tropical forests with larger diversity, and the understanding of mechanisms determining ecosystem structure and distributions is significantly lacking [6,7].

Field-based biodiversity estimates are generally used to measure species richness and abundance [8]. Trait variability within a species may not be reliably measured with field-based studies [9] as it requires collecting large samples of data over wider regions. Linking field and remote sensing-based metrics can help in getting better estimates of trait variability and species distribution, and ecological and remote sensing communities are making efforts to blend these disparate areas of research towards more effective biodiversity assessment [10]. It is reported that intra- and inter-specific variations in plant functional traits are normally reflective of environmental variation across the species' ranges of occurrence, and are indirectly observable as emergent patterns of spatial variations in trait characteristics [11]. Wide area mapping with remote sensing [12] provides a divergent approach over field methods of sampling to monitor biodiversity. Hill et al. [13] highlighted the possible application of hyperspectral remote sensing towards monitoring forest functional diversity by comparing spectral variability. Proximal and airborne hyperspectral sensors, such as the Airborne Visible InfraRed Imaging Spectrometer (AVIRIS and AVIRIS—Next Generation (AVIRIS-NG)) have shown immense promise and potential in exploring inter- and intra-species variability and divergence in trait characteristics [14–16]. However, there are limited empirical investigations, thereby restricting our ability to find general mechanisms that link traits and competition in the main functional realms of the plant world [17].

It has been observed that measures of variation in the spectral characteristics of a remotely sensed image are related to species richness [18,19]. The observed spectral heterogeneity can act as a proxy of species diversity and has been linked to variability in plant community assemblages [20] and environmental gradients [21]. As an example, Dahlin [22] has demonstrated that the convex hull volume (CHV) computed from the first three principal components of AVIRIS-NG spectral data can be used to reveal key insights on the relative importance of drivers of community assembly, even in the absence of additional data about plant functional traits. So, CHV not only reduces the dimensionality of spectral data through principal component analysis (PCA), but can also be used as a spectral diversity metric [23]. Wang and Gamon [10] observed that spectral diversity metrics can provide cues about the "dimensionality" of spectral data of a target area, and future work on these lines can bring clarity about regulations linked to the "plant phylogenetic–functional–spectral relationship" that may enhance understanding about the remote sensing of biodiversity. A myriad of factors impact plant growth, response, and performance, and reports indicate that variability in environmental settings, such as grazing, climate change [24], climatic gradient [25], and precipitation gradient [26], impact plant functional trait responses. The present study is an attempt to check how well spectral diversity, measured as CHV, can infer tropical tree species response to climate variability, such as different rainfall patterns and atmospheric temperature differences.

Some earlier studies presented a correlation between spectral diversity observed through remote sensing to species diversity and species response to climatic change in Amazonian forests [27] and North American biomes [28]. However, such studies are limited in the tropics and more so in regions with high heterogeneity and immense anthropogenic pressure. The tropics support most of the global diversity, and climate change has a significant impact on these diverse systems [1]. Climate impacted changes in the tree functional compositions of forests magnify the feedback of forest biomass [3]. Linking trait data to climate inputs can help in understanding how biodiversity and the associated functioning of the species are affected by climate change [29]. Other reports [30–32] show that tropical forests are directly affected by rainfall patterns. The forest ecosystems of India are not well studied using remote sensing data in spite of being diverse and vast [33], with a large number of areas listed as UN biodiversity hotspots [33]. Some earlier studies around the Godavari basin [34] and the

Eastern Ghats [35] reported that multispectral data can be utilized in mapping forest cover features. Gairola et al. [36] presented a review of methodologies for ecological studies and suggested that high resolution remote sensing data are better suited to study the Indian Himalayan region due to the high diversity and heterogeneity of the landscape. Wieczynski et al. [29] advocated for the inclusion of trait and abundance data of the forest cover of underrepresented regions of the world to have a better understanding of the effect of climate change on the overall aboveground biomass content. To address this ecologically important aspect, the abundant species mapping of forest cover using remote sensing becomes an important target to achieve. The tropical forests of India are exposed to highly variable rainfall dynamics and the corresponding changes in their structure, function, and distribution have not been fully studied. This study aims at filling some of the gaps in our understanding about Indian forests and the changes happening in the diversity and distribution of tree species in the context of climate variability.

Trees with bigger canopy sizes can be easily distinguished with the spatial resolution of airborne remote sensing sensors [10]. AVIRIS-NG data, over some of the forest cover of India, were used to map dominant species and diversity metrics at a community level [37]. This study builds on earlier work and explores the possibility of linking field observations and diversity metrics derived from remote sensing data to assess the variability due to rainfall and temperature (henceforth referred to as climatic gradient). To this end, we investigate the following questions:

- How species diversity correlates with spectral diversity calculated from remote sensing;
- Can measured CHVs assist in elucidating the intra- and inter-species variability of abundant species;
- To what extent climatic gradient impacts the distribution of species and the associated spectral diversity.

2. Materials and Methods

2.1. Site Description

Three forest sites in India, Shoolpaneshwar Wildlife Sanctuary (SWS), Vansda National Park (VNP) in Gujarat, and Mudumalai Tiger Reserve (MTR) in Tamil Nadu, were selected for this study (Figure 1). The three sites are designated as protected areas and experience highly variable rainfall patterns, along with increasing anthropogenic pressure (Table 1). Mean annual rainfall is different in the three zones, with the lowest record at SWS and the highest at MTR. A temperature gradient is also observed, with MTR recording the lowest mean values and the highest being at SWS, which is an opposite trend to that of rainfall (Table S1). Such variation in climate is expected to affect the species distribution, diversity, and abundance, which is of interest in this work. Different steps carried out in the study are schematically represented in Figure 2.

Table 1. Physiographic characters and decadal means of rainfall and temperature data of the three sites.

Sites	Location		Area (km^2)	Average Elevation (m)	Rainfall (mm)	Temperature (°C)
	Lat (°N)	Lon (°E)				
SWS	21.88	73.65	607.70	287	1210.66	27.00
VNP	20.82	73.44	23.99	169	1557.72	26.52
MTR	11.73	76.46	688.00	1233	1731.88	24.84

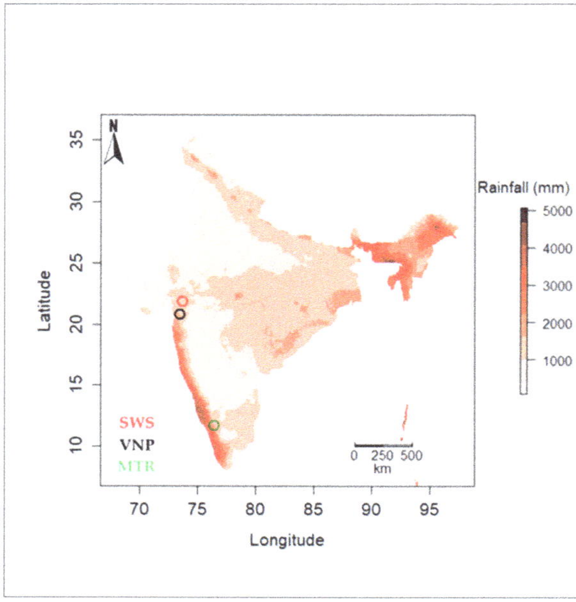

Figure 1. Map showing the location of the three study sites, Shoolpaneshwar Wildlife Sanctuary (SWS), Vansda National Park (VNP), and Mudumalai Tiger Reserve (MTR). The color gradient indicates decadal mean rainfall across the country.

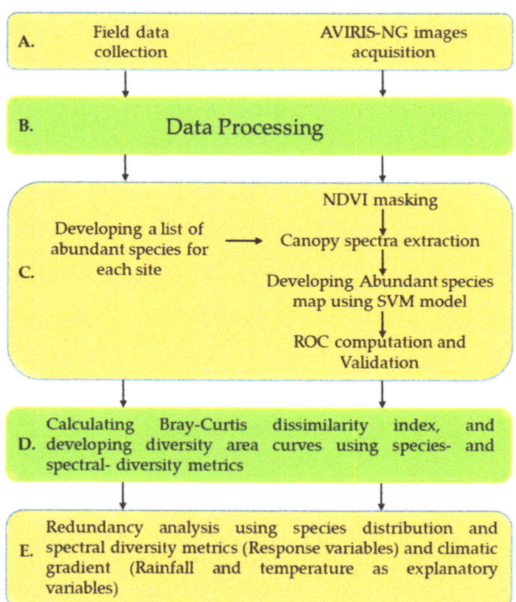

Figure 2. Schematic view of the workflow carried out.

2.2. Vegetation Characteristics

The forests of all three sites support tree species with diverse physiognomies and phenological characteristics. The tree species of SWS and VNP are largely deciduous with a few evergreens. MTR also supports both deciduous and evergreens, with a larger proportion of evergreen species (~25%). Plantations of teak and bamboo at SWS and VNP, and teak, bamboo, eucalyptus, and silver oak at MTR constitute sizable proportions of tree cover. Amongst the tree species recorded during field study at each site, some of the species showed widespread distribution (referred as abundant species), while others are sparse (referred to as lesser distributed species). Table S2 illustrates common tree species found in the three sites.

2.3. Field Data Collection

Initial field surveys were conducted ± 5 days from the image acquisition dates (Table 2) for each site. Field work includes the marking and registering of field plots, recording the phenological phases of tree species, and the collection of foliar samples from representative individuals of each species observed in these plots. Sites were revisited in subsequent months to perform detailed species inventories. Registered field plots and additional ones were laid across the length and width of the flight path at the three sites over accessible terrain (Figure 3). During the initial field visits, it was noticed that individual trees of some of the commonly occurring species were spread over > 4 m, and the expected spatial resolution from the flight campaign was 4–8 m. Keeping in view of the observed distribution of tree species in the field, and the suggestive spatial resolution of the sensor, quadrats of 8 m × 8 m were randomly laid down across trails of forest cover, some away from the trail (nearly 200 at each site). In addition to these, three to five polygons of varying sizes (500–750 m^2) were laid out. The number of quadrats at each site, and the size of the polygons considered, was primarily based on the observed vegetation characteristics and remarks of the local forest officials. All trees of more than 4 cm in diameter at breast height (dbh), falling within each quadrat and polygon, were counted. Tree species diversity, stem density, height, and canopy spread were measured and trees were identified to the species level with the help of field manuals, consultations with local residents, and forest service personnel. Tree crowns and field plots were geolocated using a handheld global positioning system (GPS) unit (Garmin Ltd. Olathe KS, USA) with < 3 m accuracy. The GPS readings were noted after three minutes of position averaging for better accuracy. Qualitative observations showed that the progression and duration of the senescent phase differed amongst species and across the three sites. The proportion of leaves fallen from the canopy also differed between species and the three sites. Few species were fully foliated with greener crowns.

Table 2. Summary of Airborne Visible InfraRed Imaging Spectrometer—Next Generation (AVIRIS-NG) image data.

Sites	Flight Elevation (km)	Image Acquisition Date	Image Acquisition Time (GMT +5:30)	Flight Footprint Length (km)	Cloud Cover
MTR	4.83	05-Jan-2016	10:62:47–12:81:39	42.22	Clear to hazy
SWS	4.15	08-Feb-2016	12:58:43–13:61:58	44.85	Clear
VNP	4.16	09-Feb-2016	11:02:17–11:57:46	13.32	Clear

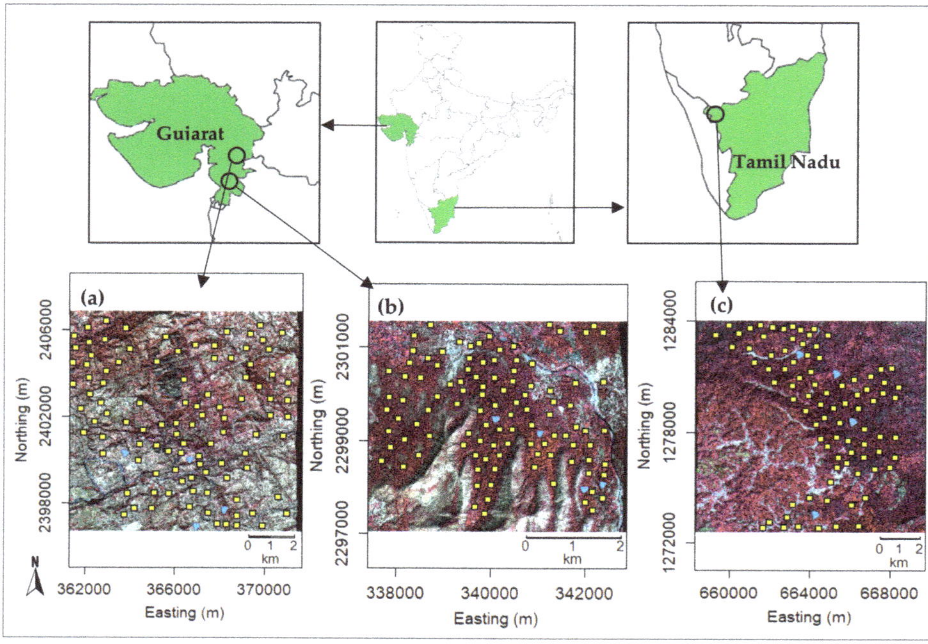

Figure 3. Sampled quadrats (yellow) and polygons (blue) established during field study. (**a**) SWS, (**b**) VNP, and (**c**) MTR.

2.4. Climatic Data

Following Diaz et al. [25], variability in the measurements of temperature and rainfall together are referred to as climatic gradient. We assessed two different datasets for sourcing temperature and rainfall data on annual, decadal, and long-term timescales from European ReAnalysis-Interim (ERA-Interim) and Climate Hazards Infrared Precipitation with Station data (CHIRPS). Data were projected to a spatial resolution of 5.12 km (0.05°) for CHIRPS data and 12.5 km (0.125°) for ERA-Interim data and extracted for all sampling locations at the three sites. The meteorological data from 1984–2016 indicated a spatial rainfall gradient from MTR (wetter) to VNP to SWS (lesser wetter, more arid). Annual, decadal, and long-term mean minimum and maximum temperatures had with minor variations at the three sites. Decadal mean rainfall and temperature at MTR are ~40% higher and ~4% lower compared to SWS (Table S1).

2.5. AVIRIS-NG Data Acquisition and Processing

As part of a joint campaign between the Indian Space Research Organization (ISRO) and the National Aeronautics and Space Administration (NASA, USA), the AVIRIS-NG sensor was flown over a large number of sites in India between October 2015 and February 2016. AVIRIS-NG acquired data at wavelengths from 380 to 2510 nm with 5 nm spectral sampling. The sensor was flown on a King Air B200 at a 4000–8000 m elevation, with resulting pixel sizes of 4–8 m, depending on flight altitude. The spatial resolution of the AVIRIS-NG data of the three sites considered in this study was 4 m. The instrument measurements were calibrated and atmospherically corrected following Thompson et al. [38]. Airborne imaging spectroscopy data were subject to significant sun–sensor surface illumination effects due to both topography and anisotropic reflectance by vegetation. These were corrected following methods in Soenen et al. [39] and Wanner et al. [40] (code available at https://github.com/EnSpec/HyTools-sandbox) resulting in topographically and bidirectional reflectance distribution function (BRDF)-corrected images. Details of the AVIRIS-NG image data acquisition are given in Table 2.

The topographically and BRDF-corrected data were mosaicked into contiguous blocks for all sites. Noisy bands (<411 nm) and water absorptions bands (1348–1428, 1778–1949 nm) were removed, resulting in final image spectral subsets comprising a total of 366 usable bands (from a total of 425). The normalized difference vegetation index (NDVI) of each image was used primarily to generate non-forest masks with thresholds of 0.4 (SWS and VNP) and 0.6 (MTR) based on visual observations. The difference in NDVI threshold values is because of the observed variability in canopy greenness and senescent phases at the time of the flight pass over the three sites. To lower the overall brightness variation, a continuum removal (CR) transformation was also applied (as in [22]). These CR-transformed data were considered for further processing.

$$NDVI = \frac{\rho_{860} - \rho_{650}}{\rho_{860} + \rho_{650}} \quad (1)$$

2.6. Data Analysis

2.6.1. Field Data Analysis

The data of the quadrats laid down in field plots were pooled (10 quadrats in each field plot), resulting in 20 field plots for SWS, 19 for VNP, and 21 for MTR. Abundance data of each site at the laid down quadrats and polygons indicated that ~20 species are abundant at each site, covering 75–80% of the forest of each site (henceforth referred to as abundant species). Species with lesser occupancy of each site are together referred to as lesser distributed species. The BiodiversityR package [41] has been used for analyses of field data. Kindt's exact method [41] was used for plotting tree species area accumulation curves using 1000 permutations for each site. These curves were developed for all the recorded species and the abundant species of each site. We tested different methods to obtain estimates of the number of tree species for each site (first- and second-order Jackknife, Chao, and Bootstrap), with an expectation that the tree species richness estimates of a site would be better characterized by using a range than a single one. Bray–Curtis dissimilarity graphs were developed to assess the dissimilarity in the composition of any pair of field plots of a site. Estimated values for the Bray–Curtis dissimilarity distance ranged between 0 and 1 depending on the 100% similarity (0) or complete dissimilarity (1) of existing species in a pair of plots. Akin to species–area curves, Bray-Curtis graphs were developed for all the recorded species and abundant species.

2.6.2. Abundant Species Mapping with the Support Vector Machine (SVM) Model

A spectral library of crown-level reflectance spectra for the abundant tree species of each site (20 species at SWS and VNP, 19 in MTR) was developed for all the three sites using field records. At many points, we could obtain abundant species having a canopy spread of > 5m or as pure patches with > 15 m spread (Table 3). These locations were utilized to extract the canopy spectra of each abundant species. Data from the canopy spectral library were used to develop species maps of each site. The classification of the images to map the abundant species required an abundant species map of each site to assess differences, if any, in these tree species distributions, and spectral diversity as CHVs, across the climatic gradient. Minimum noise fraction (MNF) transformation [42] was applied to the CR-transformed spectral dataset for feature reduction. A supervised classification of the abundant species of each site was performed using the support vector machine (SVM) method [43] using the MNF-transformed canopy spectra of each species. The parameters of the SVM model are optimized using three-fold cross-validation based on the available training data. After visual screening, the first 16 bands (for SWS and MTR) and the first 17 bands (for VNP) of MNF were selected as inputs for the SVM model to develop an abundant species map of the three sites. The field data were separated as training and testing data with a 60:40 ratio. For the SWS and VNP sites, additional field visits were carried out to validate randomly selected locations from the species map for an additional level of verification of classification accuracy.

Table 3. Measured biophysical parameters of the tree species at each site.

Sites	Height (m)			Canopy Area (m^2)		
	Mean (± SD)	Min	Max	Mean (± SD)	Min	Max
SWS	12.48 ± 4.80	2.60	25.80	55.91 ± 65.27	2.41	613.80
VNP	18.31 ± 6.30	3.50	35.00	71.89 ± 70.44	2.55	446.93
MTR	22.90 ± 7.56	6.00	39.00	75.45 ± 74.66	2.41	638.20

2.6.3. Spectral Diversity Area Curves and Metrics

Spectral diversity metrics were generated both from NDVI masked standalone images and from the abundant species maps of each site. Three different ways were followed to assess the applicability of spectral diversity metrics towards gauging tree species diversity. In the first approach, to understand the applicability of hyperspectral data in measuring diversity sans field study inputs, 175 squares (10 × 10 pixels) per site were randomly extracted from an NDVI masked image of each site. The CR-transformed spectral data of the extracted squares of each site were analyzed based on Dahlin [22]. Principal component analysis (PCA) was performed to reduce the dimensionality of the data with the resulting components accounting for most of the total variance. The first three principal components accounted for 94% of the variation in the recorded spectral data. Two spectral diversity measures, the sum of variance of the first three PCs and the convex hull volume value, were calculated [22]. The CHV quantifies the volume of the trait space occupied by species in a community regardless of the shape of the distribution and represents the multivariate equivalent of range [44]. The sum of variance of the first three PCs and CHV values derived from spectral data of this study were empirically considered as proxies towards showcasing the tree species functional trait diversity of each site, as noted in Dahlin [22]. These data were also used to verify whether the spectral diversity metric was altered in lieu of the changes in species diversity expected across the forest cover of each site.

In the second approach, to find out the relation between the field-measured species diversity of the plots and the spectral diversity of the remote sensing measurements over the same, the spectral data of areas falling within Ground Control Points (GCPs) locations of the plots (20 of SWS, 19 of VNP, and 21 of MTR) of each site were extracted and CHV values from the CR-transformed spectral data of each plot were calculated and used to make a moving average-based spectral diversity area curve. This curve represented the spectral diversity of the species recorded during field study.

In the third approach, a number of plots of similar sizes, coinciding with the field study, were selected from the abundant species maps of each site, spectral data were extracted, and CHVs were calculated. The patterns identified in this approach were considered as abundant species spectral diversity–area curves, as the spectral data were based on the derived classification map indicating abundant species in the regions of interest. To analyze the one-to-one relationship between the number of abundant species and cumulative spectral diversity, equal numbers of canopy spectra (n = 10) of each abundant species were successively added and CHVs were calculated. A linear regression line was drawn between species numbers and CHVs. In another way, equal sized plots (0.5 ha, n = 5) were chosen from subsets of abundant species maps with a varied number of species. The CHVs of extracted spectral data were calculated to observe whether the value changed with species number.

2.6.4. Intra- and Inter-Species Variability, CHV Values

The abundant species maps derived from the remote sensing data were used to test the strength of spectral data as proxies for functional trait diversity in explaining the impact of climatic gradient on tree inter- and intra-species variability. Fifty spectra of each of the common abundant species (five tree species) of three sites were extracted from each of the CHIRPS 0.05° grid cells based on the abundant species maps. The CHVs of the extracted spectral data of each of the common abundant species falling in CHIRPS grid cells at each site were calculated and plotted against CHIRPS decadal mean rainfall data. The expectation was that the CHV values of each of these five common tree species did not vary significantly without being impacted by the observed climatic gradient.

2.6.5. Redundancy Analysis

To look at the impact of climatic gradient on the tree species diversity, distribution, and spectral diversity of the three sites, redundancy analysis (RDA) was performed. RDA is a constrained ordination technique that directly relates the relative abundance of species across environmental gradients [41] and allows for establishing relationships between observed gradient of abundances vis-à-vis environmental drivers considered here as a climatic (rainfall–temperature) gradient. Analyses were performed using field plot data, the CHVs of plots for all recorded species, and abundant species (response variables) against recorded climatic gradient (explanatory variable).

3. Results

3.1. Tree Species Diversity, Distribution

A total of 123 tree species were recorded during field study at the three sites, of which 31 species were found to be common at all three sites. The recorded tree species richness was similar at VNP and SWS (68 and 70, respectively) but relatively higher at MTR (80 species). Amongst different the methods employed to estimate the species richness of each site, the Bootstrap method showed relatively closer values to the observed ones, as shown in Table S3. Estimates also showed a greater richness of species at MTR over the other two sites. A list of the common species at the three sites and the abundant species at each site are given in Tables S2 and S4. Species abundance data obtained from field plots was corroborated with observed distributions in larger polygons (Figure 2) laid down during field study, and were found to be matching. Log-based rank abundance curves of the abundant tree species, and all the recorded species of each site, showed variable width and similar slopes, indicating variability in the occurrence of these species (Figure S1). Lesser distributed species were in the lower part of the abundance curve, with many having the same rank. Site specificity was evident in the recorded tree species diversity, and the distribution of tree species, across the rainfall gradient. The most abundant and common tree species across three sites were *Ficus racemosa*, *Tectona grandis*, *Terminalia bellirica*, and *Terminalia tomentosa*. Bamboo (*Dendrocalamus strictus* at SWS and VNP and *Bambusa bambos* at MTR) was the fifth most common abundant species at the three sites. The relative distribution of common abundant species and their hierarchy, as seen in the abundant species maps, was not the same at the three sites (Figure 4). The distribution of *T. grandis* increased with an increase in rainfall, while bamboo species showed a reverse trend. *T. bellirica* showed lesser spread at VNP, while *T. tomentosa* showed higher spread, indicating intra-genus inconsistency in spread.

The phenology of deciduous and evergreen tree species showed diverse phases at the three sites at the time of the AVIRIS-NG flight pass. Differences within deciduous species were also noticed. Deciduous species at MTR appeared greener compared to the other two sites. The recorded biophysical parameters of trees, such as height and canopy spread, were higher at MTR compared to the other two sites. Marked trees with larger dimensions (Table 3), and the recorded clusters of some of the species during field study, made it possible to develop a canopy spectral library of abundant species. This effectively assisted in developing abundant species maps of each site with better accuracy.

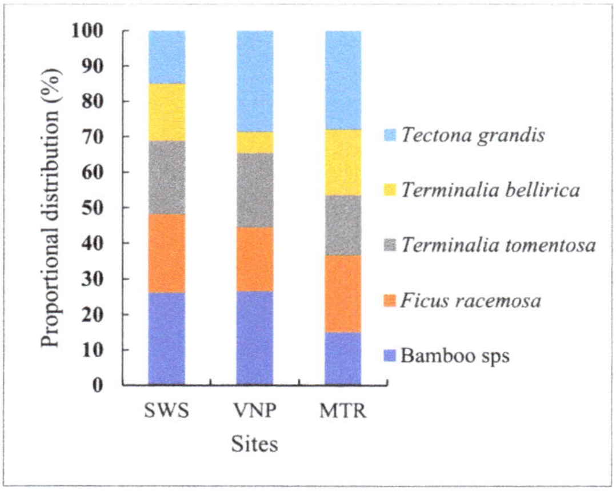

Figure 4. Proportional distribution of five common abundant species across the three sites from abundant species maps.

3.2. Abundant Species Maps

The support vector machine (SVM) classification model for the abundant species of each site performed well, with accuracies of 80.15%, 76.32%, and 78.54% and kappa 0.79, 0.75, and 0.76 for SWS, VNP, and MTR, respectively, as shown in Tables S5–S8. Accuracy measures were cross-verified with receiver operating characteristic (ROC) curves for each species, and the area under the curve (AUC) was calculated (Figure S2 and Table S9). The mean AUC value for each site was 0.99, 0.98, and 0.94 for SWS, VNP, and MTR, respectively, showing the subtlety and specificity of the accuracies of the developed species maps. These maps were used to inquire about how spectral diversity metrics derived from remote sensing data can reveal species diversity and distribution. Figure 5 shows how the abundant species were spread at each site. The spread of common species was distinctive and site-specific. Spectral diversity, calculated as CHV, increased with species number and their relationship showed high R^2 values (Figure 6), indicating that metrics of spectral data can be used/tested as a proxy for species diversity inputs for unexplored and relatively inaccessible forest cover. The CHVs of spectral data obtained from plots of 0.5 ha size selected from abundant species maps showing different species numbers increased with an increase in species numbers, as shown in Figure 6. Species specific to MTR, such as *C. tetraandra* and *F. tsjahela*, showed higher CHVs. Seed-propagated abundant species showed relatively higher CHV values compared to vegetation-propagated ones, as can be observed in Table S10. This supports the common understanding that diversity is usually greater in seed-propagated species compared to species showing vegetative propagation.

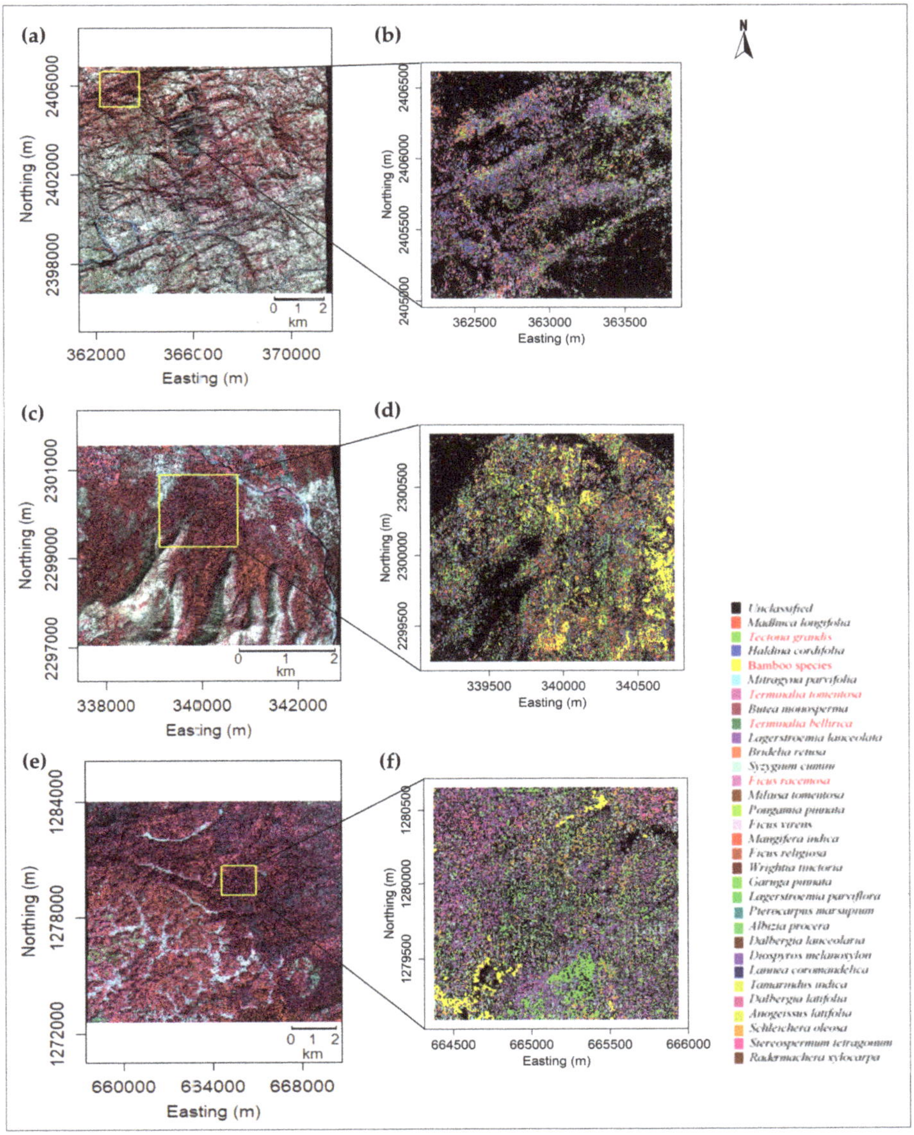

Figure 5. RGB image (R = 862 nm, G = 652 nm, B = 552 nm) and zoomed in subset of abundant species classification maps of the three sites (**a**,**b**: SWS, **c**,**d**: VNP, **e**,**f**: MTR). The abundant species color code is the same as for Figures 4 and 5. The names of common abundant species at the three sites are marked in red-colored font.

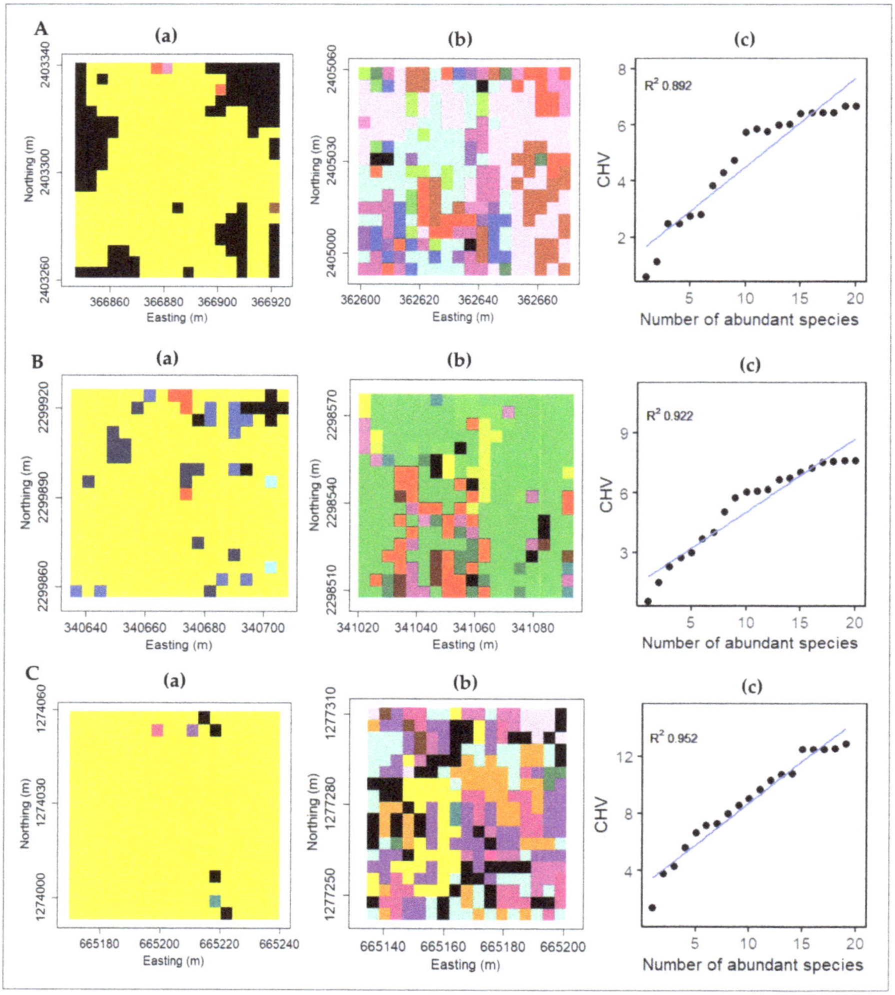

Figure 6. Convex Hull Volumes CHVs of 0.5 ha plots (n = 5) selected from abundant species maps of the three sites (**A**: SWS, **B**: VNP, **C**: MTR) with different numbers of occupant species; CHV values are: a = 2.41 (n = 4), b = 6.05 (n = 12) for SWS, a = 3.31 (n = 5), b = 5.65 (n = 12) for VNP, and a = 1.86 (n = 4), b = 6.1 (n = 11) for MTR; Regression lines are of cumulative CHVs and the number of abundant species. (**a**) represents 0.5 ha plot with lower diversity (n = 4 or 5); (**b**) represents 0.5 ha plot with higher diversity (n = 11 or 12); (**c**) Regression lines of cumulative CHVs.

3.3. Diversity–Area Curves

Species diversity–area curves revealed a higher diversity at MTR compared to SWS (Figure 7). Spectral diversity–area curves developed for the three sites by following three different approaches showed distinguishing patterns. Firstly, to test the approach given in Dahlin [22] for these lesser studied sites, the sum of variance of the first three PCs and CHV were used to compare spectral diversity within and across the three sites. This approach revealed the utility of remote sensing data in measuring biodiversity variables sans field inputs. The results of this approach account for the spectral diversity of standalone tree cover diversity in the forests of each site. Spectral diversity–area curves of the summed variance of 175 plots showed an increase in variance over the observed climatic gradient, with the lowest value at SWS and the highest at MTR (Figure 7). The CHVs of 175 plots at MTR were nearly twice the value of SWS (Table S11), indicating greater spectral diversity over the climatic gradient. Spectral diversity–area curves from the other two approaches also showed that spectral diversity increased over the observed climatic gradient. Abundant species at each site showed higher spectral diversity compared to the recorded tree species (Figure 7).

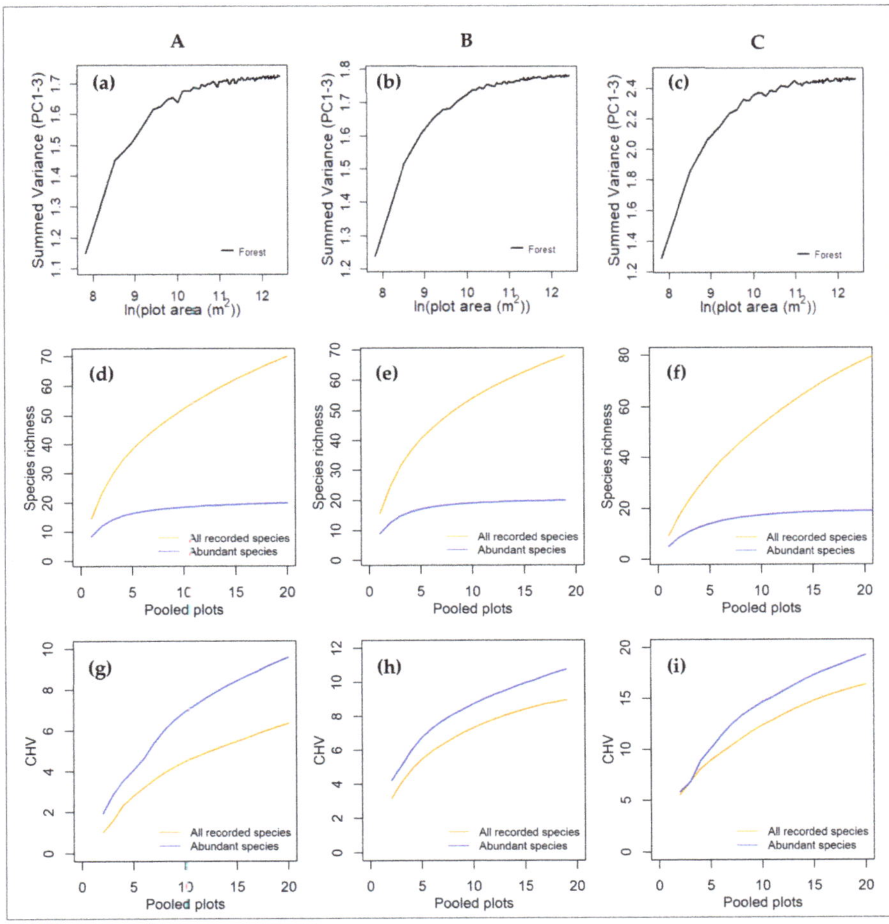

Figure 7. Diversity area curves (**A**: SWS, **B**: VNP, **C**: MTR). Spectral diversity–area curves for the summed variance of 175 plots (**a**–**c**); Species diversity–area curves (**d**–**f**); Cumulative CHVs and area curves (**g**–**i**).

Heat maps of Bray–Curtis dissimilarity graphs between plots developed for all the recorded species and abundant species showed differences in the values of species and spectral diversity (as CHVs) metrics. For the SWS and VNP sites, values measured from the spectral diversity metrics were higher compared to the ones coming from field metrics. At MTR, the values were the same for both field and spectral data metrics (Figure 8 and Figure S3).

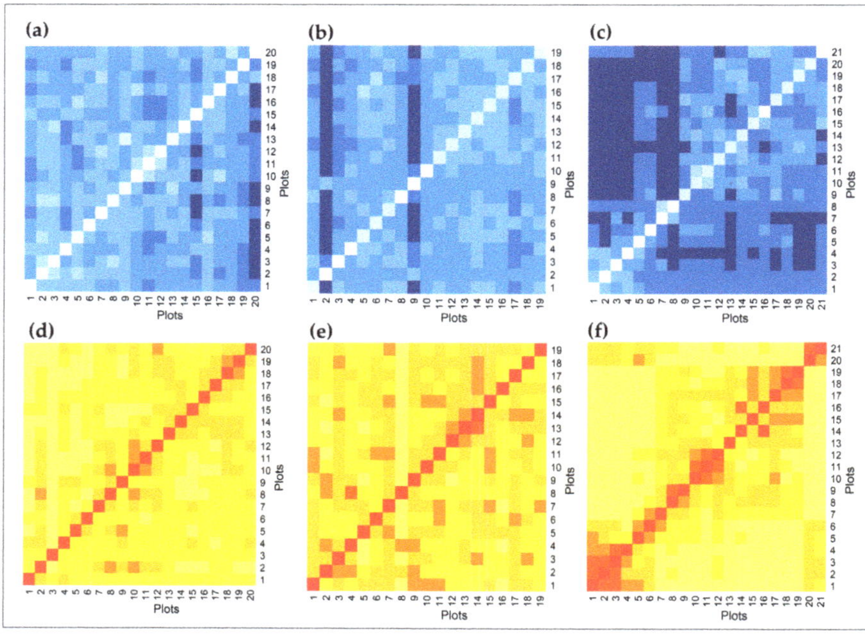

Figure 8. Heatmap of Bray–Curtis dissimilarity graphs of abundant species. The mean dissimilarity value of species abundance (blue) and that of spectral diversity as CHVs (yellow) are (**a**) 0.64 and (**d**) 0.81 for SWS, (**b**) 0.56 and (**e**) 0.68 for VNP, and (**c**) 0.81 and (**f**) 0.81 for MTR.

3.4. Intra- and Inter-Species Spectral Diversity, Rainfall Gradient

The abundant species maps developed from AVIRIS-NG datasets were found to be very useful in finding the intra- and inter-spectral variability of species spread over ~80% of cover at all three sites. The CHV values of the sets of 50 spectra of each common abundant species from each of the 0.05° CHIRPS grid cells across rainfall gradient showed that rainfall has a positive effect on intra-species CHV values, indicating a dynamic range in their functional diversity. CHVs increased over rainfall gradient consistently across the sites (R^2 0.86–0.98). Across species, CHV variability was relatively narrow for *T. grandis* (Figure 9) and substantially higher for *F. racemosa* and bamboo species. Inter-species variability was found to be relatively higher at MTR (Figure S4).

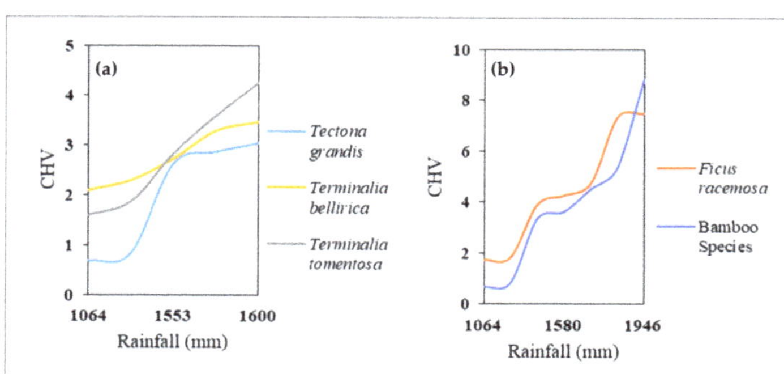

Figure 9. Common abundant species intra-species spectral variability (as seen in CHVs) across the climatic gradient of the three sites. (n = 5 in **a**, n = 7 in **b**).

3.5. Redundancy Analysis

Redundancy analysis was performed using climatic data as explanatory variables and species abundance and CHV values as response variables. RDA plots revealed that species are broadly spread across the climatic gradient. The climatic gradient explained 12–17% of the variance in the species and spectral diversity distribution of all recorded species at the three sites (Figure S5). The Spectral diversity metrics were observed to be more sensitive compared to the species abundance metrics by showing higher explained variance towards the climatic gradient. The field and spectral diversity metrics of the abundant species as response variables showed higher explained variance, as shown in Figure 10, indicating that 13–24% of the variance in the distribution of the species by site matrix was because of the climatic gradient.

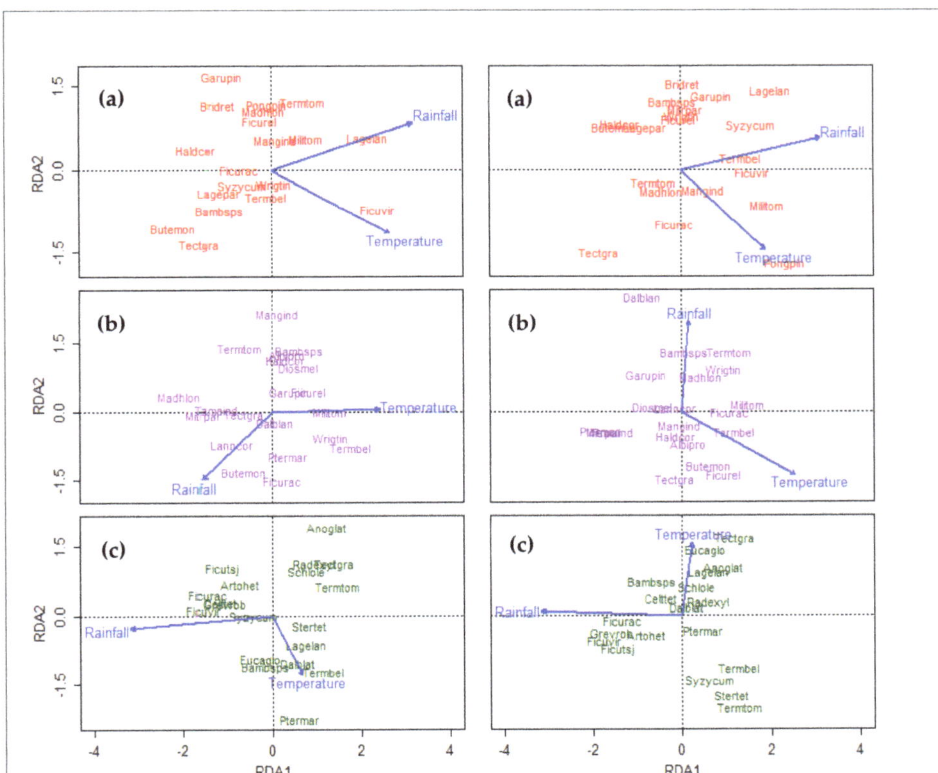

Figure 10. Redundancy analysis output of abundant species distribution values (left) and spectral diversity as CHVs (right) (considered as response variables) with climatic gradient (explanatory variable). The variance explained for species distribution ranged between 13–24%, and for spectral diversity as CHVs, varied between 14–23%. (**a**) SWS, (**b**) VNP, and (**c**) MTR.

3.6. Caveats

The level of uncertainty in spectral diversity measurements can be impacted by the classification accuracy of species maps and the available number of spectra for lesser distributed species. Our results indicate that the selected field plots numbers were reasonable for these sites. Other regions with variable species and spectral diversity metrics may require different sampling scales for correlating with the measured inputs.

4. Discussion

4.1. Tree Species Diversity Metrics

The present study focuses on measuring the tree diversity of three sites of a megadiverse tropical country, India, using field-based and remote sensing techniques, which is of importance to forest managers and conservationists [45]. Measured dissimilarity indices indicated that spectral diversity metrics can provide a better understanding of the diversity compared to sampling-based field study metrics. Amongst the tree species recorded (70–80 species), field-based metrics revealed that around a quarter of them are seen as abundant, and this pattern is referred to as an established attribute of communities [46]. Similar to previous findings [47,48], of the 31 common species of the three sites, 11 were found to co-occur as abundant species at MTR, 14 at VNP, and 15 at SWS, showing site- and climatic gradient-specific distributions, resulting in disparate inter-specific interactions. The differences in species composition, seen here across climatic gradient, are congruent to the studies reported earlier across rainfall gradients and drought [49,50]. A few of the species, such as *Gmelina arborea*, showed sparse distribution at all the three sites, indicating that some of the species inherently show restricted spread, irrespective of the observed climatic variability. These observed shifts in community composition can eventually determine the growth rates and productivity [3] of the three sites, which have a long-term impact on prevailing forest structure and function. The trends in spectral CHVs observed across the three sites of this study indicate that species assemblages may have strong local controls, and the corresponding patterns of co-occurrence are largely aligned along major climatic gradient. As observed earlier in [51], ecological and remote sensing products together, as reported in this study, can improve the verity of ecological models for effective forest management.

The congruence observed in the field- and spectra-based diversity metrics and the increase in the measured CHVs proportionate to abundant species per unit area (Figure 4) affirm the utility of spectral diversity metrics in measuring biodiversity. A similar association was reported between species richness and spectral diversity [52]. The higher rainfall observed at MTR resulted in a larger number of recorded tree species (by 15%), and significantly higher spectral diversity, as observed in the CHVs of abundant species (nearly twice that of the readings at SWS) compared to the other two sites. Spectral diversity positively increased across the climatic gradient, affirming the positive impact of rainfall on diversity. The flattening of the species–area curves appeared first for abundant species and later for all the recorded species. This pattern was reversed for spectral diversity–area curves, indicating that greater intra-species variability, as seen in the abundant species of this study, may have a surprise impact on biodiversity estimates coming from remote sensing [16], and this is to be factored in while projecting biodiversity estimates of the tropics using remote sensing. Higher intra-species variability amongst the abundant species of the three sites, as seen from CHVs, could be one of the many reasons for their expansive reach at the three sites. As mentioned by Avolio et al. [46], our study makes an effort in estimating abundant species distribution over larger spatial scales through remote sensing and conveys functional diversity through spectral diversity metrics, which may not be visible in traditional field-based studies.

Random plots of each site, selected from the forest cover maps of the three sites, showed different spectral diversity values (Figure 7) from the ones reported by Dahlin (whose study area was forests sof southwestern Michigan, USA) [22], demonstrating how effectively the spectral data of AVIRIS-NG correlates with tree species diversity observed in the tropical cover of this study. Similar to the observation of Wang and Gamon [10], our work supports and strengthens the efficacy of high resolution remote sensing data in the assessment of biodiversity variables of forest cover through spectral diversity metrics.

4.2. Abundant Species Maps

The abundant species maps of the three sites showed a higher accuracy and matching similarity with field-observed data. The corroboration of the developed species maps with additional field visits showed a > 85% match. Classification accuracy is mostly dependent on intra- and inter-species spectral variability [53,54]. This is evident in the spectra of abundant species, which resulted in developing classification maps showing > 75% accuracy in the classification of ~20 tropical tree species spread over ~80% of the forest cover of the three sites, revealing the advantage of AVIRIS-NG in tree enumeration studies over expansive regions. As reported earlier, the distribution and spread of abundant species canopies [55] and the quality of extracted endmember spectra [10] made it possible to have abundant species maps of the three sites with better accuracy. Earlier, Ferreira et al. [54] reported 81–84% user's accuracy for eight tree species of tropical deciduous forests using hyperspectral data with a 1-m spatial resolution. With seven individual tree species and an additional class of broadleaved ones in forest cover in Poland, Modzelewska et al. [56] reported 77% accuracy in classification using 5-m resolution hyperspectral data. The user's accuracy values of this study (83.54–86.70%) for ~20 tropical tree species with 4-m spatial resolution data are reasonable. The distribution maps of abundant species developed here by using high resolution sensors provide valuable inputs for forest ecological studies [57] and also for forest management.

4.3. Intra- and Inter-Species Spectral Variability

Jetz et al. [58] indicated that spectra reveal biochemical and structural aspects of vegetation, assisting in tracking plant functional diversity remotely. Schweiger et al. [14] underlined the importance of plant spectral diversity in predicting the ecosystem functionality of Cedar Creek. These reports broadly support the foundation of this study. Though transformations, such as PCA, do not directly represent measured traits, the CHV metric calculated here indirectly expresses the multivariate range of trait values captured through the variability of spectra collected via the high-fidelity AVIRIS-NG instrument, as in [44]. Patterns of intra- and inter-species variability in CHVs across the three sites (Table S10) help show the variability in canopy traits and simultaneously help elucidate the variations in traits along a major climatic gradient. The CHVs of species with restricted distributions are relatively lower when compared to the abundant species, indicating a high degree of trait conservatism.

Albert et al. [59] emphasized considering intra-specific variability in plant ecology, and Siefert et al. [60] reported that 25% of trait variation in a community is because of intra-species variability. The observed intra-species variability of abundant species in measured CHVs, and on species assemblages over the three sites, is in sync with these reports. Functional variability can be looked at from an individual perspective or at the species-based [61] and community level. These are evident from the consistent increase in the CHVs of the common abundant species across the climatic gradient over the three sites (Figure 9). Poorter et al. [61] reported 44% trait variation within the most abundant species. The amplitudes of intra-specific variability among abundant species, seen in this study, are large. Additionally, the dynamics in CHV values are species-specific. Bolnick et al. [62] suggested how intra-specific trait variation can modulate species abundance and interaction, impacting community dynamics. Similar reasoning can be given for the observed dynamics of abundant species distribution over the three sites of this study.

4.4. Climatic Gradient and Species Response

RDA, performed primarily as a constrained ordination technique, showed the impact of the climatic gradient on the spread of tree species. The impact of the climatic gradient is evident in the direction and spread of the arrows in the RDA plot (Figure 10), indicating a constrained response of species spread towards environmental variables with high positive, negative, and near zero correlations. Climatic gradient explained more variance about the abundant species distribution, revealing their explicit response, with ~80% occupancy at each site. The explained variance increased from SWS to VNP to MTR, showing positivity in the distribution dynamics of recorded tropical tree species over the climatic gradient. The measured variance values went down for the RDA of all recorded species, indicating that sensitivity towards climatic gradient could be different for abundant species compared to others. Though the percentage of variation in species distribution, explained by the observed gradient, appears smaller at the lower end, values at the upper end can match with the ones given for the three different factors of woody plants of the Panama Republic [63]. Though further confirmations are needed, it is interesting to notice how the same common species place themselves in site-specific community conformations in response to rainfall variability. Stavros et al. [64] raised the question, "do diverse communities respond to climate change differently from simpler communities?". The outputs of the multivariate analysis of this study are indicative of a similar approach by the recorded species of this study.

5. Conclusions

This study utilized a large field-sampled dataset over a climatic gradient to leverage imaging spectroscopy data from the joint ISRO-NASA AVIRIS-NG India campaign to help generate an understanding of the patterns of linkage between tree species field and spectral diversity metrics. High resolution datasets worked effectively in developing abundant species maps, finding out intra- and inter-species variability as CHVs within and over three sites in the tropics. Similar to the report of [65], which mentioned that climate primarily drives the distribution of regional species, the climatic gradient in this study was shown to impact the spread and dispersion of abundant tree species of three sites differently, leading to site-specific strategic species assemblages. The response of two of the important species, *T. grandis* and bamboo species, depict how well these species can tune their inherent functional diversity towards climatic gradient. The findings of this study are relevant in providing vital inputs to the management of these precious and vulnerable forest covers, and support the view in [66] that an integrative approach of remote sensing and vital field data has huge potential in the conservation of nature. The outputs of this study, coming from regions less represented in the global database, can contribute to global models. Observed changes in species distribution and diversity metrics over climatic gradient can help in assessing these forests' responses to the projected dynamics in rainfall over the coming decades.

Supplementary Materials: The following are available online at http://www.mdpi.com/2072-4292/12/13/2130/s1.

Author Contributions: N.S.R.K. conceived the idea and designed the experiments. A.N.C., M.G.D., and R.M.P. equally and collectively carried out field work, AVIRIS-NG data processing, and statistical analysis. P.M. assisted in remote sensing data processing. N.S.R.K. wrote the first draft of the manuscript. P.M. and A.S. made extensive edits to the first draft. All authors contributed to the final version. B.B. is the coordinator of the AVIRIS-NG India program. All authors have read and agreed to the published version of the manuscript.

Funding: This research was funded by the DST-NISA program (BDID/01/23/2014-HSRS/20) and the SAC-ISRO-AVIRIS-NG-AO program.

Acknowledgments: The authors are thankful to Mr. Rajesh Kallaje, IFS and Mrs. Aradhana Sahu, IFS for field work support. We are very much thankful to Philip Townsend, Adam Chlus, and Zhiwei Ye at the University of Wisconsin for sharing topographic and BRDF-corrected AVIRIS-NG images for the three sites. The authors are thankful to the reviewers for their critical suggestions.

Conflicts of Interest: The authors declare no conflict of interest.

References

1. Esquivel-Muelbert, A.; Baker, T.R.; Dexter, K.G.; Lewis, S.L.; ter Steege, H.; Lopez-Gonzalez, G.; Mendoza, A.M.; Brienen, R.; Feldpausch, T.R.; Pitman, N.; et al. Seasonal drought limits tree species across the Neotropics. *Ecography* **2017**, *40*, 618–629. [CrossRef]
2. Kraft, N.J.B.; Valencia, R.; Ackerly, D.D. Functional Traits and Niche-Based Tree Community Assembly in an Amazonian Forest. *Science* **2008**, *322*, 580–582. [CrossRef] [PubMed]
3. Zhang, T.; Niinemets, Ü.; Sheffield, J.; Lichstein, J.W. Shifts in tree functional composition amplify the response of forest biomass to climate. *Nature* **2018**, *556*, 99–102. [CrossRef]
4. Knapp, A.K.; Ciais, P.; Smith, M.D. Reconciling inconsistencies in precipitation–productivity relationships: implications for climate change. *New Phytol.* **2017**, *214*, 41–47. [CrossRef]

5. Babst, F.; Poulter, B.; Trouet, V.; Tan, K.; Neuwirth, B.; Wilson, R.; Carrer, M.; Grabner, M.; Tegel, W.; Levanic, T.; et al. Site- and species-specific responses of forest growth to climate across the European continent. *Glob. Ecol. Biogeogr.* **2013**, *22*, 706–717. [CrossRef]
6. Kapos, V. Seeing the forest through the trees. *Science* **2017**, *355*, 347–349. [CrossRef]
7. Poorter, L.; Wright, S.J.; Paz, H.; Ackerly, D.D.; Condit, R.; Ibarra-Manríquez, G.; Harms, K.E.; Licona, J.C.; Martínez-Ramos, M.; Mazer, S.J.; et al. Are Functional Traits Good Predictors of Demographic Rates? Evidence from Five Neotropical Forests. *Ecology* **2008**, *89*, 1908–1920. [CrossRef]
8. Pélissier, R. Tree spatial patterns in three contrasting plots of a southern Indian tropical moist evergreen forest. *J. Trop. Ecol.* **1998**, *14*, 1–16. [CrossRef]
9. Swenson, N.G. The assembly of tropical tree communities—The advances and shortcomings of phylogenetic and functional trait analyses. *Ecography* **2013**, *36*, 264–276. [CrossRef]
10. Wang, R.; Gamon, J.A. Remote sensing of terrestrial plant biodiversity. *Remote Sens. Environ.* **2019**, *231*, 111218. [CrossRef]
11. Díaz, S.; Kattge, J.; Cornelissen, J.H.C.; Wright, I.J.; Lavorel, S.; Dray, S.; Reu, B.; Kleyer, M.; Wirth, C.; Colin Prentice, I.; et al. The global spectrum of plant form and function. *Nature* **2016**, *529*, 167–171. [CrossRef]
12. Schepaschenko, D.; See, L.; Lesiv, M.; Bastin, J.-F.; Mollicone, D.; Tsendbazar, N.-E.; Bastin, L.; McCallum, I.; Laso Bayas, J.C.; Baklanov, A.; et al. Recent Advances in Forest Observation with Visual Interpretation of Very High-Resolution Imagery. *Surv. Geophys.* **2019**, *40*, 839–862. [CrossRef]
13. Hill, J.; Buddenbaum, H.; Townsend, P.A. Imaging Spectroscopy of Forest Ecosystems: Perspectives for the Use of Space-borne Hyperspectral Earth Observation Systems. *Surv. Geophys.* **2019**, *40*, 553–588. [CrossRef]
14. Schweiger, A.K.; Cavender-Bares, J.; Townsend, P.A.; Hobbie, S.E.; Madritch, M.D.; Wang, R.; Tilman, D.; Gamon, J.A. Plant spectral diversity integrates functional and phylogenetic components of biodiversity and predicts ecosystem function. *Nat. Ecol. Evol.* **2018**, *2*, 976–982. [CrossRef] [PubMed]
15. Miraglio, T.; Adeline, K.; Huesca, M.; Ustin, S.; Briottet, X. Monitoring LAI, Chlorophylls, and Carotenoids Content of a Woodland Savanna Using Hyperspectral Imagery and 3D Radiative Transfer Modeling. *Remote Sens.* **2020**, *12*, 28. [CrossRef]
16. Wang, Z.; Townsend, P.A.; Schweiger, A.K.; Couture, J.J.; Singh, A.; Hobbie, S.E.; Cavender-Bares, J. Mapping foliar functional traits and their uncertainties across three years in a grassland experiment. *Remote Sens. Environ.* **2019**, *221*, 405–416. [CrossRef]
17. Kunstler, G.; Falster, D.; Coomes, D.A.; Hui, F.; Kooyman, R.M.; Laughlin, D.C.; Poorter, L.; Vanderwel, M.; Vieilledent, G.; Wright, S.J.; et al. Plant functional traits have globally consistent effects on competition. *Nature* **2016**, *529*, 204–207. [CrossRef]
18. Palmer, M.W.; Wohlgemuth, T.; Earls, P.; Arévalo, J.R.; Thompson, S.D. Opportunities for long-term ecological research at the Tallgrass Prairie Preserve, Oklahoma. In Proceedings of the ILTER Regional Workshop: Cooperation in Long Term Ecological Research in Central and Eastern Europe, Budapest, Hungary, 22–25 June 1999; Volume 2229, p. 123128.
19. Palmer, M.W.; Earls, P.G.; Hoagland, B.W.; White, P.S.; Wohlgemuth, T. Quantitative tools for perfecting species lists. *Environmetrics* **2002**, *13*, 121–137. [CrossRef]
20. Rocchini, D.; Chiarucci, A.; Loiselle, S.A. Testing the spectral variation hypothesis by using satellite multispectral images. *Acta Oecologica* **2004**, *26*, 117–120. [CrossRef]
21. Asner, G.P.; Martin, R.E. Spectranomics: Emerging science and conservation opportunities at the interface of biodiversity and remote sensing. *Glob. Ecol. Conserv.* **2016**, *8*, 212–219. [CrossRef]
22. Dahlin, K.M. Spectral diversity area relationships for assessing biodiversity in a wildland–agriculture matrix. *Ecol. Appl.* **2016**, *26*, 2758–2768. [CrossRef]
23. Gholizadeh, H.; Gamon, J.A.; Zygielbaum, A.I.; Wang, R.; Schweiger, A.K.; Cavender-Bares, J. Remote sensing of biodiversity: Soil correction and data dimension reduction methods improve assessment of α-diversity (species richness) in prairie ecosystems. *Remote Sens. Environ.* **2018**, *206*, 240–253. [CrossRef]
24. Bello, F.D.; Lepš, J.; Sebastià, M.-T. Predictive value of plant traits to grazing along a climatic gradient in the Mediterranean. *J. Appl. Ecol.* **2005**, *42*, 824–833. [CrossRef]
25. Díaz, S.; Cabido, M.; Zak, M.; Carretero, E.M.; Araníbar, J. Plant functional traits, ecosystem structure and land-use history along a climatic gradient in central-western Argentina. *J. Veg. Sci.* **1999**, *10*, 651–660. [CrossRef]

26. Aguirre-Gutiérrez, J.; Oliveras, I.; Rifai, S.; Fauset, S.; Adu-Bredu, S.; Affum-Baffoe, K.; Baker, T.R.; Feldpausch, T.R.; Gvozdevaite, A.; Hubau, W.; et al. Drier tropical forests are susceptible to functional changes in response to a long-term drought. *Ecol. Lett.* **2019**, *22*, 855–865. [CrossRef]
27. Féret, J.-B.; Asner, G.P. Mapping tropical forest canopy diversity using high-fidelity imaging spectroscopy. *Ecol. Appl.* **2014**, *24*, 1289–1296. [CrossRef]
28. Martin, M.E.; Plourde, L.C.; Ollinger, S.V.; Smith, M.-L.; McNeil, B.E. A generalizable method for remote sensing of canopy nitrogen across a wide range of forest ecosystems. *Remote Sens. Environ.* **2008**, *112*, 3511–3519. [CrossRef]
29. Wieczynski, D.J.; Boyle, B.; Buzzard, V.; Duran, S.M.; Henderson, A.N.; Hulshof, C.M.; Kerkhoff, A.J.; McCarthy, M.C.; Michaletz, S.T.; Swenson, N.G.; et al. Climate shapes and shifts functional biodiversity in forests worldwide. *Proc. Natl. Acad. Sci. USA* **2019**, *116*, 587–592. [CrossRef]
30. Lewis, S.L. Tropical forests and the changing earth system. *Philos. Trans. R. Soc. B: Biol. Sci.* **2006**, *361*, 195–210. [CrossRef]
31. Enquist, B.J.; Enquist, C.A.F. Long-term change within a Neotropical forest: assessing differential functional and floristic responses to disturbance and drought. *Glob. Chang. Biol.* **2011**, *17*, 1408–1424. [CrossRef]
32. Lewis, S.L.; Sonké, B.; Sunderland, T.; Begne, S.K.; Lopez-Gonzalez, G.; van der Heijden, G.M.F.; Phillips, O.L.; Affum-Baffoe, K.; Baker, T.R.; Banin, L.; et al. Above-ground biomass and structure of 260 African tropical forests. *Philos. Trans. R. Soc. B Biol. Sci.* **2013**, *368*, 20120295. [CrossRef] [PubMed]
33. Roy, A.; Roy, P.S. Biodiversity information in India: Status and future scope. In *Biodiversity in Tropical Ecosystems*; Springer: Berlin, Germany, 2015; pp. 9–13. ISBN 1-55528-370-5.
34. UNNI, N.V.M.; ROY, P.S.; PARTHASARATHY, V. Evaluation of LANDSAT and airborne multispectral data and aerial photographs for mapping forest features and phenomena in a part of the Godavari basin. *Int. J. Remote Sens.* **1985**, *6*, 419–431. [CrossRef]
35. Ambastha, K.R.; Jha, C.S. Geospatial analysis of Tamil Nadu Eastern Ghats forest types at landscape level with reference to fragmentation and species diversity. *J. Indian Soc. Remote Sens.* **2010**, *38*, 453–463. [CrossRef]
36. Gairola, S.; Procheş, Ş.; Rocchini, D. High-resolution satellite remote sensing: a new frontier for biodiversity exploration in Indian Himalayan forests. *Int. J. Remote Sens.* **2013**, *34*, 2006–2022. [CrossRef]
37. Jha, C.S.; Fararoda, R.; Singhal, J.; Reddy, C.S.; Rajashekar, G.; Maity, S.; Patnaik, C.; Das, A.K.; Misra, A.; Singh, C.P.; et al. Characterization of Species Diversity and Forest Health using AVIRIS-NG Hyperspectral Remote Sensing Data. *Curr. Sci.* **2019**, *116*, 1124. [CrossRef]
38. Thompson, D.R.; Gao, B.-C.; Green, R.O.; Roberts, D.A.; Dennison, P.E.; Lundeen, S.R. Atmospheric correction for global mapping spectroscopy: ATREM advances for the HyspIRI preparatory campaign. *Remote Sens. Environ.* **2015**, *167*, 64–77. [CrossRef]
39. Soenen, S.A.; Peddle, D.R.; Coburn, C.A. SCS+C: a modified Sun-canopy-sensor topographic correction in forested terrain. *IEEE Trans. Geosci. Remote Sens.* **2005**, *43*, 2148–2159. [CrossRef]
40. Wanner, W.; Li, X.; Strahler, A.H. On the derivation of kernels for kernel-driven models of bidirectional reflectance. *J. Geophys. Res. Atmos.* **1995**, *100*, 21077–21089. [CrossRef]
41. Kindt, R.; Coe, R. *Tree Diversity Analysis. A Manual and Software for Common Statistical Methods for Ecological and Biodiversity Studies*; World Agro-forestry Center (ICRAF): Nairobi, Kenya, 2006.
42. Green, A.A.; Berman, M.; Switzer, P.; Craig, M.D. A transformation for ordering multispectral data in terms of image quality with implications for noise removal. *IEEE Trans. Geosci. Remote Sens.* **1988**, *26*, 65–74. [CrossRef]
43. Cortes, C.; Vapnik, V. Support-vector networks. *Mach. Learn.* **1995**, *20*, 273–297. [CrossRef]
44. Cornwell, W.K.; Schwilk, D.W.; Ackerly, D.D. A Trait-Based Test for Habitat Filtering: Convex Hull Volume. *Ecology* **2006**, *87*, 1465–1471. [CrossRef]
45. Fassnacht, F.E.; Latifi, H.; Stereńczak, K.; Modzelewska, A.; Lefsky, M.; Waser, L.T.; Straub, C.; Ghosh, A. Review of studies on tree species classification from remotely sensed data. *Remote Sens. Environ.* **2016**, *186*, 64–87. [CrossRef]
46. Avolio, M.L.; Forrestel, E.J.; Chang, C.C.; Pierre, K.J.L.; Burghardt, K.T.; Smith, M.D. Demystifying dominant species. *New Phytol.* **2019**, *223*, 1106–1126. [CrossRef]
47. Ricklefs, R.E.; Miller, G.L. *Ecology*, 4th ed.; W.H. Freeman & Co: New York, NY, USA, 2000; ISBN 978-0-7167-2829-0.
48. Ricklefs, R.E. Disintegration of the Ecological Community. *Am. Nat.* **2008**, *172*, 741–750. [CrossRef]

49. Scheiter, S.; Langan, L.; Higgins, S.I. Next-generation dynamic global vegetation models: learning from community ecology. *New Phytol.* **2013**, *198*, 957–969. [CrossRef]
50. Engelbrecht, B.M.J.; Comita, L.S.; Condit, R.; Kursar, T.A.; Tyree, M.T.; Turner, B.L.; Hubbell, S.P. Drought sensitivity shapes species distribution patterns in tropical forests. *Nature* **2007**, *447*, 80–82. [CrossRef]
51. Randin, C.F.; Ashcroft, M.B.; Bolliger, J.; Cavender-Bares, J.; Coops, N.C.; Dullinger, S.; Dirnböck, T.; Eckert, S.; Ellis, E.; Fernández, N.; et al. Monitoring biodiversity in the Anthropocene using remote sensing in species distribution models. *Remote Sens. Environ.* **2020**, *239*, 111626. [CrossRef]
52. Rocchini, D.; Hortal, J.; Lengyel, S.; Lobo, J.M.; Jiménez-Valverde, A.; Ricotta, C.; Bacaro, G.; Chiarucci, A. Accounting for uncertainty when mapping species distributions: The need for maps of ignorance. *Prog. Phys. Geogr. Earth Environ.* **2011**, *35*, 211–226. [CrossRef]
53. Zhang, J.; Rivard, B.; Sánchez-Azofeifa, A.; Castro-Esau, K. Intra- and inter-class spectral variability of tropical tree species at La Selva, Costa Rica: Implications for species identification using HYDICE imagery. *Remote Sens. Environ.* **2006**, *105*, 129–141. [CrossRef]
54. Ferreira, M.P.; Zortea, M.; Zanotta, D.C.; Shimabukuro, Y.E.; de Souza Filho, C.R. Mapping tree species in tropical seasonal semi-deciduous forests with hyperspectral and multispectral data. *Remote Sens. Environ.* **2016**, *179*, 66–78. [CrossRef]
55. Ustin, S.L.; Gamon, J.A. Remote sensing of plant functional types. *New Phytol.* **2010**, *186*, 795–816. [CrossRef]
56. Modzelewska, A.; Fassnacht, F.E.; Stereńczak, K. Tree species identification within an extensive forest area with diverse management regimes using airborne hyperspectral data. *Int. J. Appl. Earth Obs. Geoinf.* **2020**, *84*, 101960. [CrossRef]
57. Vaglio Laurin, G.; Puletti, N.; Hawthorne, W.; Liesenberg, V.; Corona, P.; Papale, D.; Chen, Q.; Valentini, R. Discrimination of tropical forest types, dominant species, and mapping of functional guilds by hyperspectral and simulated multispectral Sentinel-2 data. *Remote Sens. Environ.* **2016**, *176*, 163–176. [CrossRef]
58. Jetz, W.; Cavender-Bares, J.; Pavlick, R.; Schimel, D.; Davis, F.W.; Asner, G.P.; Guralnick, R.; Kattge, J.; Latimer, A.M.; Moorcroft, P.; et al. Monitoring plant functional diversity from space. *Nat. Plants* **2016**, *2*, 1–5. [CrossRef]
59. Albert, C.H.; Grassein, F.; Schurr, F.M.; Vieilledent, G.; Violle, C. When and how should intraspecific variability be considered in trait-based plant ecology? *Perspect. Plant Ecol. Evol. Syst.* **2011**, *13*, 217–225. [CrossRef]
60. Siefert, A.; Violle, C.; Chalmandrier, L.; Albert, C.H.; Taudiere, A.; Fajardo, A.; Aarssen, L.W.; Baraloto, C.; Carlucci, M.B.; Cianciaruso, M.V.; et al. A global meta-analysis of the relative extent of intraspecific trait variation in plant communities. *Ecol. Lett.* **2015**, *18*, 1406–1419. [CrossRef] [PubMed]
61. Poorter, L.; Castilho, C.V.; Schietti, J.; Oliveira, R.S.; Costa, F.R.C. Can traits predict individual growth performance? A test in a hyperdiverse tropical forest. *New Phytol.* **2018**, *219*, 109–121. [CrossRef]
62. Bolnick, D.I.; Amarasekare, P.; Araújo, M.S.; Bürger, R.; Levine, J.M.; Novak, M.; Rudolf, V.H.W.; Schreiber, S.J.; Urban, M.C.; Vasseur, D.A. Why intraspecific trait variation matters in community ecology. *Trends Ecol. Evol.* **2011**, *26*, 183–192. [CrossRef]
63. Svenning, J.-C.; Kinner, D.A.; Stallard, R.F.; Engelbrecht, B.M.J.; Wright, S.J. Ecological Determinism in Plant Community Structure Across a Tropical Forest Landscape. *Ecology* **2004**, *85*, 2526–2538. [CrossRef]
64. Stavros, E.N.; Schimel, D.; Pavlick, R.; Serbin, S.; Swann, A.; Duncanson, L.; Fisher, J.B.; Fassnacht, F.; Ustin, S.; Dubayah, R.; et al. ISS observations offer insights into plant function. *Nat. Ecol. Evol.* **2017**, *1*, 0194. [CrossRef]
65. van Ewijk, K.Y.; Randin, C.F.; Treitz, P.M.; Scott, N.A. Predicting fine-scale tree species abundance patterns using biotic variables derived from LiDAR and high spatial resolution imagery. *Remote Sens. Environ.* **2014**, *150*, 120–131. [CrossRef]
66. Hoffmann, S.; Schmitt, T.M.; Chiarucci, A.; Irl, S.D.H.; Rocchini, D.; Vetaas, O.R.; Tanase, M.A.; Mermoz, S.; Bouvet, A.; Beierkuhnlein, C. Remote sensing of β-diversity: Evidence from plant communities in a semi-natural system. *Appl. Veg. Sci.* **2019**, *22*, 13–26. [CrossRef]

MDPI
St. Alban-Anlage 66
4052 Basel
Switzerland
www.mdpi.com

Remote Sensing Editorial Office
E-mail: remotesensing@mdpi.com
www.mdpi.com/journal/remotesensing

www.ingramcontent.com/pod-product-compliance
Lightning Source LLC
Chambersburg PA
CBHW061545220926
51850CB00005B/259
* 9 7 8 3 0 3 6 5 8 5 2 6 0 *